International Conference on Environmental and Energy Engineering (IC3E 2017)

IOP Conference Series: Earth and Environmental Science
Volume 63

Suzhou, China
22 - 24 March 2017

ISBN: 978-1-5108-4038-6
ISSN: 1755-1307

Printed from e-media with permission by:

Curran Associates, Inc.
57 Morehouse Lane
Red Hook, NY 12571

CURRAN ASSOCIATES INC. proceedings .com

Some format issues inherent in the e-media version may also appear in this print version.

Printed by Curran Associates, Inc. (2017)

Institute of Physics
Dirac House, Temple Back
Bristol BS1 6BE UK

Phone: 44 1 17 929 7481
Fax: 44 1 17 920 0979

techtracking@iop.org

Additional copies of this publication are available from:

Curran Associates, Inc.
57 Morehouse Lane
Red Hook, NY 12571 USA
Phone: 845-758-0400
Fax: 845-758-2633
Email: curran@proceedings.com
Web: www.proceedings.com

Table of contents

Volume 63

2017 International Conference on Environmental and Energy Engineering (IC3E 2017)

22–24 March 2017, Suzhou, China

Accepted papers received: 27 April 2017
Published online: 9 May 2017

Preface

Energy Research and Applications

Sustainable Development

Energy and Environment

Energy and Environment

IOP Conf. Series: Earth and Environmental Science **63** (2017) 011001 doi:10.1088/1755-1315/63/1/011001

PREFACE

IC3E 2017 was held in Suzhou, China during March 22nd-24th, 2017. IC3E 2017 was organized by Asia Pacific Institute of Science and Engineering, sponsored by HKSME, supported by other organizations. The conference provides a useful and wide platform both for display the latest research and for exchange of research results and thoughts in broad Environmental, Energy Engineering, Sustainable Development and Green Buildings. The participants of the conference were from almost every part of the world, with background of either academia or industry, even well-known enterprise. The success and prosperity of the conference is reflected high level of the papers received.

The proceedings are a compilation of the accepted papers and represent an interesting outcome of the conference. This book covers 3 chapters: Energy Research and Applications; Sustainable Development; Energy and Environment.

We would like to acknowledge all people who supported IC3E 2017. Each individual and institutional help were very important for the success of this conference. Especially we would like to thank the organizing committee for their valuable advices in the organization and helpful peer review of the papers.

We sincerely hope that IC3E 2017 will be a forum for excellent discussions that will put forward new ideas and promote collaborative researches. We are sure that the proceedings will serve as an important research source of references and the knowledge, which will lead to not only scientific and engineering progress but also other new products and processes.

Prof. Dou Binlin, Dalian University of Technology, China

IOP Conf. Series: Earth and Environmental Science **63** (2017) 011002 doi:10.1088/1755-1315/63/1/011002

COMMITTEES

Conference Committee Chair

Prof. Hossein Ganjidoust, Tarbiat Modares University, Iran

Program Commitee Chairs

Prof. Michael K. H. LEUNG, City University of Hong Kong, Hong Kong

Prof. Zhi Chen, Concordia University, Canada

Prof. Dou Binlin, University of Shanghai for Science and Technology, China

Publication Chair

Prof. Enji Sun, China Academy of safety science and technology, China

International Technical Committee

Prof. Rajasekhar Balasubramanian, National University of Singapore (NUS), Singapore

Dr. Katsuyuki KAWAI, Kindai University,Japan

Dr. Liu Liangying, JI nan university,China

Dr. Sayed M. Bateni, University of Hawaii at Manoa, USA

Dr. Lim Teik Thye, Nanyang Technological University, Singapore

Dr. Kong Lingjun, Guang zhou University, China

Dr. Grzegorz Lisak, Nanyang Technological University, Singapore

Dr. Cheung Sai Hung, Joseph, Nanyang Technological University, Singapore

Dr. Huichao Chen,Southeast University, China

Dr. Chunhua LIU, The City University of Hong Kong, Hong Kong

Dr. Huaiyu SHAO, University of Macau, China

Dr. Xinjing Zhang, Chinese Academy of Sciences (CAS), China

Dr. Masoud D. Champiri, University of Houston, USA

Peer review statement

All papers published in this volume of *IOP Conference Series: Earth and Environmental Science* have been peer reviewed through processes administered by the proceedings Editors. Reviews were conducted by expert referees to the professional and scientific standards expected of a proceedings journal published by IOP Publishing.

IOP Conf. Series: Earth and Environmental Science **63** (2017) 012001 doi:10.1088/1755-1315/63/1/012001

Internet of Things Based Combustible Ice Safety Monitoring System Framework

Enji Sun

China Academy of Safety Science and Technology, Beijing, 100012, CHINA

enjisun@gmail.com

Abstract. As the development of human society, more energy is requires to meet the need of human daily lives. New energies play a significant role in solving the problems of serious environmental pollution and resources exhaustion in the present world. Combustible ice is essentially frozen natural gas, which can literally be lit on fire bringing a whole new meaning to fire and ice with less pollutant. This paper analysed the advantages and risks on the uses of combustible ice. By compare to other kinds of alternative energies, the advantages of the uses of combustible ice were concluded. The combustible ice basic physical characters and safety risks were analysed. The developments troubles and key utilizations of combustible ice were predicted in the end. A real-time safety monitoring system framework based on the internet of things (IOT) was built to be applied in the future mining, which provide a brand new way to monitoring the combustible ice mining safety.

1. Introduction

As the development of human society, more energy is requires to meet the need of human daily lives. Compared with traditional energies, new energies generally have the characteristics of less pollution and large reserves. New energies play a significant role in solving the problems of serious environmental pollution and resources exhaustion in the present world. At the same time, many of the new energies are well-distributed, which are also very meaningful to solve the war caused by energy problem.

Combustible ice is essentially frozen natural gas – a natural gas hydrate, and is one of the newest energy sources to be discovered, which has been found in high altitude frozen plateaus as well as underwater in marine sediments. Natural gas hydrates are essentially just frozen methane and water and can literally be lit on fire bringing a whole new meaning to fire and ice. One cubic meter of combustible ice contains 164 cubic meters of regular natural gas and is considered to have few impurities, meaning it can burn cleaner with less pollutant.

While reserves of combustible ice have been found around the world, China's discovery is big and estimates say that there is enough energy frozen in the tundra to provide the country with energy for 90 years. Natural gas, which is not a zero emission fuel, can only be considered a "cleaner" energy source compared to oil, but given the alternative of it just being released from melting, it might be wise to continue research on this new energy source.

This paper analysed the advantages and risks on the uses of combustible ice. By compare to other kinds of alternative energies, the advantages of the uses of combustible ice were concluded. The combustible ice basic physical characters and safety risks were analysed. The developments troubles and key utilizations of combustible ice were predicted in the end. To deal with all the safety risks during the mining, a real-time safety monitoring system framework based on the internet of things (IOT) was built to be applied in the future mining, which provide the a brand new way to monitoring the combustible ice mining safety.

2. The advantages and risks of combustible ice

2.1. The basic parameters of combustible ice

China organizes launching the previous research to natural gas hydrate since 1997. [1] In 1999, Ministry of Land and Natural Resources started the reconnoitring at sea of natural gas hydrate, launched the investigations of 18 voyages, find the intersection of north and land slope, the South China Sea of China, have very natural gas hydrate to compose and deposit the terms, and obtained a series of geophysics, geochemistry, geology, biology, etc. obvious evidence. [2-3]

At the end of December of 2003, the marine geological investigation bureau of Guangzhou was in the South China Sea area in the north for the first time, utilizing the sea floor to make a video recording, technology has found the ashen blocky deposition material of group in deep sea floor of 3000 meters, analyse that is thought, such ashen blocky material of group is after the combustible ice in the deep department stratum is decomposed, the coastal bottom of methane gas ruptures and gushes out to overflow the sea floor, take shape. [4-5]This conclusion gets the approval of the authoritative expert of international combustible ice. [6-7]This discovery verifies there is combustible ice in the sea area of the north of the South China Sea of China. Combustible ice is clean new energy with the burning results only with water. The burning combustible ice burning flame as showed in Figure 1.

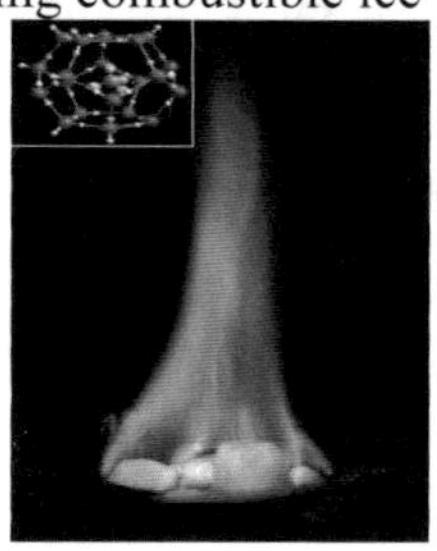

Figure 1. Combustible ice burning flame [8]

To form combustible ice there are three basic conditions. First of all, the temperature cannot be too high, may produce above the zero degree, from 0 to 10 centigrade for suitable, the most altitude limit is about 20 centigrade. The temperature will be again high to decompose combustible ice. The second pressure must suffice, but cannot be too big. In zero centigrade and above 30 atmospheric pressure conditions, it produces on the possibility. [9]

2.2. Safety risks of combustible ice

Although the development prospect of combustible ice is considerable, the mining difficulty is great. On the one hand is the combustible ice is located at the vast ocean base or on the high and cold dangerous far tundra, which are development difficultly. On the other hand, combustible ice owns some physical characters, also enables the combustible ice mining to be beset with difficulties.

The combustible ice is common product in the low temperature and the high-pressured. But the combustible ice "the hobby" may probably be much bigger than to the high pressure the low temperature. The people already discovered that under enough high pressure, combustible ice when 18℃, still might maintain "the ice" the nature, is one kind out-and-out "the high temperature ice".

This has brought the very major difficulty for combustible ice's mining. If cannot guarantee that the enough high pressure and the low temperature, the combustible ice will melt rapidly for the water and the methane, if cannot carry on to the methane collects well, the massive methane will enter the atmosphere directly, the global climate will face a disaster.

Carries on mining to the ocean deep place's combustible ice, besides faces the similar environmental protection difficult problem, but also possibly causes the seabed to soften, presents the large-scale seabed landslide, the destruction seabed project facility situation.

At present develops mainly faces three problems: First, in the technical stratification plane has the difficulty, our country mines the natural gas the technology to be quite mature, but develops combustible ice to compare to development natural gas many procedures. Moreover develops improper, will create the natural gas revelation. Second, in the cost aspect, actually to mine is worthwhile. If the oil price reduces, whether to increase the development cost, also will be worth considering. Third, environment question, if will develop will create the very big destruction to the environment, then its development will also not have the significance. In fact all energy mining can the varying degree create the harm to the environment, but along with science's and technology's development, combustible ice can in control under the condition develops is feasible. But did not have the effective solution at present, the concrete solution is exploring.

2.3. The advantages of combustible ice

Combustible ice, known as the new energy that can be used 1000 years by human is the first choice of the future alternative to oil, coal and other traditional energy sources. One cubic meter of combustible ice can be released from 160-180 cubic meters of natural gas. Its energy density is 10 times that of coal. After combustion no residue and exhaust was produced.

The recent discovery marks the first time that exposed methane hydrate deposits have been found in Japanese waters. Methane hydrate, which is normally found several hundred meters beneath the ocean floor, is a sherbert-like substance that burns when exposed to flame. It forms when low temperature and high pressure under the ocean floor causes methane molecules to become trapped inside frozen water molecules.

Scientists from the University of Tokyo, the Japan Agency for Marine-Earth Science and Technology (JAMSTEC), and the National Institute of Advanced Industrial Science and Technology (AIST) are working together to collect samples from two points located at depths of 800 to 1000 meters (2600 to 3300 feet), about 30 kilometres (19 miles) offshore. They are using unmanned submarines to collect the ice. Based on the high electrical conductivity of the ground beneath the ocean floor, the scientists suspect the existence of large underground columns of methane hydrate. [10-12]

While methane hydrate is being hailed as a potential source of fuel in the future, methane is a greenhouse gas. Methane is generated when organic matter in deep layers of sedimentary rock breaks down due to heating. The methane moves into upper layers, where it accumulates and forms methane hydrate. As the temperature rises and pressure falls, methane hydrate dissociates into methane and water. The resulting methane concentration in the surrounding seawater ranges from dozens to thousands of times higher than normal.

3. System framework

To improve combustible ice mining safety, combustible ice mining monitoring system (CISMS) aims to monitor the combustible ice mining safety anytime, anywhere, which is based on the internet of things (IOT). The entire system datum are acquired by all kinds of wireless sensors, processed by the cloud computing platform, and then broadcasted to the terminals either on PDA or the clients.

CISMS is based on the internet of things (IOT) which is under the security technology and management concept generally.

3.1. The sensor layer of CISMS

Sensor layer is the core data acquiring component and the basic layer in the CISMS. Different kinds of sensors, such as the temperature sensor, pressure sensor, position gas sensor etc., are applied to acquire the combustible ice mining data. The sensors are mostly wireless sensors, which typically consist of a large number of low-cost, low-power, and multifunctional sensor nodes that are deployed in a region of interest. These sensor nodes are small in size, but are equipped with sensors, embedded microprocessors, and radio transceivers, and therefore have not only sensing capability, but also data processing and communicating capabilities.

These sensors communicate over a short distance via a wireless medium and collaborate to accomplish a common task. Compared with traditional wireless communication networks, for example, cellular systems and MANET, sensor networks have the following unique characteristics and constraints:

- Dense Node Deployment. Sensor nodes are usually densely deployed in a field of interest. The number of sensor nodes in a sensor network can be several orders of magnitude higher than that in a MANET.
- Battery-Powered Sensor Nodes. Sensor nodes are usually powered by battery. In most situations, these sensors are deployed in a harsh or hostile environment, where it is very difficult or even impossible to change or recharge the batteries.
- Severe Energy, Computation, and Storage Constraints. Sensor nodes are highly limited in energy, computation, and storage capacities.
- Self-Configurable. Sensor nodes are usually randomly deployed without careful planning and engineering. Once deployed, sensor nodes have to autonomously configure themselves into a communication network.
- Application Specific. Sensor networks are application specific. A network is usually designed and deployed for a specific application. The design requirements of a network change with its application.
- Unreliable Sensor Nodes. Sensor nodes are usually deployed in harsh or hostile environments and operate without attendance. They are prone to physical damages or failures.
- Frequent Topology Change. Network topology changes frequently due to node failure, damage, addition, energy depletion, or channel fading.
- No Global Identification. Due to the large number of sensor nodes, it is usually not possible to build a global addressing scheme for a sensor network because it would introduce a high overhead for the identification maintenance.

3.2. The network layer of CISMS

The network layer includes the remote control, VPN, M2M wireless access, wireless sensor network, heterogeneous network integration and the fourth generation network. The network layer bridges the sensor layer and the application layer.

The network layer is responsible for routing system packets delivery including routing through intermediate routers, whereas the data link layer is responsible for media access control, flow control and error checking. The network layer provides the functional and procedural means of transferring variable length data sequences from a source to a destination host via one or more networks while maintaining the quality of service functions.

The network layer of CISMS makes it possible to monitor the combustible ice in anywhere on anytime. Its networks are partitioned into sub networks and connect to other networks for wide area communications. Networks use specialized hosts, called gateways or routers to forward packets between networks. This is also of interest to mobile applications, where a user may move from one location to another, and it must be arranged the messages. Version 4 of the internet protocol (IPv4) was not designed with this feature in mind, although mobility extensions exist. IPv6 has a better designed solution.

3.3. The application layer of CISMS

The application layer is the top level of the CISMS structure. Basically it is based on the support service platform, information platform and the cloud computing platform to accomplish the CISMS functions through public middleware, intelligent recognition and the industrial monitor.

The cloud computing platform deals with the entire sensors layer datum transferred by the network layer, which is a technology that uses the internet and central remote servers to maintain data and applications. Cloud computing allows users to use applications without installation and access personal files at any computer with internet access. This technology allows for much more efficient computing by centralizing storage, memory, processing and bandwidth. The information platform is mainly to promote, share and exchange information on the combustible ice monitoring issues in different regions. The information platform aims to illustrate the combustible ice monitored datum graphically, which are calculated after the cloud computing platform. The support service platform provides the concerned support service to the CISMS users. All the system problems can be solved online on time. Any errors or abnormal behaviours are logged in the server.

4. Conclusions

Combustible ice, also called natural gas hydrate, with giant reserves and high energy density. It is a kind of new type green energy source which has a good prospect. According to comparative analysis of two kinds of development ideas, it presents that both international cooperation and independent technology innovation can promote developing more green and low cost recovery technology of combustible ice.

This paper analysed the advantages and risks on the uses of combustible ice. By compare to other kinds of alternative energies, the advantages of the uses of combustible ice were concluded. The combustible ice basic physical characters and safety risks were analysed. The developments troubles and key utilizations of combustible ice were predicted in the end. To deal with all the safety risks during the mining, a real-time safety monitoring system framework based on the internet of things (IOT) was built to be applied in the future mining, which provide the a brand new way to monitoring the combustible ice mining safety.

5. Acknowledgments

This work was supported by National Natural Science Foundation of China (71373245, 50974109), the basic research funding of China Academy of Safety Science and Technology (2016JBKY07), and National Key Technology R&D Program of the Ministry of Science and Technology (2016YFC0801305).

6. References

[1] CSNS Project Team. China Spallation Neutron Source Project Proposal. Institute of High Energy Physics and Institute of Physics, Chinese Academy of Sciences, 2007.
[2] Lengeler H, WEI J. Spallation Neutron Source. In: Chao A and Tigner M ed. Handbook of accelerator physics and engineering: a compilation of formulae and data. Singapore: World Scienti_c Publishing Co., 1998; third Printing, 2006. Section 1.6.17.
[3] Wei J. *Reviews of Modern Physics*, 2003, **75**(4): 1383-1432.
[4] Spallation Neutron Source: The Next-generation Neutron Scattering Facility for the United States. Department of Energy, 1998. SNS website: http://www.sns.gov, 1998.
[5] Oyama Y. Nucl. Instrum. *Methods Phys. Res. A*, 2006, **562**(2): 548-552.
[6] A Second Target Station at ISIS. Council for the Central Laboratory of the Research Councils, 2000. RAL-TR-2000-032.
[7] Jorgensen J D, Veal B W, Nowicki L J et al. Physical Re-view B, 1990, 41(4): 1863-1877.
[8] Dai P, Mook H, Hayden S M et al. *Nature*, 2000, **406**(6799): 965.
[9] Mao W L, Mao H, Goncharov A F et al. Science, 2002, **297**(5590): 2247-2249.
[10] Kolesnikov A I, Zanotti J M, Loong C K et al. *Phys. Rev.Lett.*, 2004, 93(3):35503-1-4.
[11] W. S. Cooper Fact Retrieval and Deductive Question-Answering Systems. *J. ACM*, 1964, **11**(2):117-137.
[12] Liao Zhi-min, Xiong Shan. A New Green Energy Source: Combustible Ice. *Natural Gas Technology*, 2008, **2**(20):345-347.

Discussion on Boiler Efficiency Correction Method with Low Temperature Economizer–Air Heater System

Liu Ke[1], Yang Xing-sen[1], Hou Fan-jun[2] and Hu Zhi-hong[1]

[1] State Grid Shandong Electric Power Research Institute, Jinan 250003, China
[2] Zhong Shi Yi Tong Group, Jinan 250003, China

luliuke@163.com

Abstract. This paper pointed out that it is wrong to take the outlet flue gas temperature of low temperature economizer as exhaust gas temperature in boiler efficiency calculation based on GB10184-1988. What's more, this paper proposed a new correction method, which decomposed low temperature economizer-air heater system into two hypothetical parts of air preheater and pre condensed water heater and take the outlet equivalent gas temperature of air preheater as exhaust gas temperature in boiler efficiency calculation. This method makes the boiler efficiency calculation more concise, with no air heater correction. It has a positive reference value to deal with this kind of problem correctly.

1. Introduction

The heat loss due to exhaust gas lays direct impacts on the efficiency of the boiler. In the premise of corrosion in low temperature being under control, the lower the exhaust gas temperature is the better. In most of the circumstances, reformation of the utilization of flue-gas heat is not capable of reducing the temperature of the exhaust gas in boiler efficiency calculation; however, it is effective for improving the economy of the machine unit [1-5]. For example, heating condensate water is a common way to utilize residual heat [6].

Along with the development in ultra-low emission and energy conservation reconstruction work in China, a new technical reconstruction scheme has been gradually popularized, namely the modification of the low-temperature economizer`s joint heater [7-8], in which, a low temperature economizer is added in the chimney intake of the original deduster to coordinate with the modification of the electrical deduster and realize low temperature dedusting, as well as achieve the dust emission target at the electrical dedusting outlet; a heater is designed to heat up the air at the preheater inlet through the low temperature economizer-heated condensate water. The system can not only satisfy the low temperature dedusting demand, but also heat the condensate water and exhaust air; meanwhile, it has also realized the utilization of residual heat at the rear section, which could demonstrate significant economic benefits. However, such technical scheme has also brought certain interferences towards the boiler efficiency calculation; some errors may occur in terms of exhaust gas temperature determination, inlet air temperature rectification and the like. Thus, the article carries out studies corresponding to these problems by exploring and discussing methods for boiler efficiency rectification during system operation.

2. System introduction and boiler efficiency calculation boundary

2.1. System introduction

Figure 1 is Low Temperature Economizer–Air Heater System. Condensate water from the low-pressure feed water heater enters into the low temperature economizer; after being heated, the water enters into the heater in order to heat up the air; then, it returns to the low-pressure feed water heater at the subordinate level. After being heated by the heater, air temperature at the preheater inlet will be increased, and the flue-gas temperature at the preheater outlet will also be increased accordingly; however, after passing through the low temperature economizer, the flue-gas temperature could drop to 90~100℃. It is subject to a common error to consider the foregoing as the exhaust gas temperature in boiler efficiency calculation.

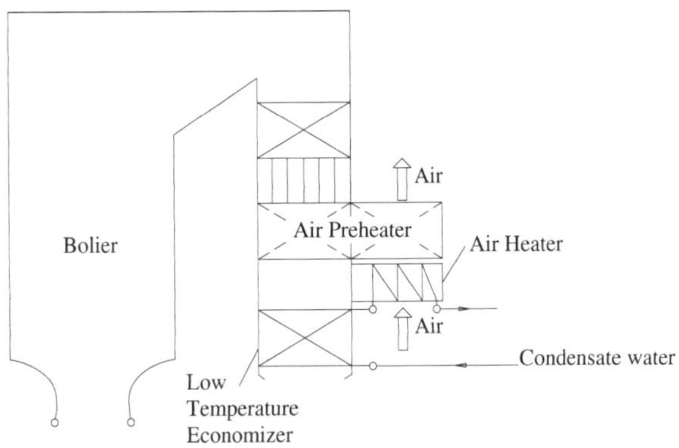

Figure 1. Low temperature economizer–air heater system.

2.2. Error Cause Analysis

Boiler efficiency calculation boundary specified in GB10184-1988 standard is shown in Figure 2, according to which, the calculation boundary shall be divided as from the preheater outlet to the deduster; heater that uses external heat has also been ruled out from the calculation boundary. Nevertheless, the low temperature economizer is located behind the preheater and before the deduster; as for whether it can be merged into the calculation boundary, it requires the analysis from the aspect of System Energy Conservation by considering the coverage of the calculation boundary as a system, in which, energy input into the system is equivalent to effectively-used heat plus the system heat loss. Assuming that the low temperature economizer`s joint heater is merged into the system and the heat absorbed by the low temperature economizer is referred to as Q_{dd}, heat absorbed by cold air in the heater is referred to as Q_{nf}; the heat of air in the low-pressure feed water heater reduced by the existence of the low temperature economizer is referred to as Q_{pj}, and $Q_{dd}=Q_{nf}+Q_{pj}$; then system heat loss has reduced by Q_{dd} through comparison to the previous value, yet the heat input into the system only increases by Q_{nf}. Although Q_{pj} is returned to the system together with water, total heat brought to the water has not been increased accordingly; Q_{pj} has merely exhausted partial air in the low-pressure feed water heater, based on which, wrong conclusion is made about the occurrence of non-conservation of system energy. It indicates that, the previous assumption of merging the low temperature economizer`s joint heater into the system is incorrect. So, the calculation boundary is still at the preheater outlet, namely the exhaust gas temperature is still the flue-gas temperature at the preheater outlet instead of at the lower temperature economizer outlet.

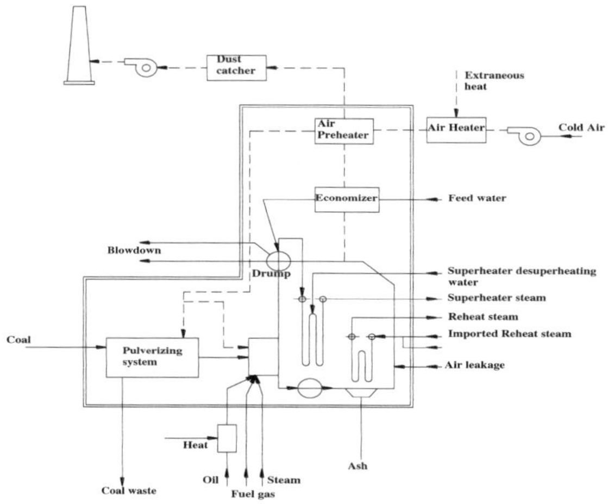

Figure 2. Thermal balance system of boiler unit.

3. Calculation of boiler efficiency

3.1. Calculation Thought

Based on the above analysis, while the low temperature economizer - air heating system is functioning, according to respective GB10184-1988 standard, the boiler efficiency calculation shall consider the flue-gas temperature at the preheater outlet as the actual exhaust gas temperature, and the heat absorbed by the heater as the external heat source to be added in the feeding heat of the boiler. In the case of no heater under normal circumstances, it is only necessary to convert the exhaust gas temperature in accordance with equation (1) so that it can guarantee the inlet air temperature. At this moment, it is required as a must to take into consideration the influences of the heater on the exhaust gas temperature. According to Part 7.2.1.2 in the GB10184-1988 standard, when it is equipped with a heater and put into operation, calculations and rectifications corresponding to three different circumstances shall be performed, analysis of which is relatively complex.

$$\theta_{\mathrm{py}}^{\mathrm{b}} = \frac{t_0^{\mathrm{b}}\left(\theta_{\mathrm{ky}}' - \theta_{\mathrm{py}}\right) + t_{\mathrm{ky}}'\left(\theta_{\mathrm{py}} - t_0\right)}{\left(\theta_{\mathrm{ky}}' - t_0\right)} \tag{1}$$

Where,

$\theta_{\mathrm{py}}^{\mathrm{b}}$——exhaust gas temperature while conversion to the guaranteed inlet air temperature, ℃;

t_0^{b}——guaranteed inlet air temperature, ℃;

t_0——Measured temperature, ℃;

θ_{ky}'——Measured gas temperature at air preheater inlet, ℃;

θ_{py}——Measured exhaust gas temperature, ℃.

As the rectification method for a heater is obscure and difficult to use, in the following section of the article, we would discuss about a boiler efficiency calculating method that doesn`t require the rectification of exhaust gas temperature in accordance with the heater.

Based on the above analysis, we discover that, there are two ultimate flow directions for flue-gas heat absorbed by the low temperature economizer; one part would be used to heat up cold air, whereas the other part would remain in the condensate water, with the former part serving as a pre-positive preheater. Thus, according to the flow direction of heat, the low temperature economizer`s heater can be decomposed into two sections, the virtual pre-positive preheater and the condensate water heater. Reduced heat of the flue-gas in the pre-positive preheater is equivalent to the increased heat of the cold air entering into the boundary. Therefore, the pre-positive preheater can be merged into the calculation boundary. The condensate water heater also absorbs partial flue-gas heat, but the feed water enthalpy entering into the boiler has not increased accordingly, which shall not be merged into the calculation boundary. The re-divided calculation boundary is shown in Figure 3. After the pre-positive preheater being merged into the calculation boundary, exhaust gas temperature of the boiler shall be equivalent to the flue-gas temperature at the pre-positive preheater outlet. Due to the fact that the pre-positive preheater doesn`t exist for real, we refer to the flue-gas temperature at the outlet as converted exhaust gas temperature. By considering the converted exhaust gas temperature as the real temperature, boiler efficiency during the operation with a low temperature economizer - heater can be acquired.

3.2. Calculation of converted exhaust gas temperature

Based on the flow direction of heat, the low temperature economizer's joint heater system can be decomposed into the pre-positive air preheater and the condensate water heater, and flue-gas temperatures and heat at various positions have been shown in Figure 4.

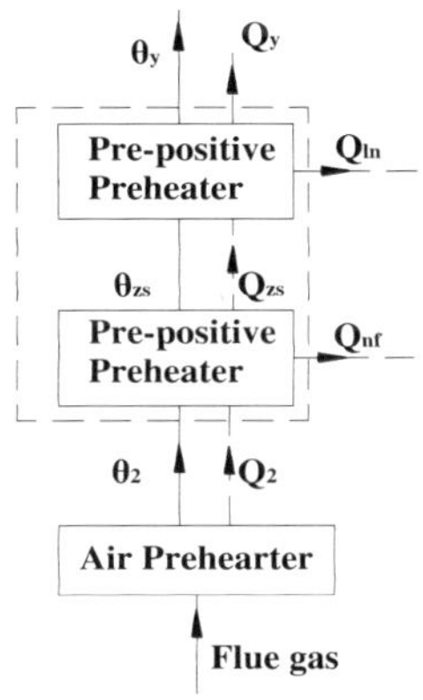

Figure 3. Schematic diagram of energy carried by flue gasθ2-exhaust gas temperatureat air preheater outlet; Q2- heat of flue gas at the outlet of air preheater; θzs-gas temperature at virtual pre-positive preheater; Qzs- heat of flue gas at virtual pre-positive preheater; Qnf-Heat absorption of air in air heater; θy-gas temperature at low temperature economizer outlet; Qy- heat of flue gas at the outlet of air preheater; Qln-Heat absorption of condensate water after heating cold air;

$$Q_{zs} = Q_2 - Q_{nf} \tag{2}$$

$$Q_{nf} = D_{ln}(h''_{ln} - h'_{ln}) \tag{3}$$

Where,

D_{ln}-Condensate water flow rate in air heater, kg/h;

h_{ln}''-Enthalpy of Condensate water at air heater outside, kJ/kg;

h_{ln}'- Enthalpy of Condensate water at air heater outside, kJ/kg;

Through equation (3), heat output of condensate water in the heater can be calculated. Assuming the heat exchange efficiency is 100%, then heat is reduced by Q_{nf} as being taken away by flue-gas in the hypothetical pre-positive preheater. According to equation (2), heat Q_{zs} taken away by flue-gas at the pre-positive preheater outlet can be calculated, based on which, the converted exhaust gas temperature θ_{zs} at the pre-positive preheater outlet can be calculated through iteration. The process of iteration is shown in Fig.4. It is required to firstly assume the flue-gas temperature at the pre-positive preheater outlet is θ'_{py}, based on which, the exhaust gas heat loss Q'_2 can be calculated. In case of the difference between it and the heat Q_{zs} taken away by flue-gas at the pre-positive preheater outlet being no more than 0.05%, the assumption holds; otherwise, the value of θ'_{py} shall be re-assumed and re-calculated, until the difference is no more than 0.05%.

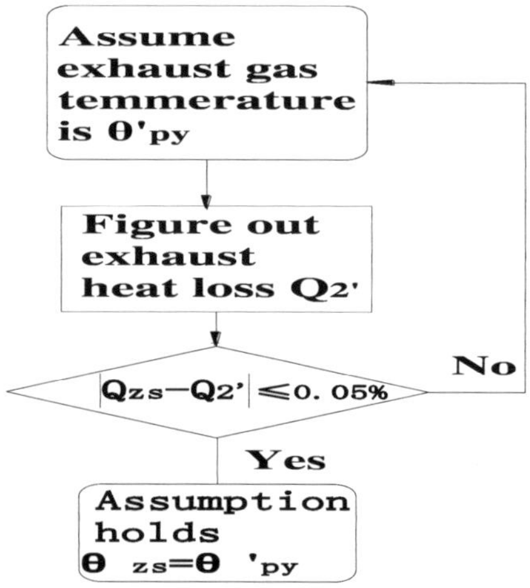

Figure 4. Flow chart of iterative calculation of converted exhaust gas temperature

4. Conclusion

While the modification of the low temperature economizer`s joint heater bringing environmental-protection and economic benefits to the thermal power unit, it also has some certain interferences with the boiler efficiency calculation.

By referring to the relevant provisions in respective GB10184-1988 standard on calculation boundary, and starting from the aspect of Energy Conservation, the article analyses corresponding error causes in boiler efficiency calculation on the basis of considering the flue-gas temperature at the low temperature economizer outlet as the exhaust gas temperature; it also points out according to the calculation methods specified in GB10184-1988 standard that, boiler efficiency calculation shall regard the flue-gas temperature at the preheater outlet as the exhaust gas temperature, with the heat from heater as the external heat and being added into the feed heat of the boiler. When rectifying the exhaust gas temperature, the influences of the heater should be considered.

In addition, by analysing the energy flow direction in the low temperature economizer - heating system, the article decomposes the system into two parts of the hypothetical pre-positive preheater and the condensate water heater while merging the previous one into the boiler efficiency calculation boundary, so as to replace the actual exhaust gas temperature with the converted waste temperature at the pre-positive preheater outlet in order to calculate the boiler efficiency. By applying this method, it is capable of avoiding the complex rectification to the heater during boiler efficiency calculation; instead, it is only necessary to rectify the (converted) exhaust gas temperature to the extent that it guarantees the feed air temperature, so that the boiler efficiency calculation can be carried out in a more concise manner.

5. References

[1] Kitto J B 1996 *Presented at Missouri valley electricassociation engineering conference* , Kansas City
[2] Guo Jianglong, Zhang Shufang and Song Zhiping 2004 *J Proceedings of the CSEE 24(3) 210-215*
[3] Jiang Feng and Wang Peihong 2008 *J Turbine Technology* 50(4) 261-264
[4] Zhang Weibin, Zhang Chunfa and Wu Zhong 2008 *J Turbine Technology* 50(4) 259-260.
[5] Lu Wanpeng, Sun Fengzhong and Shi Yuetao 2011 *J Proceedings of the CSEE* 31(11) 6-10.
[6] Huang Xinyuan, Sun Fengzhong and Shi Yuetao 2008 *J Thermal Power Generation* 37(3)56-58 .
[7] Zhao Zhijun, Feng Weizhong and Zhang Ling 2009 *J Journal of Power Engineering* 29(11)994-997.
[8] Lin Wanchao 1994 Theory of energy conservation for thermodynamic system in coal-fired power plant *Xi' an Jiaotong University Press* pp 224-239 .

IOP Conf. Series: Earth and Environmental Science **63** (2017) 012003 doi:10.1088/1755-1315/63/1/012003

Analysis on the Factors of Heat Transfer Rate and Air Volume in Tunnel

Cuifeng Du and Menglong Bian

School of Civil and Resource Engineering, University of Science and Technology Beijing,Beijing 100083,China

243519027@163.com

Abstract. In order to obtain the relationship between ventilation and heat transfer of tunnel under certain conditions, the paper analyses the convective heat transfer phenomenon under the limited conditions of driving tunnel. A simplified mathematical formula is established between heat transfer rate and air volume, which explained a phenomenon that the cooling effect is hard to be improved as air volume is increasing and in actual ventilation. According to the established mathematical formula, the ratio of heat transfer rate to air volume is analyzed in different circumstances through the field experiment and numerical simulation. A conclusion is drawn that the ratio of Φ /Q can be proposed to evaluate the ventilation cooling effect.

1. Introduction

With the depletion of shallow mineral resources, many mines have entered deep mining [1]. The heat hazard is caused by high ground temperature in deep mining [2-3]. Heat hazard is derived from compressed air, high temperature rock and equipment [4-5]. At present, the main methods to solve heat hazard are ventilation cooling, air conditioning and refrigeration, isolated heat, individual protection [6-7]. The air-conditioning system is accounted for more than 20% of all electricity costs in mine for solveing the heat hazard [8-10]. Ventilation is an economical and effective way of cooling, and it can solve the problem of heat hazard which is not serious. Grasping the relationship between ventilation and heat transfer is the basis to solve the problem of heat hazard, especially in the tunnel [11-12].

Excavated tunnel has the most serious heat hazard underground for heat is easy to accumulate due to its unique form of space. It is very important to prevent and control underground heat hazard by studying the laws of ventilation and cooling in excavated tunnel and the optimal setting of ventilation parameters.

Based on the basic theory of heat transfer, a simplified mathematical formula is established between the convective heat transfer rate and air volume, which explains a phenomenon that the cooling effect is hard to be improved as air volume is increasing and in actual ventilation. The ratio of Φ /Q can be proposed to evaluate the ventilation cooling effect through field experiment and numerical simulation.

2. Convective heat transfer model of tunnel head

There exists heat exchange in the form of thermal convection and thermal radiation between air and rock [13]. The thermal radiation can be neglected because the main components in the air are O_2 and

N_2 which neither emits nor absorbs radiation as two-atom gas [14]. Excluding the impact of water, only thermal convection needs to be considered in the cooling of surrounding rock.

After a long time of mine ventilation, the tunnel surface temperature is basically stable and the air which regarded as an incompressible fluid also has a stable temperature [15]. Without considering the heat and moisture exchange and other heat sources, it can be concluded that the heat dissipation of the surrounding rock is only caused by the thermal convection. In order to simplify the analysis process, this paper analyzes the heat convection in "Heat exchange area", as shown in Figure1.

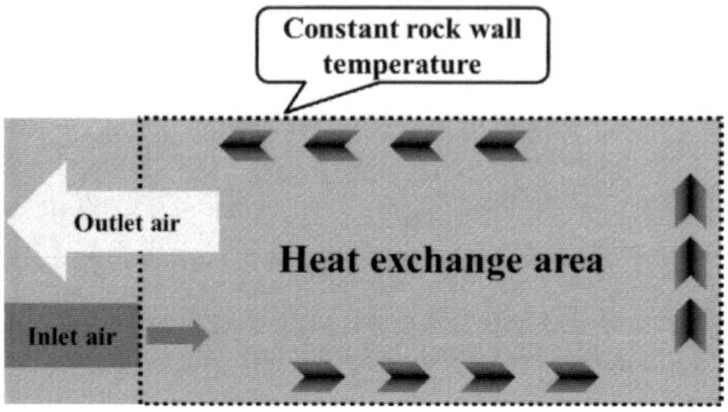

Figure 1. Sketch drawing of heat convection in tunnel.

As shown in Figure1, the fresh air temperature of the forced ventilation is set to be constant, and the air flow is heat exchange in the space between the outlet of the air duct and the head of the tunnel. The temperature of the rock wall is also constant, so only the ventilation air volume has an effect on the heat exchange. In this paper, the relationship between convective heat flow and ventilation air flow is established. According to convective heat flow calculation formula:

$$\Phi = qA \tag{1}$$

Where: Φ is convective heat transfer rate (W or J/s); q is heat flux (W/m^2); A is convective heat transfer area (m^2); The calculation formula of q:

$$q = h\left(t_w - t_f\right) \tag{2}$$

Where: t_w is the temperature of rock surface(℃); t_f is the average temperature of the fluid(℃); h is convective heat transfer coefficient(W/m^2•K). The calculation formula of h:

$$h = \frac{\lambda}{D} N_u \tag{3}$$

Where: λ is thermal conductivity of the fluid (W/m•K), and the thermal conductivity of air is 0.025 W/m•K; D is characteristic length (m); Nu is Nusselt number (dimensionless).

According to the experimental formula of forced convection tube, the calculation formula of Nu:

$$Nu = 0.023\,\mathrm{Re}^{0.8}\,\mathrm{Pr}^{0.4} \tag{4}$$

Where: Re is Reynolds number:

$$\mathrm{Re} = \frac{\rho v D}{\mu}$$

ρ is fluid density (kg/m^3); v is fluid velocity (m/s); μ is dynamic viscosity coefficient (Pa•s). Pr is Prandt number:

$$\mathrm{Pr} = \frac{C_p \mu}{\lambda}$$

C_P is specific heat at constant pressure (kJ/kg•K). The calculation formula of C_P:

$$C_p = 1.05 - 0.365\theta + 0.85\theta^2 - 0.39\theta^3$$

$$\theta = \frac{T}{1000} = \frac{273.15 + t}{1000}$$

Substituting (2), (3) and (4) into formula (1), a conclusion is obtained:

$$\Phi = 0.023 \frac{\lambda^{0.6} \rho^{0.8} v^{0.8} C_p^{0.4} A \,(t_w - t_f)}{D^{0.2} \mu^{0.4}} \tag{5}$$

According to the air volume calculation formula: $Q=v{\bullet}S$, substituting it into formula (5).

$$\Phi = 0.023 \frac{\lambda^{0.6} \rho^{0.8} Q^{0.8} C_p^{0.4} A \,(t_w - t_f)}{D^{0.2} \mu^{0.4} S^{0.8}} \tag{6}$$

When the temperatures of rock surface and inlet air air are constant, the data such as λ , ρ , C_P, A, D, μ and S are basically stable. The formula (6) is simplified as:

$$\Phi = kQ^{0.8}(t_w - t_f) \tag{7}$$

The coefficient of k in the formula is considered to be constant:

$$k = 0.023 \frac{\lambda^{0.6} \rho^{0.8} C_p^{0.4} A}{D^{0.2} \mu^{0.4} S^{0.8}}$$

The convective heat flow is directly proportional to the 0.8 th power of the air volume and the temperature difference between air and wall. The temperature difference is constant after a long time of mine ventilation, and the air temperature is falling not noticeable due to the increased air flow along with the heat flow. This is the reason that the air cooling effect is not obvious as the air flow is increasing. The field experiment and numerical simulation are used to analyze the relationship next.

3. Field experiment

In the field experiment, an arch tunnel is selected that is 5.4m^2 of sectional area and is 8.9 m of perimeter. A ventilator was set in the tunnel, and the model is JK67-2 with the power of 11kW. A flexible duct connected to the ventilator has a diameter of 0.4m and there is 10m between inlet air and tunnel head face. The inlet air temperature is 22.5℃, the rock wall temperature is 29℃, and the air temperature is 29℃ in the area before the ventilation. In the field experiment, the measuring points are arranged in several sections of the area, and the average temperature of the section is monitored. The 1 hour experimental data as shown in Table 1.

Table 1. Experimental data of temperature.

time /min	t_f /℃	time /min	t_f /℃	time /min	t_f /℃	time /min	t_f /℃	time /min	t_f /℃	time /min	t_f /℃
1	29	11	25	21	24.7	31	24.2	41	24	51	23.8
2	27	12	24.9	22	24.7	32	24.1	42	24	52	23.8
3	26.3	13	24.9	23	24.6	33	24.1	43	24	53	23.8
4	25.3	14	24.9	24	24.6	34	24.1	44	23.9	54	23.8
5	25.3	15	24.8	25	24.6	35	24.1	45	23.9	55	23.8
6	25.2	16	24.8	26	24.4	36	24.1	46	23.9	56	23.8
7	25.2	17	24.7	27	24.5	37	24.1	47	23.9	57	23.8
8	25.1	18	24.7	28	24.3	38	24	48	23.9	58	23.8
9	25.1	19	24.8	29	24.3	39	24	49	23.8	59	23.8
10	25	20	24.7	30	24.3	40	24	50	23.8	60	23.8

Table1 presents a variation of air temperature with ventilation time. The convective heat flow was analyzed after one hour of ventilation: λ = 0.025 W/m•K, ρ =1.183 kg/m^3, C_P=1006.31 J/kg•K,

A=94.4 m^2, D=2.43 m, μ =1.825×10^5 Pa•s, S=5.4 m^2, Substituting these data into equation (6) and (7), Φ =652 W, k=75.5. Due to the experimental conditions, the numerical simulation software FLUENT is used to study the heat transfer process under different air flow conditions.

4. Numerical simulation

4.1. Tunnel model

In accordance with the experimental site conditions, the establishment of model is shown in Figure 2. The total length of the tunnel is 11m, and the length of duct is 1m.

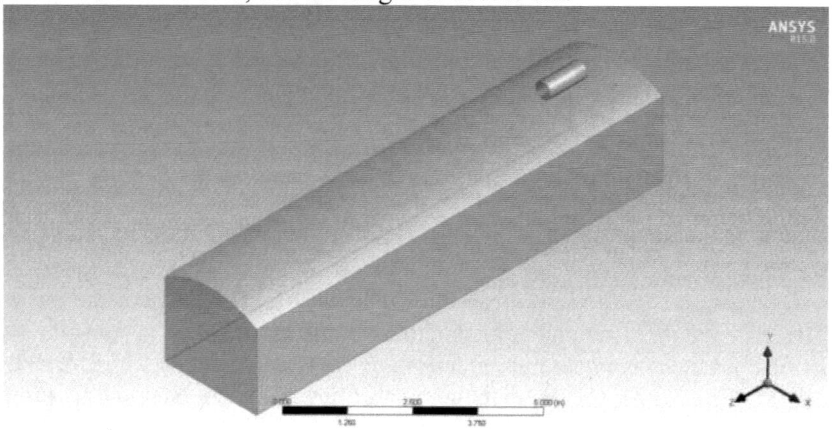

Figure 2. The tunnel model.

4.2. Meshing

The model which is shown in figure 3 is meshed with unstructured meshes, the number of mesh is 137963, and the number of nodes is 31632.

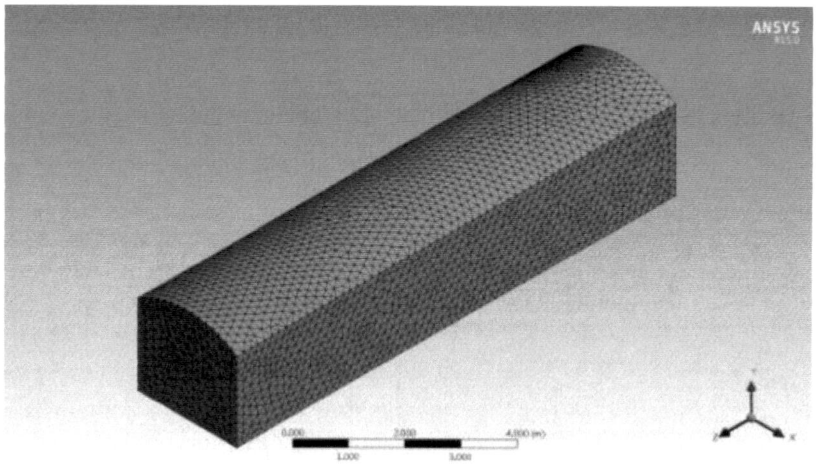

Figure 3. The meshed model.

4.3. Parameters setting

The parameters are set as table 2. Table 2 presents the performance parameters of Fluent.

Table 2. Fluent parameter settings.

name	setting information

Solver	Pressure - based solver
time	Unsteady state
Energy equation	on
model	Realizable k-ε model
Turbulence intensity	3.75% -5%
Hydraulic diameter	2.43m
Entry boundary type	velocity
Exit boundary type	outflow

4.4. Simulation results

The temperature field was simulated after 1 hour's ventilation which the air volumes are respectively as 0.38 m^3/s, 0.63 m^3/s, 0.88 m^3/s, 1.13 m^3/s, 1.38 m^3/s, 1.88 m^3/s, 2.39 m^3/s, 2.89 m^3/s, 3.39 m^3/s. Figure 4, figure 5, figure 6, figure 7 are respectively the temperature field of 0.38 m^3/s, 1.13 m^3/s, 1.88 m^3/s, 3.39 m^3/s.

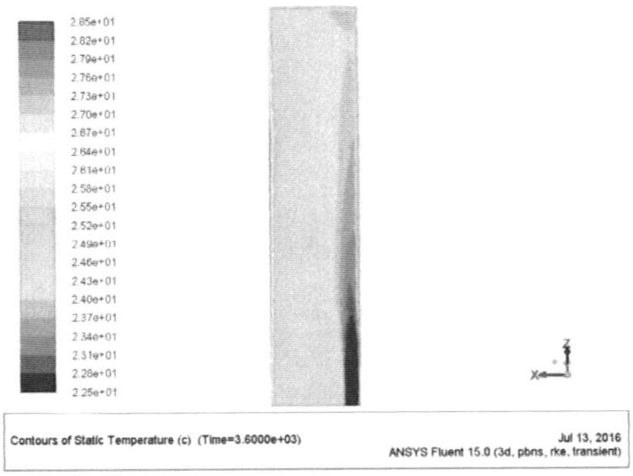

Figure 4. Temperature field of 0.38m^3/s ventilation volume.

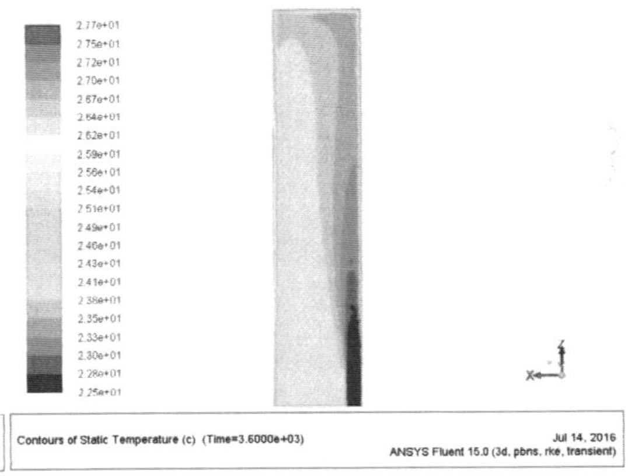

Figure 5. Temperature field of 1.13m^3/s ventilation volume.

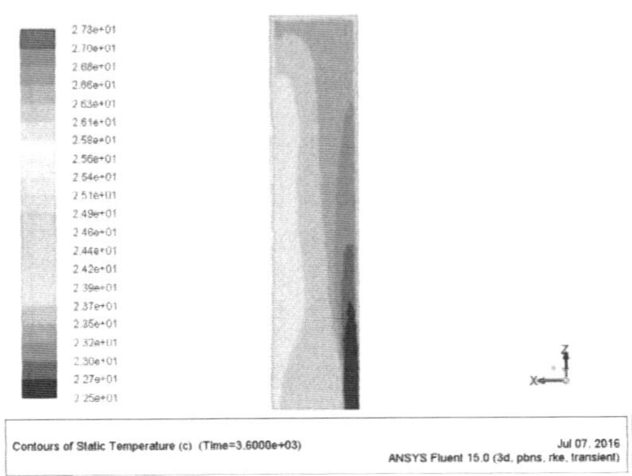

Figure 6. Temperature field of 1.88m^3/s ventilation volume.

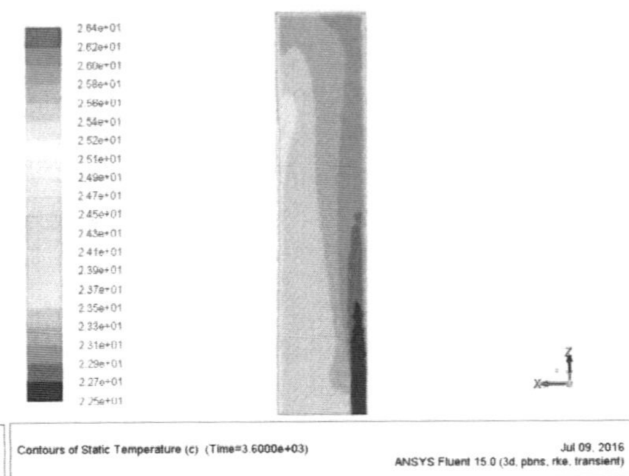

Figure 7. Temperature field of 3.39m^3/s ventilation volume.

Figures presents that the larger the air volume, the greater the influence range of the jet and the more stable the air temperature. The air volume in Figure 6 is the same as that in the experiment. The average air temperature is 23.8℃ after one hour of ventilation (Table 1), and the numerical simulation result is 23.7℃, the two values are very close.

In order to analyze the relationship between air volume and convective heat flux, the wall and air temperature values must be determined. Starting from the outlet of the duct, a cross section is selected every 2.5 m, the arithmetical mean of the five sections is taken as the average temperature t_f.

4.5. Data analysis

After processing the data, t_f which under different air volumes are obtained. According to the formula (7) , the calculated data are shown in Table 3, $\Delta t=t_w-t_f$, t_w=29℃, k=75.5.

Table 3 presents that Q is increased from 0.38 m^3/s to 3.39 m^3/s by 8.92 times, and Φ is increased by 8.23 times from 137.6 W to 1132.8 W. Combined with the data to form the air volume Q, Φ and Δt changes in Figure 8.

Table 3. Data of convective heat transfer

Q /m^3·s^{-1}	t_f /℃	Δt /℃	Φ /W	Φ/Q /J·m^{-3}
0.38	25	4	137.6	365.1
0.63	24.5	4.5	232.9	370.9
0.88	24.2	4.8	330.3	375.7
1.13	24	5	414.1	366.4
1.38	23.9	5.1	493	356.8
1.88	23.7	5.3	664.7	352.8
2.39	23.5	5.5	824.6	345.5
2.89	23.4	5.6	985.8	341.3
3.39	23.3	5.7	1132.8	334

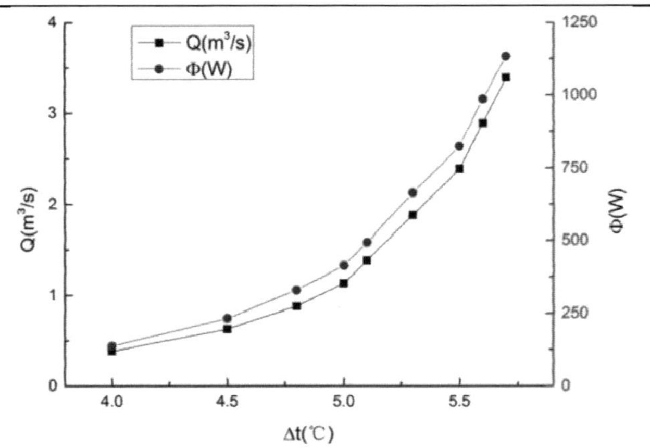

Figure 8. Ralation of Q、Φ and Δt .

Figure 8 presents that heat flow will increase with the air volume from the growth trend of Q and Φ. This is consistent with the formula (7) .The data of Φ/Q and Δt from Table 3 are formed in Figure 9.

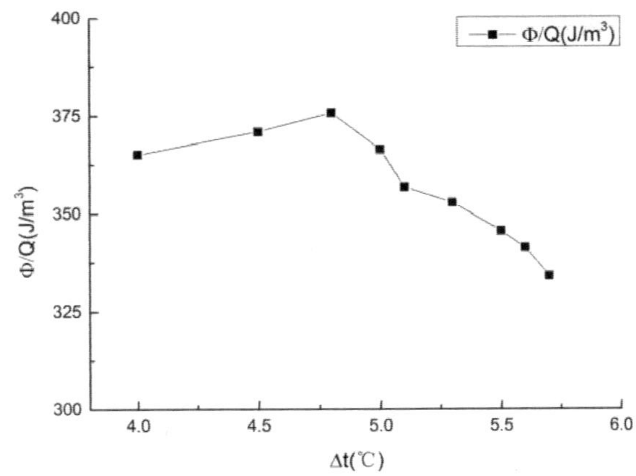

Figure 9. Ralation of Φ/Q and Δt.

Figure 9 presents that *Φ/Q* have different numbers at different *Δ t*, *Φ/Q* has maximum at *Δ t*=4.8 ℃ when *Φ/Q* = 375.7 J/m^3 and *Q* = 0.88 m^3/s. The unit of *Φ/Q* is J/m^3. *Φ/Q* can be expressed as the heat transfer in the average unit volume of air, the bigger the number, the more the heat transfer, the greater cooling effect of ventilation. Therefore, *Φ/Q* can be used as a cooling indicator to evaluate the effect of ventilation. Strictly speaking, *Φ* is a constantly changing number, but the thermal parameters of ventilation and tunnel are stable after a few years or even decades of ventilation, and the heat transfer process is stable, the heat flow is also basically stable. Ventilation and heat transfer are widely existed in mine ventilation. Whether the heat source is from surrounding rock, gushing water or equipment, there exists the maximum number of *Φ/Q* in the mine ventilation. The ratio has broad applicability to evaluate the ventilation cooling effect. Under the experimental conditions, the maximum number of *Φ/Q* appears only when *Δ t* = 4.8℃and *Q* = 0.88 m^3/s. Therefore, the ventilation can solve the problem that heat hazard is not serious in the mine.

5. Conclusion

The conclusions of this paper are as follows:

1) The problem of convective heat transfer in tunnel under limited condition is studied. A simplified mathematical formula is established between heat transfer rate and air volume, which explained a phenomenon that the cooling effect is hard to be improved as air volume is increasing and in actual ventilation.

2) Through the field experiment and numerical simulation, the temperature field data of the tunnel under different air volumes are obtained, and the calculation method of convective heat transfer rate is provided according to the formula (7).

3) The ratio of *Φ/Q* can be proposed to evaluate the ventilation cooling effect. It needs to be further verified by theory and practice.

6. References

[1] Liu W V, Apel D B, Bindiganavile V S. Cylindrical models of heat flow and thermo-elastic stresses in underground tunnels. *International Journal of Numerical Methods for Heat & Fluid Flow*, 2016, **26** (7): 2139-2159.

[2] Sasmito A P, Kurnia J C, Birgersson E, Mujumdar, A S. Computational evaluation of thermal management strategies in an underground mine. *Applied Thermal Engineering*, 2015, **90**: 1144-1150.

[3] Liu Jingxian, Mao Ning, Chen Xiaosong, Chang Deqiang. Research on geothermal distribution of Hongtoushan underground metal mine. Proceedings of The First International Conference on

Information Sciences, Machinery, Materials And Energy. *Advances in Intelligent Systems Research*, 2015, **126**: 541-544.

[4] Ghoreishi Madiseh S A, Sasmito A P, Hassani F P, Amiri L. Heat transfer analysis of large scale seasonal thermal energy storage for underground mine ventilation. Clean, Efficient and Affordable Energy for A Sustainable Future. *Energy Procedia*, 2015, **75**: 2093-2098.

[5] Hao Xiaohua, Ma Heng, Liu Yinxia, Zhang Huibo. Calculation model of airflow temperature and simulation system in coal mine. *Advanced Materials Research*, 2014, **1010-1012**:1449-1453.

[6] Flouris A D, Cheung S S. Design and control optimization of microclimate liquid cooling systems underneath protective clothing. *Annals of Biomedical Engineering*, 2006, **34** (3): 359–372.

[7] Kenny G P, Schissler A R, Stapleton J, Piamonte M, Binder K, Lynn A, Lan C Q, Hardcastle S G. Ice cooling vest on tolerance for exercise under uncompensable heat stress. *Journal of Occupational and Environmental Hygiene*. 2011, **8** (8): 484–491.

[8] Du Plessis G E, Arndt D C, Mathews E H. The development and integrated simulation of a variable water flow energy saving strategy for deep-mine cooling systems. *Sustainable Energy Technologies and Assessments*, 2015, **10**: 71-78

[9] Uys D C, Kleingeld M, Cilliers E. Converting an ice storage facility to a chilled water system for energy efficiency on a deep level goldmine. *Proceedings Of The 13th Conference On The Industrial And Commercial Use Of Energy*, 2015, 76-83.

[10] Du Plessis G E, Liebenberg L, Mathews E H. Case study:The effects of a variable flow energy saving strategy on a deep-mine cooling system. *Applied Energy*, 2013, **102**: 700-709.

[11] Yue Gaowei, Li Minmin, Xu Mengfei. Heat transfer characteristics of seasonal ventilation and surrounding rock in deep mines. *Science & Technology Review*, 2016, **34** (2): 195-199.

[12] Chen Xishan, Zhang Yongliang. Pre-cooling the inlet airflow of the mine utilizing the shallow rock low temperature resource. *Advanced Materials Research*, 2014, **868**:241-250.

[13] Wang Yijiang, Zhou Guoqing, Wu Lei. Unsteady heat−moisture transfer of wet airway in deep mining. *Journal of Central South University of Technology*, 2013, **20** (7): 1971−1977

[14] Zhang Yang, Jin Xing, Xin Chaojun. Temporal evolution of laser plasma temperature. *High Power Laser and Particle Beams*, 2009, **21** (9): 1326-1330.

[15] Loredo C, Ordonez A, Garcia-Ordiales E, Alvarez R, Roqueni N, Cienfuegos P, Pena A, Burnside N M. Hydrochemical characterization of a mine water geothermal energy resource in NW Spain. *The Science of the total environment*, 2016, **576**: 59-69.

IOP Conf. Series: Earth and Environmental Science **63** (2017) 012004 doi:10.1088/1755-1315/63/1/012004

Analysis of Short-circuit and Protection Failure Risk Considering Random Output of Distributed Photovoltaics

Yao Da, Zhang Huan and Deng Wei

Maintenance company, State Grid Hunan electric power company, Changsha 410007, China

310851982@qq.com

Abstract. Distributed photovoltaics (DPV) will increase or shunt the fault current as a branch of the power supply. The random outputs of DPV will also cause a random distribution of fault current, while the breaking capacity of the breaker and the setting value of the current protection is pre-set value, and cannot flexibly change, so DPV will bring a certain degree of influence on the breaking margin and the sensitivity of protection. This paper makes probability distribution calculating model of fault current containing DPV, and takes IEEE 33-node system as an example, simulated the probability distribution of fault current at different penetration of DPV. Finally, from the two indicators of the breaking margin of breaker and the sensitivity of protection, analysed the protection failure risk after the access of DPV.

1. Introduction

Conventional current protection is configured in a single-sided power-supply mode, while DPV will provide fault current as a branch of the power supply, and with the different distance between the fault point and the access point, fault current provided by DPV current will play a role of increase, shunt or reverse, making different influence on protection, even lead to protection misoperation or rejective-operation [1-4]. In [5-8], the fault current is simulated to analyze the protection misoperation or rejective-operation of different locations, and new protection setting methods after DPV access are proposed. However, in the above literature, DPV uses a constant power model, without considering the influence of randomness of DPV output on fault current and protection.

This paper develops the probability distribution model of fault current according to the random output of DPV, and taking IEEE 33-node system as an example, simulated the probability distribution of fault current at different penetration of DPV. Finally from the two indicators of the breaking margin of breaker and the sensitivity of protection, analyzed the protection failure risk after the access of DPV.

2. Probability Distribution Calculating Model of Fault Current containing DPV

In the system with DPV, DPV will provide part of the fault current, so it can transfer to one voltage source and reactance, as shown in figure 1. Compared with the impedance matrix of the original system, the access of DPV is equivalent to an increase of a ground branch with internal impedance of X_{DPV}[9].

The greater the output of the DPV before the fault, the greater the fault current provided by DPV after the fault, equivalent to the smaller internal impedance X_{DPV}; Conversely, the smaller the output of the DPV before the fault, the smaller the fault current provided by DPV, equivalent to the greater internal impedance X_{DPV}. Therefore, the fault current provided by DPV is different when different output; When multiple DPVs access, DPVs at different positions have different effects on fault current. Thus it's needed to calculate the probability distribution of fault current.

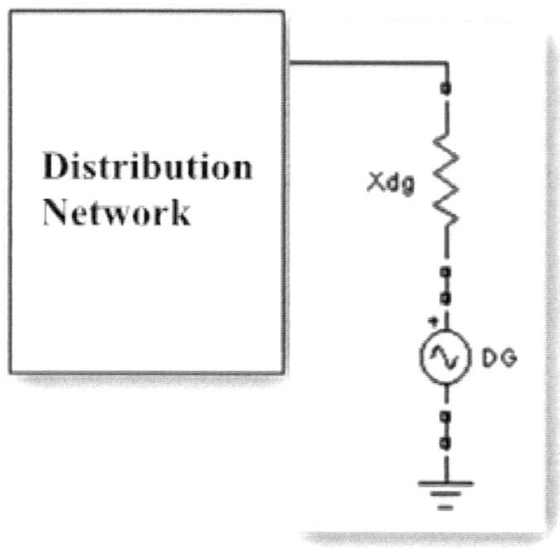

Figure 1. Fault current calculating model of DPV

3. Analysis of Breaking Margin

This paper takes IEEE33 node system as example to analyze breaking margin, the figure 2 shows its schematic, the node and branch parameter is referred in [10].

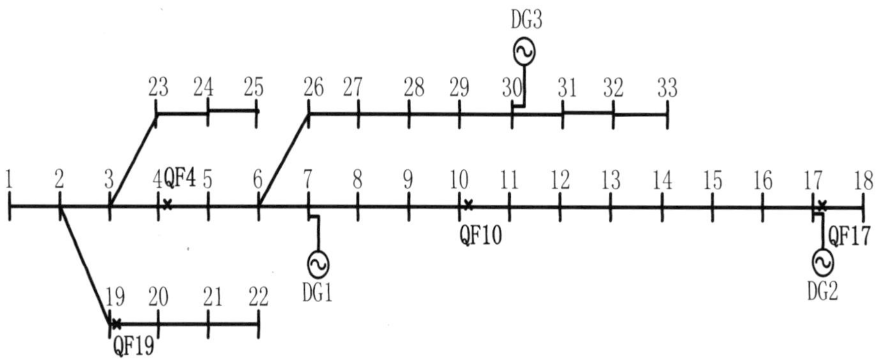

Figure 2. Schematic of IEEE33 node system

This part chooses node 7, 17 and 30 as access point and analyzes breakers QF4, QF10, QF17, QF19 on line 4-5, 10-11, 17-18, 19-20.

When three-phase short-circuit happens at downstream of circuit breaker, the fault current is largest, which can check the breaking margin of breakers. Setting penetration of DPV at 20%, 30%,40%, and three-phase short-circuit at the downstream outlet of QF4, QF10, QF17, QF19, the probability distribution of fault current is shown in figure 3.

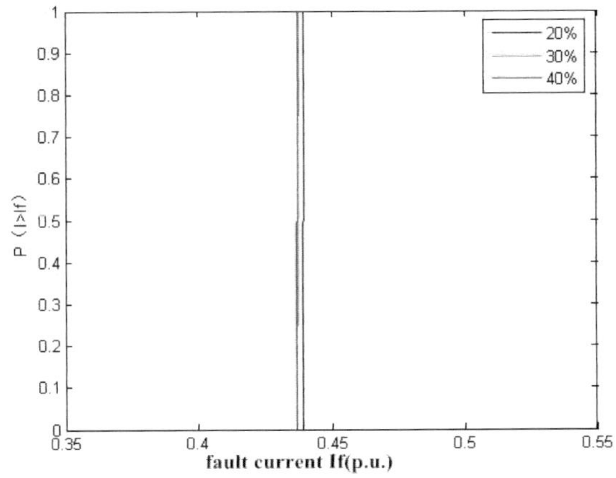

(a) Fault current of QF4

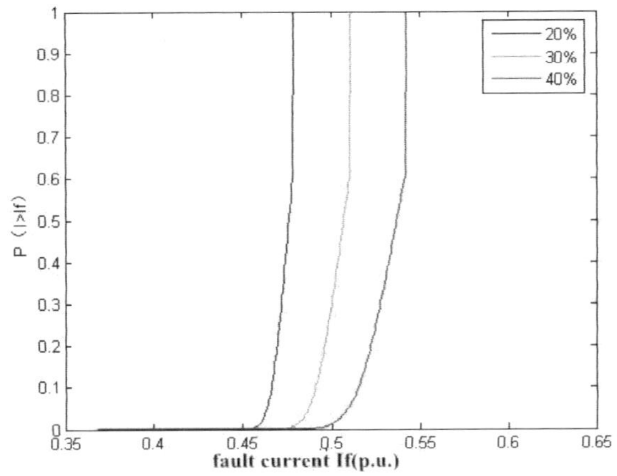

(b) Fault current of QF10

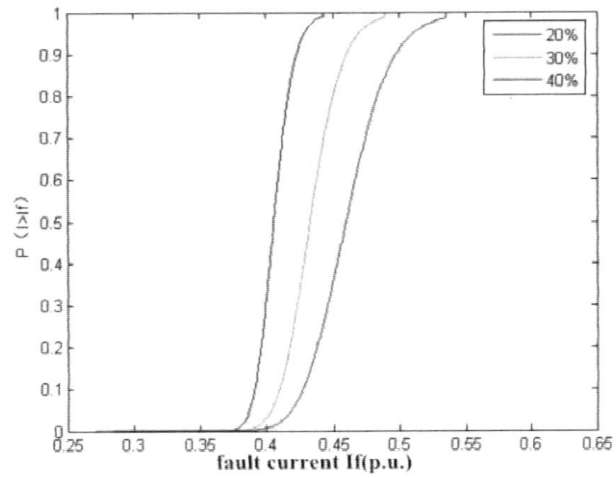

(c) Fault current of QF17

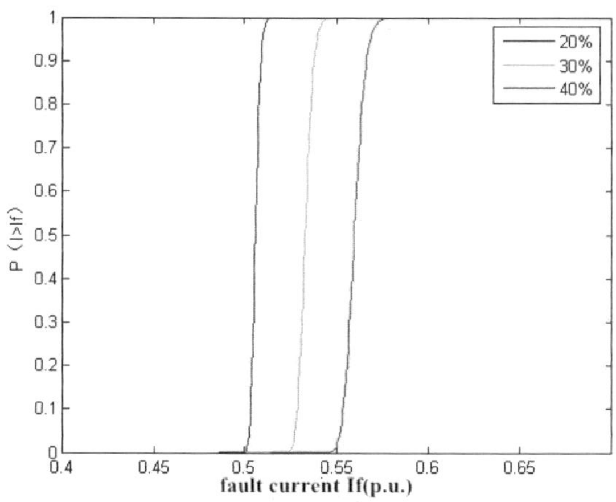

(d) Fault current of QF19

Figure 3. Probability distribution of fault current

The breaking margin (The maximum fault inrush current/maximum breaking current) of each breaker is shown in table 1.

Table 1. Breaking margin of each breaker at different penetration

	10%	20%	30%	40%	60%
QF4	55.99%	55.98%	56.03%	56.01%	56.01%
QF10	59.09%	51.87%	48.67%	45.44%	37.23%
QF17	64.93%	55.40%	47.74%	41.14%	31.34%
QF19	54.28%	48.07%	44.71%	41.40%	36.86%

It can be drawn from figure 3 and table 1:

1). The fault current provided by DPV doesn't flow through QF4, so the maximum fault current flowing through the breaker QF4 does not change with the penetration of DPV, whose breaking margin isn't affected by penetration;

2). QF10, QF17, and QF19 can flow through the fault current provided by DPV when short-circuit at the downstream outlet, so the higher the penetration, the greater the fault current, and the stronger the dispersion, leading the breaking margin becoming smaller;

3). Comparing breakers QF10 and QF17, the former only flow through the fault current provided by two of the three DPVs, while the boosting fault current from all the three DPVs can flow through QF17. Therefore, the fault current flowing through the breaker QF 17 is highly dispersive, and the breaking margin increases more rapidly with the increase of the penetration.

4. Analysis of Sensitivity of Protection

The analysis of section 3 shows that the random output of DPV makes the fault current dispersive, even the same nature of short-circuit at the same node, fault current will be different. As the current protection settings are mostly based on the maximum value of fault current as a benchmark (section 3 is based on the maximum load current). The minimum fault current can be used to check the sensitivity, so the access of DPV will bring a certain degree of influence on the sensitivity of protection.

In this part, sensitivity analysis is carried out with current protection section 2 as an example. The sensitivity coefficient of section 2 is defined as follows,

$$K^{\text{II}}_{sen} = \frac{I^{(2)}_{k.end.\min}}{I^{\text{II}}_{set}} \tag{1}$$

Where, $I^{(2)}_{k.end.min}$ is the minimum 2-phase short-circuit fault current that may occur at the end of the line; I^{II}_{set} is the setting value of current protection section 2. If the calculated $K^{II}_{sen} \geq 1.4 \sim 1.5$, indicating that the setting value in line with the sensitivity index.

According to the above principle, the sensitivity coefficient of protection at QF4, QF10, QF19 is calculated and analyzed for different penetration of DPV, whose results are shown in table 2- table 4

Table 2. Sensitivity coefficient of QF10 at different penetration

	10%	20%	30%	40%	60%
I^{II}_{set} (A)	192.35	250.33	278.08	305.84	332.62
$I^{(2)}_{k.end.min}$ (A)	278.38	309.61	321.03	331.29	340.78
K^{II}_{sen}	1.447	1.237	1.154	1.083	1.024

Table 3. Sensitivity coefficient of QF19 at different penetration

	10%	20%	30%	40%	60%
I^{II}_{set} (A)	234.37	272.44	299.52	327.20	353.78
$I^{(2)}_{k.end.min}$ (A)	311.15	342.71	357.76	371.44	385.56
K^{II}_{sen}	1.327	1.258	1.194	1.135	1.090

Table 4. Sensitivity coefficient of QF4 at different penetration

	10%	20%	30%	40%	60%
I^{II}_{set} (A)	214.12	214.32	214.15	213.98	214.07
$I^{(2)}_{k.end.min}$ (A)	299.52	299.58	299.17	299.38	299.26
K^{II}_{sen}	1.401	1.398	1.397	1.399	1.398

It can be drawn from table 2- table 4:

The random output of DPV makes the fault current dispersive, the large range of fault current may occur when short-circuit at the same node. As the current protection settings are mostly based on the maximum value of fault current as a benchmark, and the minimum value of fault current to check the sensitivity, the increase of the penetration of DPV will make the protection sensitivity decrease;

The degree which the protection sensitivity is affected is related to the relative position between the protection and the access point of DPV. If the protection is located upstream of all DPVs, the sensitivity is not affected, for example, QF4; If the protection is located downstream of the DPV, the sensitivity changes greatly, such as QF10; If the protection is located in other branches without DPV, the change of sensitivity is small as the fault current provided by DPV is small, such as QF19.

5. Conclusion

In this paper, the variance of the internal impedance X_{DPV} of DPV in the short-circuit calculation model is used to reflect the randomness of the DPV output, which can simulate the probability distribution of the fault current and analyze the protection failure risk. Taking the IEEE33 bus system as an example, the calculation results show that the bigger the DPV penetration and the closer the protection from the access point, the greater the dispersion of the fault current, leading the smaller breaking margin and protection sensitivity.

6. Acknowledgments

The project was supported by Maintenance company of State Grid Hunan electric power company.

7. References

[1] Liang C H and Duan X D 2001 Distributed generation and its impact on power system. *Automation of Electric Power System* **12** pp53-56.
[2] Pang J Y, Xia X B and Fang M 2007 Impact of distributed generation to relay protection of distribution system. *RELAY* **11** pp5-8.
[3] Liu L 2008 Researches of Distribution Network Current Protection with Distributed Generation. *Southwest Jiao tong University*.
[4] Zhu L L, Li C K, Zhang H Z 2009 Directional Overcurrent Protection for Distribution Systems Containing Distributed Generation Power System Technology **14** pp 94-98.
[5] Xie H 2006 Study of the Effects of Distributed Generation on Distribution System Safety Operation *Chongqing University*.
[6] Gomez J C and Morcos M M 2001 Coordinating overcurrent protection and voltage sag in distributed generation systems. *IEEE Power Engineering Review* **2** pp16-19.
[7] Andy G and Sukumaran B 2001 Effect of Distributed Generation on Protective Device Coordination in Distribution System. *Proceedings of IEEE PES Summer meeting*.
[8] Tang K M, Wang F S, Luo J 2010 Study on Relay Protection Scheme for Distribution Network Containing Distribution Generation. *East China Electric Power* **6** pp 876-810.
[9] Lian X L, Wu Z Q and Zhao K Y 2006 Maximum Penetration Capacity Calculation for Distributed Generation taking into account Fault Current Constrains. *Journal of Electric Power Science and Technology* **1** pp50-55.
[10] Ma X F 2013 Research on Planning and Optimal Operation of Distribution Network Including Decentralised Generation. *North China Electric Power University*

IOP Conf. Series: Earth and Environmental Science **63** (2017) 012005 doi:10.1088/1755-1315/63/1/012005

Ultra-Short-Term Wind Power Prediction Using a Hybrid Model

E. Mohammed, S.Wang and J. Yu

School of Electrical Engineering and Automation, Harbin Institute of Technology, Harbin 150001, China

moheissa183@gmail.com

Abstract. This paper aims to develop and apply a hybrid model of two data analytical methods, multiple linear regressions and least square (MLR&LS), for ultra-short-term wind power prediction (WPP), for example taking, Northeast China electricity demand. The data was obtained from the historical records of wind power from an offshore region, and from a wind farm of the wind power plant in the areas. The WPP achieved in two stages: first, the ratios of wind power were forecasted using the proposed hybrid method, and then the transformation of these ratios of wind power to obtain forecasted values. The hybrid model combines the persistence methods, MLR and LS. The proposed method included two prediction types, multi-point prediction and single-point prediction. WPP is tested by applying different models such as autoregressive moving average (ARMA), autoregressive integrated moving average (ARIMA) and artificial neural network (ANN). By comparing results of the above models, the validity of the proposed hybrid model is confirmed in terms of error and correlation coefficient. Comparison of results confirmed that the proposed method works effectively. Additional, forecasting errors were also computed and compared, to improve understanding of how to depict highly variable WPP and the correlations between actual and predicted wind power.

1. Introduction

Wind power prediction (WPP) is one of the most important issues of wind-power grid integration. An accurate WPP is suggested to be practical to operators in power systems. From the applied time-series data analysis, ultra-short-term WPP (USTWPP) is very important and contributes to the stability to the power system. The accurate prediction is not only useful for ensuring operation security for the power system, but also for enhancing the economic feasibility and performance of the power system [1].

In this paper, we developed and applied a modern hybrid method to predict high resolution by applied time-series analysis in data for wind power system, which focus on understanding the current methods to forecast on different time series (different time window and different length of prediction sample). The developed method is multiple linear regressions, and least square (MLR&LS) based a hybrid model which utilizes a new principle. As a proposal is a concern, there are two main ways to predict wind power. The first way is data transformation: convert raw data of wind power into some ratio and use this ratio as the value to be predicted. The second way is data inverse-transformation: convert the predicted ratio value into actual value and then predict total wind power. In many studied cases, there are sets of alternative forecast strategies. Forecast with a single model or method (SF) may not always lead to fitting values, while the forecast with hybrid model or method (HF) is more appropriate because the hybrid strategy can take the advantages of each SF and is more likely to get better results.

In recent years, wind power as one of the most important renewable-energy sources is growing fast. Utilization of wind energy showed rapid growth in the latest years which span many regions around the world [2, 3]. In China, the growth of wind power is revealed to be faster than accounting for a growth rate of 21.2% in 2012 [4]. Most of the wind-power projects are achieved based on national goals of reducing an emission, and producing high-output electricity.

It has been shown that, wind power output has been strong fluctuating by characteristics and only is partly controllable in nature [5, 6]. Short-term or ultra-short-term WPP is required for controlling and scheduling of power generation, for self-dispatching of power transmission market [7] and for rating improvement of the overhead line [8].

High penetrated wind power to the power grid brings many challenges to power system operators [9] if not managed properly. Accurate load forecasting is more important for electric utilities planning for the future [10]. More research on wind power technology is then necessary to supplement the conventional energy sources. However, wind power is a highly fluctuating resource. The fluctuations at large wind farms have the impact to the control and management strategies of their power output in the power grid [5].

Several studies have investigated the accuracy of short-term WPP through three approaches: physical numerical weather prediction models, statistical models based historical data and statistical models with numerical weather prediction data as additional exogenous inputs [4]. Estimating wind power output through accurate forecasting could significantly reduce the uncertainty of system operators by using incorporated meteorological data. Short-term WPP and USTWPP are required for monitoring and scheduling of power grid [7].

Power forecasting has become one of the major areas of research in engineering, particularly for short-term and ultra-short-term forecasting that increasingly become important since the rise of the competitive power system in markets. Although power forecasting is the important, great challenge is expected due to influence of several exogenous variables of the power [11]. The main purposes of ultra-short-term and short-term forecasting that ranges from minutes to hours are for real time control, operation and security evaluation of the power system [12].

2. Related work and model

2.1. Time series models in forecasting

Time series is the sequence of data points, typically measured at sequential times, predominantly with uniform time intervals [13]. There are many different modern and traditional approaches to time series modeling. In this paper, the time-series analysis is performed using the autoregressive moving average(ARMA) model, autoregressive integrated moving average(ARIMA) model and MLR&LS model.

There are two types of goals for time-series analysis. The first goal is to identify the nature of the phenomenon represented by the sequence of observations, and the second one is to forecast by predicting future values of the time-series variable. Both goals require identifying and more or less formally describing the pattern of observed time-series data. Time-series data of wind power is developed on understanding the behavior of wind power signals at different time scales. As it is well known, output power of wind farm is high variable and only partial controllable [14]. In the time-series analysis, it is assumed that, the data consist of a systematic pattern (usually a set of identifiable components) and noise (error) which usually makes the pattern difficult to be identified. Most time series analytical techniques involve some form of filtering out of noise to make the pattern more salient [11, 14-16].

2.2. ARMA model

ARMA model is considered as a basic sequential method and a practical model for comprehensive time series [17]. ARMA model is a kind of time-series analysis statistical model, historically suggested by the American G. Box and British G. Jenkins of statisticians [18]. For the random and dynamic wind power data, ARMA model with time series reflects a larger advantage. The problem for this model is to determine the order [16].

ARMA model uses autocorrelation or persistence in a time-series. The general form of ARMA model is as following:

$$w_t = c + \sum_{i=1}^{p} \varphi_i w_{t-i} + \sum_{j=1}^{q} \theta_j \varepsilon_{t-j} \qquad (1)$$

Where, t is the time to be forecasted; w_t represents the forecasting value at time t; w_{t-i} represents the historical sample data at time $t-i$; φi is the autoregressive coefficient; θj is the moving average coefficient; c is the constant value; ε_{t-j} represents error at time $t-j$, is a random variable; p is the order of the autoregressive terms; q is the order of the moving-average process.

Forecast the future values; they predict future values can be using the realized ARMA model, in equation (2) is applied to predict the minutes ahead forecasting values (m = 1; 2; 3 minutes).

$$w_{(t+m)} = \sum_{i=1}^{p} \varphi_i w_{t-i} + \sum_{j=1}^{q} \theta_j \varepsilon_{t-j} \qquad (2)$$

Where, $w(t+m)$ is the predict wind power generation at time $(t+m)$.

2.3. ARIMA model

ARIMA model is one of the most popular and frequently used stochastic time series models in forecasting. ARMA model is used only for stationary time series, but ARIMA model could be used in cases of non-stationary time series. Applying generalized mathematical formulation with parameters such as(p, d, q), ARIMA model can make a non-stationary time series become stationary. ARIMA model is based on the assumption of linearity of the predicted values [19].

The general form of ARIMA model is as following:

$$(1-\sum_{i=1}^{p} \varphi_i L^i)(1-L)^d w(t+m) = (1+\sum_{j=1}^{q} \theta_j L^j)\varepsilon_t \qquad (3)$$

Where, L represents the lag operator, and d is the order of the difference.

2.4. ANN Model

Artificial neural network (ANN) model is considered an advanced technology of short-term forecasting, so it becomes an alternative technique to time series forecasting, and gained immense popularity in last few periods. ANN has a flexible structure to model a wide variety of nonlinear problems. The advantage of ANN model is that it has the potential to approximate a large class of functions with a high degree of accuracy [19]. The ANN with multi-layer structure is very adaptable because it can incorporate several levels in a single model [20]. The goal of use ANN is to find the neural network weights that reduce the model errors.

In this paper, ANN structure used is a three-layer feed-forward back propagation neural network which includes input layer, hidden layer and output layer. Actual input data in the following simulation used two variables, recorded wind power and time for the forecasted horizon. Recorded data obtains from the historical records of offshore region wind farms. The composition of the sample set is as follows: training, validation and testing samples account for 70%, 15% and 15% respectively.

In general, ANN forecasting model can be written as follows:

$$W = F[H_1(w), H_2(w), ..., H_q(w)] + \varepsilon \qquad (4)$$

Where, W represents the dependent output variables (forecasted wind power values); w represents a set of input variables (historical wind power data); F represents output layer activation functions, H_i

represents hidden layer activation functions; ε is the error term; forecasted power will be calculated by equation (5) with time series values:

$$W(t+m)=\sum_{j=1}^{q}\theta_j\cdot g(C_{0j}+\sum_{i=1}^{p}\varphi_{ij}\cdot w_{t-1})+\varepsilon_t \quad (5)$$

Where, φ_{ij} $(i=0,1,2,...,p, j=1,2,...,q)$ θ_j $(j=1,2,...,q)$ are connected weights, p is the number of input neurons and q is the number of hidden neurons.

2.5. Multiple Linear Regressions

MLR analysis in WPP technique used the weighted least square estimation method, based on statistical analysis, can calculate the relationship between other factors [21]. The least square method is a wide-range used parameter estimating method in the MLR. The basis of this method is that it can lead to the least mean square error between the actual value and the estimated value [17].

3. Methods for USTWPP

In this section, we will describe the detail procedure of USTWPP with maximum width of the prediction time window no more than 60min.The output power of a wind farm or multiple wind farms in a power grid will be determined by historical wind power data through hybrid time series model.

3.1. Procedure of USTWPP

Wind power has a random and intermittent nature characteristic, and the output power of wind represents this characteristic. Hence it is very vital to reduce these influences for improving the accuracy of WPP. This issue will be discussed from the following aspects:

- For different width of the prediction time window, what is the most appropriate length of prediction?

In this paper, the width of the prediction time window will be assigned to 20, 30 and 60min; the length or step of prediction will be assigned to 5 and 10min.

- What's the performance difference between SF and HF?

In this paper, SF is implemented by ARMA, ARIMA and ANN respectively; HF is implemented by the MLR&LS model.

- What's the performance difference between single-point prediction and multi-point prediction?

In this paper, single-point prediction is to forecast the power or power ration of the next one length, and multi-point prediction is to forecast the power or power ration of the next few lengths within the width of the prediction time window.

- For single wind farm, what's the performance difference between direct prediction for single wind power and indirect prediction for power ration to total wind power?

Power ration to total wind power is defined as equation (6):

$$w_i=\frac{P_i}{P_{total}} \quad (6)$$

Where, P_i is the wind power of the ith wind farm, and P_{total} is the sum of wind power of multiple wind farms including the ith wind farm.

After predicting power ratios w_i and total power P_{total}, the WPP of the ith wind farm is as following:

$$P_i^{pred}=w_i\times P_{total} \quad (7)$$

- For multiple wind farms, what's the performance difference between direct prediction for total wind power and indirect prediction for single wind power?

For indirect prediction, first predict the wind power of each wind farm, and then sum them to obtain the total wind power of multiple wind farms. The wind power of each wind farm can be from the direct WPP.

Figure 1 shows the detail procedure of USTWPP designed in this paper.

In figure 1, the procedure of USTWPP collects data from SCADA systems of wind farms, and power grid dispatches center, and predicts WPP of single wind farm, total WPP of multiple wind farms, and power rations of different wind farms at different prediction conditions and with different prediction ways. The predicted value of the box②can be obtained from the box③and④according to the formula(14), and the predicted value of the box⑤can be obtained from the sum of the box①.

This paper will observe and compare the error of USTWPP respectively from the above aspects.

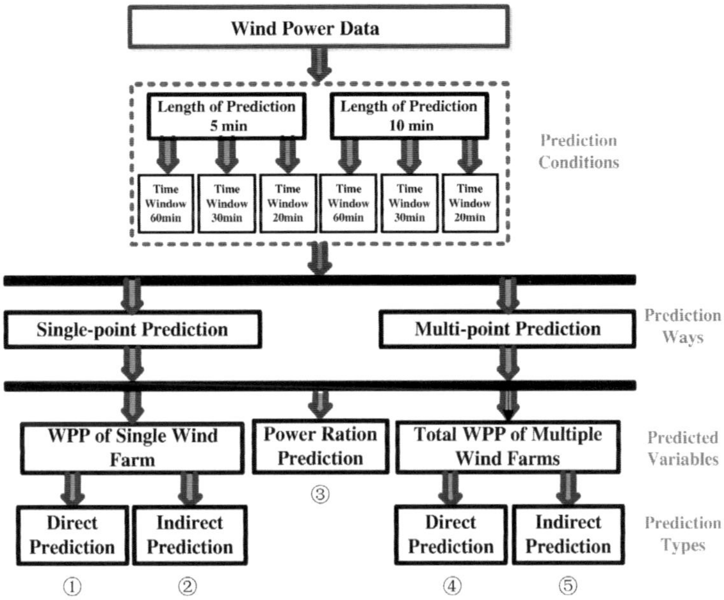

Figure 1. Detail procedure of USTWPP.

3.2. Forecasting frame-work by MLR & LS hybrid model

MLR is a new analysis tool, and has the statistical learning mechanism to explain the behavior of one of the dependent variables as a function of many other predictor variables. MLR allows finding the best linear prediction equation between dependent and other predictor variables. The basic idea of the hybrid model in this paper is to combine MLR and LR, which retains advantages to each approach and access to the largest combination advantages. The main goal of doing so is to improve the forecast accuracy.

Next, we will review the hybrid model forecasting method. The general form of MLR model is as following:

$$\beta_0 + \sum_{j=1}^{p} \beta_j w_j + \varepsilon_i = W_i \tag{8}$$

Where, W_i is the variable to be predicted; w_j is the predictor variable; p is the number of predictor variables; ε is the regression error; β is the regression coefficient that reflects the weight of each predictor variable denotes the regression error.

The regression coefficients are calculated by using the least square function according to the following formulas:

$$\sum_{i=1}^{n} (W_i - \beta_0 - \sum_{j=1}^{p} \beta_j w_{ij})^2 \rightarrow \min \tag{9}$$

The optimal solution of β is:

$$\hat{\beta} = (w^{\mathrm{T}} w)^{-1} w^{\mathrm{T}} W \tag{10}$$

The fitted values are $\hat{W} = w\hat{\beta}$, the residuals are $\hat{E} = W - \hat{W}$.

Equation (8) can be represented by the following matrix to estimate the regression coefficients.

$$\begin{bmatrix} 1 & w_{11} & w_{12} & \cdots & w_{1p} \\ 1 & w_{21} & w_{22} & \cdots & w_{2p} \\ \cdots & \cdots & \cdots & \cdots & \cdots \\ 1 & w_{n1} & w_{n2} & \cdots & w_{np} \end{bmatrix} \begin{bmatrix} \beta_0 \\ \beta_1 \\ \beta_2 \\ \cdots \\ \beta_p \end{bmatrix} + \begin{bmatrix} \varepsilon_1 \\ \varepsilon_2 \\ \cdots \\ \varepsilon_n \end{bmatrix} = \begin{bmatrix} W_1 \\ W_2 \\ \cdots \\ W_n \end{bmatrix} \quad (11)$$

When MLR model is used to WPP, the first part to the left of equation (11) showed the historical wind power or power ration data matrix with $n\times(p+1)$dimensions, the right of the equation (11) show the predicted wind power or power ration (response vector) with n dimension. Equation (11) above can be simplified to: $\boldsymbol{w\beta + \varepsilon = W}$

From equation (6) and (7), one can find that following equation holds in the matrix of equation (11) The summation of power ration values $w_{i,t}$, mast be equal one, as equation (12)

$$\sum_{i=1}^{n} w_{i,t} = 1 \quad (12)$$

Since $[W_1, W_2 .... W_n]$ is a predicted vector, the summation of $W_i$might not be equal to 1. Thus, a correction is needed, as in equation (13)

$$P_i^{pred} = \frac{W_i}{\sum_{i=1}^{n} W_i} \quad (13)$$

After predicting the total power $P_{total}$and power ration by correction equation P_i^{pred}, the indirect WPP of the i^{th}wind farm is as following:

$$\hat{P}_i = P_i^{pred} \times P_{total} \quad (14)$$

Where, $\hat{P}_i$ is the indirect WPP of the i^{th}wind farm which satisfies $\hat{P}_{total} = \sum_{i=1}^{n} \hat{P}_i$

Figure 2 shows the process of combine method for direct and indirect of USTWPP.

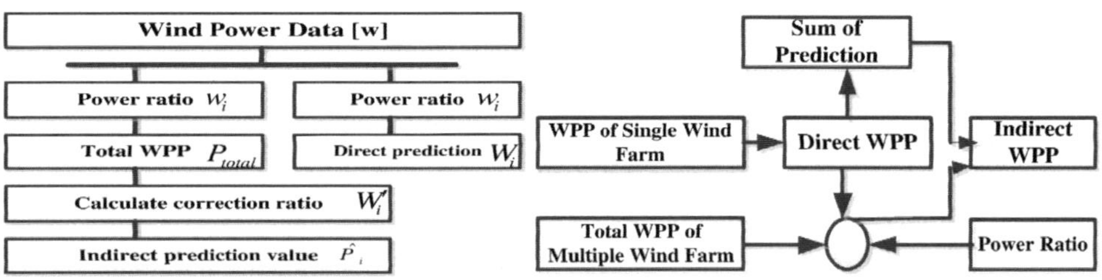

Figure 2. Process of combine method for direct and indirect of USTWPP.

4. Simulation and results

In this section, simulations are carried out for USTWPP using the time-series analysis. The test data is divided into two sets: one is the comparing results, and the other is the confirmed results.

4.1. Comparisons between single-point prediction and multi-point prediction

The prediction accuracy is a function for the difference between the predicted value and actual values. In this section, the comparison of a numerical result of the single-point and multi-point prediction ways will be displayed on the table 1. In this table, data is taken from the actual wind power data, and the width of prediction time window T is assigned to 60, 30 and 20 min; the length of prediction Δt is assigned to 5 and 10 min. MAPE is the mean absolute percentage error, and MSE is the mean square error.

Table 1. MAPE and MSE comparisons of two ways.

Error	Single-Point	Multi-Point	Single-Point	Multi-Point
	T=60min,	Δt=5min	T=60min,	Δt=10min
MAPE	2.01	2.32	2.18	2.45
MSE	2.09	2.78	2.61	3.30
	T=30min,	Δt=5min	T=30min,	Δt=10min
MAPE	1.41	1.50	1.52	2.27
MSE	1.09	1.27	1.45	2.32
	T=20min,	Δt=5min	T=20min,	Δt=10min
MAPE	0.97	1.07	1.13	1.40
MSE	0.51	0.65	0.69	1.02

Table 1 show that the single-point prediction way will give the better results, and is slightly more prone to MSE. This way could outperform others if fewer period data sets were used and fewer time intervals ahead were predicted.

Figure 3 gives the total WPP comparison of the two prediction ways, at T = 60 min, Δt = 5min.

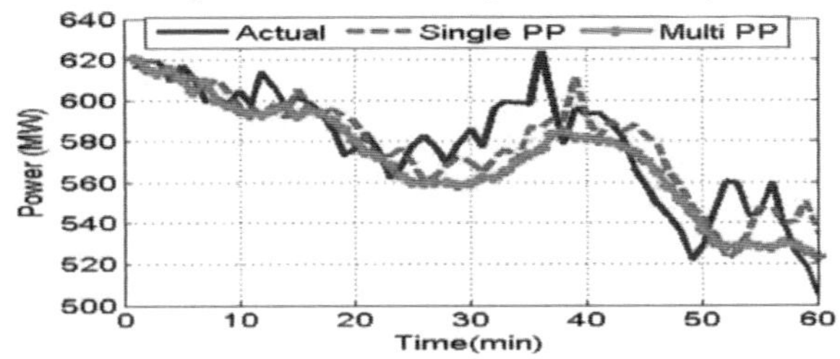

Figure 3. Total WPP with single-point & multi- point predictions.

From figure 3, the WPP curve with the single-point prediction way is closer to the actual wind power curve, and correlation coefficient between the two curves is about 0.874; while the WPP curve in the multi-point prediction way is less accurate with correlation coefficient being about 0.844.

4.2. Comparisons of multi-point USTWPP and its errors with different time

The graphical view about the multi-point USTWPP for three cases, anyone at different Δt with the same T is shown in figure4 - 6. From these figures, it is seen that the total WPP curve accurately follows the actual curve at the smaller time horizon, and at the end of T, only minor deviations are seen forΔt= 10 min.

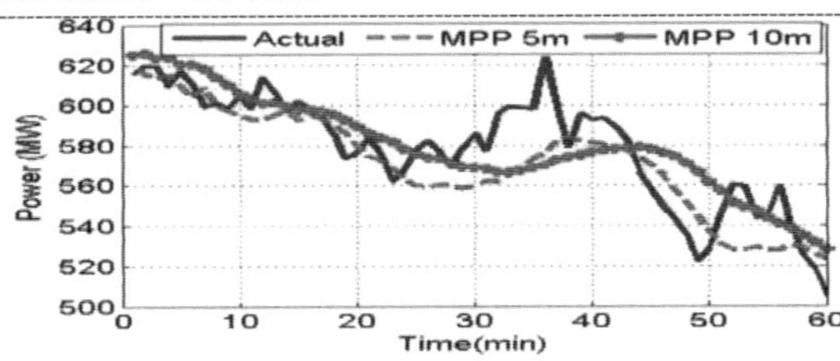

Figure 4. Multi-point USTWPP curve at Δt = 5 and 10min with T = 60min.

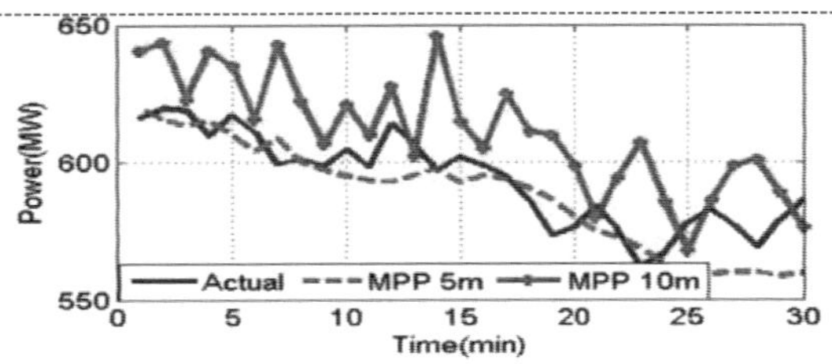

Figure 5. Multi-point USTWPP curve at Δt = 5 and 10min with T = 30min.

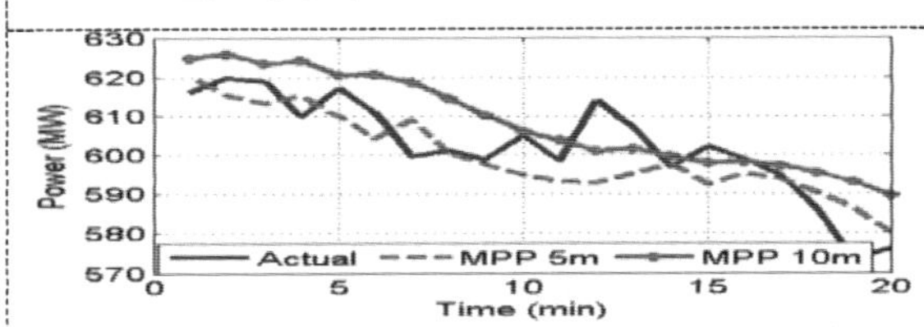

Figure 6. Multi-point USTWPP curve at Δt= 5 and

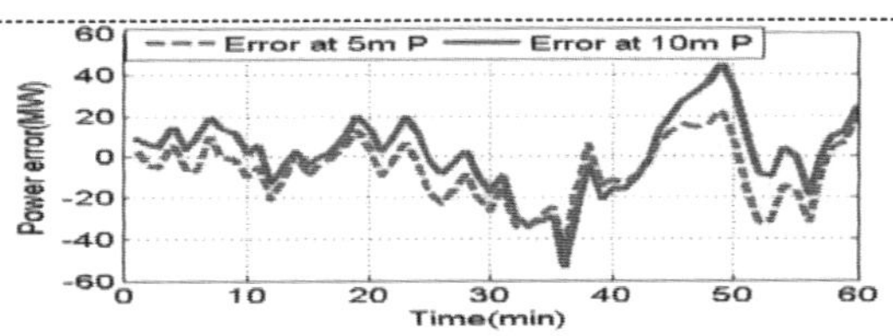

Figure 7. Multi-point USTWPP error curve at

10min with T = 20min.

Δt= 5 and 10 min with T = 60 min.

To illustrate the performance of the multi-point USTWPP in a short run, the predicted result up to 60 min are shown in table 1, and the corresponding error curves are shown in figure7 - 9. From these cases, the better accuracy for the USTWPP corresponds to the short width of prediction time window T and sampling time period.

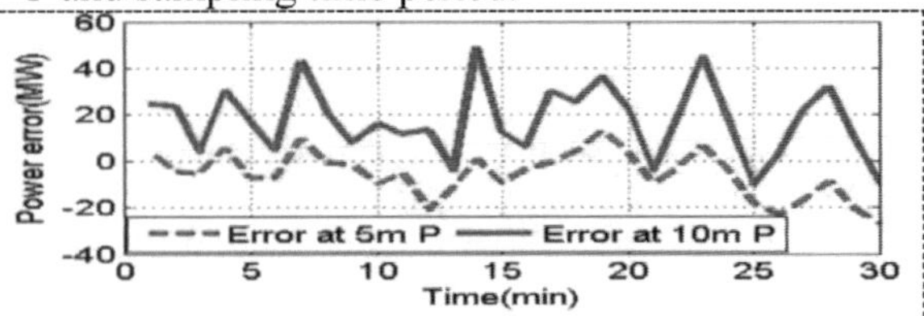

Figure 8. Multi-point USTWPP error curve at Δt = 5 and 10 min with T = 30 min.

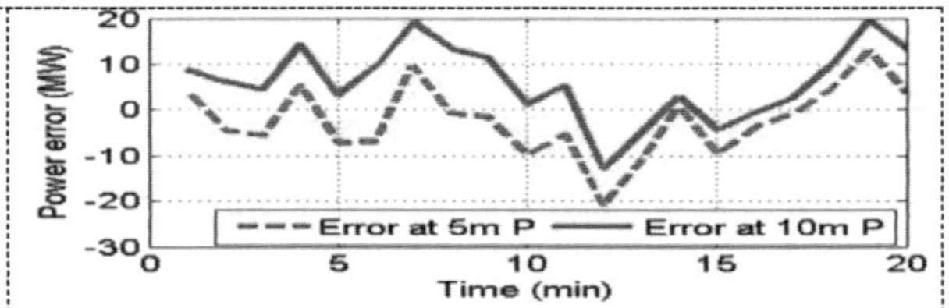

Figure 9. Multi-point USTWPP error curve at Δt = 5 and 10 min with T = 20 min.

As shown in figure9, the multi-point USTWPP error is quite low and is the smallest one, which concentrates within the prediction range.

4.3. Comparisons of two wind farms at WPP by direct and indirect prediction types

This part will show the comparisons between single WPP at two wind farm by direct and indirect prediction types. According to the values of WPP in figure 10&11, direct and indirect WPP of two wind farms respectively, WPP of two wind farms by direct type of prediction at Δt= 5 min and T = 60 min. The actual and prediction curves at two wind farm powers are shown in figure 10. It was observed that the predicted values by the direct prediction are not close to the actual values.

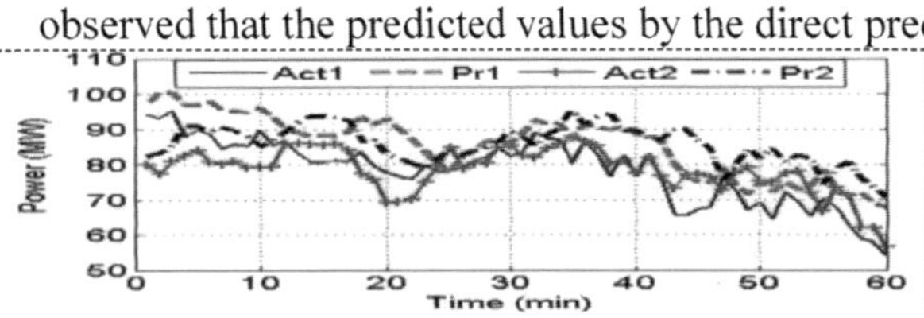

Figure 10. Actual & predicted single WPP by direct prediction type.

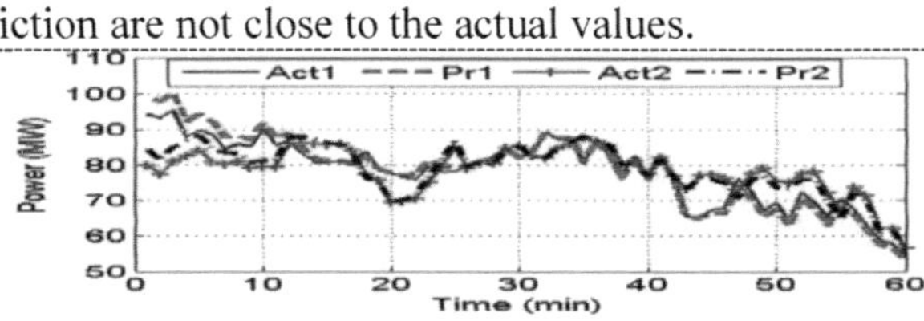

Figure 11. Actual & predicted single WPP by indirect prediction type.

The WPP of two wind farms by indirect type of prediction at Δt= 5 min and T = 60 min, as it can be seen from figure 11, the predictive power result at two wind farm is more satisfactory than the direct predict type.

As part of the comparison between direct and indirect WPP, to find out, why the prediction by the indirect type it is the high-accuracy value, in the case of indirect WPP, using the correction values to prediction wind power, causing him to better than direct prediction.

4.4. Comparisons between proposed method & ARMA, ARIMA and ANN methods

This paper studies USTWPP using four methods: ARMA, ARIMA, ANN, and MLR&LS. In this part, the simulation results for USTWPP using the statistical methods are presented.

Figure 12 gives the comparison procedures of the four methods.

IOP Conf. Series: Earth and Environmental Science **63** (2017) 012005 doi:10.1088/1755-1315/63/1/012005

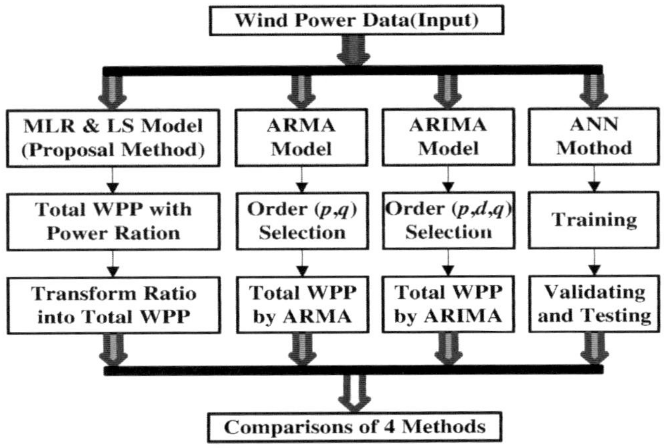

Figure 12. Comparison procedures among four methods.

The ARMA model is known to have good results, so it is used as the benchmark in this paper. Figure 13 with using the ARMA model shows that the predicted values are almost close with the actual values in the previous stage of the width of the prediction time window, and the error is very small.

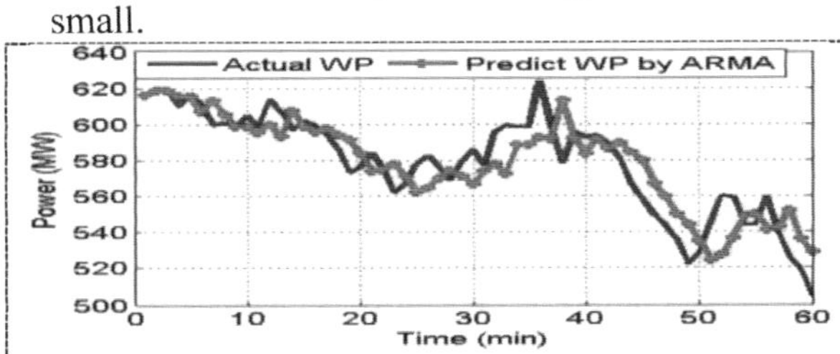

Figure 13. Actual & predicted total WP by ARMA.

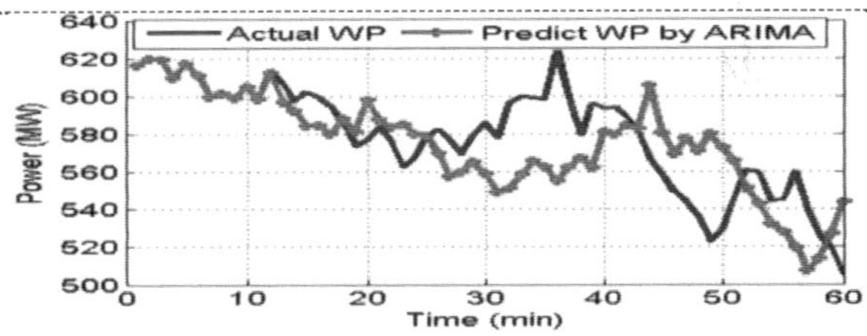

Figure 14. Actual & predicted total WP by ARIMA.

Figure 14 with using the ARIMA model shows that the predicted curve is different from the actual one. The predicted values by ARIMA model depend upon the parameter.

Simulation results of ANN model are demonstrated after many experiments with various network architectures based on BP algorithm as shown in figure 15. The accurate predicted values and the correlation between the actual values and the predicted values indicate high prediction accuracy. Devaluation expected error also indicates high prediction accuracy. ANN has been successfully used to forecast integrated wind power.

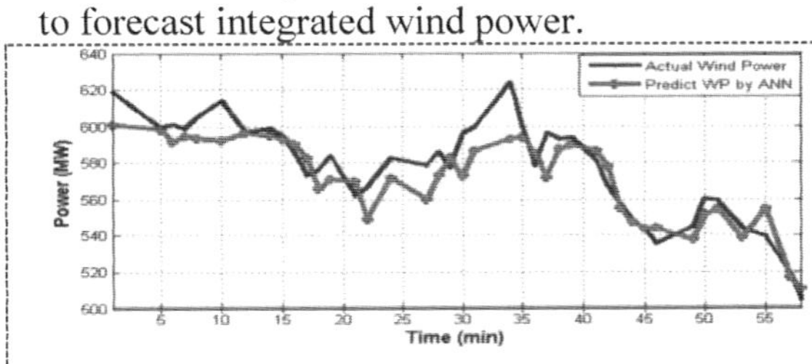

Figure 15. Actual & predicted total WP by ANN.

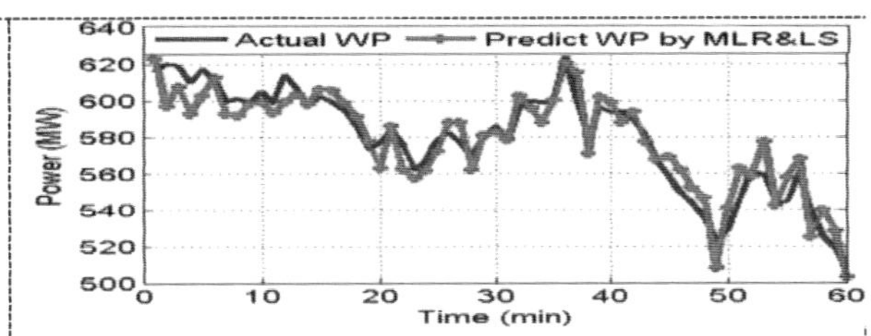

Figure 16. Actual & predicted total WP by MLR&LS.

The proposed MLR&LS hybrid model for USTWPP has been successfully implemented in this paper. Figure 16 shows a time series of WPP that is closest to the time series of actual wind power. In this study, the final predicted value by MLR&LS is more accurate than that by ARMA or other models.

4.5. Comparison of the predicted results

Comparison of ARMA, ARIMA, ANN and MLR&LS model is depicted in table 2. From table 2 and figure 13 - 16, it can be clearly observed the WPP accuracy level with ARIMA compared with ARMA is not quite significant. It can be said both methods achieved a good forecast response as errors of them are low. This result is consistent with most of previous studies. Moreover, we observed the WPP accuracy level with ANN compared with ARIMA is not quite significant. However, it is observed, the MLR & LS has a more high-accuracy level for total WPP, compared with other methods in this paper. Numerical results in table 2 and figure 17 showed the MLR&LS is best than other methods. We also noted the pattern of ARMA, and ARIMA shows the linear pattern. This finding is consistent with a majority of the studies. Finally, all analysis and simulation results prove the proposed approach gives better performance.

4.6. Forecasting accuracy comparisons of the four approaches

The forecasting accuracy is a function as the difference between the predicted values and the actual values. Table 2 gives the error of predicted values compared to actual values for the four methods. It can be inferred that the predicted values are close to the actual values when used MLR&LS. In these calculations, it is also shown that there is high correlation between the actual curve and predicted curve in all models, but the proposed hybrid model gives the better result.

Table 2. Error and correlation coefficient of the 4 methods for USTWPP at Δt= 5 min and T = 60 min.

Model	MAPE	MSE	ρ
ARMA	2.01	2.09	0.86
ARIMA	3.09	5.63	0.65
ANN	1.89	1.67	0.90
MLR &LR	1.36	0.88	0.95

It is observed from table 2 above. That WPP by the proposed hybrid model outperformed all other models. The hybrid model has both the lowest error and the maximum correlation coefficient. Compared with ARMA, ARIMA and ANN, the hybrid model has the superiority of high USTWPP accuracy. Through table 2 and figure 13-16, we evaluated the predictive performances of the four models in details. In the total WPP of multiple wind farms, it is shown that the hybrid model is better than ANN, ARMA and ARIMA.

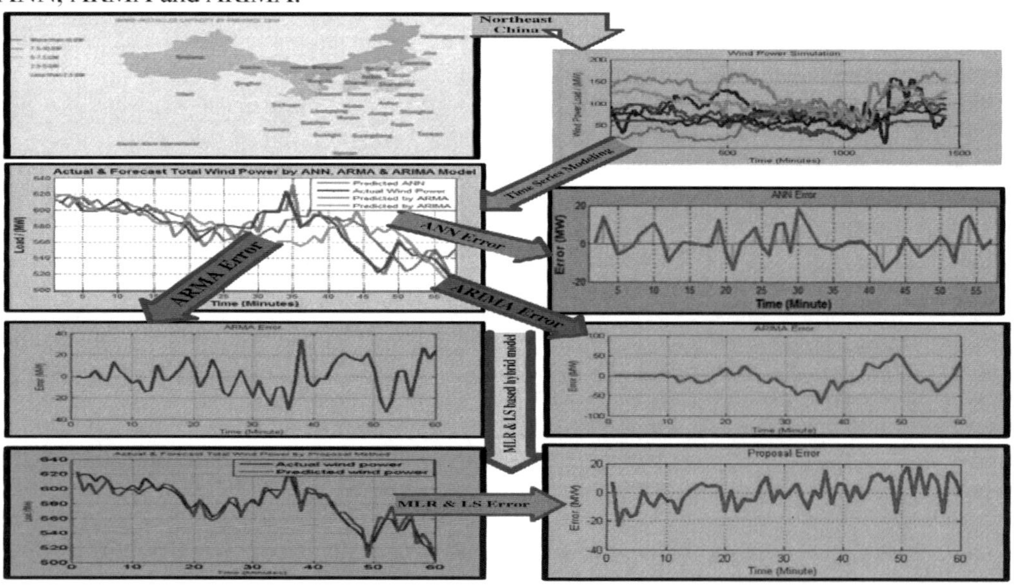

Figure 17. Simulation chart of comparison between four methods to total WPP of multiple wind farms.

Based on the simulation training about the historical data of eight wind farms, the USTWPP is reasonable. To avoid volatility due to the use of the MLR & LS hybrid model, all the forecasting were

IOP Conf. Series: Earth and Environmental Science **63** (2017) 012005 doi:10.1088/1755-1315/63/1/012005

repeated with $T = 60$ min. From table 2 and figure 17, the hybrid model clearly performs much better than ARMA, ARIMA and ANN model for single-point prediction and multi-point prediction. More accurately, in calculation of total WPP with ARMA, ARIMA, ANN and hybrid model, MAPE and MSE of the hybrid model are reduced, and ρ of the hybrid; models are increased.

5. Conclusion

For USTWPP, this paper compared the performances of four methods, including the MLR & LS hybrid model. The WPP obtained from ANN, and ARMA are accurate compared against the ARIMA model. Broadly, the hybrid model is the best one than ANN and ARMA. In the case of WPP, wind power pursues the peak and valley of actual power components of the patterns. The shape of WPP obtained from the hybrid model is more accurate than that obtained from the other methods for this study. And the USTWPP by the proposed method is close to the actual power. So, we can easily conclude that the forecasting accuracy with the combination method performs better than the individual ones in general. Our simulation results show that the proposed method has best performance than the same method for individual forecasting process and contributes for a reasonable improvement in the final WPP. After several experiments with different width as the prediction time window on the MLR & LS hybrid model, the smallest MSE was noted to give the best prediction accuracy for the test and validity data. The MSE recorded during the experiment is presented in table 2 and figures17 to show that.

6. Acknowledgment

This work is supported by the National Natural Science Foundation of China under Grant 51377035.

7. References

[1] Gargeya M A and Pabba S P 2013 *J. IJAREEIE.* **2** 1203-1212
[2] Wu Y K, LeeT C, Hsieh T Y and Lin W M 2014 *J. JPEE.* **2** 11-17
[3] Soder L, Hofmann L, Orths A, Holttinen H, Wan Y H and Tuohy A 2007.*IEEE Trans. Energy conversion.***22** 4-12
[4] Chen N, Qian Z, Nabney I T and Meng X 2014. *IEEE Trans. Power Syst.* **29**656-665
[5] Sorensen P, Cutululis N A, Vigueras-Rodríguez A, Jensen L E, Hjerrild J, Donovan M H andMadsen H 2007. *IEEE Trans. Power Syst.* **22** 958-965
[6] de Haan J E, Frunt J and Kling W L 2010.*IEEE PES.ISGT Europe.* 1-8
[7] Taylor J W and McSharry P E 2007. *IEEE Trans. Power Syst.* **22**2213-2219
[8] Albizu I, Fernandez E, Mazon A J, Sagastabeitia K J, Bedialauneta M T and Olazarri J G 2015.*IEEE 2015 ICCEP.* 382-388
[9] Chang W Y 2013 *J. Energies.* **6**4879-4896
[10] Ismail M M and Hassan M M2013. *IEEE ICMIC.* 112-118
[11] Nataraja C, Gorawar M B, Shilpa G N and Harsha J S 2012 *J. IJESIT.* **1** 45-53
[12] Jain A and Satish B 2008. *IEEERegion 10 Conference.* 1-6
[13] Botterud A, Wang J, Miranda V and Bessa R J 2010 *J. Electricity Journal.* **23** 71-82
[14] Coughlin K 2011 *J. Lawrence Berkeley National Laboratory*
[15] Chang WY 2014 *J. JPEE.* **2** 161-168
[16] Li L, Wang M, Zhu F and Wang C 2009 *J. ISCSCT.* 293-297
[17] Hongfang C and Rui S 2010. *IEEE First ACIS.International Symposium on CDEE.* 22-26
[18] Zhang J and Wang C 2013. *IEEE International Conference on.CSA.*361-364
[19] AreekulPh, Senjyu T, Toyama Hand Yona A 2010.*IEEE Trans. Power Syst.* **25** 524-530
[20] Defilippo S B, Neto G G and Hippert H S 2011 *J. 44 JAIIO - SIO 2015 - ISSN: 2451-7569*
[21] Alfares H K and Nazeeruddin M 2002 *J. International Journal of Systems Science.* **33**23-34

An Experimental Research of Biomass Crushing System

Yue Xia

College of Mechanical, Inner Mongolia University of Technology, Hohhot, China

yuexia995@126.com

Abstract. Biomass crushing is precondition for utilization and transformation of biomass energy. Compressive strength of biomass raw materials is generally small with long fiber, and multiple crushing methods like shocking, shearing and grinding should be used to realize high-efficiency crushing. Based on this, this paper researched and developed a set of biomass fine crushing system, and conducted an optimized study of main parameters influencing crushing process—revolving speed, between blade and cutter head and material moisture content. Results indicated that this set of crushing system could complete fine crushing of biomass, particle size of crushed products was controlled below 250μm, crushing energy consumption was between 70—120kwh/t, both of which were mainly decided by material moisture content and operating parameters of crusher. Revolving speed of crusher was elevated, proportion of particles (d<106μm) with fine particle size in crushed products increased, and unit energy consumption of crushing increased; narrowing space between blade and cutter head could improve crushing effect of biomass, but unit energy consumption increased; the higher the moisture content was, the less the content of fine particles in crushed products would be, and the greater unit energy consumption would be. By comprehensive comparison, the optimal operating conditions for crushing were: revolving speed was 3600r/min, space between impulse blade and cutter head was 8mm, moisture content was 10.6%, energy consumption was 98kwh/t, product particle size distribution was: those with particle size below 106μm occupied 71.5% and those between 106μm—250μm occupied 28.5%.

1. Introduction

With increasing exhaustion of fossil energy and gradual appearance of its environmental impact, biomass energy has drawn extensive global concern by virtue of its renewability and cleanness. At present, utilization methods of biomass energy include: biomass transformation method, mechanical extraction method and thermochemical transformation method, whereby thermochemical transformation method is applied the most extensively. According to chemical reaction kinetics and thermal transmission theory, crush preprocessing of biomass can improve its thermal transformation efficiency. On the one hand, reducing biomass particle size can increase specific surface area of particles, be good for heat transfer and mass transfer of chemical reaction process, reduce standing time of materials in reactor, increase processing ability of reactor, reduce temperature gradient inside particles and improve the yield. On the other hand, biomass energy density is low and range of heat value change is large, crush preprocessing can make biomass feeding even, fluctuation of heat value of stokehold feeding small, bio oil yield and components maintain stable, and then improve heat treatment efficiency of biomass and pollution control efficiency[1].

Biomass not only includes cotton stalk and branch which present brittleness with high lignin content and big relative hardness but also include straw and corn stalk with high cellulose and hemicelluloses contents with strong toughness, so it's suitable to adopt multiple methods like shocking, shearing and grinding to crush biomass[2]. Existing material crushing equipments are mostly designed for materials like industrial minerals with strong brittleness and big compressive strength such as jaw crusher, impact crusher and roll crusher, etc. Crushing force used by this kind of crushing system is not applicable to biomass crushing. Although there are some small crushers for fodder and traditional Chinese machine, such as ordinary hammer-shape crusher, water drop-like crusher, vertical spindle type crusher, etc., crushing energy consumption is high and particle size of crushed products is big, so it's not applicable to mass crushing[3].

This paper introduced a set of biomass fine crusher, investigated main factors influencing crushing performance by taking particle size of crushed products and crushing energy consumption as indexes, and the purpose of this paper was to optimize technological parameters and provide basic theory and data for follow-up industrialized application of this crushing system.

2. Overview of Crushing System

The experimental system mainly consisted of coarse crusher, fine crusher, particle collection device (cyclone dust collector and bag-type dust collector) and draught fan, etc., in the whole operating process, after materials were added coarse feeding entrance, and they were initially crushed in coarse crusher. Under the effect of draught fan, particles with small particle size （d<5 mm）were carried into fine crusher to conduct secondary crushing, particles with small particle size （d<250μm） after shocking, grinding and shearing would be drawn from exit at upper end of crushing chamber into cyclone dust collector, and those with large diameter would stay in the crusher to be crushed again. The whole process proceeded under negative pressure. Crushing system is as shown in Figure 1.

Figure 1. Biomass Crushing System

1. Coarse crusher 2. Fine crusher 3. Cyclone dust collector 4. Bag-type dust collector

Coarse crusher mainly consisted of hopper, feeding inlet, crushing chamber and discharge port. Crushing chamber was circular structure. Principal axis was fixed at two sides of crushing chamber with bearings, and electromotor drive axis through belt. Blades in crusher were uniformly distributed on axis, which ensured dynamic balance and static balance of axis and made materials uniformly distribute in crushing chamber. Some biomasses with high content of biomasses like straw and cotton stalk had hard texture, making impact load subjected by blade have certain influence on pin and axis of blade, so blades in this crusher was not designed as fixed type, and a loose welding form existed between blade and lockpin[4]. Materials in crushing chamber went shearing and grinding under high speed, collided, stroke and rubbed among blades, and then materials mutually collided and rubbed, then materials were gradually crushed into small particles and short cellosilk[5]. Coarse crusher structure is as shown in Figure 2.

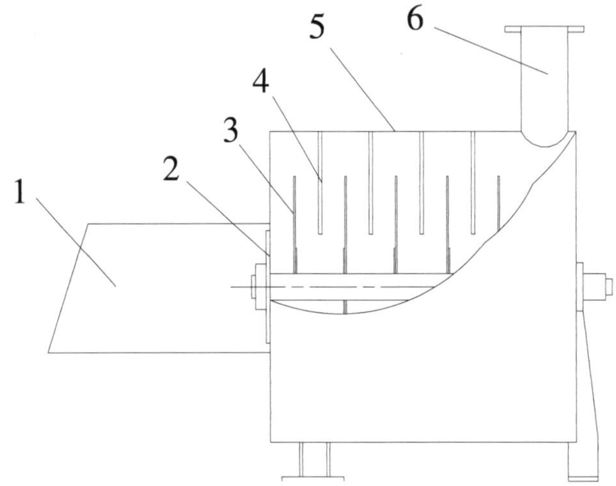

Figure 2. Coarse crusher

1. Material 2. Feeding Inlet 3. Impulse Blade 4. Fixed Blade 5. Outer Cylinder 6. Discharge Port

Fine crusher consists of feeding pipe, crushing chamber, discharging pipe and actuating unit, and there are totally 12 groups of blades in crushing chamber. When operating, as materials entering fine crushing system have passed initial crushing of previous coarse crusher, average particle size is less than 5mm, so materials in fine crusher uses high-speed revolution of motor to realize grinding between blades and blinding between materials so as to reach finer particle size[6]. Fine crusher structure is as shown in Figure 3 and operating parameters of crushing devices are as shown in Table 1.

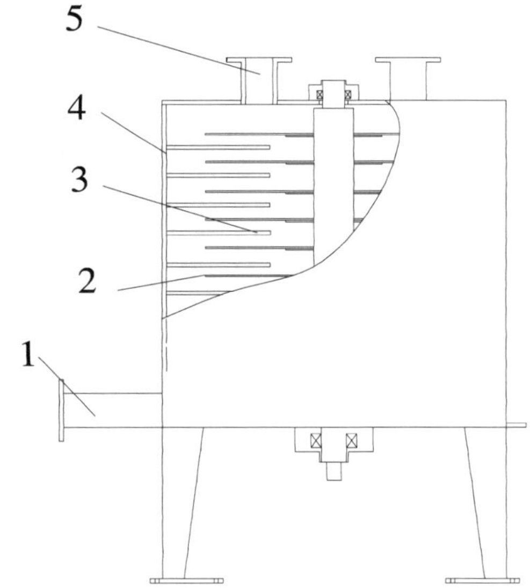

Figure 3. Fine crusher

1. Feeding Inlet 2. Blade 3. Fixed Blade 4. Outer Cylinder 5. Discharging Port

Particle collector consists of cyclone dust collector and bag-type dust collector. Draught fan is placed between cyclone dust collector and bag-type dust collector to make inside cyclone dust collector be under negative pressure status and lead flow direction of materials. After deceleration through cyclone dust collector, materials fall in receiving hopper, and a small quantity of quite fine particles, after air draft through draught fan, reach receiving cloth bag to be collected from the upper end of hopper.

Table 1. Operating Parameters of Crushing Devices

Crushing Device	Size (mm)	Motor Power (KW)	Motor Speed (r/min)	Rated Current (A)	Transmission Ratio
Fine Crusher	1000×600	15	2960	15.4	1:1
Coarse Crusher	500×800	15	2890	28.9	1:1.5

3. Experimental Research of Biomass Crushing

3.1. Experimental Method

This experiment mainly discusses about the influences of revolving speed of fine crusher, space between blades and raw material moisture content on particle size distribution and crushing energy consumption of crushed products, and experimental conditions are as shown in Table 2.

Table 2. Experimental Conditions

No.	Revolving speed (r/min)	Spacing (mm)	Moisture Content (%)
1	3000	20	10.6
2	3600	20	10.6
3	4200	20	10.6
4	4200	10	10.6
5	4200	8	10.6
6	4200	5	10.6
7	4200	10	16.1
8	4200	10	18.8
9	4200	10	24.2

3.2. Analysis and Testing Method

Particle size distribution of crushed products and crushing unit energy consumption are main standards of measuring crushing performance. Particle size distribution is screened by vibrating screen. Calculation method of crushing unit energy consumption is as following:

Wattmeter is used to conduct online measurement of motor power when crusher is operating, consumed total power (power of coarse crusher, power of fine crusher and power of draught fan) is recorded as P_N, crushing time is recorded as t_E, and mass of crushed biomass products is recorded as M. Then unit energy consumption E_N of biomass crushing is calculated through equation (1) [7]:

$$E_N = \frac{P_N t_E}{m_M} \tag{1}$$

3.3. Experimental Results and Discussion

3.3.1. The Influence of Revolving Speed on Crushing Performance.

Figure 4 and Figure 5 are respectively particle size distribution and unit energy consumption of crushed biomass products at different revolving speeds. Revolving speed of crusher is implemented through changing motor frequency.

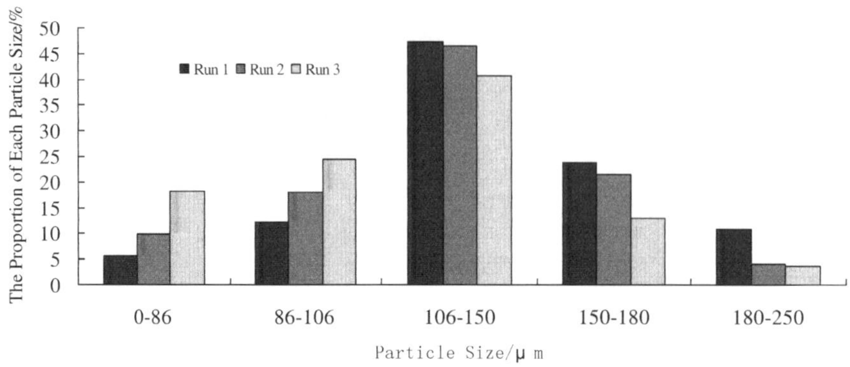

Figure 4. The Influence of Revolving Speed of Fine Crusher on Particle Size Distribution of Crushed Products

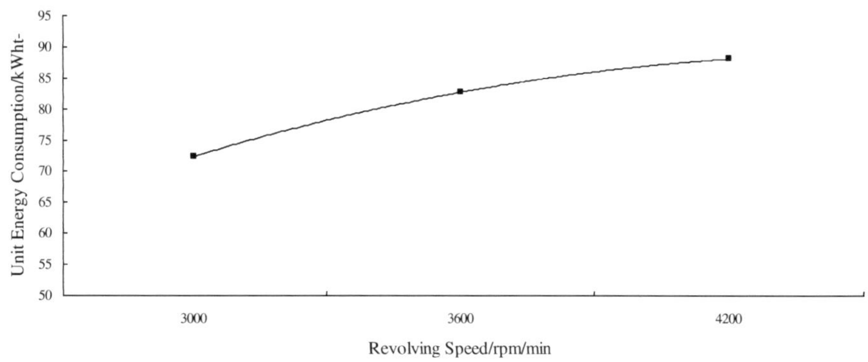

Figure 5. The Influence of Revolving Speed of Fine Crusher on Unit Energy Consumption

As revolving speed of crusher increases, content of particles with fine particle size in products presents increasing tendency. When revolving speed increases from 3000r/min to 4200r/min. Proportion occupied by particles with particle size being within 106-150 μm is the largest, particle size distribution presents normal distribution. This is because crushing methods are various of biomass in fine crusher, crushing of biomass in crushing chamber is mainly implemented through impact effect of impulse blade under high-speed revolution, meshing effect of impulse blade and sieve mesh on cutterhead, grinding effect generated by small space between impulse blade and cutterhead and mutual collision between particles of high concentration[8]. The higher the revolving speed is, the greater impact effect generated by impulse blade will be, meshing force between impulse blade and fixed blade will be enhanced, collision probability between particles of high concentration increases, crushing effect will be better, and the greater the proportion of particles with fine particle size in crushed products will be.

Variation tendency of unit energy consumption with revolving speed of crusher is as shown in Figure 5. It can be seen from Figure 5 that when revolving speed increases from 3000r/min to 4200r/min, unit energy consumption is gradually on progressive increases. This is because: when crushing is being operated, energy reaching as high as 20% is used to overcome friction energy consumption brought by high revolving speed of crusher rotor and energy consumption brought by idling of impulse blade. Large space between crushed materials and material feeding discontinuity make impulse blade be under idling status in some operating period in the crushing process, and the higher the revolving speed of rotor is, the greater the energy consumed by idling of blade will be, energy loss caused by friction will abruptly increase, which result in increasing unit energy consumption of crushed products [9].

Hence, optimal revolving speed should be selected in crushing process, which can not only meet quality requirement of crushed products but also reduce crushing energy consumption. The optimal revolving speed in this experiment is 3600r/min.

3.3.2. The Influence of Space between Blade and Cutterhead of Crusher on Crushing Performance. Figure 6 and Figure 7 are respectively variation tendencies of particle size distribution of crushed products and crushing unit energy consumption with space between blades when revolving speed of crusher is 4200rpm/min. Different spaces between blades are implemented by adding washer between blades.

It can be seen from Figure 6 that the smaller the space between impulse blade and cutterhead is, the greater the content of biomass particles with fine particle size in crushed products will be. When space between blades is 20mm, particle size of 42.8% biomass particles is below 106 μm, when space between impulse blade and cutterhead reduces to 8mm, content of biomass particles with particle size being 106 μm below increases to 71.5%. This is because in particle-type crushing system, transmission of crushing force is mainly implemented by direct contact between impulse blade and crushed materials, the smaller the space between impulse blade and cutterhead, the greater the concentration of crushing materials in crushing chamber will be, and the higher the collision probability between materials and blades will be [10]. Small space between blades is good for implementation of meshing force between blade and cutterhead, and in the meantime, it promotes grinding effect of blade and cutterhead on materials to a great extent and improves crushing effect of biomass particles.

Variation tendency of crushing unit energy consumption with space between blades is as shown in Figure 7. It can be seen from Figure 7 that with decreasing space between blade and cutterhead, crushing unit energy consumption presents increasing tendency. This is because space is small, then crushing space of particles consisting between blade and cutterhead is small, particle concentration is high, and resistance encountered by blades when doing crushing operation, which result in increasing crushing energy consumption[11].

It can also be seen from Figure 6 that the influence of space between blade and cutterhead on particle size distribution of crushed products changes not obviously within a small space scope, when the space reaches 8mm, content of biomass particles with particle size being below106 μm is 71.5%, when space reduced to 5mm, the content of biomass particles with particle size being below106 μm is 75.3%. However, when the space reduced from 8mm to 5mm, unit energy consumption increases from 98kwh/t to 110kwh/t.

Generally speaking, reducing the space between blade and cutterhead can elevate percentage content of particles with small particle size in crushed products, but will result in increasing energy consumption. Hence, in this experiment, the optimal space between blade and cutterhead is selected as 8mm.

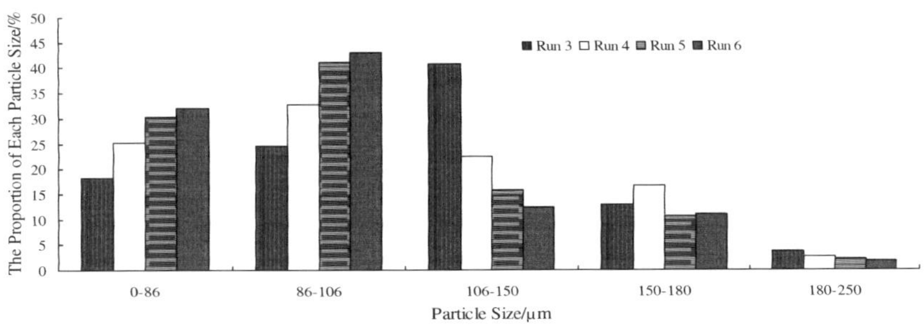

Figure 6. The Influence of Space Between Blade and Cutterhead of Fine Crusher on Particle Size Distribution of Crushed Products

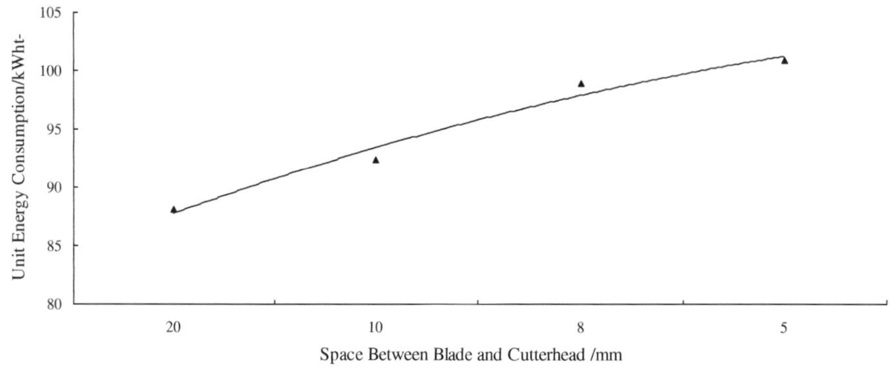

Figure 7. The Influence of Space between Blade and Cutterhead of Fine Crusher on Unit Energy Consumption

3.3.3. The Influence of Material Moisture Content on Crushing Performance. It could be seen from Figure 8 that influence of material moisture content on quality of crushed products was quite obvious, the higher moisture content was, the smaller the content of particles with small particle size in crushed products would be. This was because: increasing moisture content in materials would improve toughness of materials and reduce crushing effect to a certain degree.

The influence of moisture content on unit energy consumption of crushing was as shown in Figure 9: it could be seen that energy consumption presented increasing tendency with material moisture content. It could be seen from calculation formula (1) of unit energy consumption of crushing that: standing time of materials in crusher was one of key factors influencing energy consumption of crusher, while standing time was decided by air speed of draught fan and content of coarse biomass particles in crushing chamber. High moisture content resulted in increasing content of particles with coarse particle size, lengthened standing time of particles in crushing chamber and increased energy consumption of crushing. In the meantime, as increasing moisture content resulted in decreasing content of dry basis substances in materials, it also resulted in increasing energy consumption of crushing to a certain degree [12]. It could be seen from above analysis that it's necessary to do dry preprocessing of materials before biomass crushing, and drying conditions should be determined by comprehensive consideration of energy consumption of crushing and product quality. In this experiment, the optimal moisture content was 10.6%.

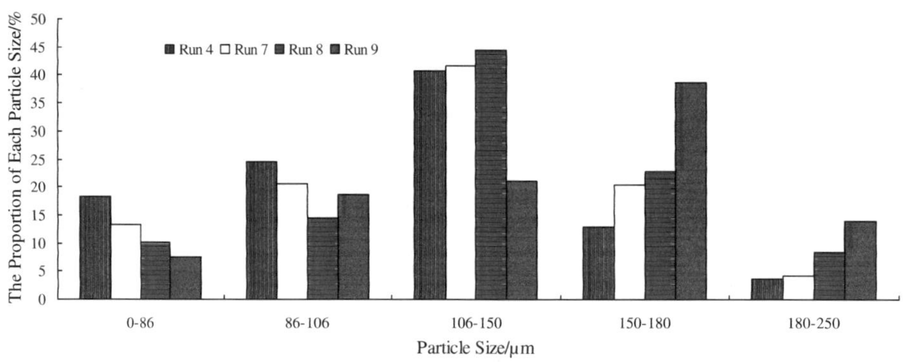

Figure 8. The Influence of Moisture Content on Particle Size Distribution of Crushed Products

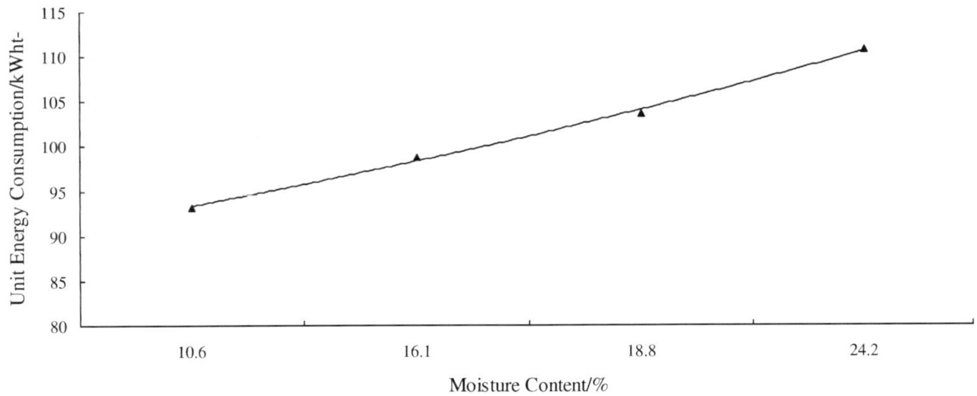

Figure 9. The Influence of Moisture Content on Unit Energy Consumption

4. Conclusion

This paper proposed a new type of crushing equipment which consisted of coarse crusher, fine crusher, particle collection device (cyclone dust collector and bag-type dust collector) and draught fan, etc., particle size of crushed biomass could reach 250μm. It made an experimental study of the influences of revolving speed, space between crusher blade and cutterhead and material moisture content on crushing performance, and then it drew the following conclusions:

(1)With increasing revolving speed of crusher, content of particles(d<106μm)with fine particle size in products presented increasing tendency, unit energy consumption increased as revolving speed of crusher increased, so optimal revolving speed during crushing process should be selected to satisfy quality requirements for crushing products and reduce crushing energy consumption. Optimal revolving speed in this experiment was 3600r/min.

(2)The smaller the space between impulse blade and cutterhead was, the greater the content of biomass particles with fine particle size in crushed products would be. With decreasing space between blades, unit energy consumption of crushing presented increasing tendency, and the influence of space between blades in particle size distribution of crushed products changed not obviously within scope of space between blades. With overall consideration, reducing space between blades could improve percentage composition of particles with small particle size in crushed products, but would result in increasing energy consumption of crushing, hence, in the experiment, optimal space between blades was selected as 8mm.

(3)High moisture content resulted in increasing content of particles with coarse particle size in crushed products, lengthened standing time of particles in crushing chamber and increased energy consumption of crushing. In the meantime, as increasing moisture content gave rise to reducing content of dry basis substances in materials and resulted in increasing energy consumption of crushing to a certain degree. Hence, it' s necessary to do dry preprocessing before biomass crushing.

5. References

[1] Zhang wudi, Song hongchuan. *Developments and Utilizations of Biomass Energy Conversion Technology* [J]. Energy research and Utilization,2000(2).

[2] Li zhongpin, Zhu jiang. *Recent Researches on Feed Grinding Technology*[J]. Feed Industry,2002(23).

[3] Zhu dewen, Chen yongsheng. *Review on Biomass Energy Grinding Technology in China*[J]. Modern Agricultural Equipments.

[4] Sun depin, *Studies and Utilization of the Hammer Crusher. Mechanical and Electrical Information*,2001(1).

[5] Lu zhengxi, Lu shoudao. *Theory and Design of Food Mechanisms and Machines*. Beijing: China

Light Industry Press,1999.

[6] Luo siyi. *Research on the design of MSW shredder and effects of particle of MSW on pyrolysis and gasification performances*[D].HUAZHONG University of Science and Technology,2010.

[7] Fan jingchun, Wang yonggang. *Economic evaluation of biomass energy technology. Energy Engineering*,2003(4).

[8] Luo S Y. *A novel biomass pulverization technology*[J].2011(2).

[9] Xiao bo, Zhou xianmei. *Study on the breakage of biomass fuel. Energy research and Information*,2006(22).

[10] Nobusuke K. *A new pulverized biomass utilization technology*. Science Direct .2007(12).

[11] Luis S. *Evaluation of different strategies for pulverization of forest biomass. Powder technology*,2006(6).

[12] Ma longlong, Wu chuangzhi. *The Utilization of Biomass Gasification Technology*. Beijjing: Chemical Industry Press.2003

Simulation of Solar Heat Pump Dryer Directly Driven by Photovoltaic Panels

H Houhou[1], W Yuan[2] and G Wang

Laboratory of Ergonomics and Environmental Control, School of Aeronautic Science and Engineering, Beihang University, Beijing 100191, China.

[1]hou.hatem@gmail.com,[2]yuanwx@buaa.edu.cn

Abstract. This paper investigates a new type of solar heat pump dryer directly driven by photovoltaic panels. In order to design this system, a mathematical model has been established describing the whole drying process, including models of key components and phenomena of heat and mass transfer at the product layer and the air. The results of simulation at different drying air temperatures and velocities have been calculated and it indicate that the temperature of drying air is crucial external parameter compared to the velocity, with the increase of drying temperature from 45°C to 55°C, the product moisture content (Kg water/Kg dry product) decreased from 0.75 Kg/Kg to 0.3 Kg/Kg.

1. Introduction

Solar energy represents the largest source of renewable energy supply. It can be converted into electrical energy using photovoltaic panels (PV), concentrating solar thermal power (CSP) and concentrating photovoltaic (CVT) [1]. It can be also converted into thermal energy using non-concentrating solar collectors such as flat-plat, hybrid photovoltaic-thermal (PVT) and concentrating solar collectors[2]. Nowadays solar thermal energy is used in great fields as generation of electricity with heat engines, domestic hot water production, heating buildings and drying food and vegetables. Solar drying system has been used for thousands years and developed basing on the good understanding of drying process and the different parameters controlling it[3] and [4], In the aim of improving its efficiency, the noticed development was the combination of the solar energy with the heat pump system[5]. In literature we cite[6], who presented a review about studies on the advances in solar assisted heat pump dryers for agriculture and marine products. Another review of heat pump systems for drying application was done by[7]; where the authors explain that improving coefficient of performance (COP) of heat pump leads to improve the efficiency of drying system with considering the drying condition. Moreover, in the design of the drying systems, we need to pay attention to the energy requirement for drying a mass of product generally characterized by the high initial moisture content. Several researchers have developed simulation models for solar convective drying systems [8]-[10]. Our study focus on building a mathematical model of the heat and mass transfer of the drying air and the product that simulates the operation of the drying process.

2. System description

The main parts of the proposed system are the dryer (drying chamber) and the heat pump system (condenser, evaporator, compressor and the expansion valve). This system contain two different cycles

2.1. The Vapor-Compression Refrigeration Cycle

The Vapor-Compression Refrigeration Cycle is comprised of four steps (figure 1):

- **Compression:** In this stage, the refrigerant enters the compressor as a gas under low pressure and having a low temperature. Then, it will be compressed adiabatically.
- **Condensation:** The high pressure, high temperature gas will go through a condensation phase and releases heat energy that will be absorbed by the drying air.
- **Throttling:** The liquid refrigerant is pushed through a throttling valve, which expands it. As a result, the refrigerant now has low pressure and lower temperature.
- **Evaporation:** The refrigerant enters the evaporator, which is in contact with the drying air. Since, low pressure is maintained the refrigerant will be able to boil at a low temperature, and then will be driven to the compressor (back to the beginning of the cycle).

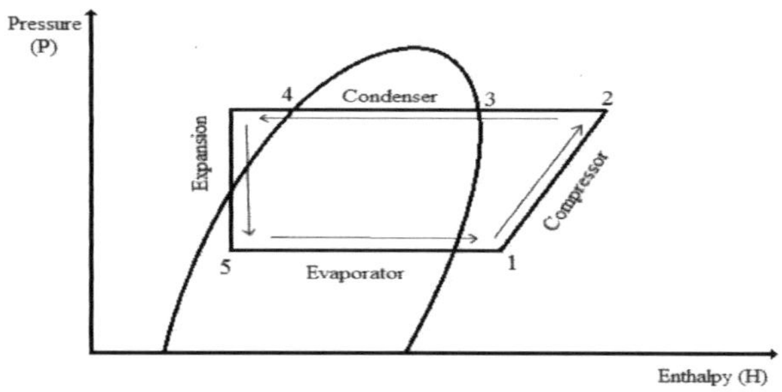

Figure 1. Vapor-Compression Refrigeration Cycle.

2.2. Drying air cycle

The drying air heated by the condenser goes into the drying chamber to heat and extract moister of the product (grape). The moist air leaving the drying chamber will be used to maintain the temperature of the evaporator to achieve a better Coefficient of Performance (COP) and finally return into the condenser. Repeatedly, the condenser heats up the drying air to reach the desired temperature of the drying process (figure 2).

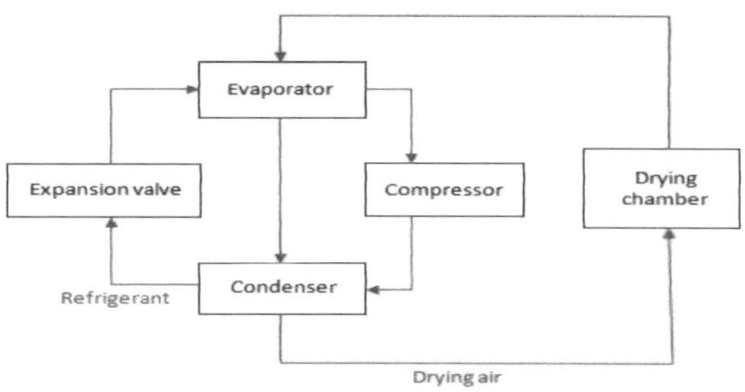

Figure 2. Description scheme of the heat pump dryer system.

3. Heat and mass transfer analyses at the drying chamber

Inside the drying chamber the various transfers are:

- Convection heat and mass transfer between the product and the air.
- Convection heat transfer between the air coming into the drying chamber and insider air.

- Convection and conduction heat transfer between internal and outsider air through the walls (thermal loose).

For what's next we give the mathematical equations describing the heat and mass transfer at the drying chamber.

3.1. Heat and mass transfer coefficients

The mass transfer coefficient can be calculated using and Chilton-Colburn analogy [11] with the dimensionless Lewis number:

$$h_m = h_{c1}\left(\rho_a c_{p,a}\right)^{-1}\left(Le\right)^{-2/3} \tag{1}$$

For the air $\left(Le \cong 1\right)$.

$h_{c1}\left(W/m^2{}^\circ C\right)$ Is the convective heat transfer coefficient between the air and the product it can be determined by the following equation:

$$h_{c1} = Nu\lambda_a D^{-1} \tag{2}$$

Nu : is the Nusselt number; we use the following correlation to calculate it [12]:

$$Nu = 0.023\ \mathrm{Re}^{0.8}\ \mathrm{Pr}^{0.4} \tag{3}$$

Re : is Reynolds number:

$$\mathrm{Re} = u\ L\ \upsilon^{-1} \tag{4}$$

Pr : Is the Prandtl number (Pr =0.7 for the air).

$h_{c2}\left(W/m^2{}^\circ C\right)$ Is the convective heat transfer coefficient between the air and inner walls, taking Nusselt number as [12]:

$$Nu = 0.036\ \mathrm{Re}^{4/5}\ \mathrm{Pr}^{1/3} \tag{5}$$

The conductive heat transfer coefficient $K\left(W/m^2{}^\circ C\right)$ between the internal air and the outside air through the walls is:

$$K = \lambda_i\ d_i^{-1} \tag{6}$$

3.2. Mass balance

The change in the amount of water inside the product is:

$$m_p(\mathrm{d}X/\mathrm{d}t) = -\dot{m}_v \tag{7}$$

Where $\dot{m}_v\left(kg/s\right)$ is mass of water evaporated from the product:

$$\dot{m}_v = A_p h_m\left(X_p\left(1+X_p\right)^{-1} + Y_a\left(1+Y_a\right)^{-1}\right) \tag{8}$$

The change in the amount of water in the air is:

$$m_a(\mathrm{d}Y_a/\mathrm{d}t) = \dot{m}_{a,i}\,Y_{a,i} + \dot{m}_v - \dot{m}_{a,o}\,Y_a \tag{9}$$

3.3. Heat balance

The energy balance equation for the product is:

$$m_p c_{p,p}(\mathrm{d}T_p / \mathrm{d}t) = -\dot{m}_v L_v + A_p h_{c1}\left(T_a - T_p\right) \quad (10)$$

The energy balance equation for the air is:

$$m_a c_{p,a}(\mathrm{d}T_a / \mathrm{d}t) = \dot{m}_{ai}((c_{p,a}T_{a,i}) + (Y_{in}(2500.9 + 1.82T_{a,i}))) - \dot{m}_{ao}((c_{p,a}T_a) + (Y_a(2500.9 + 182T_a))) + \dot{m}_v c_{p,v}T_a - A_p h_{c1}\left(T_a - T_p\right) - A_{wa}h_{c2}\left(T_a - T_{wa}\right) \quad (11)$$

3.4. Thermal looses

The energy balance equation of the wall of the drying chamber is written as:

$$\rho_{wa} c_{p,wa}(\mathrm{d}T_{wa} / \mathrm{d}t) = h_{c2} A_{wa}\left(T_a - T_{wa}\right) - KA_{wa}\left(T_{wa} - T_{am}\right) \quad (12)$$

4. Results and discussion

A MATLAB program was built to calculate the different parameters for solving the heat and mass balance equations according to the system of parameters in the table (1).

Table 1. Descriptions and values of the parameters used in the MATLAB program.

Parameters	Descriptions	Values
A_p	Surface of the product	0.5 (m^2)
A_{wa}	Surface of the wall	3 (m^2)
C_{pp}	Specific heat of the product	4180 (J/Kg°C)
C_{pa}	Specific heat of the air	1005 (J/Kg°C)
C_{va}	Specific heat of the air	717 (J/Kg°C)
C_{pwa}	Specific heat of the wall	860 (J/Kg°C)
D	Characteristic diameter of the layer of product	0.80 (m)
d_i	Thickness of the insulation wall	0.20 (m)
L	Characteristic length	1 (m)
m_p	Mass of the product	10 (Kg)
X_{in}	Initial moisture content of the product	4 (kg water/kgdryproduct)
ρ_a	Density of the air	1.06 (Kg/m^3)
ρ_{wa}	Density of the wall	2700 (Kg/m^3)
λ_i	Conductivity of the insulation wall	0.022 (W/m°C)

The temperatures and the moisture contents of the product and the air inside the drying chamber were calculated and plotted for different temperatures and velocities. The influence of these parameters in the drying process is shown in the following figures.

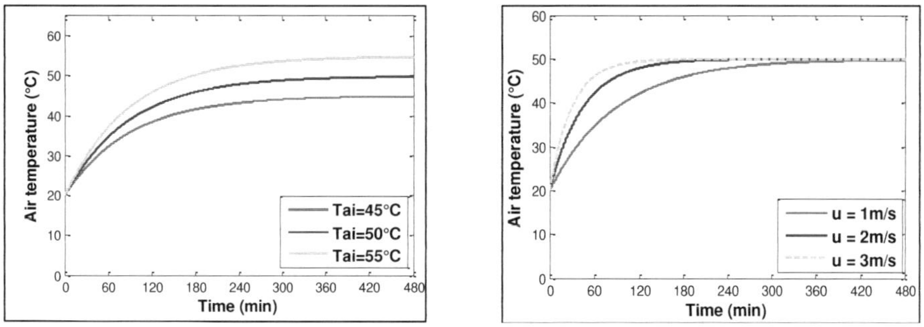

(a) (u=1 m/s) (b) (Tai=50°C)

Figure 3. Influence of drying air temperature (a) and velocity (b) on the variation of air's temperature.

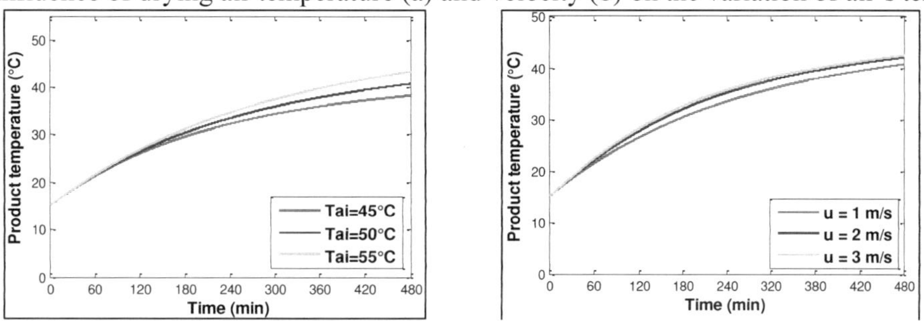

(a) (u=1 m/s) (b) (Tai=50°C)

Figure 4. Influence of drying air temperature (a) and velocity (b) on the variation of prodcut's temperature.

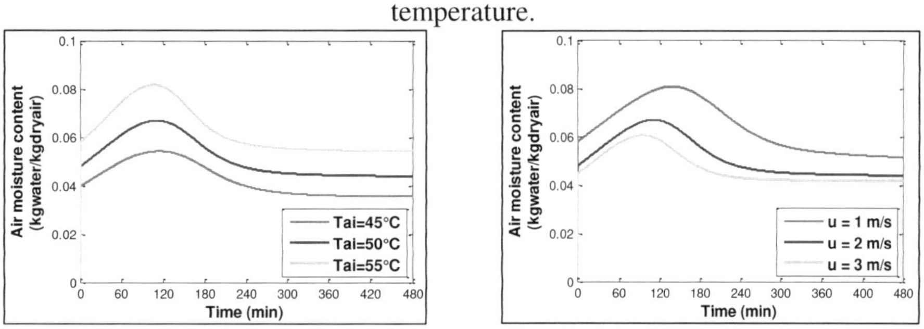

(a) (u=1 m/s) (b) (Tai=50°C)

Figure 5. Influence of drying air temperature (a) and velocity (b) on the variation of air's moisture content.

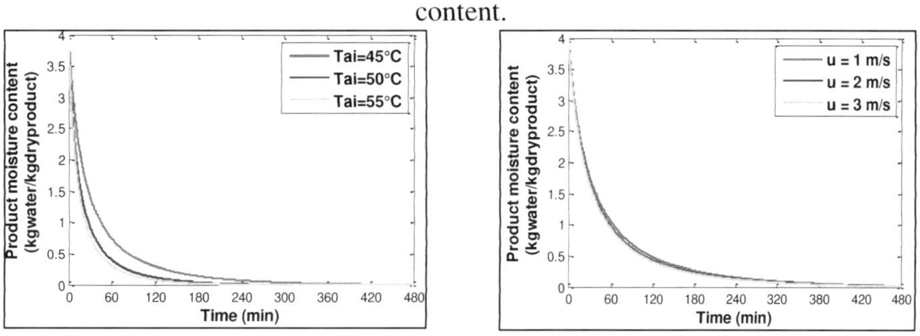

(a) (u=1 m/s) (b) (Tai=50°C)

Figure 6. Influence of drying air temperature (a) and velocity (b) on the variation of prodcut's moisture content.

Figure 3 shows the influence of the drying air temperature (a) and velocity (b) during the drying process as we can see that the influence of the temperature is more important than velocity which is the same case for the temperature of the product figure 4. This result agrees with the previous work [13] and [14]. Also for the variation of the moisture contents of the air and the product inside the drying chamber figure 5 and figure 6 the temperature of drying air is crucial parameter compared with its velocity.

5. Conclusion

In this work the simulation of the heat and mass balance equations using MATLAB software shows that the drying air temperature is an important influential parameter in the drying process. Through this simulation; the temperature and the moisture content of the hot wet air leaving the drying chamber goes into the evaporator is calculated. The influence of this hot wet air in the variation of the evaporation temperature must be handled in the design of the heat pump drying system to get higher efficiency and high value of the coefficient of performance (COP).

6. References

[1] Devabhaktuni V, Alam M, Shekara Sreenadh Reddy Depuru S, Green Ii R C, Nims D and Near C 2013 Solar energy: Trends and enabling technologies *Renewable and Sustainable Energy Reviews* vol 19, pp 555-564

[2] Tian Y and Zhao C Y 2013 A review of solar collectors and thermal energy storage in solar thermal applications *Applied Energy* vol 104, pp 538-553

[3] Bennamoun L and Belhamri A 2003 Design and simulation of a solar dryer for agriculture products *Journal of Food Engineering* vol 59, pp 259-266

[4] Krokida M, Karathanos K V, Maroulis T Z and Marinos Kouris D 2003 Drying kinetics of some vegetables *Journal of Food Engineering* vol 59, pp 391-403

[5] Amin Z M and Hawlader M N A 2013 A review on solar assisted heat pump systems in Singapore *Renewable and Sustainable Energy Reviews* vol 26, pp 286-293

[6] Daghigh R, Ruslan M H, Sulaiman M Y and Sopian K 2010 Review of solar assisted heat pump drying systems for agricultural and marine products *Renewable and Sustainable Energy Reviews* vol 14, pp 2564-2579

[7] Goh L J, Othman M Y, Mat S, Ruslan H and Sopian K 2011 Review of heat pump systems for drying application *Renewable and Sustainable Energy Reviews* vol 15, pp 4788-4796

[8] Ali I, Abdelkader L, El Houssayne B, Mohamed K and El Khadir L 2016 Solar convective drying in thin layers and modeling of municipal waste at three temperatures *Applied Thermal Engineering* vol 108, pp 41-47

[9] Singh Chauhan P, Kumar A and Tekasakul P 2015 Applications of software in solar drying systems: A review Renewable and Sustainable Energy Reviewsvol 51, pp 1326-1337

[10] Kasaeian A, Ghalamchi M and Ghalamchi M 2014 Simulation and optimization of geometric parameters of a solar chimney in Tehran *Energy Conversion and Management* vol 83, pp 28-34

[11] Tsilingiris P T 2010 Modeling heat and mass transport phenomena at higher temperatures in solar distillation systems – The Chilton–Colburn analogy Solar Energyvol 84, pp 308-317

[12] Pallet D, Fournier M andThemelin A 1987 Modélisation, identification et simulation d'un séchoir solaire à bois *Revue de physique appliquée* vol 22, pp 1399-1409

[13] Babalis S J and Belessiotis V G 2004 Influence of the drying conditions on the drying constants and moisture diffusivity during the thin-layer drying of figs *Journal of Food Engineering* vol 65, pp 449-458

[14] Yaldiz O, Ertekin C and Uzun H I 2001 Mathematical modeling of thin layer solar drying of sultana grapes *Energy* vol 26, pp 457-465

Photovoltaic Generation Data Cleaning Method Based on Approximately Periodic Time Series

J Zhang[1,3], Sh Zhang[1,3], J Liang[1,3], B Tian[1,3], Z Hou[2,3] and B Zh Liu[2,3]

[1] Electric Power Research Institute, State Grid Ningxia Electric Power Company, Yinchuan, China
[2] School of Electrical and Electronic Engineering, North China Electric Power University, Beijing, China

[3] zhangjun@dky.nx.sgcc.com.cn, 0zhangsh0nx@163.com, liangjian@dky.nx.sgcc.com.cn, tianbei@dky.nx.sgcc.com.cn, zan_hou@163.com, bzliu@ncepu.edu.cn

Abstract. Data cleaning of photovoltaic (PV) power generation is an important step during data preprocessing for further utilization, such as PV power generation forecasting. The PV power generation data can be treated as a time series. An improved data cleaning method based on approximately periodic time series is proposed. First, the abnormal data in the PV data time series is classified with three types of the outliers. Then these three types of outliers are quantified based on the physical characters of PV power generation, and the effective corresponding cleaning implementations are described considering the rate capacity of PV station and period of PV data time series. Finally, the data cleaning method is tested on the PV generation data from a certain real power grid. The results show that this data cleaning method can effectively improve the PV data quality, and provide an effective support tool for the further application of PV data.

1. Introduction

Photovoltaic power generation has the advantages of no pollution, zero emission, and few limitation of geographical resources distribution. In recent years, the installed capacity of PV power generation continues to increase. But due to the randomness and volatility of PV power generation, large-scale grid-connected PV stations have a negative impact on power quality, systemic stability and reliability. In many photovoltaic (PV) power plants, problems of communication errors, equipment failures and PV power curtailment result in high proportion of outliers in measured PV power data, which is difficult for performance analysis of PV power plants and application of power data [1]. Studying the characteristics of PV power generation and then generating its simulation sequence is very important for assessing the planning and operation of the grid-connected PV stations.

In the application of PV generation data, a short-term power forecasting model was proposed based on adaptive fuzzy time series method to forecast the power of grid connected PV power generation system [2]. Based on the historical generation data of PV power generation system, the adaptive algorithm is adopted to match the data structure with the forecasting model.

But the PV data directly collected to the gird is often incomplete, noisy and inconsistent, so the original PV data quality often cannot meet the follow-up application requirements, such as PV power generation forecasting, so it is necessary to carry out data cleaning.

In the terms of big data cleaning in power system, paper [3] studied the types of the anomalies and then proposed an iterative data cleaning method based on time sequence analysis. Paper [1] proposed a new identification methodology of the PV data with outliers, which is based on Copula theory to describe the relationship between the measured global radiation and PV power. Machine identification models are proposed to identify three typical types of outliers, which are verified using measured data of artificial data and PV power plants. The effectiveness of applying the outliers identification methods is investigated through a day-ahead PV power forecasting application. Paper [4][5] proposed an effective data cleaning methods in power system.

In this paper, an improved data cleaning method is proposed based on approximately periodic time series. The abnormal data in the PV data time series is classified with three types of the outliers. Then these three types of outliers are quantified based on the physical characters of PV power generation, and the effective cleaning implementations are described considering the rated capacity of PV station and period of PV data time series. Finally, the data cleaning method is tested on the PV generation data from a certain power grid.

2. Principle of cleaning method

2.1. Basic definition

In order to achieve the PV power generation data cleaning effectively, we choose the time series theory as the basis of cleaning analysis.

Time series refer to a series of ordinal data recorded for a time-varying phenomenon according to the order of time intervals. Time series analysis is to explore all the information contained in the data, to study such a group of real data in the long-term fluctuations in the process of statistical regularity. Time series can be divided into standard periodic time series and approximately periodic time series.

Standard periodic time series, that is fixed cycle length, and for different periods of the same position on the value is exactly the same, $X(t)=X(t\text{-T})$, T is the period.

The approximately periodic time series is a time series whose seasonal trend is an approximate periodic function. The concept of the N-order skeleton with approximate periodic function, and then the concept of approximate periodic time series is given in paper [6-7]. These two papers gave the definition as follows:

Let $\{f(t), t \geq 0\}$ be an approximately periodic function. Denoting $t_0 \geq 0$ and $t_k = t_0 + k\Delta t$, where $\Delta t \geq 0$ and $k = 1,2,\cdots$, if there exists a natural number N such that $\min\{T_k - T_{k-1}\}/\Delta t, \geq N$, where $k \geq 1$ and $T_k - T_{k-1}$ is the length of kth approximate period, $k = 1,2,\cdots$, then $\{f(t_k), k \geq 0\}$ is called an N-order skeleton of $f(t)$. Generally, N needs to be greater than 5.

The experimental data in this paper are obtained from the PV power data collected at a real PV power plant, the sample interval is 1min, and the measured PV power data curve is shown in Figure 1. It can be seen that the original PV power fluctuates greatly and cannot be directly used for the hourly average PV power prediction modelling. Moreover, there are abnormal data in the PV power generation data, such abnormal data will affect the follow-up application, such as PV power prediction. Data cleaning of these time series is an effective way to solve the problem.

According to the problem mentioned above, from the view of equipment operation, a higher failure rate may occur with the outdoor PV power generation equipment, inverters and a variety of communication equipments which are facing the various surrounding environment, weather and seasonal changes. Especially those equipments with poor installation process and production quality the failure rate will be higher. They cannot guarantee the generation of electricity and other monitoring information in real time and correct transmission.

2.2. Approximately periodic time series

State data of the PV generation under normal operation state generally show the character of periodic time series. Similarity can be observed with the observation points after s time interval. It can be fitted by a time series as ARIMA (p, ds, q).

2.2.1. Compression of PV generation data. A real PV generation curve is shown in Figure 1. This is a typical periodic time series. From the point of view of solving the problem briefly, the period is constant as 24 hours. But the PV equipment only generates active power in good sunny conditions during the day. So the PV equipment doesn't work in almost half the time of the day. In order to save storage space, simplify the workload of data analysis, a large number of zero-value data is not necessary to store and handle. The PV generation curve which the zero value has been removed is shown in Figure 2.

From Figure 1, PV power generation data has a strong randomness and volatility, also the periodicity is very apparent.

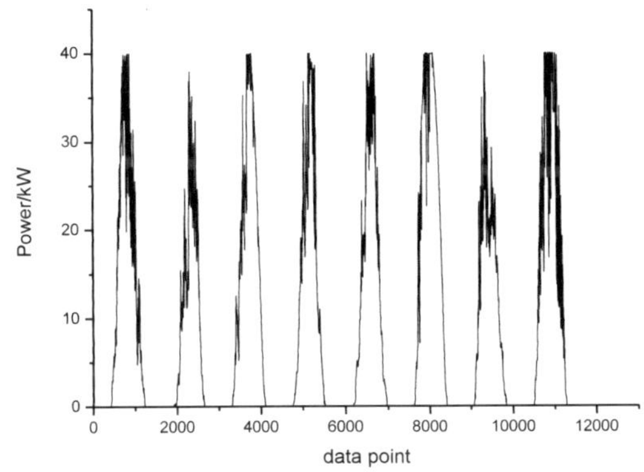

Figure 1. PV generation data under normal operation with zero-value data.

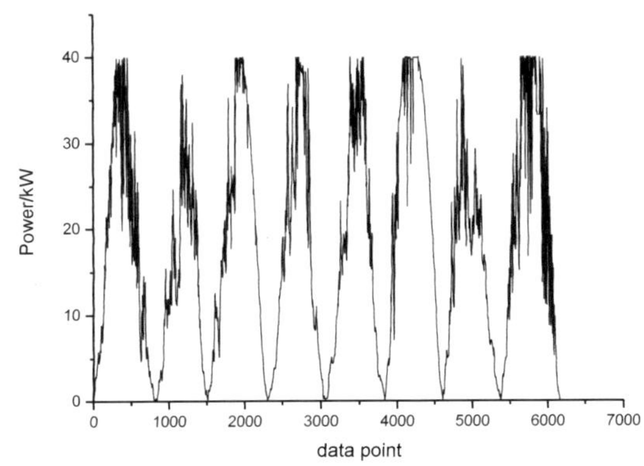

Figure 2. PV generation data under normal operation without zero-value data.

Compare the curve in Figure 1 with that of Figure 2, data storage capacity decreases by about 50%. The useful information in the data is completely retained and not missing. The compressed PV data curve is more convenient to adapt the time series analysis method.

The time series shown in Figure 2 is a typical approximately periodic time series. It exhibits certain periodic characteristics, but its period length is not a fixed value. So it is necessary to extract the approximate period.

2.2.2. Period extraction of approximately periodic time series. For an approximately periodic time series (t, X_t), $X_t=f(t)$, there is a time transformation function $t'=g(t)$, let $(t', X_{t'})$, $X_t=f(g(t))=f(t)$, this time series has the standard periodicity.

According to Figure 2, the time series $\{T_k, k = 0, 1, 2,\cdots\}$ can be calculated and shown in Table 1. The period is approximate constant.

Table 1. Time series T_k and the approximate period.

k	T_k	$T_k - T_{k-1}$
1	805	805
2	1512	707
3	2305	793
4	3059	754
5	3835	776
6	4610	775
7	5373	763

From Table 1, the approximate periodic character is obvious. If we want to know the exact value at any time in the time interval (T_k, T_{k+1}), the time transform function $g(t)$ is necessary to deduced. The function $g(t)$ can achieve the task of PV data cleaning. The specific formula of $g(t)$ can be a fitting estimation method based on paper [6-7].

3. Process of cleaning method

The existence of abnormal data will make the parameter estimation deviation of PV data time series. Therefore, the time series and the number of unknown outliers are not known. Combined with the physical properties of the PV data, different cleaning strategies are used to clean the observed PV data. The abnormal data in the time series can be divided into innovational outlier, additive outlier and level shift outlier and a combination of three types of outliers.

Suppose X_t is a time series with no outliers which obey the distribution of ARIMA (p, d, q) [2][8]. It can be expressed as

$$X_t = \frac{\theta(B)}{\varphi(B)\nabla^d}\alpha_t \tag{1}$$

$$\theta(B) = 1 - \theta_1 B - \theta_2 B^2 - \cdots - \theta_q B^q \tag{2}$$

$$\varphi(B) = 1 - \varphi_1 B - \varphi_2 B^2 - \cdots - \varphi_q B^q \tag{3}$$

Where B = delay operator; $\theta(B)$ = smoothing operator; $\nabla = 1 - B$; $\varphi(B)$ = reversible operator; α_t is independent white noise sequence with the same distribution.

When Z_t is used to represent the observed time series, the formula for the noise point at time T (pulse generation time) can be expressed as the following three outlier models [2][8].

3.1. Innovational outlier (IO)

IO affects all PV data after time T, and their effects are related to the model form of Z_t. Through the system described by the dynamic characteristics of the impact of the latter series. The IO model is expressed as

$$Y_t = \frac{\theta(B)}{\varphi(B)\nabla^d}\left(\alpha_t + \omega I_t^{(T)}\right) = Z_t + \omega\left(\frac{\theta(B)}{\varphi(B)\nabla^d}\right) I_t^{(T)} \tag{4}$$

Where $I_t^{(T)} = \begin{cases} 1, & t = T \\ 0, & t \neq T \end{cases}$.

For a certain PV data time series, this type of IO usually appears as a zero value during the daytime or a value greater than the rated power. So according to this physical character, it is easily to complete the data cleaning work.

3.2. Additive outlier (AO)

$$Y_t = Z_t + \omega I_t^{(T)} = \frac{\theta(B)}{\varphi(B)\nabla^d}\alpha_t + \omega I_t^{(T)} \tag{5}$$

AO only affects the PV data at time T where the disturbance occurs, and does not affect the sequence value after that time. The missing value in the time series can be thought of as an AO through an unknown ω.

The same cleaning strategy can be used to AO as well.

3.3. Level shift outlier (LO)

$$Y_t = Z_t + \frac{\omega}{1-B} I_t^{(T)} = \frac{\theta(B)}{\varphi(B)\nabla^d}\alpha_t + \frac{\omega}{1-B} I_t^{(T)} \tag{6}$$

Level shift outliers affect all observations after *T*, and the effect is the same.

A special strategy may be needed to handle this type of outliers. Depending on the length of time of the missing data, a suitable cleaning procedure should be used. An example shown in Section 4 is adapted to illustrate the special cleaning process.

4. Simulation case

In order to verify the practicability and effectiveness of the data cleaning method proposed in this paper, the original PV data at the moment of t =1312min is set 347.69kW as the point of failure, which is shown in Figure 3. While the rated capacity of PV station is 40kW. From t=1860min to t=6096min, the real PV data is replaced by a constant value as 34.14kW. Another type of outlier is set as IO, which begins at the time of 7103min with the value of 10.32kW. The modified curve is shown in Figure 4.

Use the process of cleaning method mentioned in Section 3, a time series can modeled and the outlier type can be recognized. The cleaning result is shown in Table 2.

Table 2. The cleaning result.

Outlier Type	Outliers Interval	
	Begin time/min	End time/min
AO	1312	1312
LO	1860	6096
IO	7103	7241

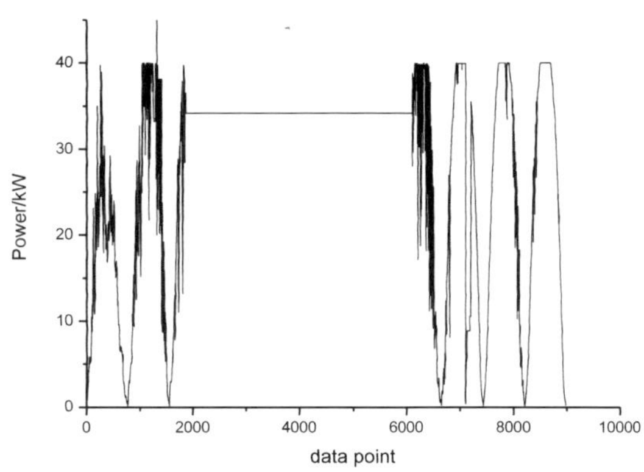

Figure 3. PV generation data with outliers.

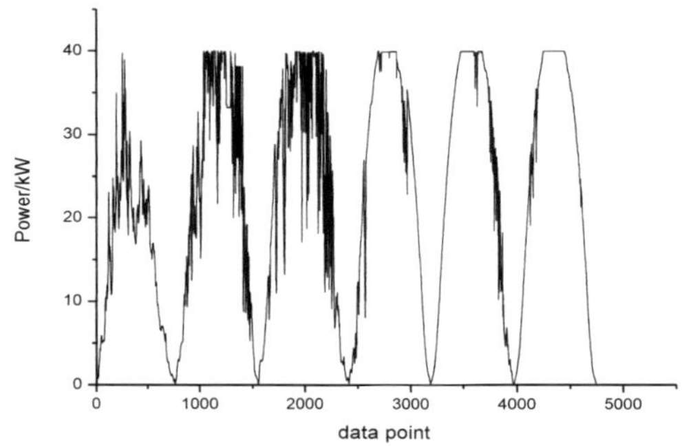

Figure 4. PV generation data without outliers after cleaning.

For the AO, it is greater than the rated capacity of the PV equipment. So the cleaning method is adapted to set this data as 40kW.

For the LO, the time interval continues for a long time, more than a few cycles. So the idea of revision or repair is likely unreasonable. The best way is to remove these large number of wrong values. During the cleaning process, the full cycle must be maintained.

5. Conclusions

Data cleaning of PV power generation is an important step during data preprocessing for further utilization. An improved data cleaning method based on approximately periodic time series is proposed. The main ideas are as follows:

The abnormal data in the PV data time series is classified with three types of the anomalies. Different types of outliers have its own cleaning steps. The approximately periodic time series use the time transform function to achieve the outliers correction considering the rated capacity of PV station. The data cleaning method is tested on the PV generation data from a certain real power grid, which show that this data cleaning method can effectively improve the PV data quality, and provide an effective support tool for the further application of PV data.

6. References

[1] Gong Y F, Lu Z X, Qiao Y, Wang Q and Cao X 2016 Copula Theory Based Machine Identification Algorithm of High Proportion of Outliers in Photovoltaic Power Data. *Automation of Electric Power Systems.* 40(9), 16-23

[2] Yang Z C, Zhu F, Zhang Ch L, Ge L and Yuan X D 2014. Photovoltaic Power Generation Short-term Power Forecasting based on Adaptive Fuzzy Sequence Method. *Journal of Nanjing Institute of Technology*, 12(1), 6-13

[3] Yan Y J, Sheng G H, Chen Y F, Jiang X Ch, Guo Zh H and Qin Sh P 2015 Cleaning Method for Big Data of Power Transmission and Transformation Equipment State Based on Time Sequence Analysis. *Automation of Electric Power Systems.* 39(7), 138-144

[4] Wang Zh L and Hu Y H 2012 *Application of time series analysis*. Beijing:Science Press

[5] Chen J Y, LI W Y and LAU A 2010 Automated Load Curve Data Cleansing in Power Systems. *IEEE Trans on Smart Grid.* 1(2), 213-221

[6] Diao Y L, Sheng W X, Liu K Y, He K Y and Meng X L 2015 Research on Online Cleaning and Repair Methods of Large-Scale Distribution Network Load Data. *Power System Technology.* 39(11), 3134-3140

[7] Wu Sh J, Zhu X Y and Yang X 2015 Analysis of Approximately Periodic Time Series. *Chinese Journal of Applied Probability and Statistics* 31(2), 199-212

[8] Hong Sh H 2015 *Periodicity Recognition and Extraction of approximately periodic time series* Shanghai East China Normal University

Numerical Simulation of Wind Pressure Coefficient and Distribution Trend of the Dust Concentration for Parabolic Trough Solar Collector

MingzhiZhao[1], XuZhang[1], XiaomingZhang[2], LuluZou[1], XiaoboKang[1]

[1]School of Energy and Power Engineering, Inner Mongolia University of Technology, Hohhot, China 010051

[2] College of Environmental and Energy Engineering, Beijing University of technology, Beijing, China 100124

zhaomingzhi2020@163.com; zczdzhangxu@163.com

Abstract. Analysis the influence rule of the changing wind angles on the distribution of the pressure coefficient at windward and lee side of collector, and adding the dust particles phase at the completion of the ideal air simulation, simulate the changing rule of the volume fraction of dust around the collector as the change of the attack angle. Under the speed of 10m/s, the conclusion of changing wind angles had no effect on the wind pressure coefficient of the value of windward ,the maximum value range is 0.6-0.7 and the changing wind angles had a great effect on the wind pressure coefficient of the value of leeward, the minimum value range is -0.35—-1.3 and the minimum value is -1.3 at collector angle θ=30°.When different angles, surface of the collector and the flow field around the dust volume fraction changes a lot, the volume fraction of dust is 2.56 times than the dust entrance at collector angle θ=0°; the volume fraction value significantly angle θ=±60°, its valume fraction is 1.4 times than dust entrance; in the rest of the angles, the volume fraction of collector surface is not much change, only slightly than the exit.

1. Introduction

The country's most abundant resource of solar radiation distributing in most areas of western China, including Inner Mongolia, Gansu, Qinghai, etc. Meanwhile, these areas are suitable for large-scale solar thermal power plants, because the land is flat and open, the cost is low [1]. However, these areas are windy and dusty the whole year. The demonstration project of 50MW pilot solar thermal power plant for Inner Mongolia Da Tang LTD, for example, measured data shows that, the average wind speed of the power station where located in Aragon is about 5m/s. Besides, these places are located in Central Asia sandstorm area which is one of the most active regions of dust activity [2-4]. The special climatic conditions like windy and dusty will have a significant impact on the operation of solar thermal power plants, especially for the solar collector.

At present, the research on wind load is mainly focused on bridges, buildings and other facilities. Lin [5] Studied the wind loads of high-rise buildings at local wind speeds. Bhumralkar C M [6] is the first person to apply numerical simulation to develop a solar collector for wind-resistant research, he used it to carry on the 2-D simulation for the California 100MW solar thermal power station, studied the influence factor on solar thermal power plants including the wind load, the temperature, etc.

Naeeni et al. [7]Simulated the wind load characteristics of a collector under 2-D steady flow field with a 250 kW pilot solar thermal power plant at Shiraz, Iran, using numerical simulation method. According to the above, the wind load of trough collector is mainly carried out by experiments; the simulation mainly focuses on 2-D flow field. This paper focuses on the influence relation of surface pressure and the pattern of sand and dust around the collector by angle of wind approach under 3-D steady flow field.

2. Forces analysis for sand in wind-blown-sand two phase flow

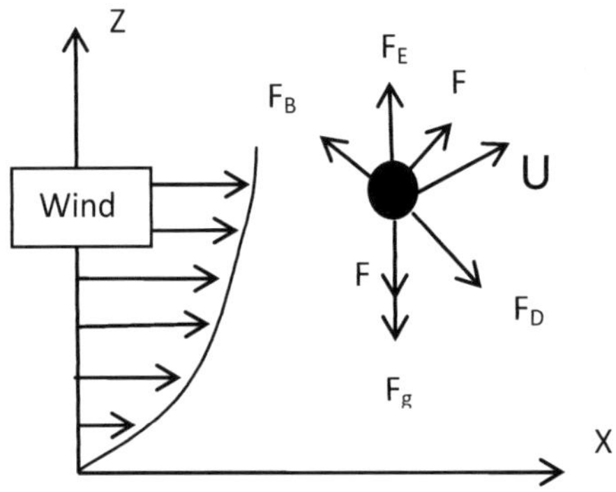

Figure 1. Sketch map of the force of sand in the air

Assuming that the sand is a uniform sphere, CFD predicts the track of a dust by integrating the force-balance equation of the particle. Figure 1 shows the forces analysis for sand in the wind-sand flow. It is assumed that the sand in the XOZ plane for two-dimension motion, the incoming wind speed is given by an exponential wind profile and the direction is along the positive X direction. With the flow in the wind-drift flow, the particle force balance

$$\frac{du_p}{dt} = F_D(u-u_p) + \frac{g(\rho_s-\rho)}{\rho_s} + F_x \tag{1}$$

equation has been given where u_p is the track speed, u is the incoming wind speed, R_p is the radius of sand and $\rho_s$is the density, ρ, g are the air density and gravitional acceleration, respectively. F_D is the grabability of sand, is the most important driving force to promote the movement of sand which mainly produced by relative movement between airflow.

In the sand grain balance equation, F_X is including electrostatic force F_E, gravity Fg, Magnuslift force F_M, Saffman lift force F_S and Basset force F_B, etc. This work ignored the influence of electrostatic forces, because there is no regularity of the influence of electrostatic force on sand movement at present.

3. Research Method

3.1Research Content

The windward angle of the collector is changing because the collector needs to track the position of the sun and the wind direction is not constant, so the windward angle is defined as the angle between the incoming wind direction and the plane where the two ends of the parabolic trough collector are connected. These changes cause the flow field around the collector to be different, thus affecting its wind pressure characteristics. In this paper, the surface pressure distribution of it was considered under the windward angle of -90 °, -60 °, -45 °, -30 °, 0 °, 30 °, 45 °, 60 °,90° . And the influence of windward angle on dust settling was studied by adding sand particles. Figure 2 shows the concept map of windward angle.

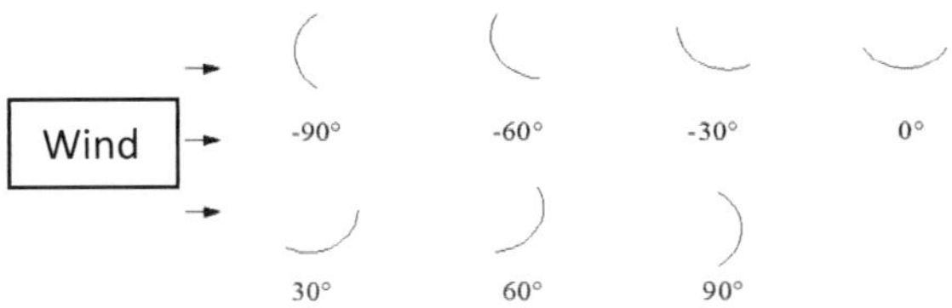

Figure2. Definition of the attack angles

Application of wind engineering research methods, taking the RP3 type trough collector [8] as the most widely used in the world as the model, the CFD software Fluent was used to simulate. The wind pressure at a height of 10 m is taken as a reference value, the mixture model of the FLUENT and the Reynolds time-averaged method were used to simulate the distribution of the wind pressure coefficients of the upwind and leeward side wind collectors with different angles at the wind speeds of 10m/s. And the influence of windward angle on the dust distribution around the collector was studied by adding sand particles.

3.2 Determination of Flow Field and Meshing

3.2.1 Determination of Flow Field. The computational model is established based on the actual model surface and the surrounding flow field. The computational domain size is large enough to ensure that the computational domain boundary does not exert a large influence on the flow field around the collector model, which makes it easier to obtain more accurate simulation results. In the numerical simulation, the blocking rate is used to represent the relationship between the model and the calculated domain size.

$$S = A_b / A_c \text{ (2)} \quad (2)$$

Where S is the blocking rate, A_b is the windward area of the model, A_c is the Cross-sectional area of the flow field.

It is pointed out that when the numerical simulation is carried out, the calculation domain blocking rate should not be more than 3% [9] [10]. The dimensions of computational domain of solar collector flow field are show in Table 1.

The computational domain in addition to guarantee the blocking rate, but also need to consider whether the flow field is fully developed while study flow around. Sun [11] suggested that, for the distance from the inlet of the flow field to the windward side of the model is five or six times as the height of the model, the distance from the outlet of the flow filed to the leeward side of the model is seven or nine times as the height of the model, in order to fully develop the flow field.

In this paper, the simulation is a single parabolic trough collector, the design parameters as shown in Table 2. The wind direction is perpendicular to the opening face of the collector, the wind direction angle is constant, whole flow field is symmetrical. In order to simplify the calculation, the computational domain was assumed to be symmetric. So, only half of the simulation was considered.

Table 1. The dimensions of computational domain of solar collector flow field

Altitude (m)	Width (m)	Length of windward (m)	Length of leeward (m)	Bloching rate (%)
43	40	40	120	0.02

Table 2. Design parameter of single parabolic trough reflector

Type of parabolic trough collector	Opening width (mm)	Focal distance (mm)	Concentration ratio	Length (m)	Degree of application
RP3	5770	1710	82	12	Widespread

*3.2.2 Meshing.*A small domain have been built in the area near the collector (as shown in Figure 3) at first, a dense tetrahedron mesh are adopted in this small domain to adapt to the complex flow conditions on surface of the collector and its surrounding, hexahedral mesh are adopted out of the small domain and its mesh quality is sparser than the mall domain.

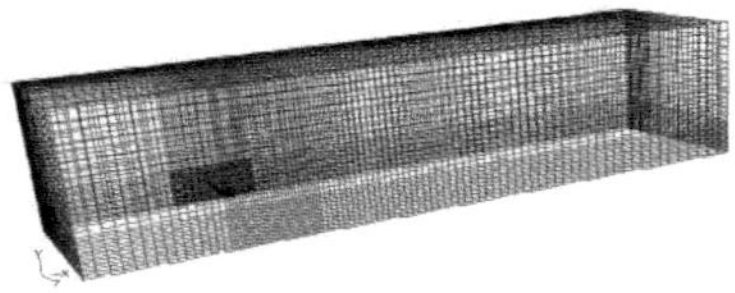

Figure 3.The mesh of computational domain

*3.2.3 Boundary Condition.*Mixture model in eulerian two phases model is adopted in two-phase flow simulation, in order to better simulate the phenomenon of dust settling. The characteristics of sand and sediment concentration are obtained from the actual statistics of sand-dust storm weather in Gansu Province in 1993[12]. Boundary conditions are as follows.

(1) The air fluid is incompressible and the density remains constant.

(2) Dust particles are uniform sphere and the average particle size is $6.44\mu m$.

(3) The particle inlet velocity is equal to the gas flow velocity, and there is no relative slip.

(4) The dust is uniformly distributed with height direction at inlet, the average volume fraction of dust is $3.2\times10^{-5}\%$.

4. Results and discussion

To change the collector angle, the distribution of the windward pressure is large different from the leeward pressure. In the horizontal wind flow, the different vertical plane formed by wind pressure and wind speed on the collector calculated the ratio of the dynamic pressure is called wind pressure coefficient.

4.1 The windward wind pressure coefficient

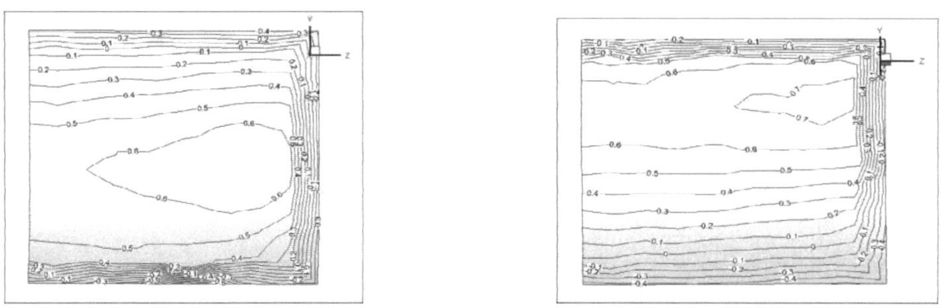

(a)the distribution of Windward pressure coefficient at θ = -90° (b)the distribution of Windward pressure coefficient at θ = -60°

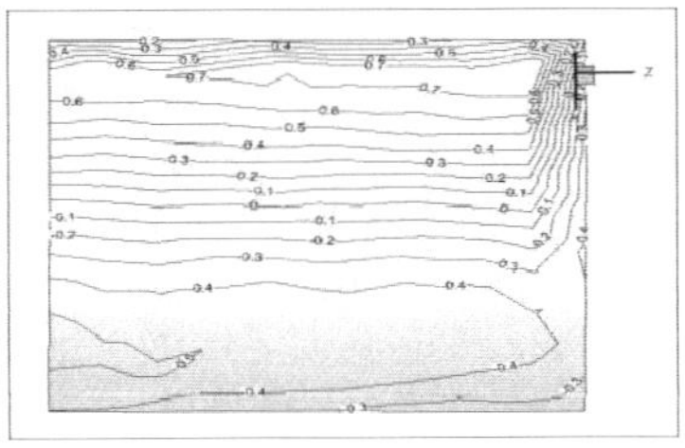

(c)the distribution of Windward pressure coefficient at θ = -30°

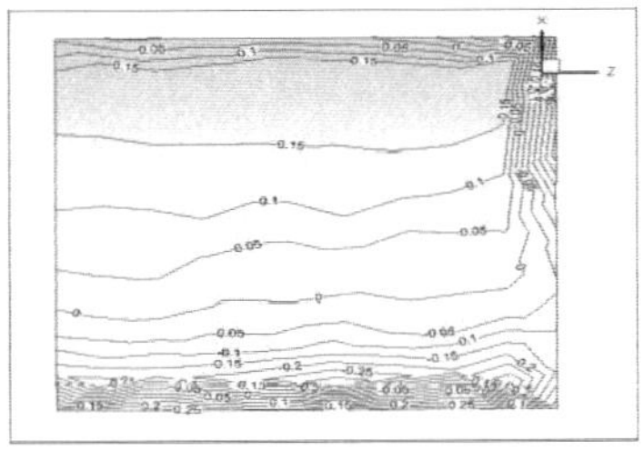

(d)the distribution of Windwardpressure coefficient at θ = 0°

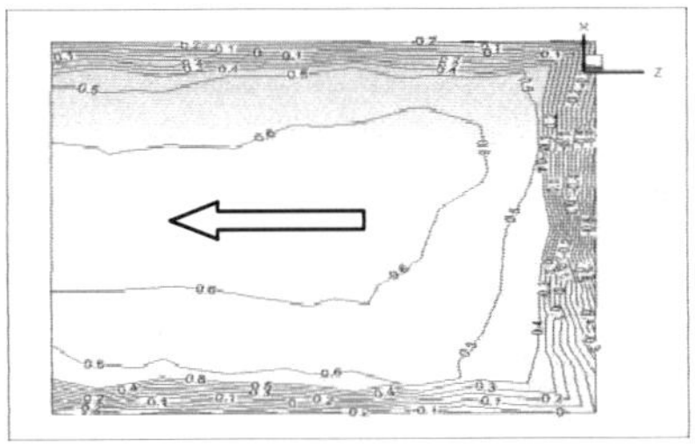

(e)the distribution of Windward pressure coefficient at θ = 30°

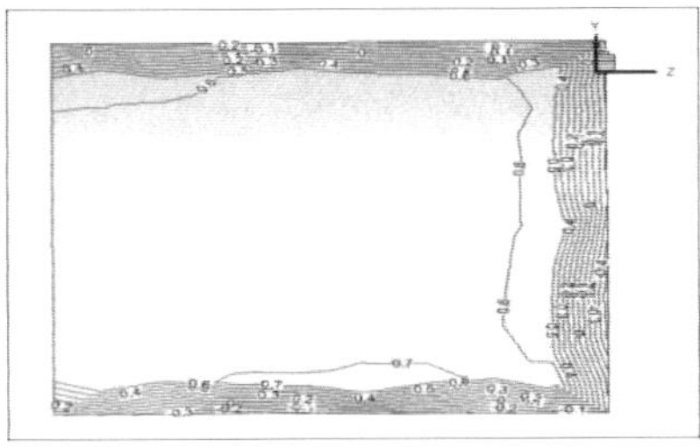

(f)the distribution of Windward pressure coefficient at θ = 60°

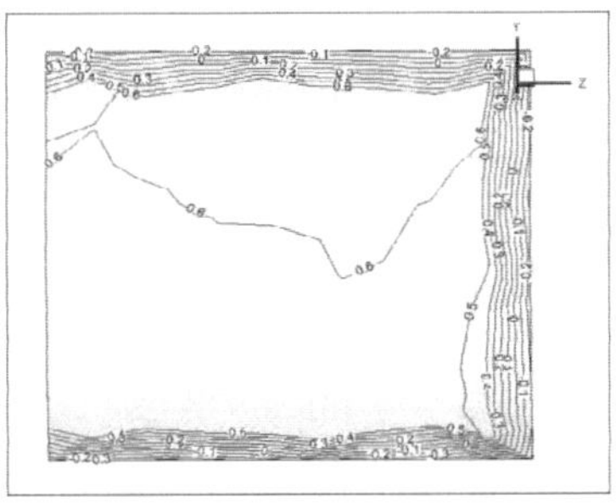

(g).the distribution of Windward pressure coefficient at θ =90°

Figure 4.The distribution of pressure coefficient at windward side of collector

*4.1.1. Flow change.*Base on the special shape of the parabolic trough solar collector, compared with the distribution of wind pressure coefficient for different collector angle from Figure 4 can be seen that changing collector angle has no effect on the windward coefficient (except collector at θ= 0°),the maximum value is 0.6-0.7.No matter how the collector angle changes collectors both ends up and down are always negative pressure, due to the collectors structure is symmetric parabola, when flow impact the surface of collectors ,part of flow rise along the collectors and through it, the other part fall along the collectors and caused eddies. At both ends of the collectors, the wind pressure coefficient of contour are most dense ,due to the distance between collector and ground are closer, the flow is not easy to separate resulting in a greater pressure gradient on the edge.

*4.1.2. Lateral change.*Along the direction of arrow at Figure 4(a), only half of the collector was calculated in the process of simulation, so the windward pressure coefficient of contour are dense in the middle of collector and are sparse at two sides. Maximum value appeared at two sides of the concave surface, is not the center. The cause of this phenomena is that when flow impact the surface of collectors, under the action of resistance the flow is separated two sides of the collector, the flow of

IOP Conf. Series: Earth and Environmental Science **63** (2017) 012009 doi:10.1088/1755-1315/63/1/012009

separation and wind blowing are push the surface of collectors cause the wind coefficient rise. So, in the process of design of collector bracket collector to consider the role of wind load on the both end.

*4.1.3.Vertical trend.*Figure 4 (a),(b) ,(c) illustrate that with the collector angle increasing, the position of the wind pressure coefficient maximum value is moving up, the wind pressure coefficient of contour are dense at the head of collector and the minimum value of the bottom of collector is invariably range -0.4—-0.5. With the collector angle increasing, the resistance for flow is decreasing, and the flow of fall along the collector cause eddies between the collector and ground.

When the collector angle at θ=0°, flow approximate flows through the surface and cause a small disturbance around the collector. The wind pressure coefficient maximum value is 0.15 and minimum value is -0.15. The wind pressure distribution on the surface is basic equal at θ>60°. Only near the edge, due to the effect of separation of the flow, the wind pressure dropped sharply.

4.2 The leeward wind pressure coefficient

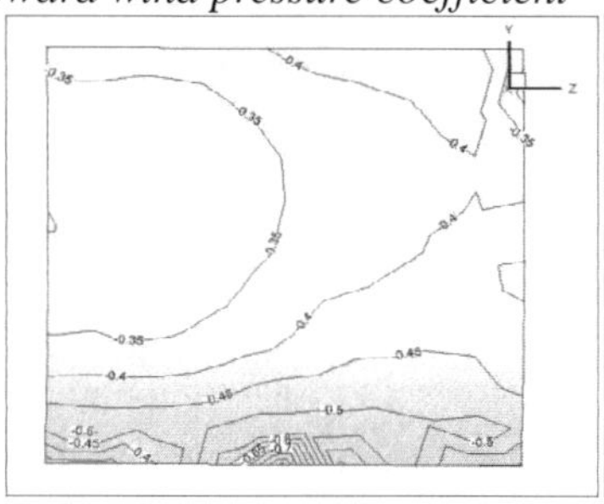
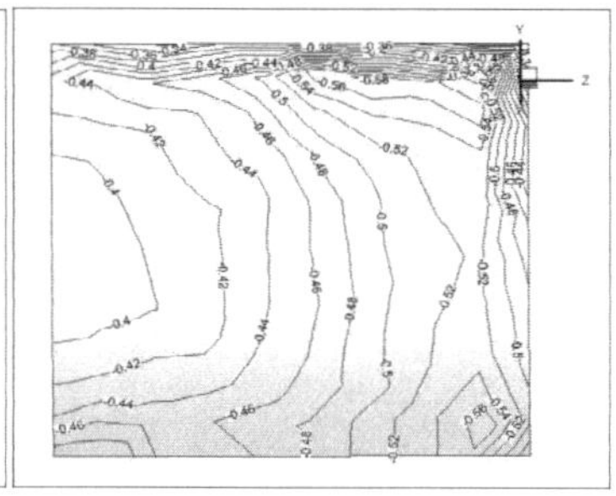

(a).the distribution of leeward pressure coefficient at θ =-90(b).the distribution of leeward pressure coefficient at θ =-60°

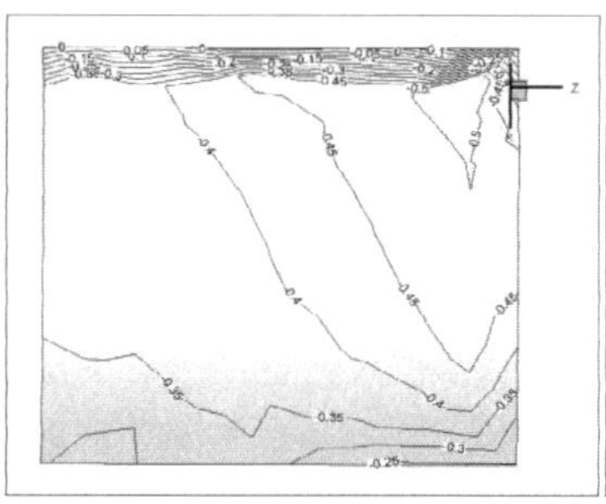
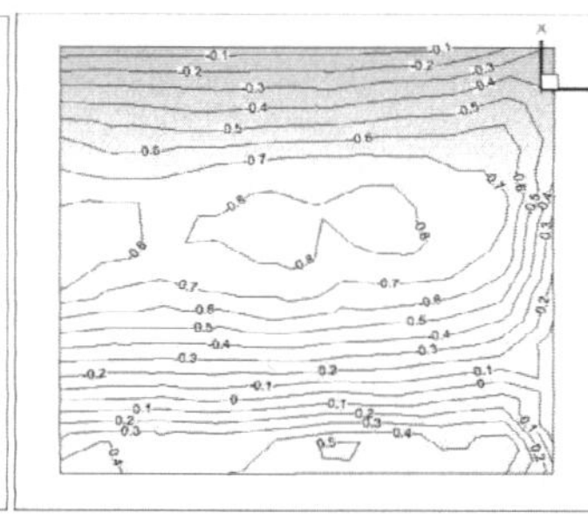

(c).the distribution of leeward pressure coefficient at θ =-30° (d).the distribution of leeward pressure coefficient at θ =0°

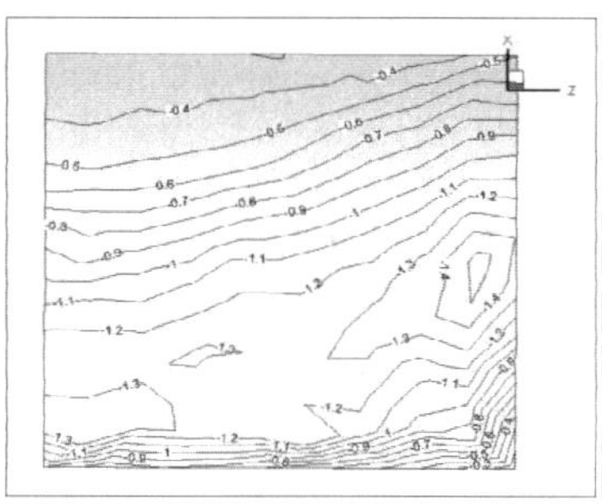
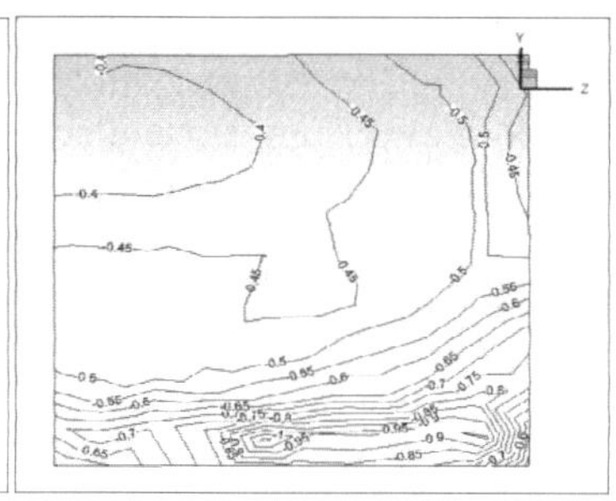

(e).the distribution of leeward pressure coefficient at θ =30° (f).the distribution of leeward pressure coefficient at θ =60°

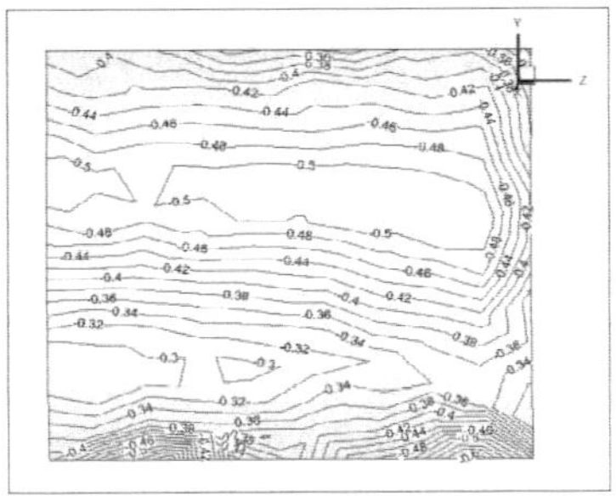

(g).the distribution of leeward pressure coefficient at θ=90°

Figure 5.The distribution of pressure coefficient at lee side of collector

The leeward wind pressure coefficient mainly negative pressure from Figure 5. Windward angle from 90° to 60°, the process of resistance increase, the separation effect of strengthening the collector edge and its maximum leeside wind pressure coefficient is lower. Collector angle θ=0°, the pressure coefficient symmetric distribution on the convex surface. The peaked of leeward wind pressure coefficient is -1.3 at θ= 30°.

4.3 Dust distribution

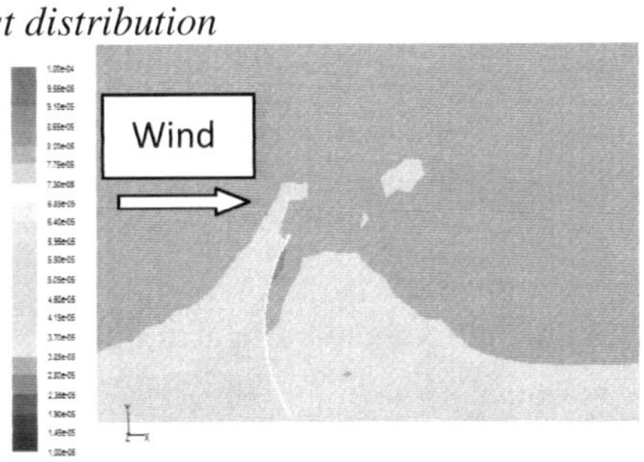

(a).the distribution of the volume fraction

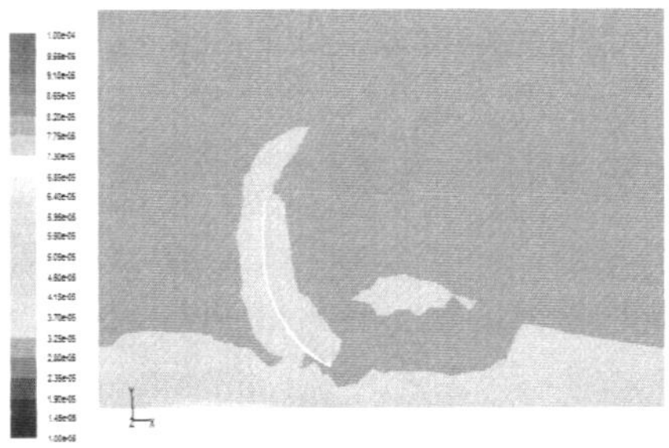

(b).the distribution of the volume fraction of dust around the collector at θ=-90°of dust around the collector at θ=-60°

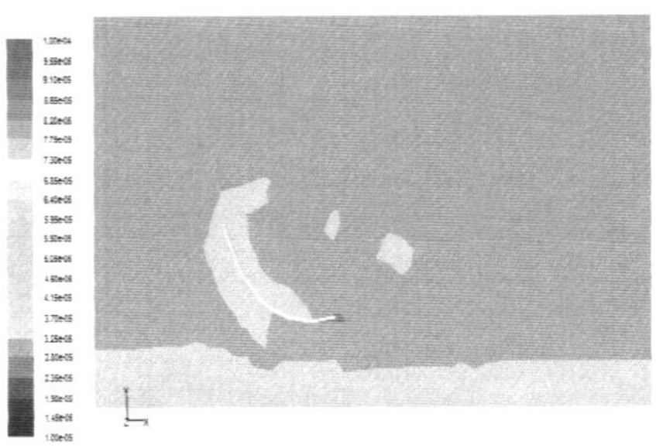

(c).the distribution of the volume fraction

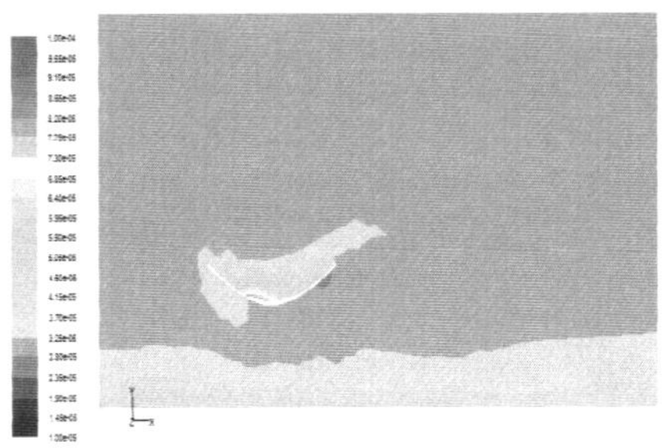

(d).the distribution of the volume fraction of dust around the collector at θ=-30°of dust around the collector at θ=0°

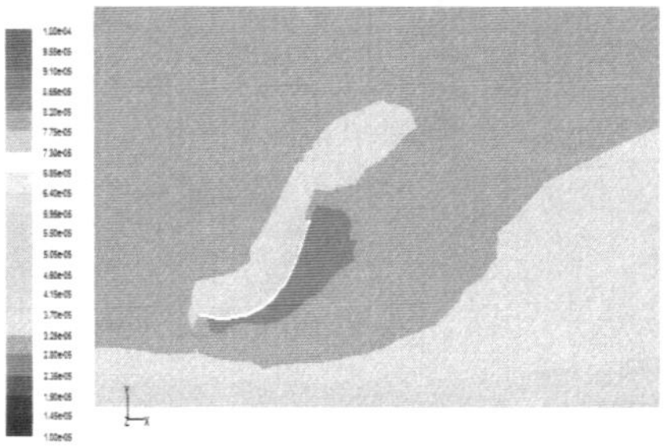

(e).the distribution of the volume fraction

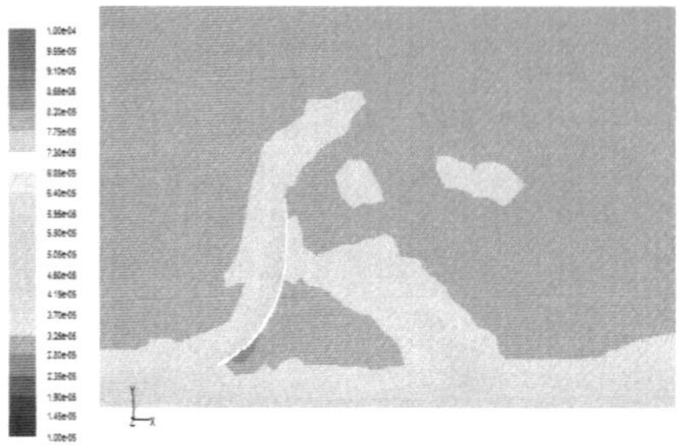

(f).the distribution of the volume fraction of dust around the collector at θ= 30°of dust around the collector at θ= 60°

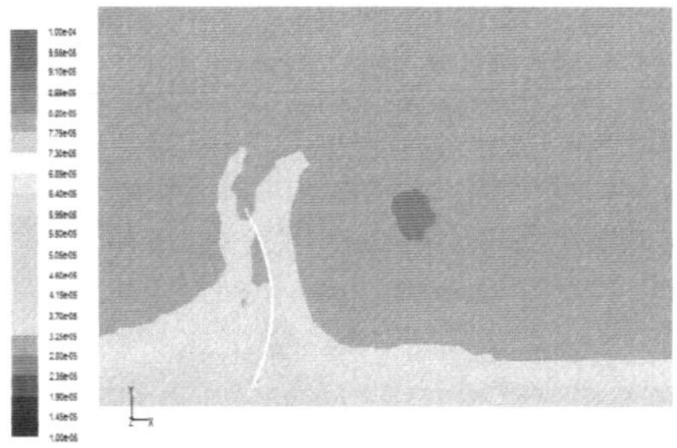

(g).the distribution of the volume fraction of dust around the collector at θ=90°

Figure 6.Volume fraction distribution of dust around collector at different attack angles

Firstly, Figure 6 illustrate the ground of windward are always appear dust-deposition area, because of the collector's resist and sand its own gravity, on the back of the collector also can appear different degree of dust-deposition area. The maximum resistance is collector angle θ=±90°, so the dust-deposition is most obvious and large range at the windward of collector. With the collector angle increase, the dust-deposition area is different. The collector angle from -90°to 0°,the dust-deposition area of the collector leeward is more and more far from the collector, when θ=0°disappear.From θ=0°to θ=90°,the dust-deposition area is appear once again.

What's more, the volume fraction of dust on collector surface is also change with the collector angle change. From Fig.8 the maximum value of the volume fraction of dust is 8.2×10^{-5} % at collector angle θ=0°, and its 2.56 times than the dust entrance. The volume fraction value significantly large angle θ=±60°, its volume fraction is 4.5×10^{-5} % and 1.4 times than dust entrance. In the rest of the angles, the volume fraction of collector surface is not much change, only slightly than the exit.

5. Conclusions

(1) Changing collector angle has no effect on the windward coefficient of the value(except collector at θ=0°), the peaked range is 0.6-0.7 ; no matter how the collector angle changes collectors both ends up and down are always negative pressure; at both ends of the collectors, the wind pressure coefficient of contour are most dense.

(2) The leeward wind pressure coefficient mainly negative pressure. Windward angle from 90°to 60°,the process of resistance increase, the separation effect of strengthening the collector edge and its maximum leeside wind pressure coefficient is lower. Collector angle θ=0°,the pressure coefficient symmetric distribution on the convex surface. The maximum value of leeward wind pressure coefficient is -1.3 at θ=30°.

(3) The volume fraction of dust on collector surface is also change with the collector angle change. From Fig.8 the maximum value of the volume fraction of dust is $8.2{\times}10^{-5}$ % at collector angle θ=0°, and its 2.56 times than the dust entrance. The volume fraction value significantly large angle

θ=±60°, its valume fraction is $4.5\times10^{-5}\%$ and 1.4 times than dust entrance. In the rest of the angles, the volume fraction of collector surface is not much change, only slightly than the exit.

6. Acknowlegement

Supported by National Natural Science Foundation of China (51466011) and Natural Science Foundation of Inner Mongolia (2014MS0511).

7. Reference

[1] Zhao Mingzhi, SongShijin, Zhang Xiaoming 2013 Method research of parabolic trough solar thermal power plant site selection *Renewable Energy Resources***3**(31):18-22.

[2] Qian Zhengan, Song Minhong, Li Wanyuan 2002 Analyses on distributive variation and forecast of sand-dust storms in recent 50 years in north China *Desert Research* **22**(2):106-11.

[3] Su Zhizhu, Dong Guangrong2002Study on the sand-dust storm disaster and prevention countermeasures in northern China *Journal of Shanxi Unversity (Natural Science Edition)***25** (4):361-5.

[4] Wang Tao, Chen Guangting, Qian Zhengan.2001Situation of sand-dust storm and countermeasures in north China *Desert Reserch***21** (4):322-7.

[5] Lin N, Letch Ford G, Tamura Y, Liang B, Nakamura O.2005Characteristics of wind forces acting on tall building *J Wind Eng Ind Aerodyn***93**:217-42.

[6] Bhumralkar C M, Slemmons A J, Nitz K C1981 Numerical study of local regional tmospheric changes caused by a large solar central receiver power plant *Journal of Applied Meteorology***20**: 660-77.

[7] N. Naeeni, M. Yaghoubi2007Analysis of wind flow around a parabolic collector fluid flow *Renewable Energy* **32** (11): 1898-916.

[8] CSPPLAZA Statistical data of the research centre*http://www.cspplaza.com/article-1830-1.html.*

[9] F Baetke, H Werner, H Wengle1990Numerical simulation of turbulent flow over surface-mounted obstacles with sharp edges and corners *Journal of Wind Engineering and IndustriAerodynamics***35**: 129-47.

[10] S A Bekele, H Hangan2002 A comparative investigation of the TTU pressure envelope-Numerical versus laboratory and full scale results. *Wind and Structures* **5** (2-4): 337-46.

[11] Sun Xiaoying, Xu Wei, WuYue 2007 The flow around a blunt body calculation domain research *The 13th national conference of structural wind engineering* Dalian.

[12] Dai XueRong, Shi YuXin, Xue Bin 1995 Granulometric characteristics and significance of the deposits from a recent extraordinary heavy duststorm in Lanzhou, Gansu, northwest China*Journal of lanzhou university(natural science)***31**(4):168—74.

IOP Conf. Series: Earth and Environmental Science **63** (2017) 012010 doi:10.1088/1755-1315/63/1/012010

Analytical Assessment of the Relationship between 100MWp Large-scale Grid-connected Photovoltaic Plant Performance and Meteorological Parameters

Jie Sheng[1,*], Qiaoming Zhu[2], Shijie Cao[1] and Yang You[1]

[1]School of Urban Railway Transportation, Soochow University, Suzhou, Jiangsu, China

[2]School of Computer Science and Technology, Soochow University, Suzhou, Jiangsu, China

jsheng@suda.edu.cn

Abstract. This paper helps in study of the relationship between the photovoltaic power generation of large scale "fishing and PV complementary" grid-tied photovoltaic system and meteorological parameters, with multi-time scale power data from the photovoltaic power station and meteorological data over the same period of a whole year. The result indicates that, the PV power generation has the most significant correlation with global solar irradiation, followed by diurnal temperature range, sunshine hours, daily maximum temperature and daily average temperature. In different months, the maximum monthly average power generation appears in August, which related to the more global solar irradiation and longer sunshine hours in this month. However, the maximum daily average power generation appears in October, this is due to the drop in temperature brings about the improvement of the efficiency of PV panels. Through the contrast of monthly average performance ratio (PR) and monthly average temperature, it is shown that, the larger values of monthly average PR appears in April and October, while it is smaller in summer with higher temperature. The results concluded that temperature has a great influence on the performance ratio of large scale grid-tied PV power system, and it is important to adopt effective measures to decrease the temperature of PV plant properly.

1. Introduction

Solar photovoltaic (PV) technology is a promising means of generating clean, sustainable energy, and the use of this technology is rapidly developing around the world [1-4]. By the end of 2015, the capacity of photovoltaic (PV) plants in China has already passed 43.18 GW, and the new additions in photovoltaic capacity is 31 GW in 2016. China has become the country with the largest photovoltaic power capacity of the world [5]. However, photovoltaic power generation output is sensitive to change of climate. With the rapid expansion of the PV plant, the intermittent and uncontrollability of the output power has brought great pressure to the safe, reliable and economic operation of the power system.

Photovoltaic power generation is a multivariable coupling nonlinear stochastic process. The main impact factor is the solar radiation, which is related to the easily determined astronomical factor and geography factor as well as many uncertain meteorological factors, including cloud cover, aerosol, sunshine hours, temperature, and humidity [6]. In recent years, many domestic and international scholars carried on a great deal of empirical research to the impact of meteorological environment factors on photovoltaic power generation, and some useful conclusions were obtained [7-10]. Most of these studies were based on pilot environmental data, the conclusions were idealistic or do not tally with the operation of the actual and commercialized photovoltaic power station.

In this paper, it is intended to research on the impact of typical meteorological factors on the photovoltaic power generation of a 100MWp grid tied photovoltaic plant. Firstly, the correlation analysis of meteorological factors to photovoltaic power generation was carried out by using the whole year's data; Moreover, the influence of meteorological parameters on energy generation capacity was analysed with monthly mean data; Last but not least, the monthly mean performance ratio was analysed to quantifies the overall effect of equipment failure and efficiency on the rated plant potential output.

2. System details

Figure 1. The 100MWp grid connected solar power plant.

The 100MWp grid connected polycrystalline PV system under performance assessment has been installed at the Yangzhou city of Jiangsu Province of China. With the theme of "ecological equilibrium, energy conservation and environmental protection" and by the pattern of "fishing and PV complementary", the solar PV plant has been installed above a fishpond with ornamental plants in it. The project makes full use of water resources in the region, organically combined solar photovoltaic power generation and aquaculture, as well as ecological agriculture and tourism, and formed a three-dimensional developed comprehensive new energy power base with special features. The schematic diagram of the system is shown in Fig. 1. The total investment of the project was one billion yuan, more than four hundred thousand pieces of polycrystalline silicon solar panels were laid in the area of 2500 mu. The grid connected PV system has been put into operation in December 2014, and realized the connection to the utility gird of the whole district in December 2015. The total energy output generated from solar PV plant was 97.588 million kWh in 2015, which amount to saving about 29000 tons of standard coal, reducing the pollutants such as SO_2, NO_x and CO_2 about 73 tons for the country.

The photovoltaic power generation data adopted in this paper was from December 2015 to November 2016 by hours throughout the whole year. The Radiation data and conventional meteorological data of the same period were taken from the real-time monitored data by hours throughout the year by environmental monitoring equipment that installed at the photovoltaic power plant.

3. Methodology

3.1. Correlation analysis

There are many meteorological factors that make great influence on the yield of photovoltaic power generation. In order to identify the importance of the impact factors, it is appropriate to measure the linear correlation degree of relationship between the electricity energy generation and various meteorological factors by using correlation analyse. The most commonly used Pearson correlation coefficient of two-dimensional variables is defined as shown in formula (1).

$$\eta_{cor} = \frac{\sum_{i=1}^{N}(x_i-\bar{X})(y_i-\bar{Y})}{\sqrt{\sum_{i=1}^{N}(x_i-\bar{X})}\sqrt{\sum_{i=1}^{N}(y_i-\bar{Y})}} \quad (1)$$

Where, N is length of sample series, $(x_i,y_i)$is the observed value of sample i of the two-dimensional variable (X,Y), and $(\bar{X}, \bar{Y})$ is the average value of the N samples of the two-dimensional variable (X,Y).

3.2. *Performance ratio (PR)*

The Performance Ratio (PR) is one of key parameter to determine efficiency of the PV systems regardless of module efficiency. The PR is useful to compare great amount of datasets graphically in order to analyse the influence of different factors. The PR indicates that how much the PV power plant is energy efficient and reliable. It provides the option for checking performance, output, seasonal oscillations and long-term degradation [9].It is given in formula (2).

$$\mathrm{PR} = \frac{Y_f}{Y_r} = \frac{E_{AC}/P_{PV}}{G_i/G_{STC}} \quad (2)$$

Where, Y_f is the yield factor, which is defined as the ratio of the AC Energy generated by PV power plant (E_{AC}) to the rated DC power of the PV power plant (P_{PV})at ideal condition (1000 W/m^2, 25℃, 1.5 AM) ; Y_r is the reference yield that can be defined as the ratio of the total in plane radiation (G_i) to module's reference in-plane irradiance(G_{STC}) which is generally taken as 1000W/m^2.

4. Results and discussion

4.1. *Correlation analysis of meteorological factors to photovoltaic power generation*

Table 1. Correlation analysis of daily photovoltaic power generation to daily meteorological factors

variables	photovoltaic power generation	average radiation	average temperature	mean wind speed	temperature range	maximum temperature	minimum temperature	hours of sunshine
photovoltaic power	1	0.937	0.248	0.058	0.659	0.282	0.106	0.522
average radiation	0.937	1	0.063	0.090	0.638	0.092	-0.077	0.300
average temperature	0.248	0.063	1	-0.150	0.058	0.988	0.966	0.782
mean wind speed	0.058	0.090	-0.150	1	-0.111	-0.143	-0.113	-0/114
temperature range	0.659	0.638	0.058	-0.111	1	0.106	-0.158	0.255
maximum temperature	0.282	0.092	0.988	-0.143	0.106	1	0.965	0.751
minimum temperature	0.106	-0.077	0.966	-0.113	-0.158	0.965	1	0.678
hours of sunshine	0.522	0.300	0.782	-0.114	0.255	0.751	0.678	1

The correlation analysis of photovoltaic power station daily output data to the daily meteorological data for the whole year (December2015-November2016) was made by using SPSS statistical software, and the results are shown in Table 1. Firstly, it is obvious that daily photovoltaic power generation is significantly positively related to daily average radiation, the Pearson correlation coefficient is 0.937; Secondly, the positive correlation coefficient of daily photovoltaic power generation to daily temperature range is 0.659; Then, the positive correlation coefficients of daily photovoltaic power generation to daily hours of sunshine, daily maximum temperature, daily average temperature, daily minimum temperature and daily mean wind speed are 0.522,0.282,0.248,0.106 and 0.058 respectively.

Because of the adoption of yearly data, there may be the phenomenon of the offsetting between positive and negative ones in variable correlation analysis. Therefore, it is need to have further detailed analysis in order to determine the influence of the key meteorological factors on photovoltaic power generation.

According to the physical process of photovoltaic power generation, solar radiation is the most direct and key impact factors that is influenced by astronomical factors, such as the revolution and rotation of the earth, and has a larger difference between each month. As shown in Table 1, the positive correlation coefficients of daily temperature range to daily average radiation is 0.638, which means a strong correlation between them. It makes little sense to analysis the impact of the temperature diurnal range on the power generation, while temperature has great influence on the efficiency of PV panels. As a result, the three key meteorological factors were confirmed as daily average radiation, daily average temperature and daily hours of sunshine.

4.2. Data analysis of meteorological parameters

By processing meteorological parameters data measured by real-time climate environment sensors in PV substation, the monthly average values of daily solar radiation, daily average temperature and daily average sunshine hours from December 2015 to November 2016 were obtained, as shown in Fig.2.

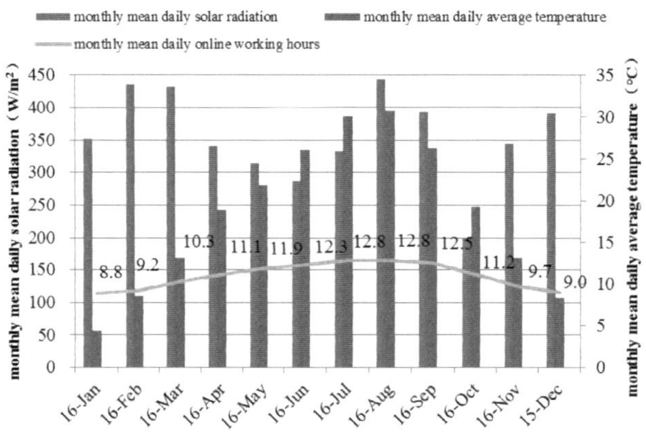

Figure 2. Monthly mean value of meteorological parameters

As can be seen from the Fig.2, the maximum value of the monthly average daily solar radiation appeared in August ,which as high as 443.72W/m^2, while the minimum value appeared in October as 199.30 W/m^2; the maximum value of the monthly average daily average temperature was also appeared in August as 30.7℃, while the minimum value appeared in January as 4.4℃; The tendency of sunshine hours was consistent with daily average temperature, the maximum and minimum values respectively appeared in August and January.

4.3. Analysis of the energy generation capacity

Based on the obtained values of meteorological parameters from the above methods, it is now possible to study the influence of meteorological parameters on the output power from PV modules during the daylight hours. Fig. 3 exhibits the daily electricity generated scatter plot of the whole year, and Fig.4 illustrates the histogram of monthly mean value of daily energy generation capacity.

As shown in the two figures, the maximum daily energy yield throughout the year was 670705.20 kWh, which appeared on 3 May 2016. The daily average solar radiation was 579W/m^2, the daily average temperature was 21.1℃, and the daily sunshine hours were 12.83h on that day. The maximum monthly value of daily energy yield was 449674.61kWh, which appeared on August 2016. The monthly average value of daily solar radiation, daily average temperature and daily average sunshine hours were 43.72 W/m^2, 30.68℃ and 12.82h respectively of August 2016.

It was mainly due to the higher sunshine intensity and the longer sunshine hours of the month, making the monthly day output is larger in August. There were two reasons that make the maximum day output appeared on 3 May. For one thing, the daily solar radiation was relatively large; and for the other more important thing, the daily average temperature was much lower on that day. The influence of the temperature on the photovoltaic power generation is complicated. On the one hand, the environment temperature affects the module operating temperature, and the performance of PV battery changes over the module operating temperature. The increase in temperature leads to the increase in conductivity of the semiconductor of solar cell, and outermost electrons acquire energy and leave the shell of the atom, as well as the reduction of the band-gap thereby increases the photocurrent slightly and decreases open circuit voltage of cell heavily. As a result, the total energy created by the solar cell decreases with increment of temperature [6,9,11]. On the other hand, there is an obvious positive correlation between temperature and solar total radiation. The solar total radiation increases, the temperatures and the corresponding photovoltaic power generation also increases. It means the temperature brings about positive effects on the photovoltaic power generation through the solar total radiation.

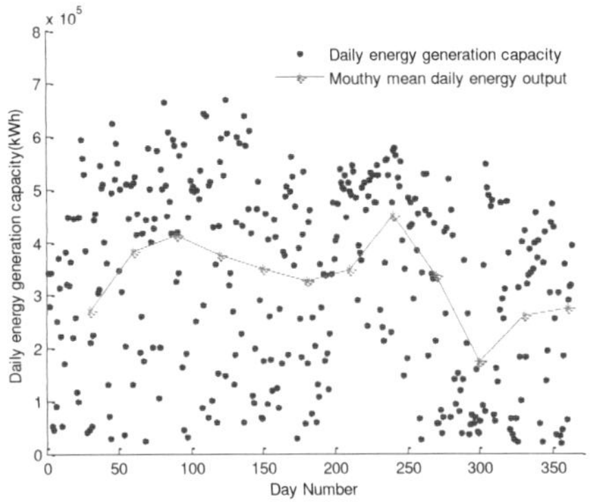

Figure 3. Daily energy generation capacity

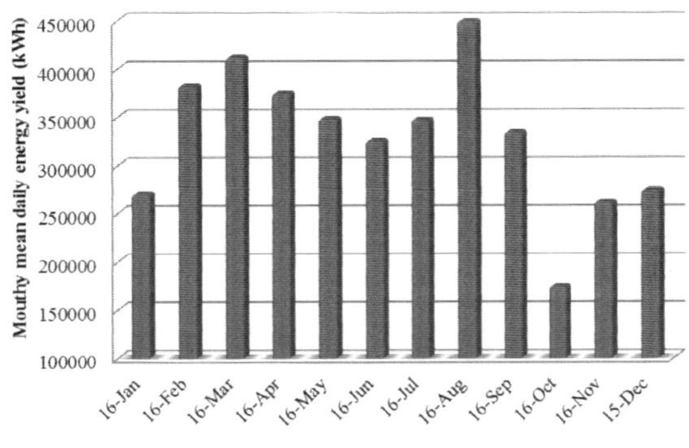

Figure 4. Monthly mean value of daily energy yield

Therefore, it is necessary to take appropriate measures to control the temperature of the photovoltaic panels. Making the PV panels work at a proper temperature ensures the efficient utilization of solar radiation, as well as the efficient control of the adverse effect on the performance of the solar PV module of temperature.

4.4. Analysis of the monthly mean PR

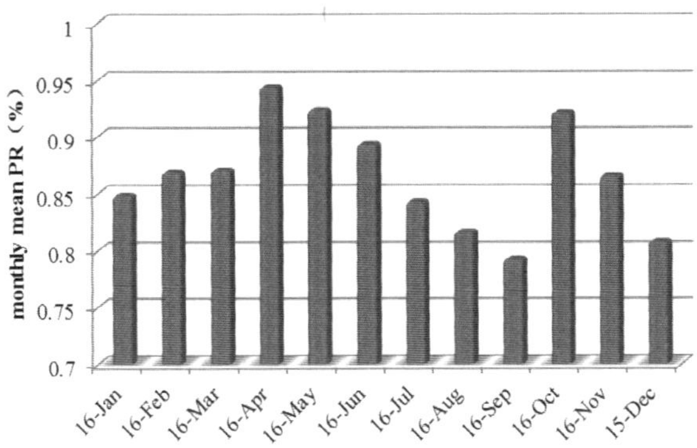

Figure 5. Monthly mean value of daily performance ratio

The monthly mean daily performance ratio is shown in Fig.5.The PR was found above 90% in April, May, and October. Among these three months, the PR were 94.4%, 92.3% and 92.1% in April, May and October respectively. The minimum value of PR is 79.2% that appeared in September. It was noteworthy that the PR of August with maximum energy output was only 81.6%, while it was as high as 92.1% in October with minimum energy generation capacity. The main reason was that although the soiling in August is higher, but the temperature was also higher which brought about the lower operating efficiency of solar cell. In October, the monthly average daily temperature was 19.2℃ that made the solar cell works with efficient utilization of solar radiation.

5.Conclusions

The impact of typical meteorological factors on the photovoltaic power generation under different months of a 100MWp grid tied photovoltaic plant has been carried out in this study. The solar PV plant has been installed above a fishpond at the Yangzhou city of Jiangsu Province of China since December 2014.

From the correlation analysis, it can be concluded that the daily PV power generation has the most significant correlation with daily global solar irradiation, followed by diurnal temperature range, daily sunshine hours, daily maximum temperature and daily average temperature. The maximum monthly average power generation appears in August, and the maximum daily average power generation appears in October. Through the contrast of monthly average PR and monthly average temperature, the larger values of monthly average PR appears in April and October with lower temperature while it is smaller in summer with higher temperature.

The results concluded that temperature has a great influence on the performance ratio of large scale grid-tied PV power system, and it is important to adopt effective measures to decrease the temperature of PV plant properly.

6.Acknowledgments

The work was supported by grants from the Natural Science Foundation of the Jiangsu Higher Education Institutions of China (15KJB510027).

7.References

[1] Li DHW, Cheung KL, Lam TNT, Chan WWH. A study of grid-connected photovoltaic (PV) system in Hong Kong. Applied Energy 2012; 90:122-127.

[2] M. Alsayed, M. Cacciato, G. Scarcella, G. Scelba. Multicriteria optimal sizing of photovoltaic-wind turbine grid connected systems. IEEE Trans. Energy Convers 2013; 28: 370-379.

[3] Su Y, Chan L-C, Shu L, Tsui K-L. Real-time prediction models for output power and efficiency of grid-connected solar photovoltaic systems. Applied Energy 2012; 93:319-326.

[4] Santos JM, Moura PS. Almeida ATd. Technical and economic impact of residential electricity storage at local and grid level for Portugal. Applied Energy 2014; 128:254-264.

[5] Energy development strategic action plan (2014-2020). China's state council general office,2014.

[6] YAN Quanquan,LI Fen,WANG Lijuan,et al. Diagnostic analysis of the impact of meteorological factors on photovoltaic power generation in different seasons. Journal of Shanghai University of Electric Power 2016;33:261-266.

[7] Nuri Gokmen, Weihao Hu, Peng Hou, et al. Investigation of wind speed cooling effect on PV panels in windy locations. Renewable Energy 2016; 90:283-290.

[8] Sharma V, Chandel SS. Performance analysis of a 190 kWp grid interactive solar photovoltaic power plant in India. Energy 2013; 55:476-485.

[9] Bhaveshkumar Dobaria,Mahesh Pandya, Mohan Aware. Analytical assessment of 5.05 kWp grid tied photovoltaic plant performance on the system level in a composite climate of western India. Energy 2016; 111:47-51.

[10] T. Khatib, A. Mohamed, K. Sopian, A review of solar energy modeling techniques, Renew. Sustain. Energy Rev. 2012; 16:2864-2869.

[11] Notton G,Lazarov V, Stoyanov L. Optimal sizing of a grid-connected PV system for various PV module technologies and inclinations, inverter efficiency characteristics and locations. Renewable Energy 2010; 35:541-554.

IOP Conf. Series: Earth and Environmental Science **63** (2017) 012011 doi:10.1088/1755-1315/63/1/012011

Finite Element Analysis for the Web Offset of Wind Turbine Blade

Bo Zhou[1], Xin Wang[1], Changwei Zheng[1], Jinxiang Cao[1] and Pingguo Zou[1]

[1] Suzhou Nuclear Power Research Institute, 1788 Xihuan Road, Suzhou, China

zhoubo_2010@cgnpc.com.cn

Abstract. The web is an important part of wind turbine blade, which improves bending properties. Much of blade process is handmade, so web offset of wind turbine blade is one of common quality defects. In this paper, a 3D parametric finite element model of a blade for 2MW turbine was established by ANSYS. Stress distributions in different web offset values were studied. There were three kinds of web offset. The systematic study of web offset was done by orthogonal experiment. The most important factor of stress distributions was found. The analysis results have certain instructive significance to design and manufacture of wind turbine blade.

1. Introduction

Wind energy is the natural resource of the earth. In recent years, as a renewable green energy, because of its vast commercial potential and environmental benefits, the wind energy is developing rapidly in the new energy industry [1]. Wind turbine converts wind energy into mechanical energy through blade, so the quality of the blade is related to the reliability of wind turbine. The stress distribution of the blade is quite complex. Generally, the structure of two webs is used in the blade, which is mainly subjected to shear and torsion. The skin provides the aerodynamic shape of the blade, which is also subjected to torsion [2-5].

The raw material for the web is a bidirectional glass fiber layer, which can improve the blade shear resistance effectively. During the production process of blade, the finished web is bonded into the blade with adhesive. Because of the local deformation of the handmade blade, it is common that the web offsets by extrusion. The blade design safety margin specified in JBT10194-2000 is 15%. The offset of the web changes the stress distribution of the blade. it is necessary to analyze the influence of blade quality. Taking into account the high cost of the blade test, research by simulation is a relatively economical and effective method. Stress distributions in different web offset values were studied by ANSYS.

2. Blade model

A blade with length of 53.8m, rated wind speed of 9.5m/s and rated turning speed of 1.49 rad/s for 2MW turbine was taken as the research object. The model of the blade was built up in the bottom-up way in ANSYS. The blade model was divided into 39 sections with 48 key points in each section. The corresponding key points were fitted into 48 curves by ANSYS command BSPLIN. The final skinning was done by the ASKIN command. The completed model is shown in Figure 1.

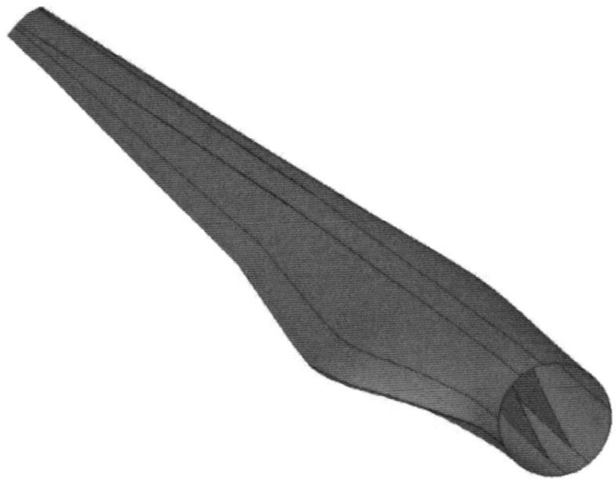

Figure 1. Blade geometry model.

The element type is SHELL99 shell element, which is an 8-node 3D shell element. Each element has 6 degrees of freedom, which is suitable for thin-medium thickness of the shell structure. Real constant was referenced by element (the direction and thickness of the composite laminates on different areas) [6]. The real constants and material properties were assigned to all corresponding areas by the MESHING function [7-9]. The finite elements were divided finally.

3. Loads and constraints

According to IEC61400-1[10], the forces on the wind turbine blade mainly include aerodynamic force, centrifugal force and gravity force, driving force and other forces such as wake force, impact force, ice force and so on[11-12]. In order to facilitate the study of the problem, the forces were simplified to aerodynamic force (twisting and bending), centrifugal force (bending, stretching and torsion) and gravity force (bending, tension and torsion).

The aerodynamic force is generated by the wind force acting on the surface of the blade. in order to achieve loading on the force surface of the blade, the method of loading along the blade's surface was selected to simulate the load condition[13].

$$\begin{cases} \mathrm{d}F_y = \frac{1}{2}\rho v^2 (C_L \cos\alpha + C_D \sin\alpha)\ \mathrm{tdr} \\ \mathrm{d}F_x = \frac{1}{2}\rho v^2 (C_L \sin\alpha - C_D \cos\alpha)\ \mathrm{tdr} \end{cases} \tag{1}$$

ρ - air density; v -rated wind speed; C_L, C_D - airfoil lift coefficient and drag coefficient; α - angle of attack of blade.

Centrifugal force is generated when the blade rotation, role in the blade center of gravity. The waving torque of the blade shifts the flexible blade in the downwind direction. The centrifugal force on the blade produces a bending moment in the waving direction that reduces this offset.

$$\begin{cases} \mathrm{d}F_z = \mathrm{m}\omega^2 \mathrm{r} \cos\beta \mathrm{dr} \\ \mathrm{d}F_y = \mathrm{m}\omega^2 \mathrm{r} \sin\beta \mathrm{dr} \end{cases} \tag{2}$$

ω - angular velocity of blade rotation; r – the distance from the centroid of the blade to the center of rotation; β - cone angle of wind wheel.

Gravity force is the force that the blade will always be subjected to during its rotation. The gravity force on each coordinate axis changes cyclically with the different azimuth of the blade rotation. Gravity force makes the bending moment of the edge vary regularly, which is the main source of fatigue when the blade level reaches its maximum and is reversed in direction from one horizontal position to the other.

$$\begin{cases} dF_x = \text{mg} \sin\theta \cos\chi \text{dr} \\ dF_y = \text{mg} \cos\theta \sin\beta \cos\chi \text{dr} \\ dF_z = \text{mg} \cos\theta \cos\beta \cos\chi \text{dr} \end{cases} \tag{3}$$

θ - the azimuth of the rotation; β - cone angle of wind wheel; χ - inclination of the wind wheel.

The root of the blade is connected with the hub by screws. The whole blade is simplified as cantilever beam model when the six degrees of freedom are fixed. The blade has the largest bending moment in the horizontal position, so this position was selected as the object of study.

4. Calculation and result analysis

There are multiple directional possibilities for the web to be offset in the blade production process. In order to simplify the research, the spacing offset, the center position offset and the verticality offset were taken as the research objects. Two webs design spacing of 53.8-type blade is 400mm. The design value of the beam width is 640mm. The limit offset of the web at the beam edge is 120mm. The middle offset value of 60mm was added as the research object. The stress of five different offset values is shown in Table 1.

Table 1. Web offset value.

Web spacing offset(mm)	Web center position offset(mm)	Web verticality offset(°)
-120	-120	-3.6
-60	-60	-1.8
0	0	0
+60	+60	+1.8
+120	+120	+3.6

The structural features of the blade are illustrated in Figure 2. It can be seen from the results that the maximum stress of the blade occurs at the connection of the main beam and the blade root[14], which is consistent with the results of the blade type test.

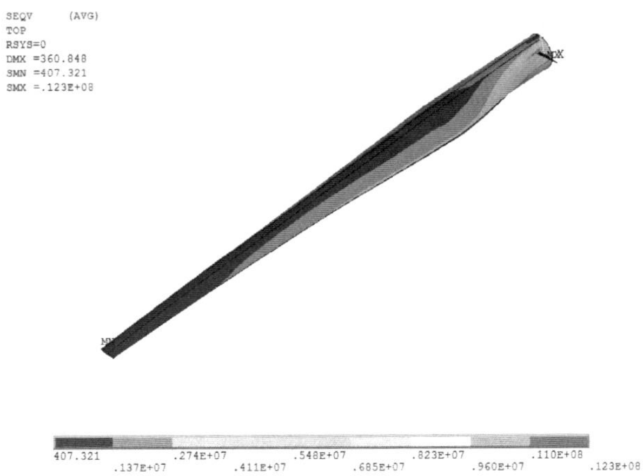

Figure 2. Blade stress distribution.

The stress calculation results of different offset of web spacing are shown in Figure 3. It can be seen that the maximum deviation of the stress from the normal value is 9.14%. There is a tendency that the stress becomes larger as the web spacing becomes smaller.

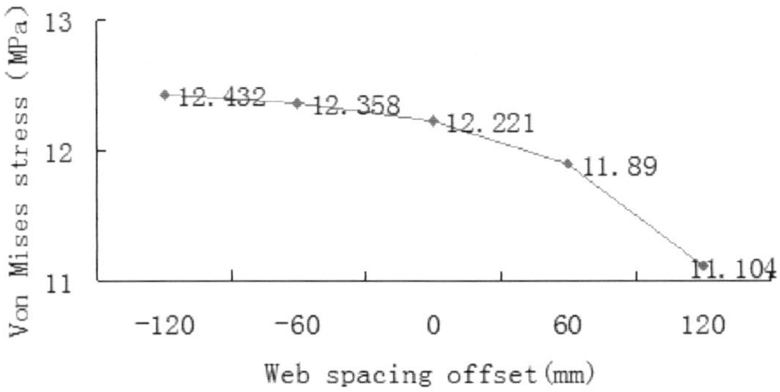

Figure 3. Stress-to-spacing variation.

The results of the stress calculation of different offset values of web center position are shown in Figure 4. It can be seen that the maximum deviation of the stress from the normal value is 5.70%. There is a tendency that the stress is becoming smaller with the web center position offset to both sides.

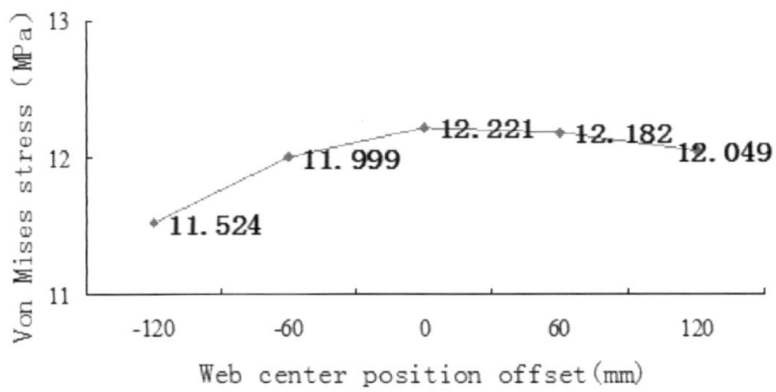

Figure 4. Stress variation with center offset.

The results of the stress calculation of different offset values of web verticality are shown in Figure 5. It can be seen that the maximum deviation of the stress from the normal value is 2.96%. There is a tendency that the stress is becoming smaller with the web verticality offset to both sides.

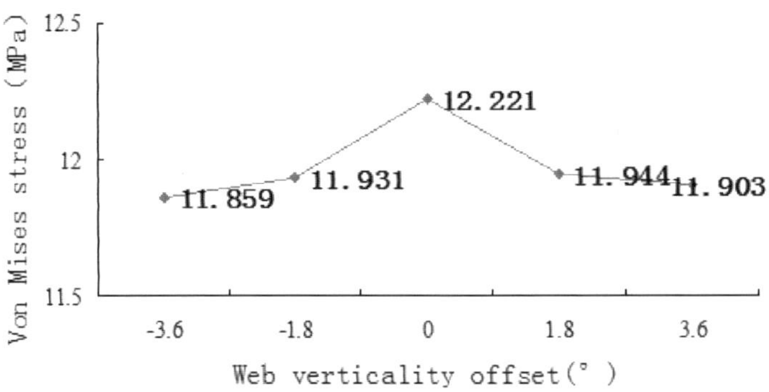

Figure 5. Stress variation with verticality offset.

Considering the above three factors, there are 125 combinations. In order to reduce the amount of calculation, the method of $L9(3^4)$ orthogonal test was used. In order to ensure that the offset value of the three factors superimposed does not exceed the beam edge, the offset value of web space was selected as 40mm, the offset value of web center position was selected as 40mm, the offset value of web verticality was selected as 1.2°. The Von Mises value of Orthogonal test is shown in Table 2.

Table 2. Von Mises value of Orthogonal test.

Web spacing offset(mm)	Web center position offset(mm)	Web verticality offset(°)	Maximum stress value(MPa)
-40	-40	-1.2	12.282
-40	0	0	12.358
-40	+40	+1.2	12.368
0	-40	0	12.05
0	0	+1.2	12.339
0	+40	-1.2	12.405
+40	-40	+1.2	11.839
+40	0	-1.2	11.668
+40	+40	0	12.007

From the stress calculation results of different web offsets, the maximum positive deviation of the stress from the normal value is 1.5%. R1=0.498, R2=0.203, R3=0.064, it can be seen that the web spacing offset has the greatest effect on blade stress, followed by the web center position offset and the web verticality offset.

5. Conclusion

The stress analysis of 2MW wind turbine blade with different web offsets showed that:

1. The blade design safety margin specified in JBT10194-2000 is 15%. From the stress calculation results of different web offsets, the maximum positive deviation of the stress from the normal value is 1.5%. The web offset plays an important role in the quality of wind turbine blade.

2. The web spacing offset has the greatest effect on blade stress, which should be avoided by more measures in the production of wind turbine blade. 1) More machines should be used during the production process to reduce the generation of offsets. 2) Larger positioning blocks can be used to secure the webs when the adhesive is not cured yet. 3) Secondary mold can be used to make the positioning of the web more accurate. Although production efficiency will be affected, it is worth it to improve the quality.

6. References

[1] Wang Yaoxian 2001 Composite structure design *Chemical Indıstry*

[2] Otero Alejandro D and Ponta Fernando L 2010 Structural analysis of wind-turbine blades by a generalized timoshenko beam model *J Sol Energ* 132:011015-1-011015-8

[3] Composites T P I 2002 Parametric study for large wind turbine blades *Sandia National Laboratories* SAND2002-2519

[4] Griffin D A 2002 Blade system design studies volume I : Composite technologies for large wind turbine blades *Sandia National Laboratories* SAND2002-1879

[5] Veers Paul S, Laird Daniel L and Sagartz Thomas G 1998 Estimation of uncertain material parameters using modal test data *Sandia National Laboratories* AIAA-0049

[6] Zhang L, Deng H and Gao J 2014 Influencing analysis on lamination parameters to static structure performance of wind turbine blade *Acta Energi Sin* 35(6):1059-1064

[7] Chen Yuqi, Wang Tiemin and Su Chenggong 2010 Research on lay-up of wind turbine blade structure *Wind Energy* (7):56-59

[8] Zehnder N and Ermanni P. A 2006 A methodology for the global optimization of laminated composite structures *Compos Struct* 72(3):311-320
[9] Lin C C and Lee Y J 2004 Stacking sequence optimization of laminated composite structures using genetic algorithm with local improvement *Comput Struct* 63(3-4):339-345
[10] IEC 61400 2003 Full-scale structural testing of wind turbine blades *IEC*
[11] Timmer W A and Van Rooij R P J O M 2003 Summary of the Delft University wind turbine dedicated airfoils *J Sol Energ* 125:488-496
[12] Yao C S, Lin J C and Allan B G 2002 Flow field measurement of device-induced embedded streamwise vortex on a flat plate *Int. Conf.1st AIAA Flow Control*
[13] Gyatt G W 1986 Development and testing of vortex generators for small horizontal axis wind turbines *NASA* CR-179514
[14] Jensen F M, Falzon B G and Ankersen J 2005 Structural testing and numerical simulation of a 34m composite wind turbine blade Development and testing of vortex generators for small horizontal axis wind turbines *Int. Conf.ICCM15*

IOP Conf. Series: Earth and Environmental Science **63** (2017) 012012 doi:10.1088/1755-1315/63/1/012012

Study on the Matching of the Wind Turbine Capacity and the Rotor Diameter in Inner Mongolia

Liang Cao[1] and Yong Zhong Wu[2]

No 128, Daxue East Road, Saihan District, Hohhot City, Inner Mongolia Autonomous Region, China.

hwzhss@126.com

Abstract: There are abundant wind resources in Inner Mongolia Autonomous Region and in order to develop the wind resources economically and reasonably, it is necessary to produce and promote wind turbine generators that fit the exploitation of the wind resources. The energy generated by the wind turbines is closely linked to the rotor swept area. If the rotor diameter doubles, the wind power will increase by four times. This paper, based on the actual wind resources, divided Inner Mongolia into different regions, and, in accordance with the present wind farm development in Inner Mongolia and the on-grid energy requirements, offered a scheme for the matching of the unit capacity and the rotor diameter in wind farms of some typical regions, which provides some scientific and theoretical references for the proper selection of the wind turbine generator parameters.

1. Introduction

Inner Mongolia Autonomous Region is located in the Mongol Plateau where there are 83,000 km^2 with the annual average wind speed of 5.5m/s, and 200,000 km^2 with the annual average wind speed between 4.5 and 5.4m/s. Its area for abundant wind energy ranks first in China and therefore this region is fit to develop wind power projects [1-4]. For the economic feasibility of the wind power projects, the operation of the selected wind turbine generators has to reach the required on-grid energy. The on-grid energy of the operated generators is closely linked to the local wind resources. However, due to the geographical features of Inner Mongolia Autonomous Region, which is extended from the northeast of China to the southwest, with a straight east-west distance of more than 2,400km and north-south distance of 1,700km, the wind resources are quite different in different regions. As a result, it is crucial to make regional divisions on the basis of the respective wind resources so as to develop the wind resources more reasonably and provide more reliable references for the producing, introducing, selecting, promoting and distributing of the wind turbines. Moreover, the amount of electricity produced by the wind turbine generators is closely linked to the rotor diameter. The rotor obtains energy from turbulence. According to the formula $P = 0.5\rho A v^3$ [5], the wind power P has a square relationship with the rotor diameter, so it is critical to choose the proper rotor diameter after defining the unit capacity based on the corresponding regional wind resources.

2. Regional Division for Wind Resources

Region A: regions with rich wind resources, including the northwest part of Xilinhot, the north central part of Ulanqab, the north part of Baotou and Bayannur. In these regions, the annual average value of effective wind power density at the height of 10m ranges from $260W/m^2$ to $320W/m^2$ and the annual hours available for operation can amount to 6100~7800h, that is, the wind turbines can work for 250~325 days in a year. The annual average wind speed stability parameter C_v values between 0.06 and 0.09 and the annual longest hours with continual ineffective wind speed are between 15 and 50 hours.

Region B: regions with adequate wind resources, including the east and south part of Xilinhot, the south part of Ulanqab, Baotou and Bayannur, the north part of Alxa, and the west part of Hulunbuir. In these regions, the annual average value of effective wind power density at the height of 10m ranges from 200 W/m^2 to $2600W/m^2$ and the annual hours available for operation can amount to 5250~6780h; in other words, the wind turbines can work for 220~280 days in a year. The annual average wind speed stability parameter C_v values between 0.06 and 0.10, and the annual longest hours with continual ineffective wind speed varies between 30 and 70 hours.

Region C: regions with available wind resources, including southwest part of Hohhot, the whole area of Xinggan League and Tongliao City, the east part of Xilinhot, most part of Chifeng City and Ordos City, and the south part of Alxa. In these regions, the annual average value of effective wind power density at the height of 10m ranges from 100 W/m^2 to $200W/m^2$ and the annual hours available for operation are up to 4380~6100h; in other words, the wind turbines can work for 180~250 days in a year. The annual average wind speed stability parameter C_v values between 0.08 and 0.13, and the annual longest hours with continual ineffective wind speed are between 40 and 100 hours.

Region D: regions with inadequate wind resources, including the northeast part of Hulunbuir, the southern part of Ordos and most part of Hohhot. In these regions, the annual average value of effective wind power density varies from 50 W/m^2 to $100W/m^2$ and the annual hours available for operation are up to 2000~4950h, which means the wind turbines can work for 80 to 200 days in a year.

3. Matching of the Unit Capacity and the Rotor Diameter

With the booming of wind power industry, Inner Mongolia, known for its abundant wind resources, is fit for constructing large-scale wind power farms. So far, many large wind farms have been built and put into operation, including Huitengxile Wind Farm, Shangdu Wind Farm, Zhurihe Wind Farm and Xilinhot Wind Farm. Now that there has been no further developing potential in Region A, the future wind farm expansion mainly focuses on areas of Region B and C. In terms of wind turbine selection, the 2MW generator system or above are not easy for inland transportation, and at the same time, as the inland wind shear is rather strong in Inner Mongolia, large diameter rotors are bound to be affected, causing low efficiency of the wind energy exploitation. As a result, it is better to choose 1.5MW and 2MW wind turbine unit models in Inner Mongolia.

Based on the current wind farm construction scheme, it can be figured out by techno-economic analysis that the optimal annual available hours of the wind turbine generator system should be up to 2,200 to 2,800 hours. If taking 2,500 hours as the standard of the annual available hours, only when the 1.5MW generator produces 5 million KW theoretical electricity in a year, and the 2.0MW generator produces 6.67 million KW, can they meet the on-grid electricity requirement. (The on-grid electricity is usually calculated as 75% of the theoretical generating power.)

Taking into account the present wind farm development, the future expansion tendency and the actual wind energy resources in Inner Mongolia, this paper chose three regions---Wulate Houqi in Bayannur City, Manchouli in Hulunbuir City, and Khorchin Youyiqianqi in Xingan League as the typical areas to determine the corresponding rotor diameter. The hut height of the selected generator model is 70m, the cut-in wind speed is 3m/s, and the cut-out wind speed is 25m/s. The rated wind speed is usually taken as 1.3~1.5 times of the mean wind speed. Choosing one rated wind speed among the value range, the rotor diameter can be worked out with the rated power. Then, it is time to test whether the wind turbine can generate enough electricity to meet the on-grid need. If it exceeds the on-grid electricity, it signifies that the rated wind speed is higher than the chosen rated wind speed value; if on the contrary, it indicates that the rated wind speed is lower than the chosen value. The iteration is repeated with different values of the rated wind speed until the theoretical electricity is satisfied.

1) Wulate Houqi lies in Bayannur City and is about 1,400m above the sea level. With a plateau continental arid climate, this region, subject to the strong cold high atmospheric pressure from Mongolia, is windy all year round, especially in spring and winter. The wind farm 70m height representative year annual average wind speed is 8.0m/s, As shown in Table 1 and the annual average wind power density is 486.6 W/m^2 while the 10m height representative year annual average wind speed is 6.2m/s and the annual average wind power density is 224.1 W/m^2. According to the wind resources category criteria in this paper, this area belongs to Region B.

Table 1. 70m height representative year wind speed frequency distribution in WuLaTe HouQi

wind speed range(m/s)	wind speed frequency(%)	wind speed range(m/s)	wind speed frequency(%)
< 0.6	1	13.0	5
1.0	2	14.0	4
2.0	4	15.0	2
3.0	7	16.0	2
4.0	8	17.0	1
5.0	9	18.0	1
6.0	9	19.0	0
7.0	9	20.0	0
8.0	9	21.0	0
9.0	8	22.0	0
10.0	8	23.0	0
11.0	7	24.0	0
12.0	5	25.0	0

The annual unit power generation is the total sum of the annual average wind speed hours at the respective wind speed scale (within the range of effective wind speed) multiplying the corresponding out-put power at the respective wind speed scale. The formula [6] is as follows:

$$G = \sum N_i W_i \qquad (1)$$

among which, G ($kW \bullet h$) means the energy generation; N_i (h) refers to the total number of hours of the wind speed at the respective wind speed scale) in a year; W_i (kW) signifies the corresponding out-put power under the respective wind speed scale.

In Wulate Houqi, the mean air pressure is 84.08 kPa, and the annual average temperature is 8.6℃. Accordingly, the average air density ρ can be calculated by the following formula:

$$\rho = P(R \times T)^{-1} \qquad (2)$$

among which, P (Pa) is the annual mean air pressure; T (t℃$+273$) stands for the average absolute temperature of Kelvin temperature scale in a year; R is the gas constant which values 287 273$J/kg \bullet K$.By calculation, the average air density can be obtained, whose value is 1.04 kg/m^3.

The following is the formula for the out-put power of the wind turbine:

$$W = 0.5 \rho A v^3 c_p \qquad (3)$$

among which, ρ is the air density, A is the rotor swept area, v is the wind speed and c_p is the rotor power coefficiency.

Through the above formula, combined with the existing market models, the respective coefficients can be determined and therefore the rotor diameter of the two selected models can be obtained to meet the on-grid electricity standards. By calculation, it is obtained that the rotor diameter of the 1.5MW wind turbine generator system is approximately 78m, and that of the 2MW generator system is approximately 90m.

2) Manchouli lies in the west of Hulunbuir, 660m above the sea level, with a rolling prairie landscape. The 70m height representative year annual average wind speed at the wind farms is 7.2m/s, As shown in Table 2 and the average wind power density is 504 W/m^2 while the 10m height annual average wind speed is 5.8m/s and the average wind power density is 252.4 W/m^2. According to the wind resources categories made by this paper, this area also falls into the category of Region B.

Table 2. 70m height representative year wind speed frequency distribution in ManZhouLi

wind speed range(m/s)	wind speed frequency(%)	wind speed range(m/s)	wind speed frequency(%)
< 0.6	2	13.0	3
1.0	6	14.0	2
2.0	6	15.0	2
3.0	8	16.0	1
4.0	10	17.0	1
5.0	10	18.0	1
6.0	9	19.0	0
7.0	8	20.0	0
8.0	8	21.0	0
9.0	7	22.0	0
10.0	6	23.0	0
11.0	5	24.0	0
12.0	4	25.0	0

The air density of the wind farms in Manchouli is $1.19\,kg/m^3$, so it can be calculated that the proper rotor diameter of the 1.5MW wind turbine generator system in this area is 80m and that of the 2MW generators is 92m.

3) Khorchin Youyiqianqi, located in the south of the Great Xingan Mountains in Inner Mongolia, is a hilly area with a cold temperate continental monsoon climate. In this area, the northwest wind prevails all year round. The 70m height annual average wind speed is 7.4m / s, As shown in Table 3 and the average power density is 446 W/m^2, whereas the 10m height annual average wind speed is 5.3m/s, and the average wind power density is 190 W/m^2. According to the category standard, this area belongs to the category of Region C.

Table 3. 70m height representative year wind speed frequency distribution in Khorchin YouYiQianQi

wind speed range(m/s)	Wind speed frequency (%)	Wind speed range (m/s)	Wind speed frequency (%)
< 0.6	0	13.0	3.67
1.0	1.08	14.0	2.51
2.0	2.64	15.0	1.29
3.0	5.58	16.0	0.79
4.0	8.14	17.0	0.48
5.0	10.33	18.0	0.22
6.0	12.09	19.0	0.14
7.0	10.9	20.0	0.07
8.0	10.05	21.0	0.08
9.0	8.9	22.0	0.01
10.0	8.1	23.0	0
11.0	7.01	24.0	0
12.0	5.91	25.0	0

By calculation, the wind farm air density in Khorchin Youyiqianqi is 1.2 kg/m^3 and following the same calculating process as the above, it can be obtained that the appropriate rotor diameter of the 1.5MW wind turbine generator system in this area is 76m and that of the 2MW generators is 88m.

4. Conclusion

The paper divided the wind resources in Inner Mongolia into three categories and then chose two typical places from category B and one typical place from category C to study the matching of the unit capacity and the rotor diameter. With meeting the on-grid electricity as the premise, the paper calculated the corresponding rotor diameters based on the essential factors of the wind resources in the three places. By calculation, the rotor diameters of 1.5MW wind turbine generator system for wind farms in Wulate Houqi, Khorchin Youyiqianqi and Manchouli are respectively 78m, 76m, and 80m. The rotor diameters of 2MW generator system in these three regions are 90m, 88m and 92m. Moreover, it can be inferred that based on the characteristics of wind resources in these areas, the rotor diameter can be determined by choosing the rated wind speed of 10~11m/s. As a result, the appropriate rotor diameter for 1.5MW wind turbine generator system is between 76~80m whereas the rotor diameter for 2MW generator system is between 88~92m. By analysis and calculation, this study offers some theoretical basis and references for the matching of the unit capacity and the rotor diameter in Inner Mongolia wind farms.

5. References

[1] Zang Rui and Feng Shouzhong 1997 *China. J. Inner Mongolia wind resource assessment and wind power development . International Electric Power for China.* **4** L17-20
[2] Wang Yuan-yang and Huang Yong 2012 *China. J. Wind Energy Resource Assessment and Derivation of Characteristic Wind Speed In Wind Field of Inner Mongolia. APPLIED ENERGY TECHNOLOGY.* **9** L1-6

[3] Chen Xin and Song Lili 2011 *China. J. Study on Characteristics of wind energy resources in two tipical areas in China. ACTA ENERGIAE SOLARIS SINICA.* **32** L 331-7
[4] Zhu Rui-zhao and Xue Heng 1983 *China. J .Division of Wind Energy in China. ACTA ENERGIAE SOLARIS SINICA.* **4** L 123-32
[5] Tony Burton 2011 M *Wind Energy Handbook* (Beijing: Energy Press)
[6] Wang Cheng Xu and Zhang Yuan 2003 M *Wind power* (Beijing: China Electric Power Press)

IOP Conf. Series: Earth and Environmental Science **63** (2017) 012013 doi:10.1088/1755-1315/63/1/012013

Solar Radiation at Surface for Typical Cities in the Arid and Semi-Arid Area in Xinjiang, China Based on Satellite Observation

Fang Sihua[1, 2], Liu Haichen[1, 3], Huang Jiamin[1], Zhang Yunqi[1], Hu Jun[1], ChenYonghang[1], Kang Yanming[1], WangXue[1] and Huang Chengjie[1]

[1]College of Environmental Science and Engineering, Donghua University, Shanghai 201620, China
[2]Baruch College, City University of New York, New York 10010, USA
[3]Tandon School of Engineering, New York University, New York 11201, USA

yonghangchen@126.com; ymkang@dhu.edu.cn.

Abstract: Xinjiang, a region of China with arid and semi-arid areas, has abundant solar incidence with 166.5×104 km^2 and diverse underlying surface. The meager number of surface radiation observatories cannot meet the need for efficient exploration of solar energy. In this study we classified Xinjiang into three regions: southern Xinjiang, northern Xinjiang and Tu-Ha region and applied satellite data to provide the surface solar radiation's temporal distribution for 10 typical cities. The study is focused on seasonal, annual and variations of all sky downward shortwave radiation flux at surface based on 24-year satellite dataset GEWEX-SRB from the WCRP/GEWEX (World Climate Research Program/Global Energy and Water Cycle Experiment) from 1984 to 2007. The results are as follows. In general, the monthly average solar radiation flux for the cities in the Tu-Ha region was the largest followed by the south Xinjiang and northern Xinjiang. The solar radiation in the most northern cities were less than 150.0 W/m^2 in winter, the minimum is 138.7 W/m^2, while the other cities were greater than 150.0 W/m^2. The maximum of monthly solar flux for the Tu-Ha region, southern and northern Xinjiang was 400.0 W/m^2.

1. Introduction

Excessive depletion of fossil fuels results in severe global energy crisis with a large amount of greenhouse gas from combustion emissions leading to global warming and serious threats to the environment and human health. Solar radiation has been internationally recognized as one of the preferred sources of clean energy due to its wide distribution, in situ availability, no need for transport and small impact on the environment [1-4].

Scholars in China and overseas have utilized global surface observation, numerical simulation and satellite remote sensing to estimate solar radiation reaching the earth's surface for research areas including global and cities [5-6], focusing on distribution of solar radiation [7-10], its long-term change trend in time and space [11-12], and its influencing factors [13-15].

Xinjiang Uygur Autonomous Region is located in northwestern China. It accounts for about 1/6 of the total land area of China. It is in the arid and semi-arid climate zone with abundant solar energy. Its annual total radiation is up to 5430-6670 MJ/m2, and the theoretical annual solar energy is 1450-1720 kWh/m2, which suggests a powerful potential for future development of electricity [16].

Although there has been some research on solar radiation in Xinjiang [17-20], it was mainly based on a few surface observations that limited the results. Therefore, in this study SRB satellite data from 1984 to 2007 for 10 typical cities of Xinjiang was used to study seasonal, annual and inter-annual variations of shortwave radiation flux in order to provide better scientific understanding of solar radiation in Xinjiang.

2. Materials and Methodology

In this paper, the northern area to the Tianshan Mountains in Xinjiang is defined as the northern Xinjiang and the southern area is defined as the southern Xinjiang. The Turpan and Hami area is defined as the Tu-Ha region (Figure 1). Because of differences in geographical locations, climate and terrain conditions, we selected representative cities in each region for research: Altay, Yining, Urumqi for the northern Xinjiang; Korla, Aksu, Kashgar, Qiemo, and Hotan for the southern Xinjiang; Turpan and Hami for the Tu-Ha region. Seasonal, annual and inter-annual variations based on monthly solar radiation flux are presented in this paper.

Radiation measurements often contain erroneous and spurious values due to technical failures and/or operation-related problems. As a result, lack of adequate and high-quality solar radiation data has become a universal problem [21, 22].GEWEX-SRB is a series of long-term Surface Radiation Budget (SRB) dataset provided by NASA's WCRP/GEWEX (World Climate Research Program/Global Energy and Water Cycle Experiment). An improved (by Pinker/Laszlo) algorithm on shortwave to calculate shortwave radiation was used in the dataset [23]. The algorithm was used to calculate the broadband solar radiation for every time indicated. The Double-flow Eddington approximation model was used to map broadband reflected radiation flux from the top of the atmospheric radiation to ground surface. Furthermore, total reflected radiation flux at the top of the atmosphere was calculated through conversion of narrow-band to broadband in the visible radiation. The angle distribution model was from Earth Radiation Budget Test (ERBE). The satellite observation dataset, SRBREL3.0 shortwave 3 hourly data product [24-26] in the time period from January 1984 to December 2007 was adopted in this research. The spatial resolution is 1°× 1°, and the temporal resolution is 3 hours. All Sky Downward shortwave radiation flux was analyzed.

Figure 1. Locations of the ten typical cities in Xinjiang

3. Results and Discussion

Comparison of annual variations of shortwave radiation among the three regions. Generally, in Fig.2, the monthly average solar radiation in northern cities was the smallest among the three region sand those in winter were below 200.0 W/m^2, with a minimum of 138.7 W/m^2; it was greater than 150.0

W/m^2 in the other cities, also the values were relatively concentrated. For both southern and northern cities, the maximums were all close to 400.0 W/m^2. Differences among cities were significant in spring and summer; it could reach to 150.0 W/m^2 from maximum to minimum. In autumn and winter, however, radiation value lines of each city almost overlap due to the relatively small differences among cities.

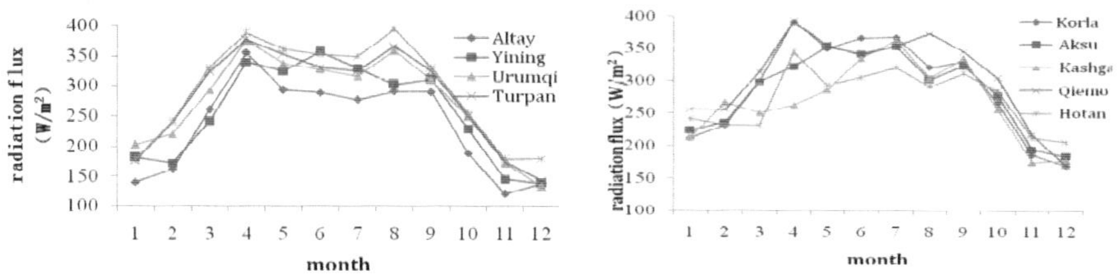

(a) (b)

Figure 2. (a) Annual variations of solar radiation flux in 5 typical cities in northern Xinjiang and the Tu-Ha Region averaged for the period from 1984 to 2007;
(b) Annual variations of solar radiation flux in 5 typical cities in the southern Xinjiang averaged for the period from 1984 to 2007.

Figure 2 (a) shows, in general, Hami had the maximum monthly values followed by Turpan, Urumqi, Yining, Altay. This is mainly related to the cities' latitudes; the higher the latitude, the smaller the city's radiation, generally. In June, for example, the largest value appeared in Yining. Maximum values for each city were in different months: April for Turpan, Altay and Urumqi, while August for Hami, and June for Yining.

From Figure 2 (b), the five southern cities in order of radiation values from larger to smaller were Qiemo, Korla, Aksu, Hotan and Kashgar. The gap between Hotan and Kashgar was especially small. In contrast with the values for northern Xinjiang, the radiation values for southern cities were hardly related to their latitudes. Korla has the highest latitude, but its average radiation value ranked second in the five cities, and even first in April, May and June. Hotan has the lowest latitude, but the largest radiation values. Because Hotan is located in the Tarim Basin, under the influence of high concentration desert aerosols, a large part of radiation will be scattered, so the part that reaches the ground will decrease accordingly. Maximum values for two cities appeared in April, Kashgar in July, and Aksu in May.

The monthly average radiation flux for the whole Xinjiang area for the 24 years was 281.0 W/m^2, with monthly average radiation flux values from 116.78 to 358.97 W/m^2 shown in Table 1. The maximum appeared in June and minimum in December.

Table 1. Monthly solar radiation flux averaged for 10 cities in Xinjiang from 1984 to 2007

Months	Jan	Feb	Mar	Apr	May	Jun
Average radiation flux (W/m^2)	131.24	180.06	242.55	304.84	346.23	358.97
Months	Jul	Aug	Sep	Oct	Nov	Dec
Average radiation flux (W/m^2)	342.68	315.70	273.49	211.84	150.60	116.78

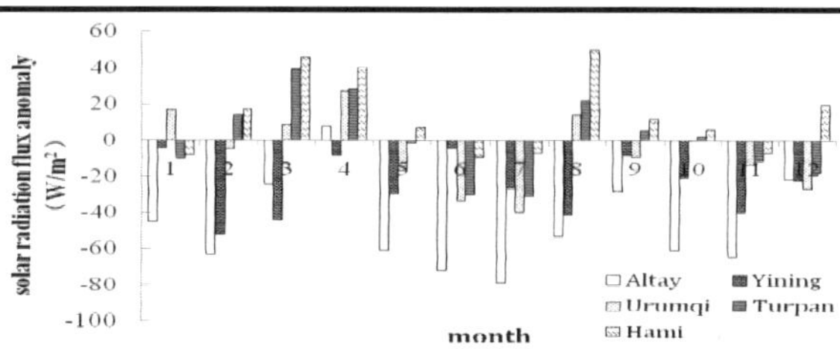

Figure 3. The annual variations of solar radiation flux anomalies for the 5 cities in northern Xinjiang and the Tu-Ha Region averaged for the period from 1984 to 2007

The monthly anomalies of solar radiation flux for the five cities in northern and the Tu-Ha Region in Figure 3. In general, radiation flux values for northern cities for most months are lower than that for the whole-Xinjiang average; in southern cities the situation is quite the opposite. Yining and Altay in northern Xinjiang are below the average in all months for the whole of Xinjiang. Radiation flux anomaly in July for Altay in particular, located in most north part of Xinjiang and in the Altai Mountains, can reach up to 78.7 W/m^2.

The result of five southern cities are shown in Figure 4. in May, June and July, almost all their radiation values are lower than the averages for the whole of Xinjiang, which is similar to the pattern for northern Xinjiang. However, in autumn and winter, namely the period from September to February, basically all southern cities' shortwave radiation values are higher than the averages for the whole of Xinjiang, while most of the northern cities have lower-than-average values. This shows that in autumn and winter solar radiation flux in the southern cities was higher than that in the northern cities.

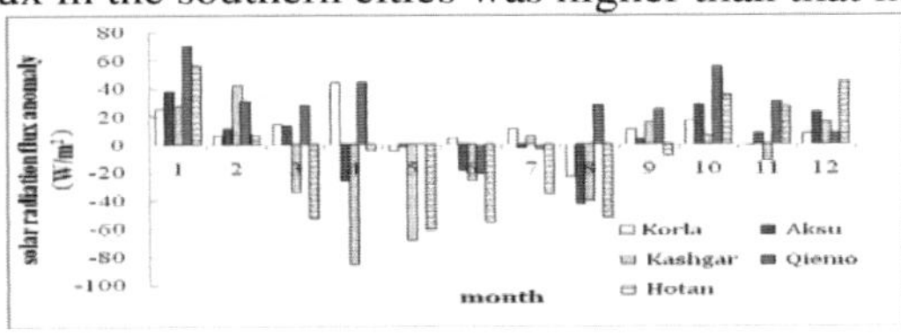

Figure 4. The annual variations of solar radiation flux anomalies for the 5 cities in the southern Xinjiang averaged for the period from 1984 to 2007

Specific values of climate tendency rates for each city for each month are shown in Table 2 below, each months' maximum was shown in bold. We can see that the changing trend in Urumqi was the most obvious with several months' maximum rate values among all the cities.

Table 2. Climate tendency rates (W/m^2/10) of solar radiation for the ten cities in each months

	Jan	Feb	Mar	Apr	May	Jun
Altay	0.313	**5.123**	5.08	-4.233	-2.872	-6.465
Yining	-1.058	3.781	1.552	-0.126	-4.845	-10.467
Urumqi	**-3.325**	-0.675	1.965	**-9.652**	**-19.464**	**-15.478**
Turpan	-1.811	-1.908	2.815	-2.408	-10.342	-4.103
Hami	-0.465	0.263	1.012	-0.858	-0.582	4.316

	Jul	Aug	Sep	Oct	Nov	Dec
Altay	-7.978	0.96	3.967	3.19	-0.463	-1.59
Yining	-2.1	3.513	2.803	1.148	-2.296	-2.088
Urumqi	**-13.779**	-11.001	-4.928	-3.319	-3.14	-1.985
Turpan	-8.804	-3.622	0.481	1.946	1.756	0.171
Hami	4.525	5.366	5.789	5.94	2.871	0.873
Korla	-8.024	-3.51	**-5.847**	0.544	0.145	**-3.542**
Aksu	-7.097	-3.458	-0.985	4.441	0.848	2.867
Kashgar	-0.193	-1.2	3.984	**7.974**	0.666	1.296
Qiemo	-6.67	-3.495	-1.288	5.471	**3.411**	0.289
Hotan	-8	**-14.828**	-5.225	-0.943	-1.935	0.431

4. Conclusions

As the lack of surface observations limited the scientific basis for exploration of solar energy, in this study GEWEX-SRB dataset from the WCRP/GEWEX (World Climate Research Program / Global Energy and Water Cycle Experiment) from 1984 to 2007 was used to help fill in the information gap over the data-sparse area. The seasonal, annual and inter-annual variations of all sky downward shortwave radiation flux at surface for 10 typical cities in Xinjiang, which is an arid and semi-arid area, were studied. The main results are summarized as follows.

(1) The monthly solar radiation flux average for the 24 years for the 10 cities in Xinjiang was from 116.8 W/m^2 to 359.0 W/m^2 with the average value for the whole Xinjiang area was 281.0 W/m^2. The maximum appeared in June and minimum in December. In general, among the three regions it was the largest for the cities in the Tu-Ha region, the second largest in the south Xinjiang and the smallest in northern Xinjiang.

(2) For annual variation, in winter (November, December and January) the solar radiation flux was the smallest. In the northern cities the monthly solar radiation flux was less than 200.0 W/m^2 with the minimum 138.7 W/m^2 which occurred in Altay and greater than 200.0 W/m^2 in the other cities. The maximum of monthly solar flux was between 340.0 W/m^2 and 400.0 W/m^2 which most occurred in April, Jun and August.

5. Acknowledgements

We would like to express our thanks to the Atmospheric Science Data Center in NASA Langley Research Center for the data support.

6. References

[1] Fthenakis V, Mason J E, Zweibel K. The technical, geographical, and economic feasibility for solar energy to supply the energy needs of the US. Energy Policy, 2009, 37(2): 387-399.

[2] Byrne J, Kurdgelashvili L, Poponi D, et al. The potential of solar electric power for meeting future US energy needs: a comparison of projections of solar electric energy generation and Arctic National Wildlife Refuge oil production. Energy Policy, 2004, 32(2): 289-297.

[3] Stanhill G, Moreshet S. Global radiation climate changes: The world radiation network. Climate Change, 1992,21, 57–75.

[4] ShenYanbo, Zhao Zongci, Shi Guangyu. The Progress in Variation of Surface Solar Radiation, Factors and Probable Climatic Effects. Advanced in earth science,2008,23(9):915-923.

[5] Russak, V. Trends of solar radiation, cloudiness and atmospheric transparency during recent decades in Estonia.Tellus, 1990, 42B: 206-210.

[6] Ohmura A, Gilgen H. The GEBA database: Interactive applications, retrieving data. Global Energy Balance Archive (GEBA), 1993: 44-60.

[7] Li Xiaowen, Li Weiliang, Zhou Xiuji. Analysis of the solar radiation variation of China in recent 30 years. Quarterly Journal of Applied Meteorology, 1998, 9(1): 24-31.

[8] Augustine J A, DeLuisi J J and Long C N.SURFRAD-A national surface radiation budget network for atmospheric research. Bulletin of the American Meteorological Society,2000,81(10):2341-2357.

[9] Wu Qizhong, Wang Zifa, Cui Yingiie. Evaluating the solar radiation resources of China in recent 20 years by meteorological model. Journal of Applied Meteorological Science, 2010, 03: 343-351.

[10]Stokes G. The Atmospheric Radiation Measurement (ARM) Program: Programmatic background and design of the cloud and radiation test bed.1994.

[11]Xia Lu, Zhang Qiang, Yue Ping. Plateau Land Surface Radiation Components and Their Annual Fluctuations in response to Climate Change.S18 The Fourth Annual Graduate,2013.

[12]Zha Liangsong. A Study On Spatial And Temporal Variation Of Solar Radiation In China. Scientiageographicasinica, 1996(03): p. 41-46.

[13]Zhang Wenyu, Feng Guanghong, Miao Yuzhi. The Discussion on The Changing Relation Between The Solar Radiation And The Atmosphere Pollution In Lanzhou [J]. Gansu Meteorology, 1997(01): p.45-46+56.

[14]Deng Zuqin, Han Yongxiang, Bai Huzhi, Zhao Tianliang. Relationship between dust aerosol and solar radiation inGobi Desert in North China [J].China Environment Science, 2011, 31(11): 1761-1767.

[15]A. Kazantzidisa, K. Eleftheratos, C.S. Zerefos. Effects of cirrus cloudiness on solar irradiance in four spectral bands [J]. Atmospheric Research, 2011, 102: 452-459.

[16]Chen Zhihua, Shi Guangyu, CheHuizheng. Analysis of the solar radiation of Xinjiang Uygur Autonomous Region in recent 40 years [J].Arid Land Geography, 2005, 28 (6): 734-739.

[17]Liu Jia, He Qing, Liu Rui, et al. Solar radiation characteristics and solar energy resource in Xinjiang [J]. Arid Meteorology, 2008, 04: 61-66.
[18]Xin Yu, Zhao Yizhou, Mao Weiyi, et al. Homogeneity test of the total solar radiation data series and further research on climatological calculation over Xinjiang [J]. Plateau Meteorology, 2011, 04: 878-889.
[19]Liu Yu，Zhang Xianfeng, Jonathan Li, et al. Estimation and seasonal variation of solar radiation in Xinjiang based on a multi-sensor data approach [J]. Journal of Arid Land Resources and Environment, 2015, 03: 161-165.
[20]Chen Zhihua, Shi Guangyu, Che Huizheng. Analysis of the solar radiation of Xinjiang Uygur Autonomous Region in recent 40 years [J]. Arid Land Geography, 2005, 28 (6): 734-739.
[21]Moradi I. Quality control of global solar radiation using sunshine duration hours. Energy, 2009, 34: 1-6
[22]Iziomon M G, Mayer H. Assessment of some global solar radiation parameterizations. J Atmos Solar-TerrPbys, 2002, G4: 1631-1643 [21] Pinker R. T. and Laszlo I. Modeling surface solar irradiance for satellite applications on a global scale[J]. Journal of Applied Meteorology, 1992, 31(2):194-11.
[23]Pinker R. T. and Laszlo I. Modeling surface solar irradiance for satellite applications on a global scale [J].Journal of Applied Meteorology,1992,31(2):194-11.
[24]Taiping Zhang, Paul W. Stackhouse Jr., Hoell, J. M., Westberg, D. J..Application of A Global-To-Beam Irradiance Model to the Satellite-Based NASA GEWEX SRB Data and Validation of the Results against the Ground-Based BSRN Data. American Geophysical Union, Fall Meeting 2012.
[25]Taiping Zhang, Paul W. Stackhouse Jr., William S. Chandler, David J. Westberg. Application of a global-to-beam irradiance model to the NASA GEWEX SRB dataset: An extension of the NASA Surface meteorology and Solar Energy datasets. Solar Energy, 2014, 110: 117–131.
[26]Taiping Zhang, Paul W. Stackhouse Jr., Shashi K. Gupta, Stephen J. Cox, J. Colleen Mikovitz, Laura M. Hinkelman. The validation of the GEWEX SRB surface shortwave flux data products using BSRN measurements: A systematic quality control, production and application approach. Journal of Quantitative Spectroscopy and Radiative Transfer, 2013, 122: 127–140.

Diatomite Modified Immobilized Delftia sp. for the Bio-Abiotic Removal of Antibiotics Amoxicillin in the Aqueous System

Lijuan Gao[1], Jing Sun[1], Kai Guan[1], Tingting Shen[1]* and Xikui Wang[2]*

[1]College of Environmental Science and Engineering, Qilu University of Technology, Jinan, Shandong, 250353, P.R. China,
[2] School of Chemistry and Pharmaceutical Engineering, Qilu University of Technology, Ji'nan, 250353, P. R. China.

stthunanyt@163.com
xkwang2015@163.com

Abstract. Diatomite modified sodium alginate (Si/SA) immobilized *Delftia*sp. *A2(2011) (STT01)*was applied to degrade amoxicillin. The immobilized pellets provided a direct and visual probe for the degradation process due to their intrinsic bright colour. The results demonstrated that 100% of amoxicillin and 68.5% of COD_{cr} removal were achieved after 72 h, comparing with the cases of sodium alginate (SA) system (81.2%, 46.9%) and the free cells system (60.5%, 35.5%). The degradation kinetics was in good agreement with Michaelis-Menten equation. The maximum rate (V_m) and Michaelis constant (K_m) were calculated as 9.09 mg L^{-1} h^{-1} and 228 mg L^{-1}, respectively. The results further revealed that diatomite not only acted as immobilization support to improve the mechanical strength and lifetime of the pellets but also as absorbent to promote the treatment efficiency. Therefore, both enzymatic catalysis and chemisorption were responsible for the removal of amoxicillin.

1. Introduction

Even though antibiotics has been widely used for treating and preventing diseases in humans and veterinary practice since 1940s, not until 2010, did it attract an increasing worldwide concern in the increasing potential risk for ecosystem and the bacterial resistance [1]. It has been reported that the rates of antibiotics-resistant isolates are increasing over time, not only in hospital wastewaters and animal production wastewaters, but also in sewage, wastewater treatment plants, and even in drinking water [2-4]. Therefore, the ecological risk of antibiotics in the aquatic environment should be urgently estimated.

A variety of techniques have been applied to pre-treat or directly degrade antibiotics wastewater, such as ozonation or O_3/H_2O_2 [5], Fenton or Fenton-like process [6], biological process [7] and combined chemical and biological oxidation process [8,9]. However, the direct biotic process is more attractive because the microbial or enzymatic degradation is an eco-friendly and cost-competitive alternative to chemical decomposition process, in addition, any pretreatment would increase the cost as well as the energy consumption.

Recently, diatomite has been applied to handle a number of wastewater due to the high chemical stabilities and specific surface areas [10-12]. These studies have demonstrated the promising outlook of diatomite that directly used as an absorber for wastewater treatment. Yet, up to date, few studies have been focused on the utilization of diatomite to immobilize *Delftia* sp. for the treatment of antibiotics wastewater.

Therefore, the objectives of this study are to identify the feasibility of using Si/SA system as an alternative to SA system and the free cells system for the direct degradation of antibiotics. As one of the most commonly used antibiotics in China, amoxicillin was selected as the model contaminant. The comparisons among these systems were established with amoxicillin concentration and COD_{cr} removal efficiency. The Si/SA process was further characterized by scanning electron microscopy (SEM), infrared spectra (IR), and Michaelis-Menten equation.

2. Experimental

2.1. Materials
Amoxicillin, NaOH, H_2SO_4(AR), diatomite with a surface area of ca. 100 m^2 g^{-1} (AR), ammonium acetate and acetonitrile (HPLC grade) were purchased from Sinopharm Chemical Reagent Co., Ltd., China. Reagents used for the bacterial media were purchased from Sigma. Deionized water was provided by a Milli-Q system (Millipore). Bacterium *STT01* was self-isolated from the activated sludge of urban sewage treatment plant by our research group in 2011.

2.2. Cells immobilization
The process of immobilization of bacteria (*STT01*) was according to the previous research [13]. The resulted pellets were soaked into 4% (w/v) $CaCl_2$ and held for 24 h at 4°C to immobilize completely, and then were washed with sterile water twice and then for degradation experiment. The control experiments were conducted under the same conditions.To achieve the optimal immobilization conditions, the diatomitedose (1.0, 2.0, 3.0, 4.0 and 5.0, g L^{-1}), and the mass ratio of wet cells to immobilization support (C:IS, 1:10, 1:20, 1:30, 1:40, and 1:50, w/v) were investigated.

2.3. Optimal conditions investigations
All degradation experiments were conducted in 250-mL tapered bottle and statically incubated in an adjustable illumination incubator (250D, Jiangsu Fuhua). As for the Si/SA system, based on large investigations, the basic operation conditions were chosen as follows: amoxicillin 600 mg L^{-1} ($COD_{cr,0}$ = 825 mg L^{-1}); pH 7.0, illumination intensity 3000 lx, temperature 30°C, DO2.0 mg L^{-1} and treatment time 72 h. Here were focused on the dosage of diatomite and the ratio of C: IS, which play the prerequisite role in the work. During the degradation, samples were collected at certain intervals for analyses. Selected samples had been repeatedly analyzed in order to validate the produced results and they were found within acceptable analytical error (±5%). The results were the means of triplicate determinations.

2.4. Analysis
The samples used for COD_{cr} test were collected by centrifuging at 10000 rpm for 20 min to remove bacteria cells. COD_{cr} was tested according to APHA Standard Methods (1995) [14]. IR spectrum was performed on Perkin Elmer FTIR spectrophotometer (USA) in KBr pellet. HPLC-MS were carried out on Thermo Finnigan LCQ-Advantage (USA) according to the methods presented by Nägele and Moritz [15].

2.5. Calculations
The removal efficiency (R_e) was calculated by the following equation:

$$Re = \frac{(C_0 - C_e)}{C_0} \times 100\% \tag{1}$$

where C_0 is the initial concentration of amoxicillin or COD_{cr}, and C_e is the detected concentration of amoxicillin or COD_{cr} in the removal process.

3. Results and discussion

3.1. Optimal parameters

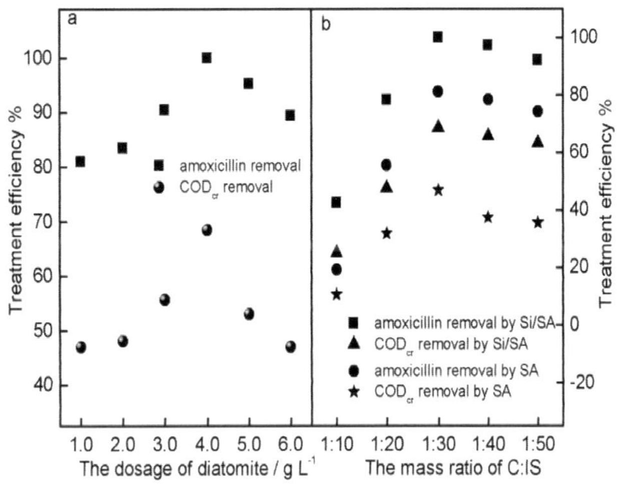

Figure 1. The treatment efficiency with varying dosages of diatomite (a) and mass ratios of C: IS (b) in the Si/SA and SA system correspondingly. Noted: C: IS is the mass ratio of the wet cells to immobilization support.

It was suggested that the appreciable treatment efficiencies were achieved at the dosage of diatomite of 4.0 g L^{-1} (Fig. 1a) and the mass ratios of C: ISof 1:30 (Fig. 1b). When the load of diatomite was more than that of 4.0 g L^{-1}, it made the immobilization support too hard to be pelletized; whereas when the load of diatomite was less than that of 4.0 g L^{-1}, which was found to do little effect on the improvement of treatment efficiencies (Fig. 1a). Similarly, when C: IS was increased to 1:20 or 1:10, higher cell loads might be an effect of substrate and/or light limitation of the cells inside the pellets, decreasing the degradation efficiencies. In addition, the less mass ratio of C: IS couldn't immobilized the cells completely that a great amount of cells was shed off in $CaCl_2$ and then led to lower biomass of *STT01*. On the contrary, the excess of Si/SA or SA load would bring up a rich nutrition environment, which had an inhibitory effect on cells growth, reducing the efficiency. The results might be ascribed that the proper dosages and mass ratio of C: IS could facilitate the preferential mass transfer, sufficient biomass and high light harvest, which was in favour of the performance of photo biological ability of *STT01*[13]. Under the optimal condition, 100%of amoxicillin and 68.5% of COD_{cr} removal efficiency were carried out in the Si/SA system compared with the cases of SA (81.2%, 46.9%), and free cells (60.5%, 35.5%). Additionally, we also found that the biodegradability of immobilization cells were not readily inhibited by the variations of external conditions, which might be due to the preferential retention of bioactivity and the avoidance of product inhibition [16].

3.2. Visual degradation efficiency

Even though amoxicillin could inhibit the growth of isolated free cells of *STT01*, yet it seems to be negligible to *STT01* in Si/SA system due to the increasing bright colour and cell density of the pellets (Fig. 2). It was noted that the removal of amoxicillin and COD_{cr} was not appreciable since the process of hydrolysis might require longer duration for digestion and the feed of amoxicillin influent was more toxic for *STT01* at the first stage, with the time on, the bacteria were gradually accommodated to it and the biodegradability was increased correspondingly. The appreciable treatment efficiencies were gained in various treatment time, i.e. 100% amoxicillin removal and 68.5% COD_{cr} removal were achieved in 72 h, 48 h, 48 h, 72 h, and 84 h when the pellets were reused five times, respectively. The decreasing treatment time could be ascribed to the increasing biomass and bioactivity of the pellets, leading to higher degradability in shorter time, whereas, when the amount of cells was increased to the maximum, it was potential to lead to a poor nutrition environment in the pellets, and the endogenous metabolism of cells might occur, resulting in the prolonged treatment time. As shown in Figs. 2e-f, the relatively darker colour of the pellets indicated that some cells were autolysis to death, the pellets gradually dissolved out of the pellets, which was also a hint that it was the time to add new active biomass for the continuous degradation.

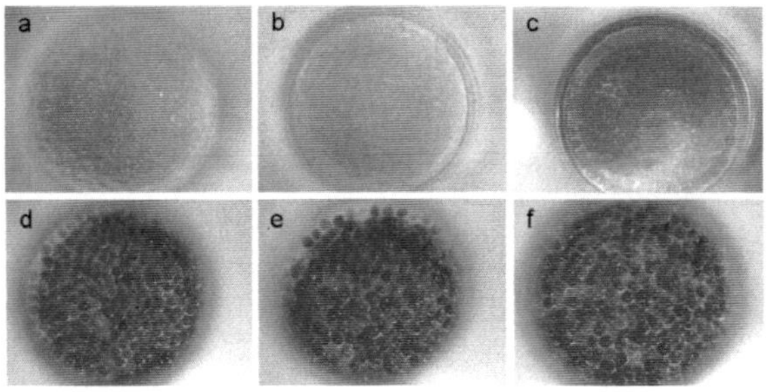

Figure 2. Optical images of the pellets immobilized with Si/SA under various treatment time of before treatment (a), once (b), twice (c), three times (d), four times (e), and five times (f).Noted: Images were taken from the top side when the effluents were emptied out after treatment

3.3. Mechanisms investigation

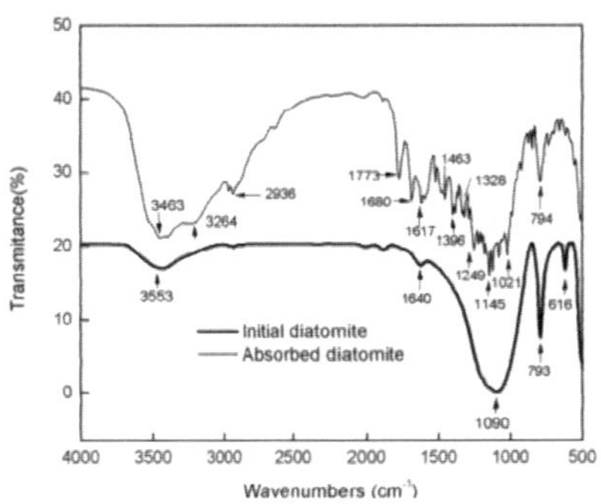

Figure3. IR spectra of the initial and absorbed diatomite

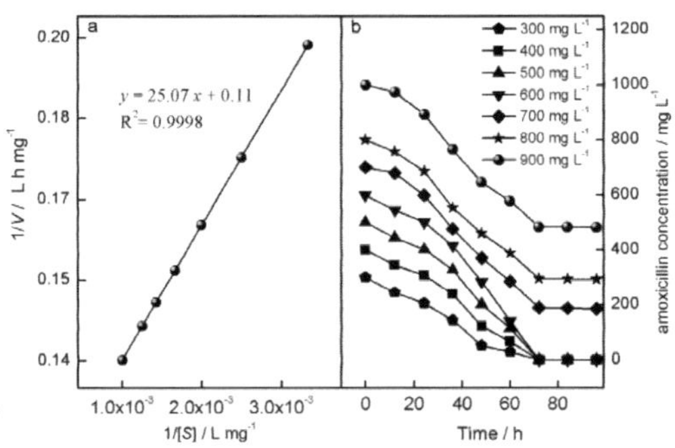

Figure4. Degradation kinetics ofamoxicillin: Lineweaver-Burkplot of 1/V and 1/[S] (a), various initial concentration (b).

3.3.1. Diatomiteabsorption mechanism. As well known, the surface of diatomite is covered by silicon hydroxyl groups with hydrogen bonds. These silicon hydroxyls the surface activity and adsorption properties of the diatomite. Adsorption ability is influenced by the action between adsorbent and absorbate in solution, as well as the pore structure [11, 12, 17].

It could be assumed that diatomite adsorbed amoxicillin in the way that its surface hydrogen bond and surface electric charge reacted with –OH, $-NH_2$, C=O group of amoxicillin to form new hydrogen bonds. To confirm this hypothesis, diatomite was thus alone selected to removal amoxicillin under the same conditions except for the contribution of *STT01*. The IR spectra of the initial and absorbed diatomite were presented in Fig. 3. It was shown that the Si-OH bond of initial diatomite shows bands at $3553cm^{-1}$and 793 cm^{-1}, Si-O bond at 1090 cm^{-1}, whereas the Si-OH bond of absorbed diatomite took shift to 3463 cm^{-1} and 794$cm^{-1}$and Si-O bond was changed into 1249 cm^{-1}, 1145$cm^{-1}$and 1021cm^{-1}.Furthermore, the peak around 1773 cm^{-1} might attribute to the C=O of β-lactam ring of amoxicillin; the vibration at 1680 cm^{-1}might assign to the C=O stretching vibration of secondary amido; the peaks at 3463 cm^{-1},3264 cm^{-1}, 2936 cm^{-1},1396 cm^{-1}, 1328 cm^{-1} could assign to the C-H bending or vibration of $-CH_2$ and $-CH_3$orN-H of β-lactam ring, and the special absorption at 1617cm^{-1} and 1464 cm^{-1} belong to benzene stretching [19]. The results showed that the physical-chemical adsorption might occur simultaneously on the surface of diatomite. Moreover, 20% of amoxicillin and 18% of COD_{cr} removal achieved in the process further demonstrated that the diatomite was responsible for the improvement of the treatment efficiency. In addition, the silicon hydroxyl is dissociated to ≡Si-O- and H^+ in aqueous solution, so the surface of diatomite carried negative charge with surface ζ potential[11,12], whereas amoxicillin is cationic at neutral condition, so it further deduced that the absorption action could be potentially carried out.

3.3.2. Degradation kinetics. The degradation kinetics were modeled with Michaelis-Mentenequation and were mathematical regression analyzed by Line weaver-Burk method, plotting the inverse of substrate concentration of amoxicillin against the inverse of the initial velocity due to the Michaelis-Mentenequation (2) and (3),

$$V = \frac{V_m[S]}{K_m + [S]} \tag{2}$$

$$\frac{1}{V} = \frac{K_m}{V_m} \bullet \frac{1}{[S]} + \frac{1}{V_m} \tag{3}$$

Where K_m: Michaelis constant/mg L^{-1}, [S]: substrate concentration/mg L^{-1}, V: enzymatic reaction starting velocity/ mg L^{-1} h^{-1}, V_m: enzymatic reaction maximum velocity/mg L^{-1} h^{-1}.

According to the Michaelis-Mentenequation [18], the average degradation velocity within 48 h was selected as the initial velocity of the experiments. The Line weaver-Burk plot (Fig. 4a) was clearly showed that there was a positive relationship between $1/V$ and 1/[S], which suggested the degradation velocity and initial amoxicillin was in line with Michaelis-Mentenequation due to the high R square (0.9998). So, according to Eq. 3, it could be deduced that $1/V_m$ is the intercept of y axis (0.11), and K_m/V_m is the slope (25.07), and thus the maximum degradation velocity (V_m) and Michael is constant (K_m) could be gained as 9.09 mg L^{-1} h^{-1} and 228 mg L^{-1}, respectively. So the enzymatic catalysis was responsible for the removal of amoxicillin in the Si/SA system. In addition, the results demonstrated that the optimal parameters in this experiment are also available for the cases of higher concentrations by scaling up the dosages of diatomite and the amount of the pellets correspondingly (Fig. 4b).

4. Conclusions

This investigation has been confirmed that the immobilized *STT01* is negligible to the toxicity of amoxicillin due to the appreciate removal efficiency. The results further demonstrate that the fast degradation of amoxicillin is due to the synergistic abiotic adsorption of diatomite and the biodegradation of *STT01*. The work opens up a direct visual platform for the degradation process due to the intrinsic bright colour of bacteria, and offers a promising entry for developing novel immobilization techniques of diatomite materials. It is expected to shed a new light on the future investigations on antibiotics-related wastewater.

5. Acknowledgements

This work was supported by National Natural Science Foundation of China (Grant no. 21507067), Shandong Natural Science Foundation (Grant no.Y2008B14) and Key lab of Paper Science and Technology, Ministry of Education (Shandong Province) (Grant no.0803135).

6. References

[1] A.Gulkowska, H.W.Leunga, M.K.So, S.Taniyasu, N.Yamashita, L.W.Y.Yeung, B.J.Richardson, A.P.Lei, J.P.Giesy, P.K.S.Lam, Removal of antibiotics from wastewater by sewagetreatment facilities in Hong Kong and Shenzhen, China. Water Res. 42 (2008) 395-403.

[2] Y.H. Yang, S.G. Fu, H. Peng, A.D. Shen, S.J. Yue, Y.F. Go, L. Yuan, Z.F. Jiang, Abuse of antibiotics in China and its potential interference in determining the etiology of pediatric bacterial diseases. Pediatr. Infect. Dis. J.12 (1993) 986-988.

[3] A. Wang, S. Yu, K. Yao, W. Zhang, L. Yuan, Y. Wang, J. Wei, X. Shen, Y. Yang, Antimicrobial susceptibility of Haemophilusinfluenzae strains and antibiotics usage patterns in pediatric outpatients: results from a children's hospital in China (2000-2004), Pediatr. Pulumonol. 43 (2008) 457-62.

[4] X.X. Zhang, T. Zhang, H.H. Fang, Antibiotic resistance genes in water environment, Appl. Microbiol. Biotechnol. 82 (2009) 397-414.

[5] R. Rosal, A. Rodriguez, J.A. Perdigon-Melon, M. Mezcua, M.D. Hernando, P. Leton, E. Garcia-Calvo, A. Aguera, A.R. Fernandez-Alba, Removal of pharmaceuticals and kinetics of mineralizationby O_3/H_2O_2 in a biotreated municipal wastewater, Water Res. 42 (2008) 3719-3728.

[6] H. Tekin, O. Bilkay, S.S. Ataberk, T.H. Balta, I.H. Ceribasi, F.D. Sanin, F.B. Dilek, U. Yetis, Use of Fenton oxidation to improve the biodegradability of a pharmaceutical wastewater, J. Hazard. Mater. 136 (2006) 258-265.

[7] M.Khademi, G.D.Najafpour,B.NavaeiNia, A.A.Zinatizadeh, R.RezaeiKalantary,Biological Treatment of Antibiotic Plant Effluent in an UASFF Bioreactor, World Appl. Sci. J. 5 (Special Issue for Environment) (2009) 1-8.

[8] L.A. Alaton, S. Dogruel, E. Baykal,G. Gerone, Combinedchemical and biological oxidation of penicillin formulation effluent, J. Environ. Manage. 73 (2004) 155-163.

[9] C. Sirtori, A. Zapata, I. Oller, W. Gerniak, A.Agüera,S. Malato,Decontamination industrial pharmaceutical wastewater bycombining solar photo-Fenton and biological treatment, Water Res. 43 (2008) 661-668.

[10] M.A.Al-Ghouti, M.A.M.Khraisheh, S.J.Allen,M.N.Ahmad, Theremoval of dyes from textile wastewater: a study of the physicalcharacteristics and adsorption mechanisms of diatomaceous earth, J. Environ. Manage. 69 (2003) 229-238.

[11] M.A.M.Khraisheh, Y.S.Al-degs, W.A.M.Mcminn,Remediationof wastewater containing heavy metals using raw and modifieddiatomite, Chem. Eng. J. 99 (2004) 177-184.

[12] Z.L. Wang,Y.X. Yang, X.P. Qu, J.B. Zhang, Y.R. Chen,L.X. Nie, Decolouring mechanism of Zhejiang diatomite, Application to printing and dyeing wastewater,Environ. Chem. Lett. 3 (2005) 33-37.

[13] T. Shen, C. Jiang, C. Wang, J. Sun, X. Wang, X. Li,A TiO_2 modified abiotic-biotic process for the degradation of azo dye methyl orange, RSC Adv. 5(2015) 58704-58712.

[14] APHA (American Public Health Association), Standard Methods for theExamination of Water and Wastewater, 19th ed (1995), Washington, DC, USA.

[15] E. Nägele, R. Moritz, Structure elucidation of degradation products of the antibiotic amoxicillin with ion trap MS^n and accurate mass determination by ESI TOF, J. Am. Soc. Mass Spectrom. 16 (2005) 1670-1676.

[16] C. Zhang, X. Zhu, Q. Liao, Y. Wang, J. Li, Y. Ding, H. Wang, Performance of a groove-type photobioreactor for hydrogen production by immobilized photosynthetic bacteria. Int. J. Hydrogen Energy 35 (2010) 5284-5292.

[17] J.L. Wu,Y.S.Yang,J.H. Lin, Advanced tertiary treatment of municipal wastewater using raw and modified diatomite, J. Hazard. Mater. B127 (2005) 196-203.

[18] L.A.Segel,M. Slemrod,The Quasi-Steady-State assumption: acase study in perturbation,SIAM Rev. 31 (1989) 446-477.

[19] Zhu H.W. Spectral Analysis of Structure of Organic moleculars, Chemical IndustryPress, Beijing, 2005.

The Measurement and Analysis of Microphysical Capabilities of the Condensation Nucleus of Artificial Warm Fog

DAI Meng-yan, CHEN Chun-sheng, LIU Jiang-hai, ZHANG Liang and HE Hai-yan

State Key Laboratoty of NBC Protection for Cilivian, Research Institute of Chemical Defence, Beijing, 102205 China

daidecai0558@163.com

Abstract. In this paper, the microcosmic capabilities of the condensation nucleus and fog droplets of artificial fog were observed and sampled by ELPI, SEM, XRD and LPSA. The chemical composition, number concentration, volume concentration and nucleation efficiency of the artificial condensation nucleus were gained. The condensation nucleus was mainly made up of chloride. The size distribution scope was about 0.291～1.59μm and the peak particle size was 0.609~0.942μm. The number concentration of the condensation nuclei was 3.98×10^5~4.82×10^5/cm^3, the volume concentration was 2.02×10^5~2.30×10^5μm^3/cm^3, and the maximum nucleation efficiency was 3.07×10^{12} /g. Actually, most condensation nuclei did not participate in the actual fog performing process, and the nucleation activation proportion was 13.57%. The condensation nucleus played an important role in the microphysical performance of fog droplets. However, the growing velocity of fog droplets was not identical as the catalytic performance difference of nuclei and the uneven distribution of watervapor in air. That was the reason why the fog droplets size distribution behaved the characteristic of bi-spectrum converting into uni-spectrum gradually.

1. Introduction

The cloud coagulate nucleus (CCN)in nature is the nuclei of cloud droplets and fog droplets in the environment of oversaturation, with the particle size range of about 10^{-3}～10μm. The coagulate nucleus is indispensable for the coagulation of water vapor. The meteorological change relies on the soluble nuclei such as NaCl and $(NH_4)_2SO_4$ [1-3]. The artificial cloud catalyst widely used in weather modification contains cold cloud catalyst and warm cloud catalyst, and the effective components including AgI, NaCl, $CaCl_2$ and so on [4]. The function of artificial cloud catalyst in atmosphere depends on its physicochemical characteristics [5-6], including particle size, microstructure, constituent and optical property.

We have made artificial fog based on the stimulation of formation mechanism of nature fog by fog aerosol [7]. The hygroscopic catalyst is transformed into artificial condensation nucleus with feat size distribution by the combustion of fog aerosol. Then water vapor coagulates on the condensation nucleus and the fog droplet comes into being. The condensation nucleus made up of hygroscopic inorganic salt (such as chloride) is used as warm cloud catalyst. The microphysical characteristics of condensation nucleus including constituent, number concentration and nucleation efficiency are closely related to its cloud catalytic performance [8-9]. Thus the microphysical characteristics of condensation nuclei have decisive effect on the growth process of fog droplets and would make very different attenuation of electromagnetic waves through artificial fog [10]. In the research of cold cloud catalyst, air filled with catalyst particles is usually injected into a cloud chamber with oversaturation condition, then catalyst particles will grow into ice crystals and the nucleation efficiency could be calculated with the help of microscope [11]. Inevitably test error caused by the wind tunnel dilution system and manual operation exists in this method. More importantly, this method is inapplicable to the warm cloud catalyst research because of the entirely different mechanism of nucleation. In this work, in order to establish the inner correlation between the microphysics characteristics of condensation nucleus and fog droplet of artificial fog, they were studied by several analytical methods including the electrical low pressure impactor, scanning electron microscope, X-ray diffraction and laser particle size analyzer. The number concentration, volume concentration and nucleation efficiency of the artificial condensation nuclei and droplets were obtained. It was expected that such work would help to reveal the nucleation control and catalysis mechanism of artificial condensation nucleus, and also have important significance for the optimum of the optical extinction effect of artificial fog.

2. Experimental

2.1. Experimental instrument

The microphysical characteristics of condensation nucleus was sampled and tested by the electrical low pressure impactor (Finland DEKATI Company). Electrical low pressure impactor (ELPI) classifies the particles into 12 size channels in the size range 0.03~10μm according to the aerodynamic diameter, such as 27.6-55nm(Channel 1), 55-93.7nm(Channel 2), 0.0937-0.155μm(Channel 3), 0.155-0.261μm(Channel 4), 0.261-0.38μm(Channel 5), 0.38-0.609μm(Channel 6), 0.609-0.942μm（Channel 7）, 0.942-1.59μm(Channel 8), 1.59-2.38μm(Channel 9), 2.38-3.97μm(Channel 10), 3.97-6.64μm(Channel 11）, 6.64-9.85μm(Channel 12). Scanning electron microscope (SEM) image of the condensation nucleus was obtained on a Hitachi S-4800 microscope (Japan). X-ray diffraction (XRD) spectrogram was tested by a Rigaku X-ray diffraction D/MAX2200 (Japan). The microphysical characteristics of fog droplets were tested by Winner311-XP Laser Particle Size Analyzer (LPSA, Jinan Winner Particle Instrument Company).

2.2. Experimental

The directions of fog aerosol contains oxidant (45~75%), bond (8~22%), combustible matter (0~10%), hygroscopic catalyzer (5~20%) and additive (1~5%) [7].

(1) Experimental method of sampling and testing of condensation nuclei

These experiments were done in a small smoke chamber with the volume of 6.375 m^3, the temperature was controlled at 14～20℃ and the relative humidity was controlled below 40%. The flow of sampling probe of the ELPI was 10L/min. Figure1(a) shows the schematic diagram of experimental set-up. 1 gram of fog aerosol was ignited by Mg rod or the high temperature gun (American Bernzomatic) respectively in each experiment. The fan stirred for 20s after combustion of fog aerosol, and the condensation nuclei were sampled and tested.

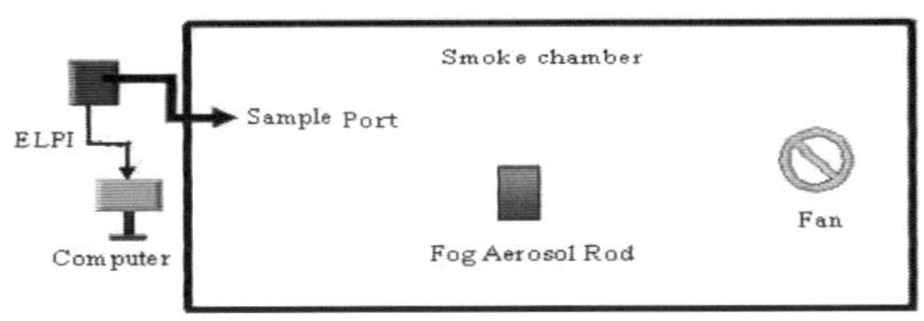

(a)Measurement of condensation nucleus

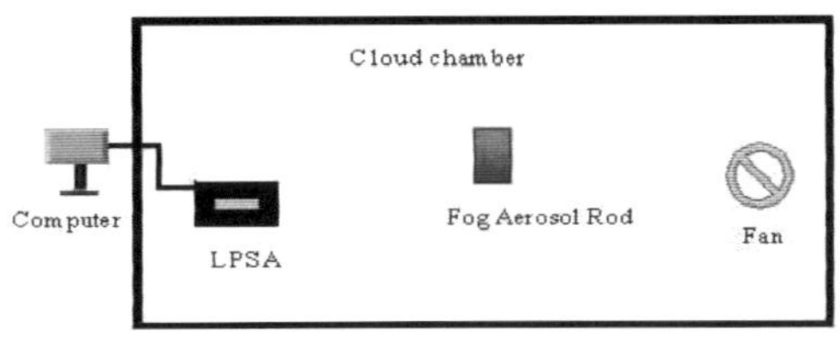

(b)Measurement of fog droplets

Figure 1. Schematic diagram of experimental set-up for the measurement of condensation nuclei and fog droplets.

(2) Experimental method of testing of fog droplets

The experiments were done in a cloud chamber with the volume of $50m^3$(4.4m×3.3m×3.4m), the temperature was controlled at 14~20℃ and the relative humidity was 75~80%. Figure1 (b) shows the schematic diagram of experimental set-up. 60 grams of fog aerosol was used in each experiment. The fan stirred for 20s after combustion of fog aerosol, and the fog droplets were tested.

3. Results and discussion

The condensation nuclei produced by combustion of fog aerosol would absorb water vapor and grow into fog droplets as soon as possible in suitable humidity conditions. If the microphysics performances of condensation nuclei and fog droplets are tested in the same experiment, it will be difficult for neither LPSA nor ELPI to distinguish between solid particles and fog droplets accurately. Thus the subsection measure method of condensation nuclei and fog droplets was designed to reduce the adverse effect of different dispersion space, environmental conditions and instruments.

3.1. The combustion of fog aerosol and dispersion of condensation nucleus

As shown in figure 2 the experiment results indicated that the hot air mass produced by combustion of fog aerosol was mixture of numerous condensation nuclei and manifold gas including CO_2, H_2O and O_2 calculated from the description of the fog aerosol(as shown in figure 2(a)). In the beginning, the hotair mass diffused around freely with high velocity. As a result of the difference in temperature and pressure between the hot air mass and atmospheree, the movement of hot air mass changed mostly attributed to air resistance, energy exchange and mass diffusion in the ascending process of hot air. Then the flow velocity of air mass decreased and diffused in the chamber gradually.

The relative humidity was controlled at below 40% in order to prevent the moisture absorption of condensation nucleus. In consequence, it was supposed all aerosol particles kept solid phase and there was no liquid droplets existed in the chamber. The condensation nuclei firstly suspended on the top of the chamber(as shown in figure 2(b)). It took about 5~6 minutes for the nucleus particles to sink, diffuse and evenly disperse in the chamber(as shown in figure 2(c)).

(a)Combustion process (b)10s after combustion (c)6mins after combustion

Figure 2. Photoes of combustion of fog aerosol and diffusion of condensation nucleus.

3.2. The microphysical performance of condensation nucleus

Figure3 gives the number concentration curve and volume concentration curve of condensation nuclei produced by fog aerosol ignited by Mg rod and high temperature gun respectively. The microphysics characteristics of condensation nuclei behaved similar trend of variability in the three conditions. In the time of 0~5 minutes after combustion, the number concentration and volume concentration of condensation nuclei increased rapidly(as shown in figure3(a)). In the time of 5~15 minutes the curves of number concentration and volume concentration demonstrated a tendency toward stabilization before slowly decline. Based on these results, the dispersion process of condensation nucleus particles could be divided into two steps. The first step was the freely diffusion process of particles propelled by gas expansion, and the second step was the settlement process due to gravity. The particles flowed fast as a result of instability of diffusion process, and this leaded to the uneven distribution of condensation nuclei in the whole chamber. Accordingly, the curve of number concentration and volume concentration of the particles displayed larger fluctuation in 0~5minutes. In 5~10minutes, the freely diffusion process of particles already finished and no obvious settlement phenomena had taken place yet. Therefore, the test results of that time could be thought as the average concentration of condensation nuclei produced by fog aerosol.

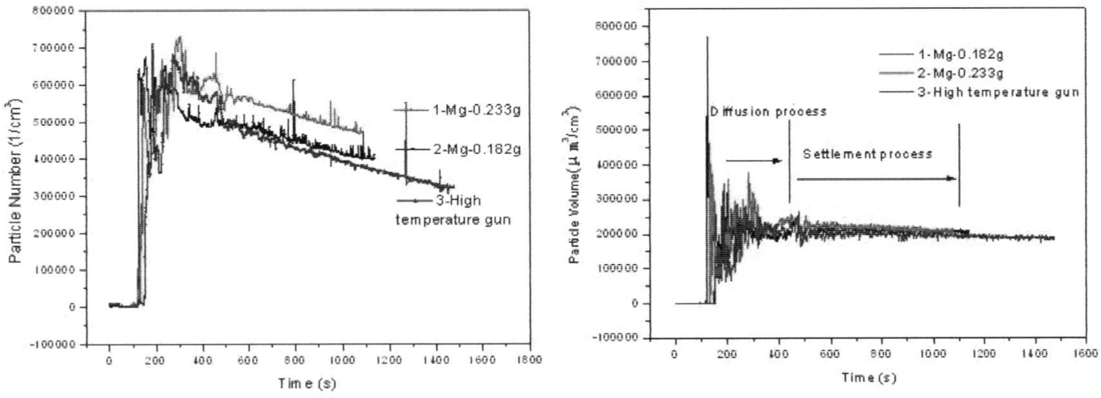

(a)Number concentration curve (b)Volume concentration curve

Figure 3. Number concentration curve and volume concentration curve of condensation nuclei.

It was indicated that the number concentration of nuclei though different ignition methods varied obviously. The number concentration of nuclei ignited by Mg rod(with the mass of 0.233g) was about 20% more than that ignited by high temperature gun. A great of deal of MgO solid particles would be generated by the combustion of Mg rod, and that caused the number concentration tested much bigger. While the combustible of high temperature gun was butane, there were no solid products that would probably interfere with the measurement of condensation nuclei. It was clearly that the results obtained from the high temperature gun ignition would be more accurate.

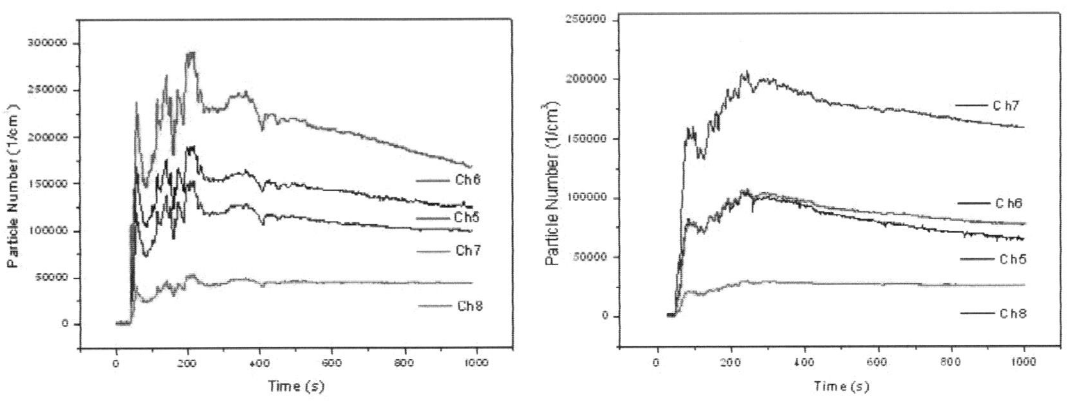

(a)Mg ignition (b)High temperature gun ignition

Figure 4. Number concentration curves of condensation nuclei tested by ELPI.

Figure 4 gives the number concentration curves of condensation nuclei produced by fog aerosol ignited by Mg rod and high temperature gun respectively. The particle size scope of condensation nuclei gained by different ignition methods was almost in the same size channels of ELPI. They were Channel 5(0.261-0.38μm), Channel 6(0.38-0.609μm), Channel 7(0.609-0.942μm), and Channel 8(0.942-1.59μm)). However, the peak particle size and characteristics of distribution were different. The condensation nuclei obtained by Mg ignition methods were smaller and the peak particle size was 0.38~0.609μm. The number concentration of the size channels of ELPI in a descending order was Channel 6, Channel 5, Channel 7, and Channel 8. The condensation nuclei obtained by high temperature gun ignition methods were comparatively bigger and the peak particle size was 0.609～0.942μm. The number concentration in a descending order was Channel 7, Channel 6, Channel 5, and Channel 8. Combined with figure 3 and figure 4, it was deduced that the MgO particles were mostly fine particulate matter from the fact that the main effect of MgO particles was a marked increase on the number concentration of Channel 6 and Channel 5.

Meanwhile, it was worth noting that the volume concentration curves of different ignition methods was similar as shown in figure 3(b). It was because the key factors impacting the volume concentration of aerosol particles comprised of number concentration and size distribution. According to the calculation equation of the volume concentration (V):

$$V = \int_0^\infty N_r \cdot \frac{4\pi}{3} \cdot (\frac{D_r}{2})^3 \qquad (1)$$

As in equation (1), N_r was the number concentration of the particles with the diameter of D_r. The volume concentration of particles was directly proportional to the cube of diameter, so the particle size contributed much more to the volume concentration than the number concentration. Since the MgO particles were all fine particulate matter, the Mg ignition method did not affect greatly on the volume concentration of condensation nuclei.

Table 1 gives the data of the microphysical characteristics of condensation nuclei in different time. The number concentration of concentration nuclei was 3.98×10^5~4.82×10^5 /cm^3, and the volume concentration was 2.02×10^5~2.30×10^5μm^3/cm^3. The particle size distribution range was 0.261~1.59μm, and the peak particle size was 0.609~0.942μm. The number concentration of the size channels in a descending order was Channel 7, Channel 6, Channel 5, and Channel 8. The volume concentration in a descending order was Channel 7, Channel 8, Channel 6, and Channel 5. The test results just verified our analysis above that the particles of bigger size contributed much more than smaller ones on the volume concentration, and this made the distribution feature of the number concentration not consistent with that of the volume concentration. For example, the particles of channel 8 were the biggest in all. In the time of 5 minutes, the number concentration of channel 8 was only 30% of that of channel 6, but the volume concentration of channel 8 was 4 times more than that of channel 6.

Table 1. The data of the microphysical characteristics of condensation nuclei.

Time	ELPI Channel	Ch5	Ch6	Ch7	Ch8	Ch1-Ch12
	Size（μm）	0.261-0.38	0.38-0.609	0.609-0.942	0.942-1.59	0.03-10
5 minutes	Number concentration(/cm^3)	9.60×10^4	9.93×10^4	1.93×10^5	2.85×10^4	4.82×10^5
	Volume concentration（μm^3/cm^3）	0.15×10^4	0.59×10^4	4.35×10^4	2.72×10^4	2.30×10^5
15 minutes	Number concentration(/cm^3)	6.56×10^4	7.67×10^4	1.61×10^5	2.50×10^4	3.98×10^5
	Volume concentration（μm^3/cm^3）	0.10×10^4	0.46×10^4	3.66×10^4	2.53×10^4	2.02×10^5

3.3 The influence of condensation nuclei on the microphysical performance of fog droplets

Figure 5 gives the SEM image and XRD spectrum of condensation nuclei. The morphology of the condensation nuclei was in the shape of near-sperical orcubic and composed of chloride such as KCland YL. Defined the nucleation efficiency as the number of the condensation nuclei per gram of fog aerosol in the unit of /g:

$$\lambda_0 = \frac{c_n \times V}{m} \tag{2}$$

As in equation (2), C_n: the number concentration of condensation nuclei, in $/cm^3$; V: the volume of the chamber, in cm^3; m: the mass of fog aerosol, in gram. As the maximum number concentration was 4.82×10^5 $/cm^3$, the nucleation efficiency could be calculated as $3.07\times10^{12}/g$.

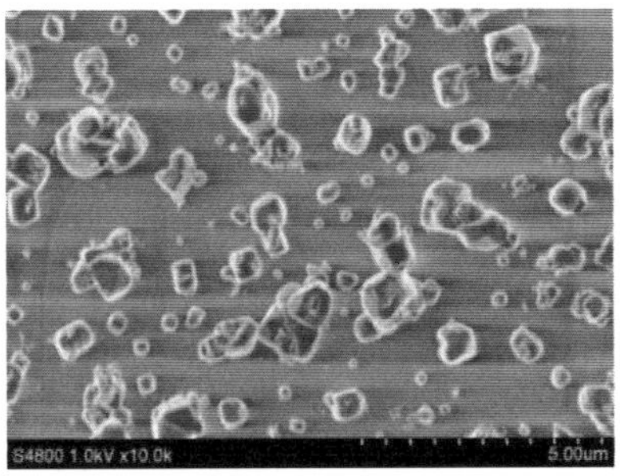

(a)SEM image(×10000)

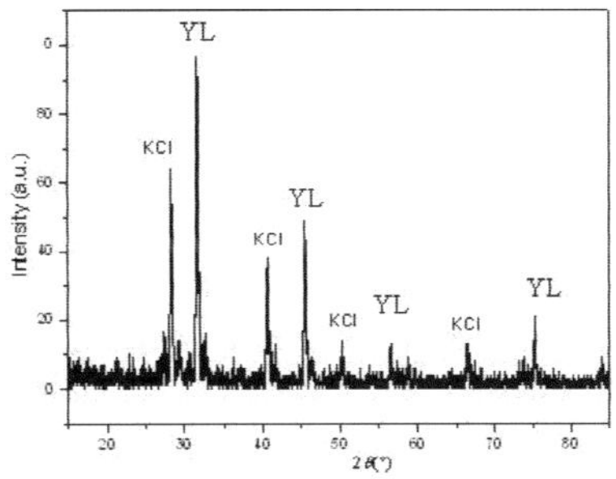

(b)XRD spectrum

Figure 5.SEM image and XRD spectrum of condensation nucleus.

Figure 6 gives the number concentration, water content and size distribution of fog droplets. The life circle of fog droplets could be divided into 4 stages: formation stage, development stage, stabilization stage and elimination stage. The formation stage was in the time of 0～3minutes, the condensation nuclei started growing into fog droplets instantly in the humidity controlled higher than the deliquescence relative humidity of chloride. The number concentration and particle size of fog droplets increased rapidly in this stage. The development stage was in 3～10 minutes, the increment speed of number concentration and particle size reduced a little than before, however, the number concentration and water content reached their maximum value in that time. The size distribution of fog droplets behaved characteristic of bimodal distribution and the peak size scope narrowed distinctly. The stabilization stage was in 10~20 minutes, the number concentration, water content and particle size became stable and decreased slightly. At the same time, the double-peak spectrum of fog droplets converted into single-peak spectrum gradually. After 20 minutes was the elimination stage, the droplets size, water content and number concentration all began to reduce. The fog would disperse little by little as the acceleration of evaporation, coalescence and settlement of fog droplets.

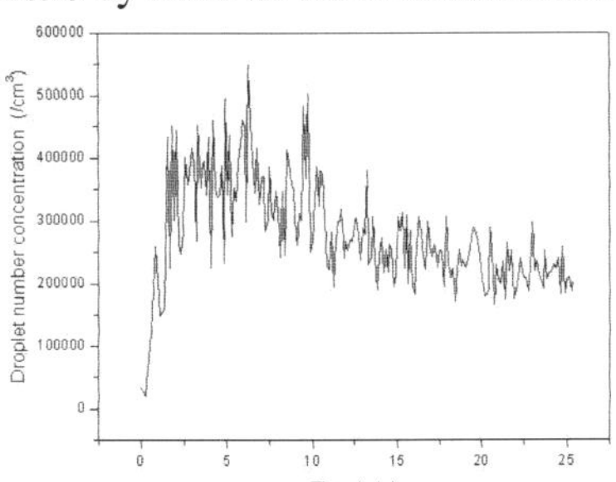

(a)Number concentration curve

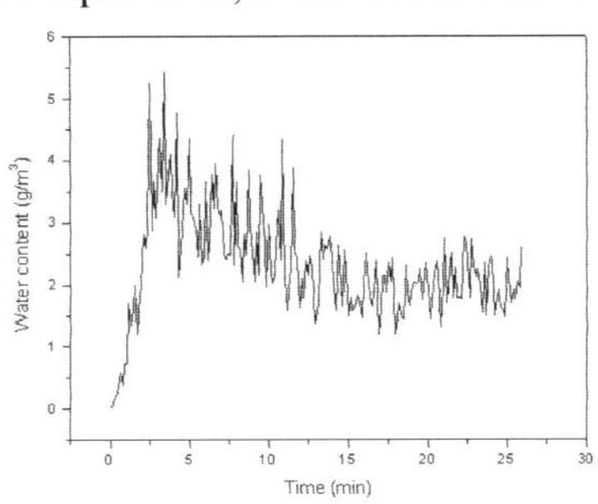

(b)Water content curve

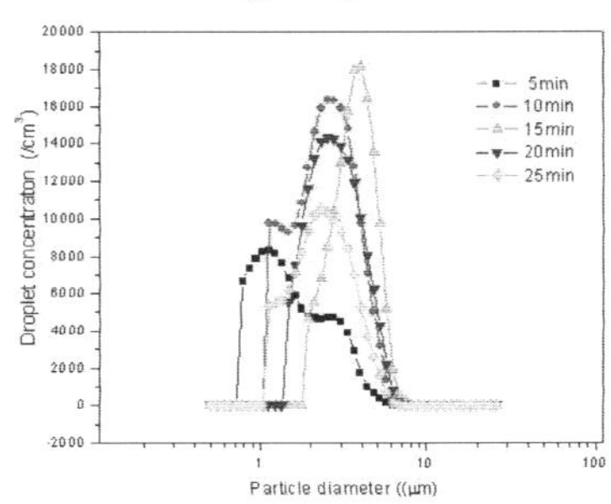

(c)Size distribution

Figure 6. Number concentration curve, water content curve and size distribution of fog droplets.

The fog droplets were grown up from condensation nuclei with moisture absorption. The microphysical characteristics of fog droplets were closely related to the catalytic performance of condensation nuclei. There was significant linear correlation between the number concentration of fog droplets and nuclei, and the size distribution and water content were important representation of the catalytic performance of condensation nuclei and the growth speed of fog droplets. Meanwhile, the watervaporplayed a crucial role on the formation of fog droplets. It was interesting to find that after the fog forming process the environmental humidity always reduced 3～5% than before. It meant that thecoagulation of condensation nuclei and formation of droplets would consume a great deal of water vapor in air and lead to decrease of humidity. Watervapor was sufficient in the formation stage and development stage of fog, and the influence of condensation nuclei on the fog droplets was also showed more obviously in these two stages. In the stabilization stage and elimination stage, the environmental humidity began to decrease and the watervapor was not enough to sustain the growth of droplets.

Although a great number of condensation nuclei were produced by fog aerosol firstly, there was fierce competition among them. Because of the uneven distribution of watervapor and the individual difference of nuclei [12], only a little part of the numerous nuclei would be able to activate in oversaturation condition and grow into fog droplets finally. Those lucky condensation nuclei were called as activated nuclei. In the formation stage of fog droplets, the number concentration of activated nuclei was impossible to measure. As the activated nuclei would transformed into fog droplets in the instant, the number concentration of fog droplets was approximately equal to the number of activated nuclei. Defined the activation proportion as the ratio of the activation nuclei to the total number of the condensation nuclei:

$$\eta = \frac{N_a}{N_0} \times 100\% \approx \frac{N_f}{N_0} \times 100\% = \frac{N_f}{\lambda_0 \times \frac{m}{V}} \times 100\% \tag{3}$$

As in equation (3), λ_0: the nucleation efficiency of fog aerosol, in /g, N_f: the number concentration of fog droplets, in /cm^3; V: the volume of the cloud chamber, in cm^3; m: the mass of fog aerosol, in gram. According to the results above, the nucleation efficiency of fog aerosol was 3.07×10^{12} /g, and the maximum number concentration of fog droplets in the formation stage was 5×10^5/cm^3. Supposed no error brought by the subsection measure method or nuclei loss, the activation proportion was calculated as follows:

$$\eta = \frac{N_f}{\lambda_0 \times \frac{m}{V}} \times 100\% = \frac{5 \times 10^5}{3.07 \times 10^{12} \times \frac{60}{50 \times 10^6}} \times 100\% = 13.57\% \tag{4}$$

It was calculated that the activation proportion was only 13.57% in ideal condition. This result suggested that most condensation nuclei did not participate in the fog forming process and the activated nuclei were real effective condensation nuclei which showed better catalytic performance.

To sum up, the condensation nuclei were of crucial importance for the microphysical characteristics of fog droplets. Owing to the microphysical performance difference of condensation nuclei and the random distribution of watervapor in the environment, the activation proportion of condensation nuclei was low and the growth rate of fog droplets was inconsistent. The size distribution of fog droplets behaved as a double-peak spectrum in the formation stage. Then the fog droplets would grow into the equilibrium size in the development stage and stabilization stage. The size distribution of fog droplets turned to a single-peak spectrum ultimately.

4. Conclusion

The microcosmic capabilities of the condensation nuclei and fog droplets of artificial fog were observed and tested by a subsection measure method. The conclusion was as follows:

(1)The microcosmic capabilities of the condensation nuclei of artificial fog were researched by ELPI, SEM, XRD and LPSA. The number concentration of concentration nuclei was $3.98\times10^5 \sim 4.82\times10^5$ /cm^3, and the volume concentration was $2.02\times10^5 \sim 2.30\times10^5$ μm^3/cm^3. The particle size distribution range was 0.261~1.59μm, and the peak particle size was 0.609~0.942μm. The nucleation efficiency was about 3.07×10^{12} /g.

(2)Most condensation nuclei did not participated in the fog forming process because of the competition for watervapor. The activation proportion was only 13.57% in the ideal condition. The increase of activated nuclei was of more significance for the artificial fog than the improvement of nucleation efficiency.

(3)The condensation nucleus was of crucial importance for the microphysical characteristics of fog droplet. Owing to the microphysical performance difference of condensation nuclei and the random distribution of watervapor in the environment, the growth rate of fog droplets was inconsistent and the fog droplets size distribution appeared the characteristic of bimodal spectrum converting into unimodal spectrum gradually.

5. Acknowledgments

This paper is supported by the National Natural Science Foundation of China (Grant No: 41205097，51103176) and the National Key Technology Research and Development Program of China (Grant No: 2013BAK03B08).

6. References

[1] Johnson G R，Ristovski Z and Morawska L2004*J. Aerosol Sci.* **35**443-55
[2] Abderrahim D and Abdelfetah M 2003 *Fluid Phase Equilibria.* **206**13-25
[3] Kreidenweis S M，Koehler K and DeMott P J 2005 *Atmospheric Chemistry and Physics Discussions.* **5**1357-70
[4] Rosenfeld D，Khain A and Lynn B 2007 *Atmospheric Chemistry and Physics Discussions***7**3411-24
[5] SU Zheng-jun, GUAN Li-you and SHI Ai-li 2010 *Meteorologyical Science and Technology.* **38**259-62
[6] HU Shuai，GAO Chang-tai, LIU Leiand LIU Zhi-tian 2013*Chinese Journal of Light Scaterring.***25**338-46
[7] DAI Meng-yan, HU Bi-ru and Wu Wen-jian 2009 *Chinese Journal of Energetic Materials.* **17**708-12
[8] Kulmala M，Korhonen P and Laaksonen A 1997 *J. Aerosol Sci.***28**749-50.
[9] Cubison M J，Ervens B and Feingold G 2008 *Atmospheric Chemistry and Physics Discussions.* **8**5629-81
[10] Kurt B and Hans G 2004 *Aerospace Science and Technology.* **8** 63–71
[11] JIN De-zhen，ZHANG Jing-hong，JINANG Zhong-han，ZHAO Ze-hui and XU Le. 2012 *Climatic and Environmental Research.* **17** 666-70
[12] SHENG Pei-xuan, MAO Jie-tai and LI Jian-guo 2005*Atmospheric Physics.* Beijing: Peking University Publishing House P290-327

Novel Local Calibration Method for Chemical Oxygen Demand Measurements by Using UV–Vis Spectrometry

Hu Yingtian, Liu Chao and Wang Xiaoping*

State Key Laboratory of Modern Optical Instrumentation, College of Optical Science and Engineering, Zhejiang University, Hangzhou 310027, China

xpwang@zju.edu.cn

Abstract. In recent years, ultraviolet–visible spectroscopy has been widely used for chemical oxygen demand (COD) measurements of water. However, chemical compositions of substance in different water samples can cause measurement deviations, so a local calibration is needed. In this study, a novel local calibration method is proposed. The absorption spectra of COD standard solutions and wastewater samples taken from four factories were collected. We analyzed the impact of chemical compositions of substance in different water samples and extracted the morphology features of their absorptive spectra for recognition models. Furthermore, we calculated the local calibration parameters of the four categories of real water samples by specific modification based on the ability of light absorption in various water environments. After the process of local calibration, the root mean square errors (RMSEs) of the predictions were very small, which highlights the potential of this method for improving the accuracy and adaptability of COD measurements based on ultraviolet–visible spectrum.

1. Introduction

Water contamination is an important environmental concern, and there is a great need for reliable information on contaminant concentrations in natural waters. Chemical oxygen demand (COD) is an important parameter that reflects the concentrations of organic pollutants in water. In recent years, ultraviolet–visible (UV–Vis)spectroscopy, as an indirect method, relies on the regressing model standards provided by chemical methods [1]. It has been widely used for chemical oxygen demand and other parameters measurements of water, because its procedure is easy and can achieve fast detections with no secondary pollution [2].

Langergraber's research group [3, 4] developed a submersible UV–Vis spectrometer for in-situ real-time measurements and put forward the concepts of global and local calibration. Lourenço ND et. Al [5] researched the water quality monitoring method and applied it in a fuel park wastewater treatment plant. Zhao [6, 7] developed an online spectrometer with an open flow pool based on the UV–Vis absorption spectrum for water quality monitoring. Mi [2] predicted COD by using machine learning methods and the result was better than that of the least squares method. Wang [8] established the prediction models for seawater COD by using partial least squares and the results were consistent with real values.Hu [1] and Tang [9] studied algorithms to compensate the turbidity in the COD measurement.

In spite of this research, these systems and algorithms suffer from lower accuracyin different application environments, in other words, the adaptability is poor. Actually, the chemical component determines the shapes of absorption spectra, i.e., absorbance response position and degrees; so, the detection of conventional parameters such as COD, total organic carbon (TOC), total suspended solids (TSS), and dissolved organic carbon (DOC) is seriously impacted by the presence of diverse chemical components in water [10]. Therefore, a local calibration is needed in various water environments. However, algorithms for the local calibration in COD measurements have not been studied in detail and intensively.

In this study, we analyzed the impact of chemical compositions of substance in different water samples and proposed a local calibration methodtheoretically and experimentally. Detection results suggest that this local calibration method can be used toimprove the measurement accuracy and adaptability, to some extent, during UV–vis-based COD measurements.

2. Materials and methods

2.1. Experimental setup

The UV–Vis spectral acquisition system consisted of a light source, a sample cell, a spectrometer, and a microcomputer (figure 1). The pulsed xenon light source produced 220–750 nm light, which was collimated by a lens and then passed through the sample cell and air respectively, as the measurement beam and the reference beam. The emergent light was collected into the spectrometer by the converging lens. An optical fiber was used to connect the xenon lamp and the sample cell, and the sample cell and spectrometer as well. The data processing was realized by an embedded system.

For the convenience of online and in situ measurement, a groove was designed in the submerged structure. In addition, the present state of light source can be caught in the double light path. The instrument was shown in figure 2, with the length of 550 mm, the diameter of 110 mm, and the light path of 35mm.

2.2. Water sample

In this study, the COD standard solutions were prepared by dissolving potassium hydrogen phthalate indeionized water supplied by a MilliQ water purification system (Millipore, Billerica, MA, USA). The concentrations included 5, 7.5, 10, 15, 20, 30, 60 mg/L.

The real wastewater samples were taken from two dyeing and printing plants and two paper mills in Zhejiang Province, China. We took water samples near their sewage-draining exits respectively and diluted with deionized water to expand the concentration range. The dilution multiples included 2, 3, 4, 5, 6, 8, 10, 12, 16, 18, 20, 25, 30, 32. Moreover, the real COD values were detected by chemical titration for the verification experiment.

A series of spectral experiments of these samples were carried out with the instrument shown in figure 1. The intensity spectrum of each water sample was measured 15 times and the spectra were averaged to reduce random error from 2.67% to 1%. We measured and deducted the dark spectrum before each measurement.

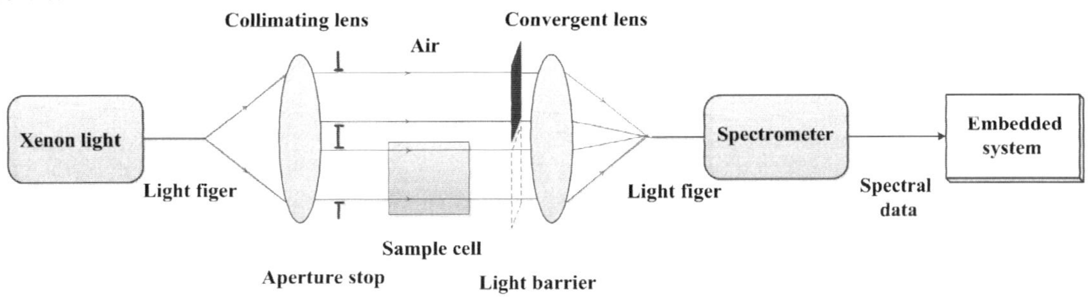

Figure 1. The schematic diagram of the spectral data acquisition system with double paths.

Figure 2. The actual object.

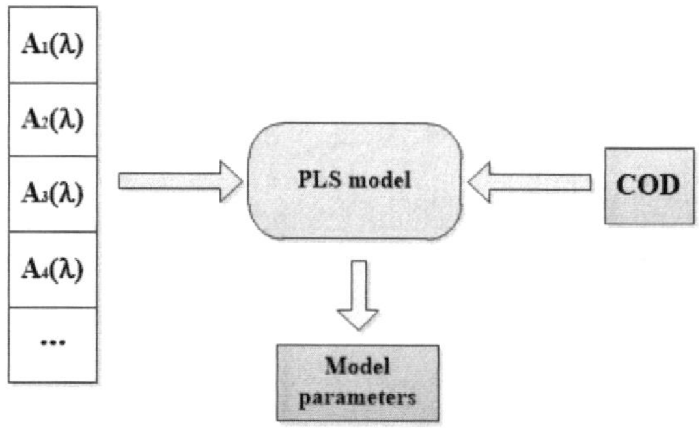

Figure3. The process of global calibration.

2.3. Global and local calibration

Ultraviolet–visible (UV–vis) spectroscopy, as an indirect method, relies on the regression model standards provided by chemical methods. The regression process includes global calibration and local calibration, which are used for a universal and a targeted regression model respectively. In this paper, the partial least squares (PLS) was used because of the mature theory and simple calculation. The process of global calibration was shown in figure 3. The COD values of standard solutions and the corresponding absorbance at eight wavelengths close to the absorption peaks were put into the PLS model to train the regression parameters. When applied in different water environments, the COD can be roughly calculated using the global calibration model.

Actually, chemical compositions of substance in water samples varied, which impacts the detection of COD because of the variation in spectral shapes. In order to adapt the variation in different water environments, i.e. to insure the accuracy even for different applications, a local calibration is needed. As a premise, the classification and recognition of water samples should be carried out. In a local calibration, the results calculated by using the global calibration model should be specifically modified based on the ability of light absorption in each water environment. In theory, absorbance follows the Beer–Lambert law:

$$A = \lg(I_0 / I) = kCL \tag{1}$$

where k is the absorption coefficient, C is the substance concentration, L is the optical path, $I_0$and Iare the original and emergent light intensity, respectively, and A is the absorbance.

As for a real water sample, the absorption coefficient of COD standard solutions was used in the global calibration rather than the real absorption coefficient of the sample. Therefore, it was the fundamental cause of measurement deviation. The local calibration parameter was defined as following:

$$\tau = k_0 / k_1 = C_1 / C_0 \tag{2}$$

where τ is the local calibration parameter, k_1 and k_0 are the absorption coefficient at a characteristic wavelength of the real water sample and COD standard solution respectively, C_0 is the COD value calculated by using the global calibration model, and $C_1$is the real COD detected by chemical titration.

In measurements, we calculated the COD by using the global calibration model and identified the category and the corresponding local calibration parameter first, and then multiplied the COD by the local calibration parameter to modify it.

3. Results and discussion

3.1. Impact analysis of chemical compositions of substance in water samples

Through research and investigation, the organic contaminants in water come mainly from sanitary sewage and industrial wastewater. In theory, unsaturated functional groups absorb ultraviolet light, such as C=O bonds, C=C bonds, and benzene rings. Apparently, the location of absorption peaks and the strength of absorption are various because of the chemical components in different water samples. As shown in figure 4, the absorptive characteristics of COD standard substance and wastewater samples were different. The COD standard substance had two distinct absorptive peaks at about 230 and 280 nm; the two absorptive peaks of wastewater samples in dyeing and printing plants were vague; the wastewater samples in paper mills only had one absorptive peak. Actually, dyeing and printing plants mainly contained dyes, phenol, sorts of benzene and its derivatives, so the second absorptive peaks around 280 nm existed. On the contrary, the mono-peak absorbance curves of paper mill wastewater were caused by the lignin-like substance. Although the sewage discharged by the same kind of factories had similar spectral shapes, there were slight variations among different factories.

In addition, the actual COD values of wastewater samples were detected by means of dichromate titration (table 1). In order to compare with the COD standard solutions, the absorbance at 280 nm and the absorption coefficient per unit length and unit concentration were also listed. The table 1 showed that the light absorbing ability of different water samples varied from the COD standard substance. The dyeing and printing plant sewage with more aromatic compounds had a stronger ability of light absorbing than potassium acid phthalate, while the ability of lignin-like substance in paper mills was weaker. Therefore, the different chemical compositions of substance in water samples affect the accuracy of prediction results, if the global calibration of COD standard substance was only used.

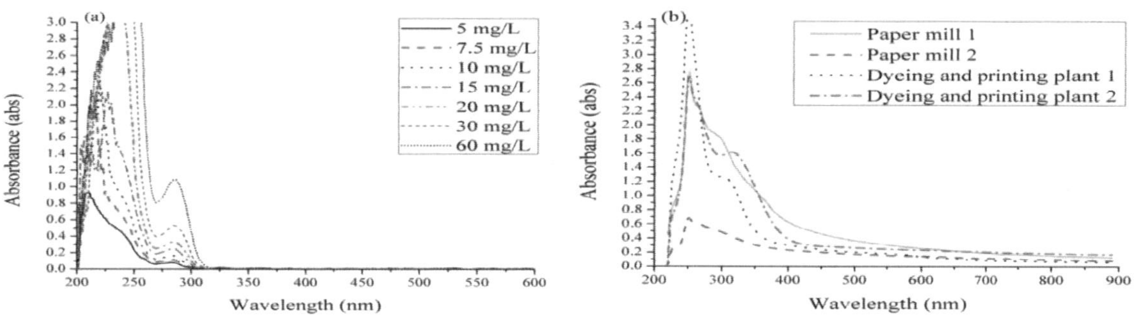

Figure 4. (a)Absorption spectra of COD standard solutions with different concentrations. (b) Absorption spectra of wastewater in two paper mills and two dying and printing plants.

Table 1. The concentration, absorbance at 280 nm, and the absorption coefficient at 280 nm of four wastewater samples and a COD standard solution.

Water samples	COD(mg • L^{-1})	Absorbance at 280nm	Absorption coefficient at 280nm(L • mg^{-1} • m^{-1})
Paper mill 1	238.5	1.92	0.23
Paper mill 2	39.4	0.53	0.38
Dyeing and printing plant 1	57.8	1.36	0.67
Dyeing and printing plant 2	62.9	1.72	0.78
COD standard solution	20.0	0.33	0.47

Table 2. The morphology features and output values of 5 categories of water samples. The peaks or valleys of each water sample were taken the first three, and the difference of absorbance was the latter one subtracted by the prior one.

No.	Water samples	Peak 1 (nm) or valley	Peak 2 (nm) or valley	Peak 3 (nm) or valley	Difference 1	Difference 2	Output
1	Paper mill 1	256	0	0	0	0	1
2	Paper mill 2	255	0	0	0	0	0
3	Dyeing and printing plant 1	257	289	307	-1.92	0.02	0
4	Dyeing and printing plant 2	256	300	316	-0.92	0.05	0
5	COD standard solution	230	268	282	-1.93	0.23	0

3.2Classification and recognition of water samples

In order to reduce the impact from chemical compositions of substance in different water samples, it is necessary to classify the water samples based on morphology features of their absorptive spectra and construct effective recognition models before local calibration. For better recognizing, the morphology features included the positions of absorptive peaks and valleys (the first three), and the difference of absorbance between two adjacent peaks and valleys. In addition, back-propagation algorithm was used in the artificial neural networks training to establish the recognition models. In the process of sample training, 28 water samples of four factories with seven dilution multiples and seven COD standard solutions were taken as the training set. The dilution multiples included 1, 2, 4, 8, 10, 16, 20. In table 2, the morphology features of every category of water samples were listed and only one sample with the largest concentration of each category was shown here. When training one category of the water samples, the output values of the category were set to 1 and the others were set to 0. For following categories of water samples, the same progress was done.

In the validation process, the rest samples of the four factories were taken as the validation set and the sample size was 32, including 8 dilution multiples of 3, 5, 6, 12, 18, 25, 30, 32. The experimental result was that the correct recognition ratios reached94%, which proved that the models were rational and effective to identify the category of water samples.

3.3Parameters of local calibration

A local calibration of wastewater samples belonging to four factoriescarried outusing the method in 2.3 on the premise of correct recognition. Only the original concentration of samples taken from every factory was selected in the calibration process to streamline the operation. The characteristic wavelengths were set at the second absorbance peaks or 280 nm if without the second peaks. The calibration parameters were calculated and listed in table 3.

In practical application, the calibration parameter can be determined after classification recognition and the prediction results can be modified by multiplying the local calibration parameters on the basis of using global calibration models. Therefore, the proposed local calibration method efficiently broadens the range of application through a simple operation, in which only one sample needs chemical titration and spectrum collection.

Table 3. The characteristic wavelengths and calibration parameters of four wastewater samples.

Water samples	Characteristic wavelengths (nm)	Calibration parameters (τ)
Paper mill 1	280	2.03
Paper mill 2	280	1.24
Dyeing and printing plant 1	303	0.75
Dyeing and printing plant 2	315	0.64
COD standard solution	280	1

3.4. Results for predictions using local calibration model

In order to validate the the effectivenessof this method for improving the accuracy in various water samples,we predicted the COD of seven standard solutionsand 60 wastewater samples of four factories with 15 dilution multiples by using the global calibration model only and both global and local calibration models.After 67 samples were predicted, the results for two types of processing methods were plotted in figure5, and their correlation coefficient (R^2) and root mean square error (RMSE) values are shown in table 4. The results for the proposed method of local calibration are generally consistent with the actual values and the RMSE values are very small.

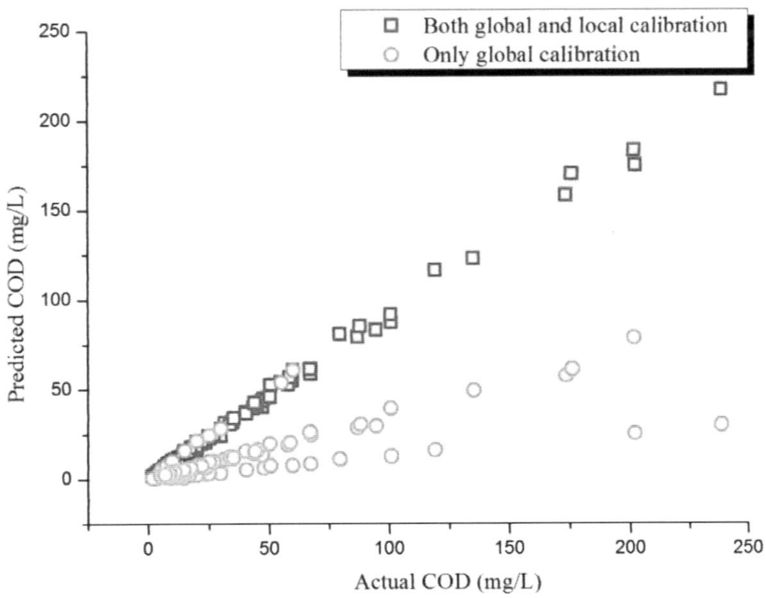

Figure 5. Prediction results for two types of processing methods.

Table 4. Prediction results of different processing methods for water samples.

Processing methods	Relative error (%)	R^2	RMSE(mg/L)
Only global calibration	42.94%	0.77	44.72
Both global and local calibration	8.43%	0.99	5.62

4. Conclusion

In this study, we sampled the absorption spectra of COD standard solutions and wastewater samples taken from four factories and proposed a novel local calibration method during COD measurements employing the UV–vis spectrum. This method improved the measurement accuracy and adaptability by specific modification for various water environments. We analyzed the impact of chemical compositions of substance in different water samples and extracted the morphology features of their absorptive spectra for recognition models.The dyeing and printing plant sewage with more aromatic compounds had a stronger ability of light absorbing than potassium acid phthalate, while the ability of lignin-like substance in paper mills was weaker.Furthermore, we calculated the local calibration parameters of the four categories of real water samples. Moreover, 67 samples of standard solutions and real wastewater with 15 dilution multiples were predicted and the results by using local calibration were consistent with the actual. In conclusion, this local calibration method can be used toimprove the measurement accuracy and adaptability, to some extent, during UV–vis-based COD measurements.

5. Acknowledgements

The study was supported by the National High Technology Research and Development Program of China (863 Program) (Project No. 2014AA06A504).

6. References

[1] Hu YT, Wen YZ and Wang XP *2016 Novel method of turbidity compensation for chemical oxygen demand measurements by using UV–vis spectrometry*Sensor Actuat. B-Chem.**227** 393-8
[2] Mi YP, Wang XP and Jin X *2008 Water COD prediction based on machine learning*J. Zhejiang Univ. Sci. A**42(5)** 790-3
[3] Langergraber G, Weingartner A and Fleischmann N *2004 Time-resolved delta spectrometry: a method to define alarm parameters from spectral data*Water Sci. Technol. **50(11)** 13-20
[4] Rieger L, Langergraber G, Thomann M, Fleischmann N and Siegrist H *2004 Spectral in-situ analysis of NO2, NO3, COD, DOC and TSS in the effluent of a WWTP*Water Sci. Technol.**50(11)** 143-52
[5] Lourenço ND, Chaves CL, Novais JM, Menezes JC, Pinheiro HM and Diniz D *2006 UV spectra analysis for water quality monitoring in a fuel park wastewater treatment plant*Chemosphere**65(5)** 786-91
[6] Zhao YQ, Li YC, Guo Y, Gu BJ and Yang Z *2012 A Novel Technology for Water Quality Testing Based on UV Spectral Analysis*Spectrosc. Spect. Anal.**32(5)** 1301-5
[7] Zhao YQ, Wang HM, Liu ZY, Li YC and Fan SF *2010 Novel method for on-line water COD determination using UV spectrum technology*Chinese Journal of Scientific Instrument **31(9)** 1927-32
[8] Wang LL, Liu XH, Shi XX, Lu YR, Qi Y and Wang MY *2013 Study on Real-Time Monitoring of Seawater COD by UV-Vis Spectroscopy*Adv. Mat. Res.**726-731** 1534-7
[9] Tang B, Wei B, Wu DC, Mi DL, Zhao JX, Feng P, Jiang SH and Mao BJ *2014 Experimental Research of Turbidity Influence on Water Quality Monitoring of COD in UV-VisibleSpectroscopy*Spectrosc. Spect. Anal. **(11)** 3020-4
[10] Hu YT and Wang XP *2017 Application of surrogate parameters in characteristic UV–vis absorption bands for rapid analysis of water contaminants*Sensor Actuat. B-Chem.**239** 718-26

IOP Conf. Series: Earth and Environmental Science **63** (2017) 012017 doi:10.1088/1755-1315/63/1/012017

The Interactive Effect of Diversification and Farming Scale on Productivity of Family Farm:Taking Rice Cultivation as An Example

Zhou Wei

Business school, Hohai University, Nanjing, China

zhouweihhu2017@163.com

Abstract. Based on the diversification and cultivation scale, the rice cropping data of rural fixed observation points in 2011 were selected and the effect of diversification degree on rice productivity was analyzed by the Tobit model. The empirical results of the model show that diversification of sample farm will lead to loss of rice production efficiency. With the increase of rice planting scale, the loss of rice production efficiency will need to be further increased by diversification. Thus, we should stick to the family farm of specialized production operation. The transfer of land, the price and quantity of leasing, respecting the law of the market; the raising of funds can be considered non-subsidized capital market financing to help, while maintaining a certain degree of diversification, to avoid idle assets, low resource efficiency loss.

1. Introduction

With the gradual development and deepening of agricultural modernization, as an emerging form of agricultural production and management, family farms are adapted to social and economic development. Due to the weak nature of agriculture and the special nature of the production process, making agriculture throughout the reproduction cycle is facing many risks. In addition, agriculture also has the risk of an economic market and social behavior shared by other industries. So agriculture is a case in high-risk industries [1]. To this end, family farms must diversify their operations to resist and mitigate the effects of natural and market risks on farmers' incomes, diversify farmers' income channels and increase farmers' incomes [2-3]. Then, the diversification of agricultural production in the form of family farms can be expected to result in a significant increase in income for farmers. Will diversification of family farms have an influence on current grain production? Can guarantee China's food security? These issues are available to question and study.

Family farm as the main body of new agricultural management, in recent years, for family farm research is very popular, the main research focused on the status qua, the significance of family farms and other production and management of the relationship between the main problems and perfect system [4-6]. For example, Huang Xinjian, etc. about Jiangxi rice cultivation, for example, should be transferred from the contract right to improve the financial subsidies to promote family farm progress [7]. Sci Guoqing and other family farms that China is an important form of agricultural micro-management, he achieved the land family business and commercial production of the organic integration. In the developed countries, he argues, there are basically three categories of workers in agriculture: those who occupy the means of production and who are only responsible for farming; and those who own the means of production and who are engaged in most of the work themselves

Producers; third is not the possession of means of production, employment of agricultural machinery operators, they have actually become industrial workers [8-9]. For its significance, Chi-hung, etc. (2013) that this is a global phenomenon. Family farms have an efficient efficiency advantage that is unmatched by other forms of business organization. Lin (2008) argues that family farms are the most appropriate system for agricultural growth in developing states. On the other hand, Wang Chunlai (2014) argues that the advantages of family farms lie in agricultural production, while the benefits of farmers' cooperatives and industrial and commercial enterprises lie in production services and market linkages. Three should seek advantages Complementarity and common development [10]. In terms of the plight of family farms, Shi Guoqing et al. (2014) argue that the plight of land transfer at this stage restricts the realization of moderate scale management of family farms in China [11-16]. Depending on Wang Jianhua et al. (2014), the construction of the sustainable development mechanism of family farms is a systematic and holistic project. It is made up of five complementary developmental stages: quasi-quasi- Into the recognition mechanism, support management mechanism, cooperation and incentive mechanism, social service mechanism, the mechanism of vocational peasant cultivation.

Most existing studies on family farms focus on the theoretical analysis of family farm progress. However, there is poorly research on the empirical analysis of the effectiveness of family farms. For instance, there are relatively few empirical studies on the relative efficiency of diversified family farms and diversified family farms, and whether the diversification has significant effects on the efficiency of production and management of family farms. Rice is the most important nation in China, more than 65% of the population of rice-based foods, its development is directly related to China's food security. Based on this, this paper chooses the family farm which is mainly engaged in rice planting as the research theme of this paper, trying to answer the question of the influence mechanism and the effect of diversification on its production efficiency.

2. Theoretical analysis and hypothesis

2.1. Behavioral decision theory

The basic view of behavior decision-making theory is that farmers as producers are not entirely rational, and their production and management behavior cannot be fully explained by rational decision-making theory.

For a long time, agriculture as a vulnerable industry has been put in a high input, low output of the embarrassing situation. Coupled with imported rice and other bulk imports of agricultural products continue to increase domestic grain prices continue to decline, leading to a sharp decline in domestic grain production income. On the other hand, a large number of farmers abandon the cultivation of the traditional rice and other traditional food crops, and change the vegetables into the second or third industry. On the other hand, a large number of farmers choose to ", Fruits and other high economic crops [17]. Farmers, as limited "rational decision makers", use "cost-benefit" comparisons in the agrarian production process to select good-benefit crops for planting. Compare to the traditional small-scale farmers, the agricultural income accounts for the proportion of the total family income of the newly-emerged agricultural management households, such as family farms. Therefore, the family farmers pay greater attention to the "cost-benefit" comparison in production and management decision-making.

In view of this, when the income of traditional grain crops such as rice is declining, the farmer is bound to look for other crops to grow crops, and even replace the existing rice cultivation, and thus showing the diversification of agricultural production situation The pursuit of higher agricultural production efficiency [18].

2.2. hypothesis

Therefore, this paper presents the first hypothesis:

Hypothesis 1: (1) Diversification of family farms helps improve rice production efficiency.

(2) The impact of family farm scale on diversification and agricultural performance

With the increase in scale of family farms, family farms can derive more economies of scale benefits from diversification and economies of scale in brand building. Therefore, opposed to small-scale rice farming family farms, large scale rice farming family farm diversification gains relatively significant. With the continuous expansion of agricultural production and management, family farms can gradually reduce the negative impact of labor transfer on pastoral production, reduce agricultural production costs by large scale mechanized production, and provide agricultural production efficiency. However, it should be pointed out that the scale of the economy there is "moderate" restrictions, not the scale of the operation the better. Once the business scale is too large, beyond the optimal scale, there may be management level lag, management ineffective and additional constraints. In addition, as family farms are still mainly family members as the main labor force, so too large scale will reduce the level of specialization of farmers. There will be diminishing returns to phenomenon. In view of this, this paper proposes:

Hypothesis 2: The scale of family farms has a positive effect on the relationship between diversification and agricultural performance.

3. Model Construction

3.1. The specific settings are as follows:

DIV denotes the independent variable, for example, the degree of diversification of the sample family farm. In order to improve the credibility of the model, the Herfindahl index (HI) is used to measure the degree of diversification of the sample farm. Xi, Zj, Pk and Wm are control variables, including personal and family traits variables, labor force variables, external environment variables and land characteristics variables.

Although all of the rice production and management efficiency of the sample family farms is observable, their distribution ranges are limited to 0-1, and the explanatory variable TE is compressed to a certain interval. In this case, In the traditional least-squares regression model, the estimation results may be inconsistent in both the global sample and the sub-sample in which the discrete points are deleted.

Compared with the traditional least squares regression model, the Tobit model is applicable to the case where the part of the explanatory variable is positive and partly 0. Specifically, if the observed value is, the unobserved value is:

3.2. conditional expectation

The conditional expectation can be written as follows for a sample satisfying the condition:

Therefore, the traditional least-squares regression regression of the above samples, ignoring the non-linear terms have been included in the reality of the random disturbance, resulting in interpretation of the variable Xi and random disturbance items are related, leading to the traditional least-squares regression The results are not consistent. In response to this situation, Tobin proposed the use of MLE to re-estimate the model, you can model the probability density function rewritten as:

Where, for the demonstrative function, if true then the function value is 1, if false then the function value is 0. Depending on the above formula, we can further write the likelihood function of the whole sample, and then estimate it by MLE to get the final estimation result.

Built on the model of analyzing the effect of diversification on rice production efficiency, this paper investigates the effects of diversified management on the diversification of rice production by analyzing the interaction between diversification and rice planting scale. And rice planting efficiency, the specific settings are as follows:

TE represents the dependent variable of the model, ie, the rice production efficiency of the sample family farm; the independent variable, which is the interaction item between the diversification degree of the sample family farm and the rice planting scale, in order to improve the credibility of the model, the degree of diversification Xi, Zj, Pk, Wm are control variables, including personal and family traits variables, labor characteristics variables, external environment variables and land characteristics variables (EI) .

4. Empirical Analysis

4.1. Data sources
A total of 9667 valid family farm samples were screened. Among them, 4245 households in Northeast China (Heilongjiang Province, Liaoning Province and Jilin Province) were divided according to the principal producing areas of rice. The Yangtze River Basin (Hunan, Henan, Hubei, Jiangsu, Anhui, Zhejiang, Jiangxi, Yunnan Province, Sichuan Province, Chongqing City, Guizhou Province, Guangxi Zhuang Autonomous Region) a total of 5422 households. The distribution of the entire sample is relatively uniform, can better represent the main producing areas of China's actual production of rice.

4.2. the status of the sample family farm operations
Because the Standard Industrial Classification does not have detailed industry segmentation criteria for the agricultural sector and the limitations of rural fixed observation data design, this paper divides the scope of operation of rice-growing family farms into four categories (That is, N = 4): pure grain cultivation industry, including rice-based traditional food crop cultivation; livestock breeding industry, including pig, sheep, chicken, duck and other poultry farming; economic crop industry, including vegetables, Crop cultivation; other industries, including leisure tourism, farmhouse and other tertiary industry activities.

Based on the formula of the Herfindahl index (HI) and the entropy index (EI), we can obtain the index of diversification of the sample family farms. On the whole, at present, the diversification of the principal rice family farms in China is relatively low. Traditional rice cultivation is still the core content of family farms.

Built on the above analysis, the traditional BCC model and Bootstrap-DEA model were used to estimate the rice production efficiency of all sample farms. In this paper, after testing the Bootstrap-DEA model and the traditional DEA model, we find that there are significant differences in the efficiency values of the two models under the assumption of equality of variance and variance, and the Bootstrap-DEA. The model is consistent with the conclusion that small sample analysis is more advantageous (Kneip, 2003). Results of Bootstrap-DEA model are as follows:

Results of model estimation show that the technical efficiency of the sample farms is relatively low under Bootstrap-DEA model and traditional DEA model. The estimated technical efficiency of the modified Bootstrap-DEA model is only 0.1817, and the lowest mechanical efficiency Farm only 0.014, and the standard deviation of 0.1512, showing that the difference between the technical efficiency of the sample farm farms. In the aspect of scale efficiency, Bootstrap-DEA model is changed to estimate the sample family farm scale efficiency value of 0.5891, of which the scale efficiency is 0.9549 and the lowest is 0.3355. In addition, in the pure technical efficiency, the Bootstrap-DEA model modified to estimate the sample family farm efficiency is relatively low, in which the sample family farm pure technical efficiency of the minimum is only 0.0234, the standard deviation of 0.2882, the overall situation and family farms Of the technical efficiency of the same. It can be seen that the pure technical efficiency and scale efficiency of the rice production in the family farms still have much room for improvement at present, but the principal factor restricting the improvement of household agricultural efficiency lies in the pure technical efficiency.

4.3. model estimation results and analysis
In the sample farm, the rice production efficiency value is between 0 and 1, which are a limited variable. Therefore, the traditional least square regression is no longer utilized, so this paper carries out regression analysis through the Tobit model. The Probability of Probability of the model was greater than the confidence level of 1% confidence level, and the R-squared was 0.6211 And 0.5621, respectively. The results show that the model has a great fit with the model.

The effect of diversification on rice production efficiency shows that all the variables have passed the significance test. The empirical results show that both the Herfindahl index (HI) and the entropy index (EI) is negatively correlated with the rice production efficiency of household farms. With the diversification of family farms, Traditional rice production efficiency will be still decline. The reason may be that, on the one hand, although diversification can improve the ability of resisting risks to some degree, the overall management level and management ability of the family farms are relatively weak, leading to the shortage of rice farms On the other hand, due to the special nature of agricultural production assets, assets between the different production activities cannot be universal.

Built on the analysis of the effect of diversification level on the rice production efficiency, the effect of diversification scale on rice production efficiency was analyzed by constructing the interaction item between the degree of diversification and the rice planting scale. Depending on the results of the regression model, the probability values of the diversified degree represented by the Herfindahl index (HI) and the entropy index (EI) were all significant at the confidence level of 1% R-squared are 0.5781 and 0.5861 respectively. Despite the fact that the fitting degree of the model is slightly different from that of the previous model, the overall regression result is still relatively good and the reliability is high.

The regression results showed that the coefficients corresponding to the interaction between the degree of diversification and the rice planting scale were both significant at the 1% confidence level and negatively correlated with the rice production efficiency, indicating that with the increase of rice planting area and diversification The rice production efficiency will decrease gradually. It is worth noting that, compared with the previous section, the coefficient of diversification is considerably higher than that of paddy rice, ie, it has a greater impact on production efficiency. It can be seen that family farms in the degree of diversification is bound to part of the management of energy and production funds from the rice production in the withdrawal, and gradually shifted to other industries, leading to rice production materials and technology, resulting in the efficiency of diversification loss. With the increase of rice planting area, the diversification of the loss effect is increasing, has slightly offset the large scale cultivation of the scale effect.

5. Conclusions and Recommendations

Based on the improvement of the traditional data envelopment analysis, this paper calculates the rice planting efficiency of domestic firms by Bootstrap-DEA model. The results demonstrate that the efficiency of rice production is relatively low, and the pure technical efficiency is the main Constraints. In addition, it is found that the rice production efficiency is relatively low on the family farms with a higher degree of diversification, and the loss of rice production efficiency will be further increased with the increase of rice planting scale.

First, adhere to the family farm of specialized production operation. Land as the most important and basic factor of production of family farms, the transfer and lease prices and quantity, to respect the laws of the market, and should not be policy-based means to determine factor prices and the number of transactions, which can improve the productivity of family farms. The problem of capital can be considered for family farms to help non-subsidized capital market financing to the market way, so that the market "foot voting", so that family farms can achieve the effect of survival of the fittest, so as to improve the efficiency of family farms .

Second, to maintain a certain degree of diversification. According to the results of the present study, the diversification level of the main rice family farms in China is relatively low. The increase of diversification will bring loss of production and management efficiency to family farms. However, complete specialization also has the malpractice, will bring the asset specificity low resources idle and the waste, also can cause the efficiency loss. Some researchers believe that diversification cannot only bring diversification of operational risks to the family farms, improve the efficiency of resource allocation and other benefits, but also conducive to the realization of environmental protection, more suitable for China's special national conditions. In China, the successful examples of family farms in the initial stage include both specialized grain farms, specialized vegetable farms, specialized fruit farms, specialized chicken farms and specialized fish farming farms; There is no shortage of diversified family farms engaged in both cultivation and cultivation. The practice shows that at this stage, diversification is suitable for China's national conditions, but also to maintain an appropriate degree, cannot be too scattered managers production and management of energy resulting in loss of efficiency.

6. References

[1] Yue Zheng-hua. Status and Problems of Family Farms in China and Policy Suggestions [J].Acta Moderna Sinica, 2013 (7): 420-424.
[2] Wang Guangquan.Chinese family farm model [J]. Theory and Contemporary, 2013 (4): 39-42.
[3] Wang Jianhua et al. Development of Family Farm in China: Impetus and Dilemma and Construction of Sustainable Development Mechanism [J]. Journal of Agricultural Modernization Research, 2013 (5): 552-555.
[4] LI Yi-xue, PENG Chao, KONG Xiang-zhi.Research on the Current Situation and Mode of Rural Land Contracting Operation Right [J].Pricing Theory & Practice, 2009 (3): 42.
[5] Liu Jinbao, Liu Hong. Weak correlation between agricultural technological progress and peasant's agricultural income growth [J]. China Rural Economy, 2009 (9): 28-34.
[6] Liu Shijin et al: between tradition and modern - growth mode and the choice of new industrialization road [M]. Beijing: China Renmin University Press.2006.
[7] Ansoff, H., Strategies for Diversification [M]. Harvard Business Review. 1957 (35): 113-124.
[8] Li Zhong, Liu Weibo. Pastoral land circulation should pay attention to several issues [J]. Economic and Trade, 2013 (3): 96.
[9] Lewis. Dual economic theory [M]. Beijing: Beijing Economic College Press.1989: 26-36.
[10] MA Wen-qiu.Experiences of American family farms and its enlightenment in China [J]. Agriculture and Technology, 2013 (7): 203.
[11] Meng Zhaochun, Chen Ping. Modern family farm is the basis of agricultural modernization in Canada to study a little inspiration [J]. Farm Economics and Management, 2003 (11): 42-45
[12] MU Xiang-li, GONG Qian-wen.Family farms: concept definition, identification criteria and development strategies [J].Rural Business Administration, 2013 (8): 23-25
[13] NI Kun-xiao, SHEN Yue-qin.Research on the development of family farms in Cixi City of Zhejiang Province [J]. Zhejiang Journal of Agricultural Sciences, 2012 (11): 27-35
[14] Ouyang Xuchu. Agricultural Economics [M]. Beijing: China Financial and Economic Publishing House.2000.
[15] Pan Shujuan. Family farm financial services to be taken [N]. China urban and rural financial news, 2012-08-29.
[16] PENG Jian-liang.Farmers, Enterprises, and Companies: A Way of Reforming State Farms [J]. China Rural Economy, 1994 (10): 46.
[17] CHEN Mei-qiu, DENG Ai-zhen, ZHOU Bing-juan, et al. Empirical Study on Farmers' Cultivated Land Protection Mentality of Different Groups [J] China Soft Science, 2005 (09): 16-22.
[18] Deng Mengzhi, Luo Renfu, Zhang Linxiu.Construction of Road Infrastructure and Non-agricultural Employment of Rural Labor Force: Based on the Survey of 2000 Farm Households in 5 Provinces [J].Technology in Agriculture Technology, 2011 (02): 4-11.

A New Type Hi-Speed BLDC Control System Base on Indirect Current Control Strategy

D P Wang[1], Y C Wang[1*], F G Zhang[1] and S Jin[1]

[1] Shenyang University of Technology, 118 ShenLiao West Road, SY 110870, China

919245539@qq.com

Abstract. High speed BLDC has the characteristic as larger air gap smaller armature inductance, traditional PWM modulation will produce a great number of high frequency current harmonics which led problem like large torque ripple and serious motor heat. In the meantime traditional PWM modulation use the diode rectifier which cause harmonic pollution in electric power net. To solve the problem above, proposes a new motor controller topology. Using the IGBT device to replace the diode on frequency converter rectifier side, apply the power factor correction technology, reduce the pollution on the grid. Using busbar current modulation on the inverter, driving bridge-arm use 3-phase 6-state open as driving Mode, realize the control on a 10000r/min,10kw BLDC. The results of Simulation on matlab show the topological structure as proposed can effectively improve the network side power factor and reduce the motor armature winding harmonic and motor torque ripple.

1. Introduction

Compare with general motors, high speed BLDC has smaller volume;smaller inertia, faster dynamic response; higher efficiency of the transmission system and precision,which make it widespread used in the area of servo control,numerically-controlled machine tool,robotics,etc.high speed BLDC get few number of turns which led very small terminal inductance [1].When use the traditional PWM modulation will produce the great number of high frequency current PWM harmonics which led a great number of non-conducting phase follow current and motor torque ripple. Literature [2] use motor drive circuit based on the buck converter PAM modulation to solve this problem, which can effectively suppressed forward and reverse follow current cause by terminal voltage zero-crossing which cause by lagging commutation conducting phase shut-before winding current peak, but because of traditional PWM modulation use the diode rectifier, the grid side current will produce harmonic and reduced the power factor, that will cause the pollution of the grid. Literature [3] proposed a PWM rectifier base on 3-phase VSR and BUCK circuit, it has the advantage of current waveform sinusoidal, little harmonic, high power factor, but it has further research on the applicability of high speed motor. Given on the basis, proposed a motor drive circuit topological structure base on 3-phase controlled rectifier bridge indirect current modulation method and then computer simulation is obtained [4].

2. The foundation of BLDC mathematical model

Take the example of the 3-phase 6-state square-wave motor BLDC with two mosfets, to analyze BLDC mathematical model and electromagnetic torque characteristic, suppose [5] :

(1)Each phase winding is completely symmetrical, air gap field is square-wave, stator current, rotor magnetic field distribution is symmetrical, corresponding electrical time constant is neglected;
(2)Magnetic circuit is unsaturated, regardless of the eddy current and hysteresis losses;
(3)Ignore the influence of cogging groove, reversing process and the influence of armature reaction, etc.
Three-phase winding voltage balance equation can be expressed as:

$$\begin{bmatrix} U_a \\ U_b \\ U_c \end{bmatrix} = \begin{bmatrix} r & 0 & 0 \\ 0 & r & 0 \\ 0 & 0 & r \end{bmatrix} \begin{bmatrix} i_a \\ i_b \\ i_c \end{bmatrix} + \begin{bmatrix} L & M & M \\ M & L & M \\ M & M & L \end{bmatrix} p \begin{bmatrix} i_a \\ i_b \\ i_c \end{bmatrix} + \begin{bmatrix} e_a \\ e_b \\ e_c \end{bmatrix} \quad (1)$$

In the equation: U_a, U_b, U_c is the phase winding of the stator voltage; i_a , i_b , i_c is the phase winding of the stator current; e_a , e_b, e_c is the stator winding back emf; L is the one phase winding inductance; M is the mutual inductance between every two phase winding; p is the differential operator $p=d/dt$.
Suppose 3-phase winding is wye-connected and there is no central line, there are:

$$i_a + i_b + i_c = 0 \quad (2)$$

$$Mi_a + Mi_b + Mi_c = 0 \quad (3)$$

Simultaneous equation (1) (2) (3), get voltage equation:

$$\begin{bmatrix} U_a \\ U_b \\ U_c \end{bmatrix} = \begin{bmatrix} r & 0 & 0 \\ 0 & r & 0 \\ 0 & 0 & r \end{bmatrix} \begin{bmatrix} i_a \\ i_b \\ i_c \end{bmatrix} + \begin{bmatrix} L-M & 0 & 0 \\ 0 & L-M & 0 \\ 0 & 0 & L-M \end{bmatrix} p \begin{bmatrix} i_a \\ i_b \\ i_c \end{bmatrix} + \begin{bmatrix} e_a \\ e_b \\ e_c \end{bmatrix} \quad (4)$$

Electromagnetic torque equation is:

$$T_e = \frac{1}{\omega}(e_a i_a + e_b i_b + e_c i_c) \quad (5)$$

In the equation: T_e is electromagnetic torque; ω is electrical machinery rotational speed.
Mechanical movement equation is:

$$T_e - T_L = J \, {d\omega}/{dt} \quad (6)$$

In the equation: T_L is the load torque; J is the rotational inertia of motor.
Basic on the equation (4) (5) (6), the BLDC mathematical model can be found. At the same time the parameters of motor used in this paper is given: rotational inertia J=0.0018164kg·m^2, damping coefficient B=0, back EMF coefficient K_e=0.021, winding single-phase resistance r=0.02Ω, one phase winding inductance L=245μh,mutual inductance between every two phase winding M=-40μh, number of pole-pairs p=2, rated speed ω=10000r/min, rated power P=10 kW. The carrier frequency of system is 10kHz, input load torque 10N·m launch after 0.01 seconds input.

3. traditional PWM control

PWM control is the technique that modulated the pulse width, it means through a series of pulse width on modulation, to equivalent to obtain waveform needed [6]. Figure 1 is the typical low speed double closed-loop BLDC control block diagram.

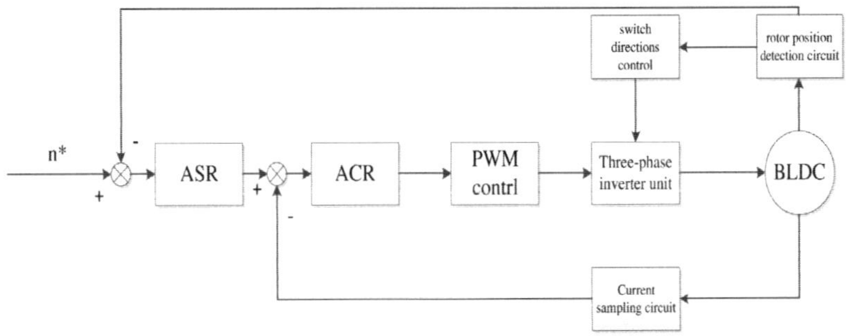

Figure 1. Double closed-loop BLDC control block diagram

According to literature [7], when carrier frequency of system is 10kHz, the inverter output voltage waveform show up significant harmonic function, not only appear the burr, but also amplitude will change over time in constant ups and downs, thus the output current and electromagnetic torque of the motor will show greatly influence. According to the parameters of BLDC show in the first quarter, the figure 2 (a) (b) (c) is the motor single-phase current waveform, the electromagnetic torque waveform and the current/voltage waveform on the grid side can be created [8].

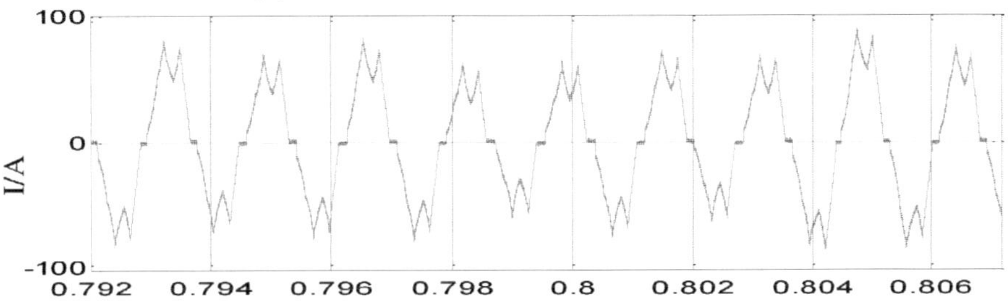

(a) single-phase current waveform
t/s

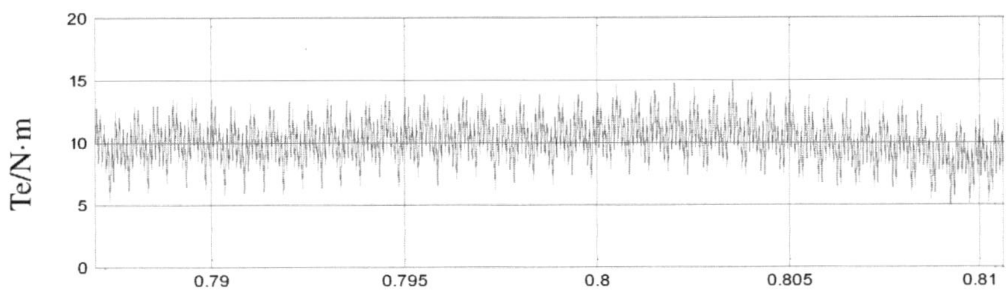

(b) electromagnetic torque waveform
t/s

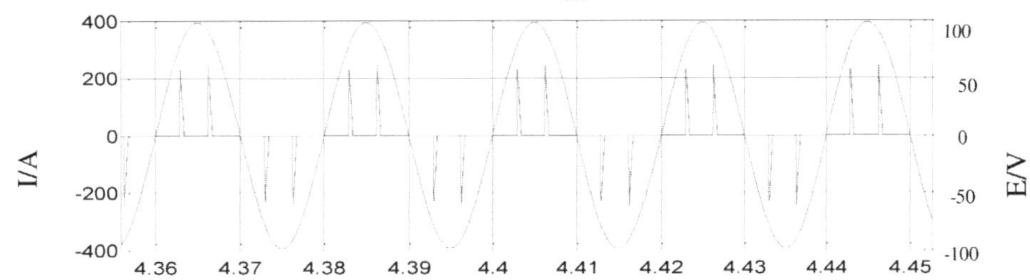

(c) current/voltage waveform on the gridside
t/s

Figure 2. BLDC waveform under the traditional PWM control

4. A new type of BLDC control system

4.1. The compitition of CSR and VSR

According to the difference of topological structure, rectifier could be divided into voltage source rectifier (VSR) and current source rectifier (CSR) [9]. If using 3-phase VSR, in order to realize voltage matching between the motors, general increase one level of BUCK circuit in the rectifier output. Compare with this, 3-phase CSR can directly control the DC current, and the output voltage from 0-1.5V_m is controllable. According to literature [10] calculation, When the rated output is 10.5Kw and below, from the point of system efficiency and power density, 3-phase CSR circuit structure has an advantage over 3-phase VSR, and VSR-BUCK is two stage topological structure, which more complex on the control system. In conclusion, choose current source rectifier as frequency converter rectifier part.

4.2. topological structure principle

Converter topological structure shown as Figure 3, add 3-phase current source rectifier to the front of the 3-phase inverter bridge. Different from the traditional inverter topology, using IGBT to replace the diode-bridge rectifier, thous to realized the controllable on recifier can greatly improve the power factor of netside. Because of use CSR, filter capacitance is added on the AC side, composite with the filter inductance on the netside become LC filter which can effectively suppressing The high frequency harmonic components on the netside current [11]. 3-phase inverter apply on 3-phase 6-state open, by this time the sampling current value on DC bus in every conduction zone, the impulse will not appear, could truely reflect the reality of the conduction phase current [12].

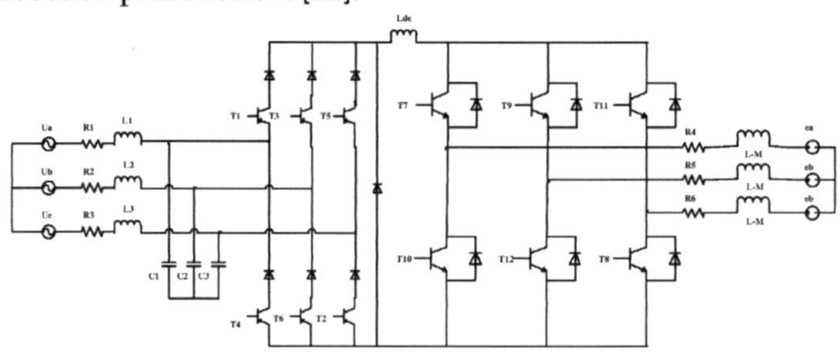

Figure 3. topological structure

4.3. The analysis of simulation result

Show the validity of this method by comparing the simulation result.Figure 4 is the current/voltage waveform on the grid side by using new BLDC control system. Figure 5(a) (b) is the motor single-phase current waveform and the electromagnetic torque waveform. From the figures show the system can realizes the operation of unity-power factor and the current on grid side sinusoidal. Compare with traditional PWM modulation show it has the advantage of low harmonic distortion and fine torque ripple suppression.

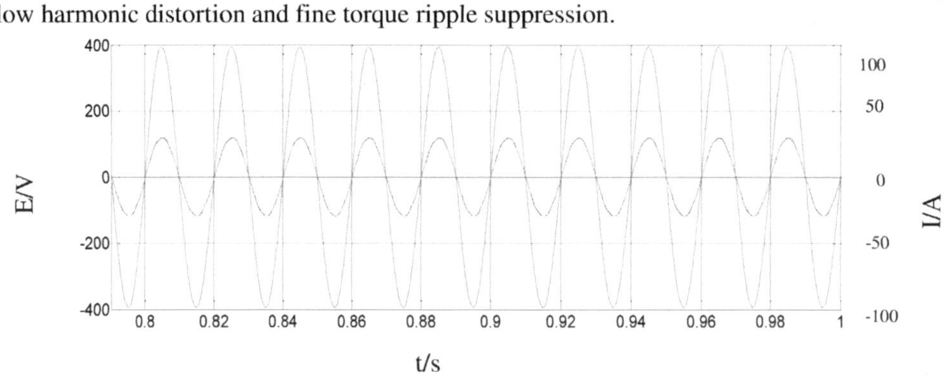

Figure 4. Current/Voltage waveform on the grid side

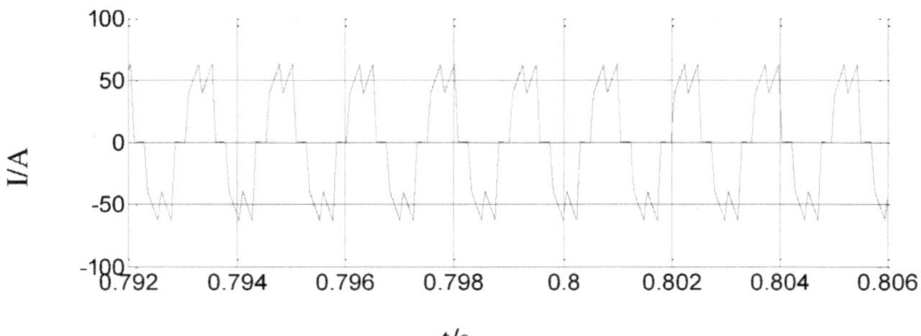

(a)

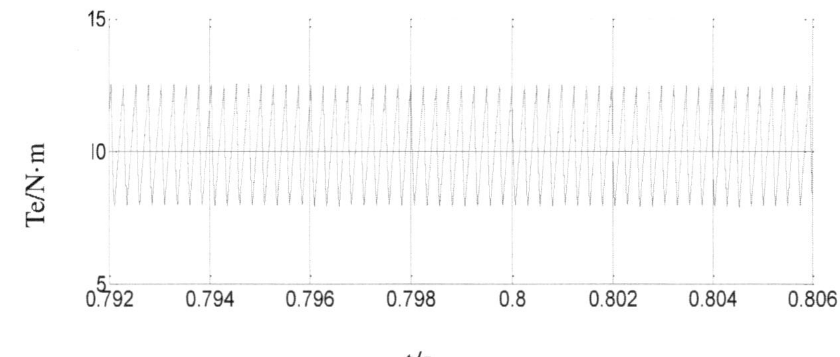

(b)

Figure 5. Single-phase current and electromagnetic torque waveform using new control system

5. Conclusion

On the basis of analyze the character of the high speed BLDC and traditional control method, improve the traditional control method and the topological structure, and made then appropriate simulation investigations. The result show when use the proposed topology can effectively improve the network side power factor and reduce the motor armature winding harmonic and motor torque ripple.

6. Acknowledgements

This research was financially supported by the phD Start-up Fund of Natural Science Foundation of Liaoning Province. China (20141072).

7. References

[1] Yue S and XiaoPeng l 2010 Simulation of Brushless DC Motor Control System Based on Matlab.*2010 3rd International Conference on Power Electronics and Intelligent Transportation System* 552
[2] Fangzheng G and Bangcheng H 2009 Research on suppression for diode freewheeling of inactive phase in high-speed BLDCM using PAM strategy.*small and special electrical machine* **42** 42
[3] Shuai L, Huaishi L, Wenyan L and Kefeng H 2012 Fuzzy-PI control algorithm for three-phase step-up/dowm PWM rectifier.*proceedings of the CSU-EPSA* **24** 121
[4] Yiongxiang P Juan D ShouZhi L and DaWei Z 2000 A novel control algorithm for Three-phase Voltage-fed PWM reversible rectifier.*Power Electroncial Technology* **6** 44
[5] Jenwei C 2010 New tactic of BLDCM simulation model based on simulink.*shipboard electronic contermeasure* **33** 112
[6] Zhaoan W and Jinjun L 2009 *Power electronics,addition 5th.*(China:China machine press) 162

[7] Haimei Y 2009 Analysis and design on high frequency inverter system. (China:Tianjing university) Chapter 2 12-17

[8] Hua J and Ping Y 2005 Quick modeling and simulation for permanent magnet brushless DC motor based on Matab/Simulink.*Journal of Yunnan University* **27** 151

[9] DaoRu Y 2007 Study of PWM Rectifier in Transducer.(China Hefei University of Technology) chapter 3 16-20

[10] Qiang G 2015 Research on control strategies and application techniques of current source PWM rectifiers.(China Chongqing university) chapter 2 13-26

[11] Bingxin Q 2012 *Development of current source PWM rectifier.*(China:North china electric power university) 12

[12] Que L,Zhiqiang W and Jiqiang T 2011 Simulation of control system for buck circuit of PAM of high speed and permanent magnet brushless DCmotor based on pspice.*aviation precision manufacturing technology* **47** 25

IOP Conf. Series: Earth and Environmental Science **63** (2017) 012019 doi:10.1088/1755-1315/63/1/012019

Enhanced Photodegradation of Paper-making Wastewater by Two-dimensional Heterojunction of Titania Plates with Exposed {001} Facets-Graphene Composite

Yiming Liu[1], Chunxiao Wang[1], Tingting Shen[1], Qing Liu[1], Jing Sun[*1] and Xikui Wang[*2]

[1]Department of Environmental Science and Engineering, Qilu University of Technology, Jinan 250353, China
[2]Chemistry and Pharmaceutical Engineering, Qilu University of Technology, Jinan 250353, China

[*]Jing Sun and Xikui Wang authors are corresponding authors contributed to the work.
jingsunok@163.com, xk_wang@qlu.edu.cn

Abstract. As a typical industrial wastewater, paper-making wastewater contains a number of toxic compounds which is harmful to the environment. In this report, TiO_2 plates/graphene composite based 2D–2D heterojunctions were synthesized via one-step solvothermal method. The structure and morphology of the as-prepared composite were characterized by Transmission electron microscope (TEM), X-ray diffraction (XRD), and Raman spectra. The results indicated that sheet-like anatase TiO_2 with exposed {001} facets was anchored on the graphene nanosheet. In the degradation of papermaking wastewater under UV irradiation, the as synthesized TiO_2 plates/graphene composite exhibited higher photocatalytic performance than the commercial Degussa P25. The effects of the amount of catalyst, H_2O_2 dosage, pH and the time of light irradiation on the photocatalytic efficiency were also discussed. At the optimum condition of 1.0 $g \cdot L^{-1}$ RGO/TiO_2 with the dosage of 0.3% H_2O_2 at pH=3, the removal rate of COD_{Cr} and chroma could reach 86% and 100%, respectively in 8 h. This work could provide new insights into the wastewater treatment with a well-defined nanocomposite.

1. Introduction

Semiconductor photocatalysis has been receiving extensive attention on account of its good application in application of solar energy and environmental pollutant treatment [1-3]. The potential of TiO_2 as a photocatalyst is well recognized because of its good biological non-toxic, low price and chemical stability [4]. It is very attractive for industrial water treatment and the degradation of Persistent Organic Pollutants. However, there are still two serious problems in pure TiO_2, which affect its photocatalytic efficiency in practical application. The first is that TiO_2 photocatalysts can only be excited under UV light due to its large band gap energy (3.2 eV); the second is the higher recombination rate of photogenerated electron-hole pairs, which seriously affected the photocatalytic reactions efficiency [5, 6].

Last several years, the photocatalytic activity of TiO_2 with exposed {001} facets has been extensively studied [3, 7, 8]. It is generally known that the performance of the photocatalytic materials was closely related to the crystal structure and crystallinity. It has been reported that anatase TiO_2

nanomaterials with exposed {001} facets showed higher photocatalytic performance than commercial Degussa P25, because of its higher surface energy. But simply changing the crystal structure of the catalyst to improve the catalytic properties is limited [9, 10].

The researchers adopted a variety of methods to improve the titanium dioxide photocatalysis. Carbon/TiO_2 composites have been proved that the photocalytic activity can be improved effectively [11, 12]. The heterojunctions between carbon materials and TiO_2 improved the separation efficiency of electron-hole pairs. Carbon doping can effectively transfer photogenerated electrons, which is an effective method to solve the recombination problem of photogenerated electron-hole pairs.

Compared with traditional carbon materials, graphene as a new material has been extensively concerned. Because of its excellent chemical stability, high electron transfer properties (2×10^5 cm^2/Vs) and adsorption capacity for organic pollutants [13]. It is a simple and effective method to improve the photocatalytic activity by combining graphene with conventional photocatalyst. It is reported that the composite materials of graphene and photocatalytic materials can not only reduce the composite of electron-hole pairs, but also make the absorption wavelength red shift of the composite material [14, 15]. Thus, the improvement of the interfacial charge transfer processes between graphene and TiO_2 is important both theoretically and in practical applications.

Papermaking wastewater is mainly produced by pulp and paper making industry [16]. A major feature of wastewater discharge is containing a number of toxic compounds, such as chlorinated lignins, dyes and organic sulfur compounds [17, 18]. These organics cannot be completely removed by traditional coagulation/flocculation and sedimentation process or anaerobic/aerobic reaction [16]. Once discharged, it may cause slime growth, color problems and an increasing environmental risk. Previous researchers demonstrated that advanced oxidation technologies, especially semiconductor photocatalysis have been successfully applied to the complete mineralization of refractory organics derived from the lignin degradation [19, 20]. Among them, TiO_2 has been extensively investigated because of its stability and high efficiency in oxidizing various pollutants in papermaking wastewater.

In this paper, a Graphene/TiO_2 system with 2D-2D heterojunction interface by a solvothermal reaction was constructed. The photocatalytic properties of composite photocatalyst materials were investigated by the degradation of papermaking wastewater under UV irradiation. The affecting factors of the synthesized materials were also discussed in detail.

2. Experimental

All materials were analyzed and without further purification. Graphite powder was purchased from Bodi Chemical Co. Ltd (Tianjin, China). Sulfuric acid(H_2SO_4), hydrochloric acid (HCl), hydrogen peroxide (30 wt%), potassium permanganate($KMnO_4$), ethanol(C_2H_6O) and tetrabutyl titanate were obtained from Sinopharm Chemical Reagent Co., Ltd. (Shanghai, China), Degussa P25 and hydrofluoric acid were purchased from Sigma-Aldrich (St.Louis, MO, USA). Paper-making wastewater was obtained from the Key Laboratory of Pulp and Paper Science & Technology of the Ministry of Education in Qilu University of Technology. Deionized water used in the experiment was prepared by pure water meter.

2.1. Synthesis of graphene oxide (GO)

Graphite oxide was prepared by modified Hummers' method [10]. In a typical reaction, 1.0 g of graphite and 1.0 g of $NaNO_3$ were added into 23.0 mL of H_2SO_4, stirred in the ice water bath. Then, 3.0 g of $KMnO_4$ was added into the above solution slowly. The mixed solution was sonicated for 5 h to form a thick paste. Then 46 mL of ultrapure water was added gradually and stirred for 10 min. Finally, another 140 mL of ultrapure water and 10 mL of H_2O_2 (30 %) was added into the mixture solution. The above solution was filtered and washed by 5% HCl solution and ultrapure water for several times. The final sediment was dried by vacuum freeze drying.

2.2. Synthesis of TiO_2 with exposed {001} facets

In a typical experimental procedure, 2.5 mL of HF (40 wt%) was added into 55 mL methanol solution under stirring. Then 8.5 mL of $Ti(OBu)_4$ was added dropwise. Finally, the viscous solution was transferred into a dried 100 mL polytetrafluoroethylene autoclave and kept at 180 ℃ for 24 h [3, 21]. After let the reactor natural cooling, open the reaction vessel to collect the samples and washed with ethanol and deionized water several times, then freeze-drying.

2.3. Synthesis of RGO/TiO_2 nanocomposites.

A certain amount of GO was mixed with methanol solution, ultrasonicated for 60 min to form a homogeneous GO (1 $g{\cdot}L^{-1}$) solution. Then, 8.5 mL of $Ti(OBu)_4$ and 2.5 mL of HF (40 wt%) were added into 55 mL the above solution slowly drop by drop under stirring. Continue stirring the above solution for 1 h. Finally, the mixed solution was transferred to a 100 mL polytetrafluoroethylene autoclave and kept at 180 ℃ for 24 h. After let the reactor natural cooling, open the reaction vessel to collect the gray-white samples and washed with ethanol and deionized water several times, then freeze-drying. With the same preparation method, a series of composite photocatalytic materials were prepared by changing the amount of GO, which are marked as 0.125%, 0.25%, 0.5%, and 1%, respectively.

2.4. Characterization

The samples structure was characterized by powder X-ray diffraction (XRD) using a scanning rate of $0.05\,°/s$ within a 2θ range of $10° - 80°$ (Bruker D8 Advance; $Cu\ K\alpha = 1.5404\ Å$). The microstructure of the samples was observed by high transmission electron microscopy (HRTEM, JEM-2100) with an acceleration voltage of 200 kV. The samples were tested by loading on the micro grid copper network.

2.5. Photocatalytic activity measurements

The photocatalytic activity was verified by photocatalytic of a certain concentration of papermaking wastewater in a photocatalytic reactor. The experiments were used the internal irradiation type of 300 W high-pressure mercury lamp as a light source. The light source was placed in the center of the quartz cold trap, and a 250 mL glass bottle was nested as a reactor to load reaction solution. The distance between the center of the light source and the inner wall of the reaction bottle was about 4 cm. The experiment was carried out by loading a magnetic stirrer at the bottom of the reactor to ensure that the reaction solution and the catalyst were mixed homogeneous during the experiment. The experiment by opening the flow of water to control the reaction solution temperature at 30 $\pm$ 0.5 ℃. 0.15 g photocatalyst was added into 150 mL of paper wastewater to form suspension and homogenized by sonication. Before the photocatalytic experiment, the mixed solution was stirred under the condition of no light for 30 min, in order to make the mixed solution reach the adsorption equilibrium. After adding a certain amount of hydrogen peroxide to the solution and starting the UV irradiation, and 10 mL of the mixed solution was sampled at regular intervals to measure the absorbance at 350 nm after filtration with a microporous filter. COD_{Cr} was detected by potassium dichromate method.

3. Results and discussion

3.1. Characterization of the TiO_2 nanosheet/graphene composite.

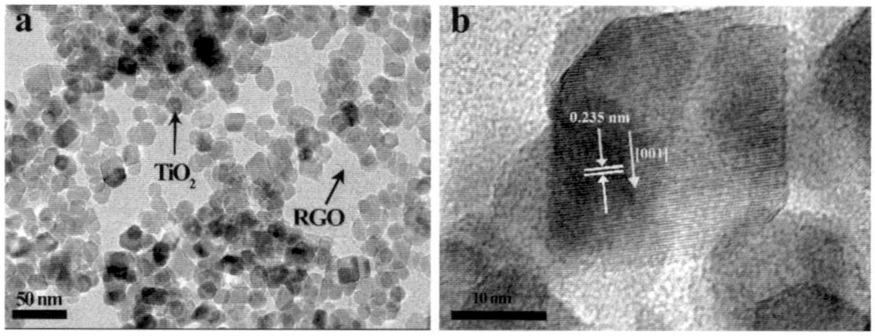

Figure 1. (a) TEM and (b) HRTEM images of 0.5% RGO/TiO_2.

Transmission electron microscopy (TEM) has been used to analyze the morphology and structure of 0.25% RGO/TiO_2. It can be clearly observed from the picture (Figure 1a) that RGO has a thin lamellar folds and exhibits 2D shape. It is seen obviously that 2D TiO_2 nanosheets evenly covering on the surface of RGO. According to the measurement of the lattice fringes of the HRTEM image (Figure 1b) shows that the lattice fringe width is about 0.235 nm, which proves that the sample is (001) exposed planes of anatase phase [9, 21]. These atomic planes prove that the TiO_2 crystals exhibit anatase facets exposed with {001}. It also further confirms the 2D-2D interface heterojunction between RGO and TiO_2 is successfully constructed.

In order to analyze the crystal shape and crystallinity of the prepared samples, the samples were scanned by XRD. XRD patterns of GO, pure TiO_2, 0.125% RGO/TiO_2, 0.25% RGO/TiO_2, 0.5% RGO/TiO_2 and 1% RGO/TiO_2 composite catalysts are shown in Figure 2. Notably, in Figure 2a, the peak at 2θ = 9.8° corresponds to the (002) reflection of stacked GO sheets. However, in the composites in Figure 2b, no diffraction peak at the same angle, confirming the graphene oxide has been completely reduced [22]. The distinctive peaks at 2θ=25.6°, 38.8°, 48.1°, 53.9°, 55.0°, 62.8°, 69.4°, 70.5° and 75.1° can be indexed to (101), (004), (200), (105), (211), (204), (116), (220) and (215) planes of pure anatase phase of TiO_2, respectively. It is obviously showed that the as prepared TiO_2 in the compositions was anatase phase (JCPDS card No: 73-1764). Sharpness of all the peaks indicates that TiO_2 is highly crystalline.

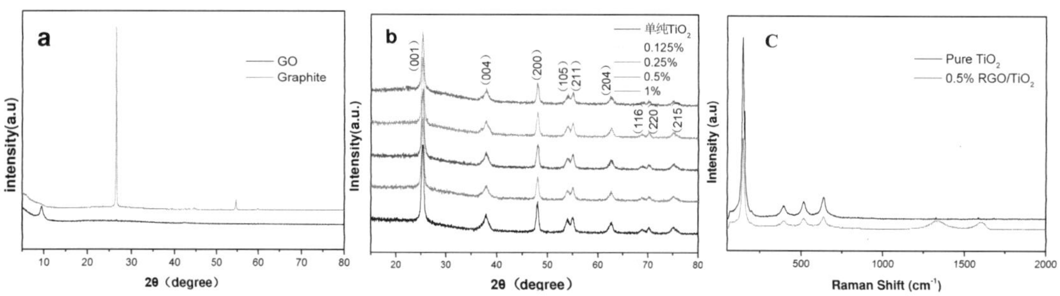

Figure 2. XRD patterns of (a) GO, Graphite, (b) TiO_2, 0.125% RGO/TiO_2, 0.25% RGO/TiO_2, 0.5% RGO/TiO_2, and 0.1% RGO/TiO_2 composite catalysts and (c) Raman spectra of TiO_2 nano-sheets and 0.5% RGO/ TiO_2 composi

The prepared catalysts were characterized by Raman spectroscopy (Figure 2c). The typical Raman bands at 153 cm^{-1}(E_g), 400 cm^{-1} (B_{1g}), 517 cm^{-1} (A_{1g}), and 640 cm^{-1} (E_g) further confirm the presence of anatase phase of TiO_2 in the composites. In addition, the D band centered at 1320 cm^{-1} (disorder band) and the G band at 1594 cm^{-1} (tangential vibration band) attributed to the graphene substrate are also present, which further proves that the catalyst issuccessfully combined with RGO [10]. The observed TEM pictures, XRD patterns and Raman spectra demonstrated that the sheet-like anatase TiO_2 with exposed {001} facets are successfully anchored on the graphene nanosheet.

3.2. Photocatalytic activity and reaction mechanism.

The photocatalytic activity of the catalyst was verified by photocatalytic degradation of paper-making wastewater. Degussa P25 and pure TiO_2 were examined as a reference for photocatalytic performance. Figure 3a displays the trend of COD_{cr} concentration of paper-making wastewater with the change of illumination time. The control experiment displayed that COD_{cr} were scarcely removed under UV light without photocatalyst or with pure H_2O_2. Upon UV irradiation, the highest photocatalytic activity was observed on 0.5% RGO/TiO_2 samples, and approximately 86% of COD_{cr} were degraded within 8 h. It is noted that the photocatalytic performance of titania with different G doping ratio is better than that of pure titania. It can be demonstrated that the 2D-2D heterojunction interface between graphene and TiO_2 nanosheet contributed significantly to the photocatalytic degradation of COD_{cr} due to the constructive to the carrier transfer. Moreover, TiO_2 nanosheets with exposed of {001} facets in papermaking wastewater showed a better effect than the commercial P25. Figure 3b displays the removal rate of chroma for three hours irradiation under different catalysts, consistent with the tendency of COD_{cr} elimination.

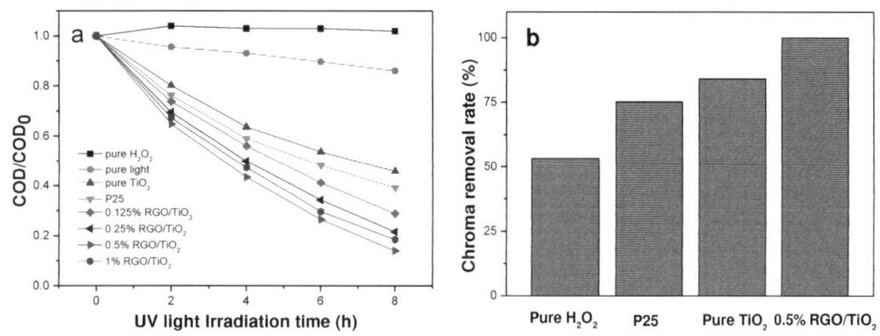

Figure 3. (a) Photocatalytic activities of different samples for paper-making wastewater degradation under UV light and (b) the removal rate of chroma for three hours irradiation.

The effects of the amount of catalyst, H_2O_2 dosage and pH on the photocatalytic degradation efficiency were studied in detail in Figure 4. The amount of the catalyst is closely related to the surface active site of the catalyst system, and the amount of the catalyst directly affects the optical transmittance of the reaction system. Although the number of active sites or the free radicals increases with the catalyst dosage, the penetration rate of light decreases and further influence COD_{cr} removal rate because of shielding effects. In Figure 4a, the optimal catalyst concentration is 1.0 $g \cdot L^{-1}$.

In Figure 4b, with the increase of H_2O_2 dosage, COD_{cr} removal rate also has a tendency to increase first and then decrease. The possible reason is the hydrogen peroxide can well capture electrons and then produce hydroxyl radicals. So adding hydrogen peroxide can enhance the oxidation ability of the system, and then improve the removal rate of COD_{cr}. However, when the hydrogen peroxide concentration continues to rise, the hydrogen peroxide will inhibit the adsorption of organic compounds on the surface of titanium dioxide, resulting in the reduction of COD_{cr} removal rate.

Generally, the pH of the suspension exhibits dissimilar effects on the COD_{Cr} removal rate of wastewater because of the different hydroxyl radical amount under various pH values [23]. At pH=3, the removal rate of COD_{cr} reached the highest level in Figure 4c. The removal rate of COD_{cr} is clearly lower at higher pH (pH>3) and lower pH (pH=2). The possible reason is that, in higher pH condition, alkaline solutions are disadvantageous at the lignin molecules adsorbing on the surface of TiO_2. Meanwhile, too strong acidity of the solution is not conducive to •OH production.

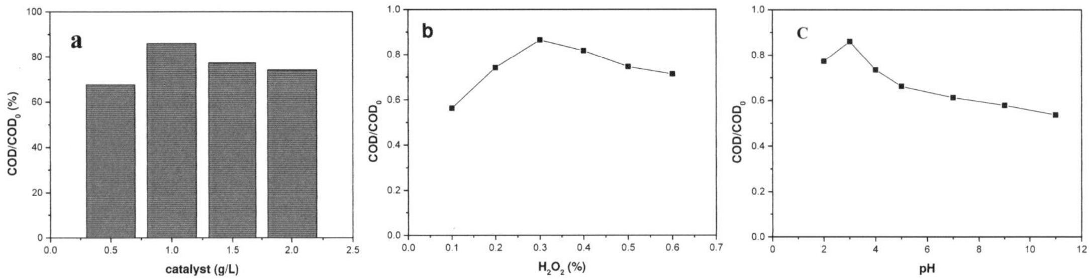

Figure 4. Effects of (a) catalyst dosage, (b) H_2O_2 and (c) pH on the removal rate of COD_{cr}.

4. Conclusions

2D-2D heterojunctions interface composed of anatase TiO_2 plates with exposed {001} facets and graphene nanosheets were constructed by a simple one-step solvothermal method. Due to the heterojunction interface nature of these hybrids, the as synthesized TiO_2 plates/graphene composite exhibited higher photocatalytic performance than the commercial Degussa P25 and pure TiO_2. The 2D-2D heterojunction system gives a new insight into the photocatalytic materials assembly. This work broadens the potential application of the photocatalytic system in wastewater treatment.

5. Acknowledgment

This work was supported by the National Nature Science Foundation of China (No. 21507067) and Open Fund of the Key Laboratory of Pulp and Paper Science & Technology of the Ministry of Education in Qilu University of Technology (No. KF201416).

6. References

[1] Li K, Chen T, Yan L, Dai Y, Huang Z, Xiong J, Song D, Lv Y and Zeng Z 2013 *Physicochem. Eng. Aspects* **422** 90-9
[2] Wang W, Lu C, Ni Y, Peng F and Xu Z 2013 *APPL SURF SCI* **265** 438-42
[3] Jiang B, Tian C, Pan Q, Jiang Z, Wang J, Yan W and Fu H 2011 *J. Phys. Chem. C* **115** 23718-25
[4] Sun J, Guo L, Zhang H and Zhao L 2014 *ENVIRON SCI TECHNOL* **48** 11962-8
[5] Luan X, Gutierrez Wing M T and Wang Y 2015 *MAT SCI SEMICON PROC* **30** 592-8
[6] Yu J, Low J, Xiao W, Zhou P and Jaroniec M 2014 *J AM CHEM SOC* **136** 8839-42
[7] Wang W, Wang D, Qu W, Lu L and Xu A 2012 *J. Phys. Chem. C* **116** 19893-901
[8] Gu L, Wang J, Cheng H, Zhao Y, Liu L and Han X 2013 *ACS APPL MATER INTER* **5** 3085-93
[9] Diak M, Grabowska E and Zaleska A 2015 *APPL SURF SCI* **347** 275-85
[10] Sun J, Zhang H, Guo L and Zhao L 2013 *ACS APPL MATER INTER* **5** 13035-41
[11] Kim H, Moon G, Monllor-Satoca D, Park Y and Choi W 2012 *J. Phys. Chem. C* **116** 1535-43
[12] Shi J, Ai H, Chen J, Cui H and Fu M 2014 *J COLLOID INTERF SCI* **430** 100-7
[13] Zhang Y, Fugane K, Mori T, Niu L and Ye J 2012 *J. Mater. Chem.* **22** 6575
[14] Zhang Y, Han C, Nadagouda M N and Dionysiou D D 2015 *Appl. Catal. B: Environ.* **168-169** 550-8
[15] Xu Y, Mo Y, Tian J, Wang P, Yu H and Yu J 2016 *Appl. Catal. B: Environ.* **181** 810-7
[16] Wang B, Gu L, Ma H 2007 *J. Hazard. Mater.* **143** 198-205.
[17] Zhang Q and Chuang K T 1998 *Ind. Eng. Chem. Res.* **37** 3343-49.
[18] Lacorte S, Latorre A, Barcel'o D., Rigol A, Malmqvist A and Welander T 2003 *Trends Anal. Chem.* **22** 725–37.
[19] Sun B, Zhou G W, Shao C W, Jiang B, Pang J L, Zhang Y 2014 *Powder Technology* **256** 118-25
[20] Marciana C, Geoffroy R, Malpass P, Artur M J, 2006 *Appl. Catal. B: Environ.* **62** 193-200.
[21] Han X, Kuang Q, Jin M, Xie Z and Zheng L 2009 *J AM CHEM SOC* **131** 3152-3
[22] Jiang B, Tian C, Pan Q, Jiang Z, Wang J Q, Yan W, Fu H 2011 *J. Phys. Chem. C* **115** 23718-25.
[23] Yang E L, Shi J J, Liang H C, Cheuk W K 2011 *Chem. Eng. J.* **174** 539-45.

Research on Influencing Factors of Biological Filtration Tower Treating Toluene Gas

Changping Zhang, Ziqing Cao, Yuqi Lu and Linggai Du

School of Energy and Environmental Engineering, Hebei University of Technology, Tianjin, China

zhchp@hebut.deu.cn ; sylviasuperme@163.com

Abstract. Through the orthogonal experimental design, the optimal combination of Triton X-100, nitrogen source, Fe^{2+}, temperature, concentration of antibiotics, pH and spray quantity was determined with surfactants, nitrogen and iron elements as additive, by which the key influencing factors were determined. In the test, the removal efficiency of the second groups was higher than that of the eighth groups, which were 89% and 87%, respectively. The best combination of a group of removal was as follows: nitrogen source concentration was 2 $g{\cdot}L^{-1}$, antibiotic concentration was 300 $U{\cdot}mL^{-1}$, the concentration of Triton X-100 was 0.05 $mL{\cdot}L^{-1}$, Fe^{2+} concentration was 14 $mL{\cdot}L^{-1}$, pH was 7, the temperature was 34℃, spray amount was 6 $L{\cdot}h^{-1}$.The antibiotic concentration was the most important factor on the removal efficiency of the toluene. The concentration of gas in each layer of toluene was detected; the curve of the outlet concentration in the optimal combination and the average state was obtained. The removal efficiency of the optimal combination was much better than the average, and it was found that the removal rate decreased with the increase of the height of the filling layer. The change of oxygen content in each layer was detected which was no significant change. It showed that oxygen was not the limiting factor of the removal of toluene by microorganisms.
Keywords: surfactants; biological filtration tower; toluene; orthogonal test

1. Introduction

Volatile Organic Compounds (VOCs) are a class of chemical compounds which are extremely complex. They are not only easily lead to photochemical smog which is harm to human and plant health, but also, after entering the atmosphere, they are participate in a series of chemical reactions to produce ozone and destroy the atmosphere, thus affect the global climate environment. Toluene is the most common volatile organic compounds, belonging to the benzene series, and is a very important chemical raw material. Benzene has been identified as a strong carcinogen by the World Health Organization (WHO) [1], and long-term exposure may cause anemia or leukemia [2-3].

There are many kinds of methods to be used to remove or reduce VOCs levels, the conventional physical and chemical methods include: catalytic incineration, adsorption, absorption, condensation, etc [4-5]. Among the new methods such as non-equilibrium plasma method, photo catalytic method and biological method, biological method because of its good effect, simple facilities, low cost, environmental protection, easy to manage and so on, it has been paid more and more attention by scholars both at home and abroad. With the characteristics of strong and fast adaptation to the pollutants, biological method domesticates the microorganisms. Pollutants are used as microorganisms' carbon source and energy, and finally been transmuted into harmless substances (H_2O, CO_2, etc.).

The process of biological treatment of volatile organic compounds mainly has three kinds of methods: biological washing, biological filtration and biological trickle filtration. This study mainly aimed at the biological trickle filtration method in which the gas is from the bottom of the tower. In the process of flow, gas is purified by contacts the biological membrane. The recycled spraying cleaning solution enters the tower from the top of the packing layer, then sedimentation at the bottom of the tower through the biological membrane.

In this study, the activated carbon was used as filler together with the domesticated strain in the treatment of toluene. Using of additives to improve the water solubility of hydrophobic organic compounds and microbial activity. Therefore, we can provide scientific basis for the application of the biological filter in the future.

2. Device and method

2.1. Experimental installation

The biological trickle filtration is made of transparent organic glass and in the type of sectional detachable. Tower diameter is 120mm the height of each layer is 150mm, and total of 4 layers. The activated carbon was used as filler. The experiment uses gas-liquid countercurrent operation so the nutrient solution flows from the top of the tower and flows out from the bottom of the tower. The waste gas, on the contrary, flows from the bottom to the top.

2.2. Experimental method

2.2.1. Detection method. Toluene detected by People's Republic Of China environmental protection standard of "Ambient air—Determination of benzene and its analogies by activated charcoal adsorption carbon disulfide desorption and gas chromatography".

The Bruke 400-GC gas chromatograph was used with BR-SWax (Carbowax® polyethylene glycol) (30 m×0.32 mm×0.5 μm) capillary column, injection port temperature was 150℃, split ratio 1:100; kept the temperature of column heater at 50℃ for six minutes, then heated it at a rate of 25℃•min-1 up to 125℃ and kept 3 minutes at this temperature; the flow rate of column was kept steady at 0.70 mL•min^{-1}; the detector temperature was 270℃ and the sensitivity of the detector was 12. Nitrogen flow rate: 25 mL•min^{-1}. Hydrogen flow rate: 30 mL•min^{-1}. Air flow rate: 300 mL•min^{-1}.

2.2.2. Additive selection. Surface active agent was chosen, nitrogen element and iron element as additives. Because the surface active agent has the hydrophilic base and the oil base, so a small quantity of it can reduce the mass transfer resistance between gas and liquid. As the result, the solubility of hydrophobic gas such as toluene in aqueous solution will be increased [6-7].

The non-ionic surfactant named Triton X-100 was selected as a solubilizer in order to increase the contact of toluene and microorganisms which was advantageous to improve the efficiency of removing [8]. At the same time, the study of Berti had discovered that by adding 1% of methyl oleate into Triton X-100 can form the oil-in-water emulsion and capacity of micellar solution [9].The emulsion increased the gas-liquid contact area of organic phase [10]. So the effects would be more obvious if we add 1% of methyl oleate.

Trace metal ions was the activator or active group of enzyme [11-13]. Nitrogen and Fe^{2+} were used to improve activity of microorganisms. Carbon source and nitrogen source were essential nutrient elements for microbial growth, so microbes absorbed carbon source by capturing toluene in the liquid and improved the biological activity by adding external nitrogen source. After Fe^{2+} dissolving into water, hydrated ions with water molecules such as $Fe(H_2O)_2Cl^-$ was formed. When they met the surface active agent in the water, the metal ion chelate may be formed, which can accelerate the dissolution of organic matter [14]. At present, there are few studies about the effect of adding Fe^{2+}, so the concentration research of Fe^{2+} had innovative.

2.2.3. Experiment design scheme. Choose L18 (3^7) orthogonal table which used orthogonal scientific to arrange experimental scheme with multi-level and multi-factor as a design table [15]. The suitable combination of various factors was found out through the analysis and treatment of the experimental results [16].

In addition to choose the above three kinds of additives, we choose the antibiotic concentration, temperature, pH, spray volume to explore the main factors that affect the efficiency of removal. Xu Lei's study discovered that when degrading toluene's concentration lower than 100 $mg \cdot m^{-3}$, fungus was about 20% higher than bactria [17].Add antibiotics to inhibit the growth of bacteria could benefit the growing of fungi. Suitable temperature, pH, spray volume could improve the efficiency of microbial degradation as well. Through the combination of various factors at different levels, we can find out the group with highest removal efficiency as the best combination.

The experimental designed with this conditions: the inlet concentration of toluene was 1000 $mg \cdot m^{-3}$, nitrogen source concentration was 2~6 $g \cdot L^{-1}$, antibiotic concentration was0~600 $U \cdot mL^{-1}$, Triton X-100 concentration was 0.02~0.08$mL \cdot L^{-1}$, Fe^{2+} concentration was 8~20 $mL \cdot L^{-1}$, pH was 5.5~8.5, temperature was 29~39℃, spray rate was 2~10 $L \cdot h^{-1}$. The air intake method was countercurrent. The experimental scheme was designed according to the table 1.

Table 1. Orthogonal experimental design

Group	A	B	C	D	E	F	G
1	2	0	0.02	8	5.5	29	2
2	2	300	0.05	14	7	34	6
3	2	600	0.08	20	8.5	39	10
4	4	0	0.02	14	7	39	10
5	4	300	0.05	20	8.5	29	2
6	4	600	0.08	8	5.5	34	6
7	6	0	0.05	8	8.5	34	10
8	6	300	0.08	14	5.5	39	2
9	6	600	0.02	20	7	29	6
10	2	0	0.08	20	7	34	2
11	2	300	0.02	8	8.5	39	6
12	2	600	0.05	14	5.5	29	10
13	4	0	0.05	20	5.5	39	6
14	4	300	0.08	8	7	29	10
15	4	600	0.02	14	8.5	34	2
16	6	0	0.08	14	8.5	29	6
17	6	300	0.02	20	5.5	34	10
18	6	600	0.05	8	7	39	2

A—Nitrogen/g•L-1
B—Antibiotic concentration/U•mL-1
C—Surface active agent/mL•L-1
D—Fe2+/mL•L-1
E—pH
F—Temperature/℃
G—Spray volume/L•h-1

3. Experimental results and analysis

The text of your paper should be formatted as follows:

The experimental scheme was designed according to the table 1; the results obtained were shown in Figure 1 and table 2.

From the Figure 1, in the design of orthogonal experiment, the removal efficiencies of the second and the eighth groups were 89% and 87%. Therefore, the conclusion was as follows.

Condition 1: nitrogen source concentration was 2g·L^{-1}, antibiotic concentration was 300 U·mL^{-1}Triton X-100 concentration was 0.05mL·L^{-1}, Fe^{2+} concentration was 14 mL·L^{-1}, pH was 7.0, temperature was 34℃, spray rate was 6 L·h^{-1}.

Condition 2: nitrogen source concentration was 6g·L^{-1}, antibiotic concentration was 300 U·mL^{-1}Triton X-100 concentration was 0.08mL·L^{-1}, Fe^{2+} concentration was 14 mL·L^{-1}, pH was 5.5, temperature was 39℃, spray rate was 2 L·h^{-1}

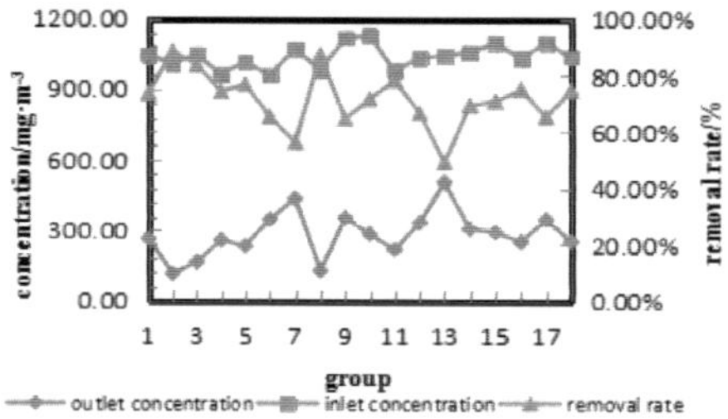

Figure 1. Operating results of the biological filtration tower

Table 2. Removal efficiency of orthogonal experiment

Group	Removal rate/%	Group	Removal rate/%
1	73.60	10	71.59
2	88.59	11	78.23
3	83.71	12	66.64
4	74.20	13	49.26
5	76.73	14	69.41
6	65.37	15	70.77
7	56.28	16	74.99
8	87.27	17	65.31
9	64.58	18	74.72

Through the date analysis the results of the removal rates in table 2, it can be concluded that the range of each factor called R. The bigger the range of substances, the bigger the effect for toluene removing, on the contrary, the smaller the range of substance, the smaller the effect for toluene removing[18].

Methods of computation: the test date obtained from the first group were added together and average the results, named the summation as K_i, the average as $k_i = \frac{K_i}{6}$, R_i was the range of the i factor. The calculated mode is:

$$R_i = \max[k_{i1}, k_{i2}, k_{i3}] - \min\,[k_{i1}, k_{i2}, k_{i3}] \qquad (1)$$

Take the nitrogen source as an example, after into the above formula, we can draw a conclusion:

$$K_{A1} = y_1 + y_2 + y_3 + y_{10} + y_{11} + y_{12}$$

$$= 0.7360 + 0.8859 + 0.8371 + 0.7159 + 0.6664 = 4.6236$$

$$k_{A1} = \frac{K_{A1}}{6} = 0.771 \quad (2)$$

In the same ways, $k_{A2} = 0.676\,, k_{A3} = 0.705$

$$R_A = \max[k_{A1}, k_{A2}, k_{A3}] - \min[k_{A1}, k_{A2}, k_{A3}]$$

$$= \max[0.771\,, 0.676\,, 0.705] - \min[0.771\,, 0.676\,, 0.705] = 0.095 \quad (3)$$

The calculation results are shown in Table 3.

Table 3. Experimental data and calculation analysis table

Number	A	B	C	D	E	F	G
k_1	0.771	0.667	0.711	0.696	0.679	0.710	0.758
k_2	0.676	0.776	0.687	0.771	0.738	0.697	0.702
k_3	0.705	0.710	0.754	0.685	0.735	0.746	0.693
R	0.095	0.109	0.067	0.086	0.059	0.049	0.065

A —Nitrogen/g•L-1
B—Antibiotic concentration/U•mL-1
C—Surface activeagent/mL•L-1
D—Fe^{2+}/mL•L-1
E—pH
F—Temperature/℃
G—Spray volume/L•h-1

According to the range R from table, we could identify the primary and secondary order of the factors affecting the removal efficiency of toluene:

B-A-D-C-G-E-F

The order of influencing factors on removal efficiency of toluene is: antibiotic concentration> nitrogen>Fe^{2+}> surface active agent> spray volume> pH> temperature.

The concentration of toluene in each layer of the outlet was detected through the sampling port, and then we could obtain the concentration of toluene in the optimal combination and the average state of all combinations. The concentration of toluene in each layer of the filter is shown in Figure 2. Removal efficiency $= \frac{iletr\ concentration - outlet\ concentration}{inlet\ concentration}$, the removal of toluene in the unit time of tower was called removal rate. Removal quantity was the main factor affecting the removal efficiency because the structure and size of each layer were the same and the gas in each layer had the same residence time. The removal quantity and removal rate of each layer of the tower are shown in Table 3.

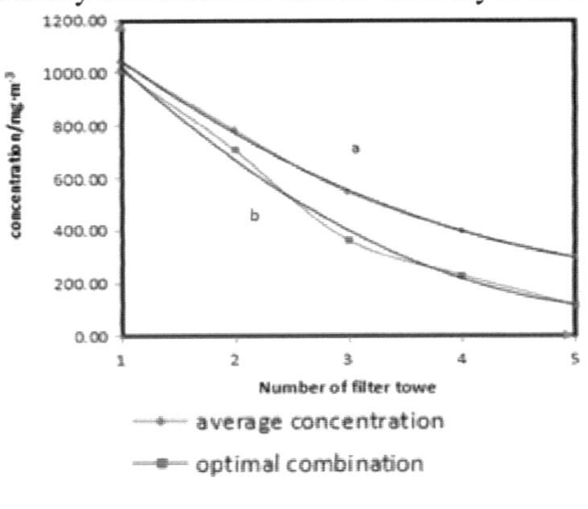

Figure 2. Concentration change of each layer in the biological filter tower

a—average combined polynomial
b—optimal combined polynomial

Two regression curves could be obtained by fitting the curve in Figure 2. Using regression calculation, regression curve equation for the average concentration of all combinations and the number of layer of the filter tower could be obtained as:

$$y = 20.008x^2 - 361.64x + 1377.7 \quad (4)$$

The curve equation between the optimal combination concentration and the number of layers is:

$$y = 41.849x^2 - 477.68x + 1457.5 \quad (5)$$

Table 4. Removal rate and removal rate of each layer

Name / Number	Average removal	Optimal combined removal	Average removal rate/%	Optimal combination removal rate/%
1	259.07	302.46	24.89	29.97
2	236.69	343.39	30.28	48.58
3	149.50	134.16	27.43	36.92
4	99.62	114.13	25.1	49.79

The fitting coefficient of regression curve equation R^2 were 0.9994 and 0.9942 respectively which were close to 1. Conclusions can be drawn from Table 4. Along with the microorganisms remove the toluene from waste gas; the toluene partial pressure in the exhaust gas was getting lower and lower, so as it was in the liquid phase, as the result, removal quantity of toluene decreased with the reaction. It was believed that this is due to the resistance of hydration film in the process of toluene in gas-liquid mass transfer. The average removal rate of each layer was relatively stable.

There was no significant change in the oxygen content of each layer, so it could be determined that oxygen will not be the limiting factor for the removal of toluene by microorganisms.

4. Conclusions

1) In the design of orthogonal experiment, the removal efficiency was significantly different among the combinations, the removal efficiencies of the second and the eighth groups were 89% and 87%.The optimal condition was: nitrogen source concentration was 2g•L^{-1}, antibiotic concentration was 300 U•mL^{-1}Triton X-100 concentration was 0.05mL•L^{-1}, Fe^{2+} concentration was 14 mL•L^{-1}, pH was 7.0, temperature was 34℃, spray rate was 6 L•h^{-1}.

2) According to the range R from table, we could identify the primary and secondary order of the factors affecting the removal efficiency of toluene: antibiotic concentration> nitrogen>Fe^{2+}> surface active agent> spray volume> pH> temperature. The effect of antibiotic concentration on the removal efficiency was largest.

3) Fitting curve of the concentration of toluene in the optimal combination and the average state of all combinations was obtained, removal quantity of toluene decreased with the increase of the height of the layer. There was no significant change in the oxygen content of each layer, so it could be determined that oxygen will not be the limiting factor for the removal of toluene by microorganisms.

4) When the inlet toluene concentration was 1000 mg•m^{-3}, the outlet concentration was about 400 mg•m^{-3}, which had not reached the National emission standard (below 60 mg•m^{-3}). Therefore, it can be connected to two stage treatment equipment in the treatment of high concentration of waste gas in order to meet the requirements of emission standards.

5. Acknowledgement

This work is supported by Science and Technology Research Project of Hebei Province, grant 402114.

6. References

[1] Li L，Liu J X. Removal of xylene from off-gas using a bioreactor containing bacteria and fungi[J]. *International bio deterioration& biodegradation*, 2006, 58(2): 60-64.

[2] Mathur A K，Majumder C B，Chatterjee S. Combined removal of BTEX in air stream by using mixture of sugar cane bagasse，compost and GAC as bio filter media[J]. *Journal of Hazardous Materials*,2007,148(1-2):64–74.

[3] Maliyekkal S M, Rene E R, Philip L,et al. Performance of BTX degraders under substrate versatility conditions[J]. *Journal of Hazardous Materials*, 2004, 109(1-3): 201-211.

[4] Devinny J, Deshusses M, Webster T. Bio filtration for Air Pollution Control[M]. *Washington: Lewis Publishers*, 1999.

[5] Zhu S Y. Environmental and industrial gas purification technology [M]. *Beijing: Chemical Industry Press*, 2001.

[6] Yang C P, Chen F Y, Luo S L, et al. Effects of surfactants and salt on Henry are constant of n-hexane. *Journal of Hazardous Materials*, 2010, 175(1-3): 187-192.

[7] Zhou X X, Removal of xylene from waste gases by strengthen bio-trickling filter. [D]. *Zhejiang University*. 2012.

[8] Hassan A A, Sorial G A.A comparative study for destruction of n-hexane in trickle bed air biofilters[J].*Chen Eng J*, 2010, 162(1): 227-233.

[9] Wang Z L. Extractive microbial transformation [M]. *Beijing: Chemical Industry Press*, 2012: 71-72.

[10] Shahna F G，Golbabaei F，Hamedi J，et al.Treatment of benzene,toluene and xylene contaminated air in a bioactive foam emµLsion reactor[J].*Chinese Journal of Chemical Engineering*, 2010, 18(1): 113-121.

[11] Meng X Z, Lai Z H, Long T R. Effects of mental ions on aerobic activated sludge activity [J]. *Journal of Safety and Environment*, 2004, 4(6):43-45.

[12] Wang X M, Yan Z C, Shi J. A review effect of the mental ions on activated sludge microbe [J]. *Petrochemical Industry Application*, 2008, 27(2): 1-3.

[13] Kou M X, Liu Q Y, Researching process of the influence of metallic ion on active sludge microbe [J], *Shanxi Architecture*, 2007, 33(5): 176-177.

[14] Xiao K J, Dai J L, He Q, et al. Characterization of Iron Nanoparticles Prepared with Non-ionicSurfactants [J], *Modern Food Science and Technology*, 2015, 31(9): 138-144.

[15] Yan L N, Li M, Chen L X. Application of Orthogonal Design Method in the Biochemistry Design Experimental Teaching [J], *Guangzhou Chemical Industry*, 2013, 41(21): 156-158.

[16] Ren L Q, Optimization design and analysis of experiment [M], *Beijing: Higher Education Press*, 2003: 10-19.

[17] Liu Q, Studies of the VOCs' purification in tricking bio filters [D], *Xi'an University Of Architecture And Technology*. 2007.

[18] Li Y Y, Hu C R, Experiment design and date processing [M], *Chemical Industry Press*, 2008: 128-132.

Numerical Simulation on the Combustion Characteristic of GO_2/GCH_4 Splash Platelet Injector

Yin Liang and Liu Weiqiang

Science and Technology on Scramjet Laboratory, National University of Defense Technology, Changsha, Hunan 410073, People's Republic of China

yl88222@126.com

Abstract. The platelet injector flow is quite different from other injectors due to the platelet technology and complex injection method. This work is aimed at assessing the effect of characteristic chamber length and middle-plate thickness for single-element splash platelet injector. Numerical simulation was adopted to investigate the gaseous methane and gaseous oxygen flow and combustion characteristics. Both characteristic velocity and combustion efficiency are examined for injector design. The results indicate that, with a 0.6mm middle-plate thickness the injector can reach the maximum combustion efficiency and shortest combustion length, but the differences are not obvious for all cases. The characteristic chamber length can be shortened effectively by using splash platelet injector.

1. Introduction

Platelet technology has been identified as a promising technology since 1964, which provides solutions to difficult thermal and fluid flow problems [1]. It has been used to produce heat exchangers, hypersonic lending edge, platelet injector, platelet mixer, micro structured reactor, and other applications [2-5]. Platelet injector is a new type of liquid rocket injector designed by Aerojet in the 1970s. It was widely used in low thrust attitude control engine because of several unique features including excellent response, simple structure, potential high performance with reduced chamber lengths [6,7].

Over the past several decades, considerable efforts have been dedicated to model in platelet injector to understand and predict the flow, mixing and combustion characteristic, many experiences have been accumulated in experiment and technology. Platelet injector exhibits many differences from traditional injector such as atomization characteristic, flow characteristic, design method, combustion stability, and the performance analysis of platelet injector is only limited to experimental analysis, the theoretical analysis is still in blank.

The effect of main configuration parameters, such as spray angle and relative thickness, expending angle have been studied, flow coefficient and mixing efficiency factor were obtained mainly by experiments [8]. Six fundamentally different injection elements were conceived and tested by Aerojet Liquid Rocket Company (ALRC), as a result of the unielement testing; splash platelet injector has proved to be the most promising candidate element with a higher performance, excellent compatibility, and insensitivity to propellant temperature [9,10]. Fanpei Lei et al. investigated the atomizing and mixing characteristics of splash platelet injector with a typical triangular lumen, the results showed that, flow coefficient and mixing efficiency factor increase firstly and then decrease with the increase

IOP Conf. Series: Earth and Environmental Science **63** (2017) 012021 doi:10.1088/1755-1315/63/1/012021

of expending angle [11]. Zhihui Guo et al. analyzed the atomization process and special spray feature of splash platelet injector using Particle Dynamics Analyzer (PDA) by experimental study [12]. These series of experiment results suggest that the injector geometry could be optimized in order to achieve a good mixing and combustion characteristics. Numerical simulation holds great, largely unfulfilled promise to be able to meet these requirements. This study is focused on obtaining baseline performance statistics of a gas-gas splash platelet injector which will be a reference for the future injector designs.

2. Numerical methods and Models

2.1. Numerical methods

2.1.1. Governing Equations Time iterative methods are employed for solving the gas-gas combustion field; the governing equations were discretized by control volume integration method.

$$\frac{d}{dt}(\rho\phi V)_{P_0} + a_0\phi_{P_0} = \sum_{j=1}^{N} a_j\phi_{P_j} + b_0 \tag{1}$$

For the steady-state process, $\frac{d}{dt}(\rho\phi V)_{P_0} = 0$. Here, P_0 is the node of the control volume, P_j is peripheral node, V is control volume, j is subscript of the interface of the control volume, N the number of interface, a_0 and a_j are the operators coefficients for P_0 and P_j, b_0 is the source term of the discrete equation.

2.1.2. Turbulence Model A compressible $\kappa-\varepsilon$ was used for the turbulence model. The turbulence kinetic energy, κ, and it rate of dissipation, ε, are obtained from the following transport equations:

$$\frac{\partial(\rho k)}{\partial t} + \frac{\partial(\rho k u_i)}{\partial x_i} = \frac{\partial}{\partial x_j}\left[\left(\mu + \frac{\mu_t}{\sigma_k}\right)\frac{\partial k}{\partial x_j}\right] + G_k - \rho\varepsilon \tag{2}$$

$$\frac{\partial(\rho \varepsilon)}{\partial t} + \frac{\partial(\rho \varepsilon u_i)}{\partial x_i} = \frac{\partial}{\partial x_j}\left[\left(\mu + \frac{\mu_t}{\sigma_\varepsilon}\right)\frac{\partial \varepsilon}{\partial x_j}\right] + C_{1\varepsilon}\frac{\varepsilon}{k}G_k - C_{2\varepsilon}\rho\frac{\varepsilon^2}{k} \tag{3}$$

The turbulent viscosity, μ_t, is computed by combining κ and ε as follows:

$$\mu_t = \rho C_\mu \frac{\kappa^2}{\varepsilon} \tag{4}$$

where, $C_{1\varepsilon} = 1.44$, $C_{2\varepsilon} = 1.44$, $C_\mu = 0.09$, $\sigma_\kappa = 1.0$, $\sigma_\varepsilon = 1.3$

2.1.3 Chemical kinetics model. The CFD code used for calculating the combusting flow field solves the fluid dynamic N-S equations incorporating a compressible $\kappa-\varepsilon$ turbulence model, the chemical kinetics model was calculated by a 9-specie, 1-reaction GO_2/GCH_4, which can be seen as follow. The interaction between turbulence and reaction was considered by EDC model [13,14].

$$CH_4 + 1.52O_2 \rightarrow 0.33CO_2 + 0.662CO + 1.454H_2O + 0.37H_2 + 0.134H + 0.216OH + 0.047O$$

IOP Conf. Series: Earth and Environmental Science **63** (2017) 012021 doi:10.1088/1755-1315/63/1/012021

2.2 Geometric Model

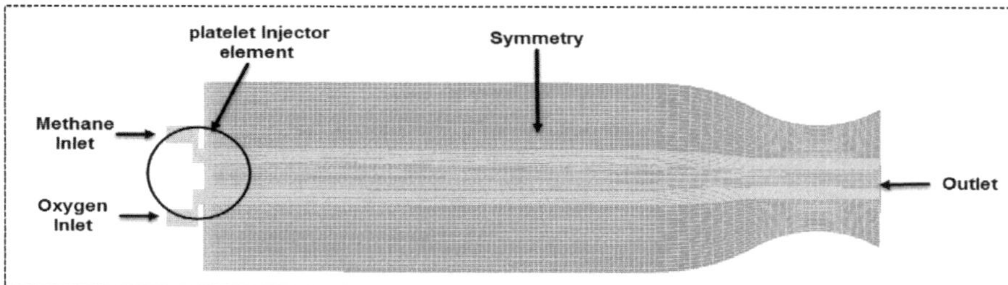

Figure 1. Grid of the calculation domain

Figure 1 shows the grid of the calculation domain, which contains domains of the platelet injector element, symmetry plane, combustion chamber, propellant inlets and pressure outlet. The inlets of methane and oxygen were both fixed mass flow rate, and the inlet turbulence intensities were both set to be 5%. The inlet temperatures are taken as 300 K. The nozzle exit was specified as a supersonic outlet, adiabatic nonslip wall boundaries are enforced at the chamber walls [15, 16]. The parameters of the splash platelet injector and the chamber can be seen in table 1.

Table 1. The parameters of chamber

Parameters	Values
Chamber pressure/MPa	0.85
Mixture Ratio(O/F)	3.2
Mass flow rate /(g/s)	5.5
Inlet gas temperature/K	298.15
Chamber diameter /mm	8.6
Throat diameter /mm	3.84
Theoretical characteristic velocity /m/s	1833.9
Ground Specific impulse /m/s	2270.2

Figure 2 shows the schematic diagram of the splash platelet injector. The key parameters which may control the flow and combustion process may contain: oxidizer orifice (D_o), fuel orifice (D_f), nozzle space (D), expending angle (γ), length of trapezoidal orifice (L_1), thickness of middle-plate ($H2$) and faceplate ($H3$). In this work the length of triangular lumen (L) was fixed to be 1.5mm. Individual middle platelet thickness can range from as thin as 0.3 mm to 1.2 mm. The detail values can be seen in table 2.

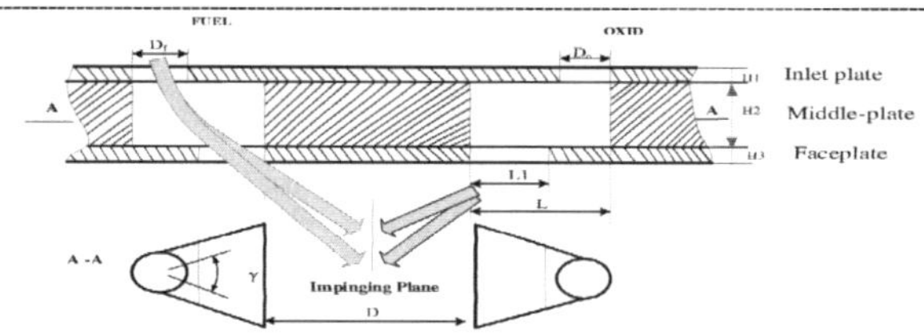

Figure 2. Schematic diagram of splash platelet injector

Table 2. Design parameters of the splash platelet element

Parameters	H_3/mm	D_O/mm	D_f/mm	D/mm	L/mm	L_1/mm	γ
Values	0.3	1	1	1.2	1.5	0.75	60°

3. Validation of numerical methods

Table 3. Design parameters of the splash platelet element

Propellant	$\dot{m}$/(g/s)	MR	P_C/MPa	D_C/mm	D_t/mm	L^*/mm	R_v
GCH4/GO2	5.5	3.5	3	26	14.8	1180	1.77

To demonstrate the validity of numerical methods and chemical kinetics model, simulation was performed on a shear-coaxial GO_2/GCH_4 injector, where P_C is the chamber pressure, MR is mixture ratio, D_c is chamber diameter, D_t is throat diameter, L^* is the characteristic length, R_v is the velocity ratio. Parameters of the shear-coaxial injector are listed in table 3, and the detail design values can get from Ref. [17]. Figure 3 shows the temperature distribution of the oxidizer post, which is consistent with the Ref. data; it can exactly describe the temperature distribution although there is little difference. This testified the accuracy of numerical methods and chemical kinetics model. Detailed chemical reaction model should be adopted in the future works.

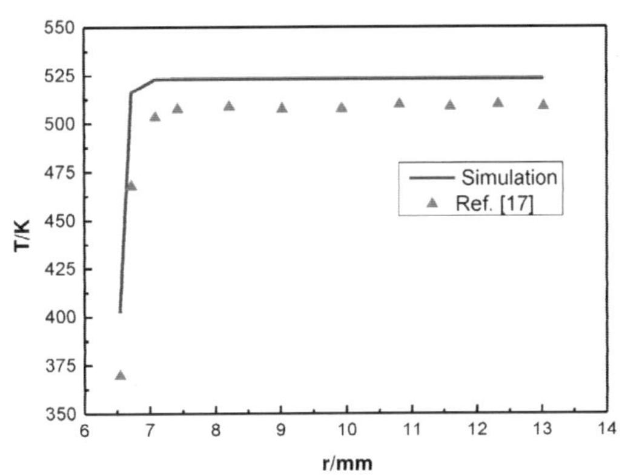

Figure 3. Temperature distribution of the oxidizer post

4. Results and discussion

4.1 Effect of the characteristic chamber length on characteristic velocity

Characteristic chamber length has an important effect on combustion efficiency. With the characteristic chamber length increasing, the volume and weight of the thrust chamber will be increased, and the surface area and thermal resistance will also be increased. For defining the GO_2/GCH_4 platelet injector thrust chamber length, the design methodology of traditional coaxial injector mostly used cannot be directly applied to platelet injector. For the GO_2/GCH_4 platelet injector, there is no mature engine type, so it is always rely on the similar experience of propellant and engine size. Four different characteristic chamber lengths L^* with 300mm, 400mm, 500mm, 600mm were simulated by Hongyan et al. These results show that when $L^* = 600mm$, which offer the maximum combustion efficiency and minimum heat load of the chamber wall. Gao Yushan el al. studied the effect of characteristic chamber length of a coaxial injector, the results show that under the same condition the characteristic chamber length of GO_2/GCH_4 is about 1.5 times that of GO_2/GH_2 [17].

Splash platelet injector is the most commonly used injectors, the splash injector involved mechanical atomization of the two propellants by means of direct impingement on the lip of the injection cup and subsequent impingement of the resulting unlike propellant fans [18]. The reason we select the splash platelet injector including ease of manufacture and high performance with reduced chamber length. In this paper, characteristic chamber length L*=300mm, 400mm, 500mm, 600mm, 700mm have been chosen for the simulation.

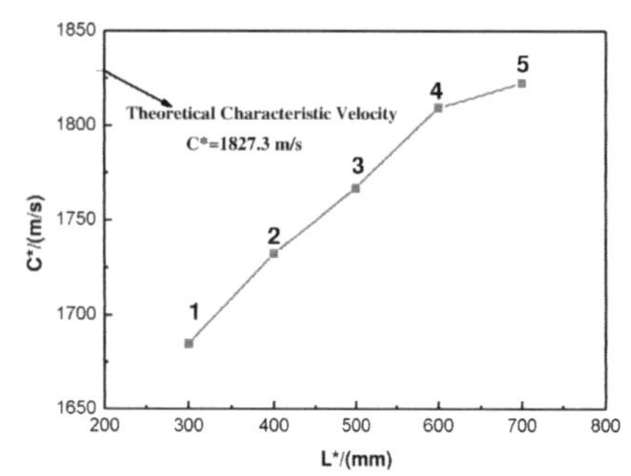

Figure 4. Characteristic velocity vs characteristic chamber length

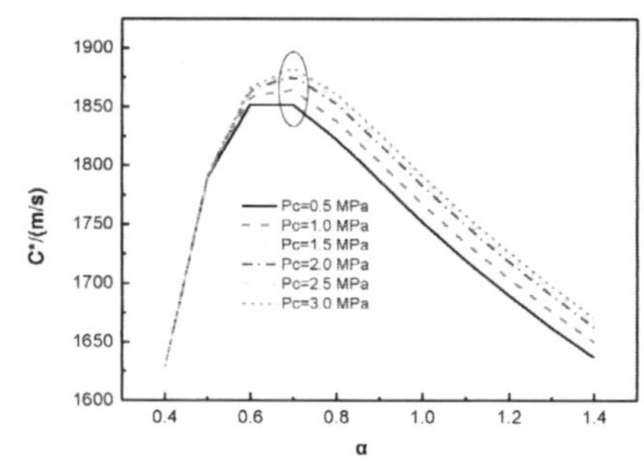

Figure 5. Characteristic velocity vs excess oxidizer coefficient, chamber pressure

Figure 4 shows the effect of characteristic chamber length on characteristic velocity. For $L^* = 300mm$, the value of characteristic velocity is about $1823.7m/s$, indicating that the characteristic velocity efficiency of GO_2/GCH_4 platelet injector can reach 92.2% when characteristic chamber length is about 300mm. With the increase of characteristic chamber length, the characteristic velocity efficiency increased significantly; when $L^* = 600mm$, characteristic velocity efficiency can reach above 99%, the corresponding values of chamber length is about 120mm. The value of $L^* = 600mm$ will be adopted in the next simulations. Figure 5 shows that, the characteristic velocity can be improved by applying a high pressure. And there is a parabola relation between characteristic velocity and excess oxidizer coefficient. In this paper, excess oxidizer coefficient $\alpha = 0.8$ has been used for the thermodynamic calculation.

4.2 Temperature

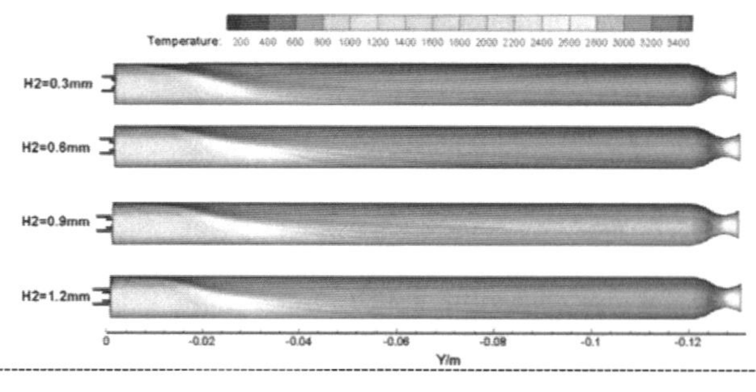

Figure 6. Temperature contours on the symmetrical plane

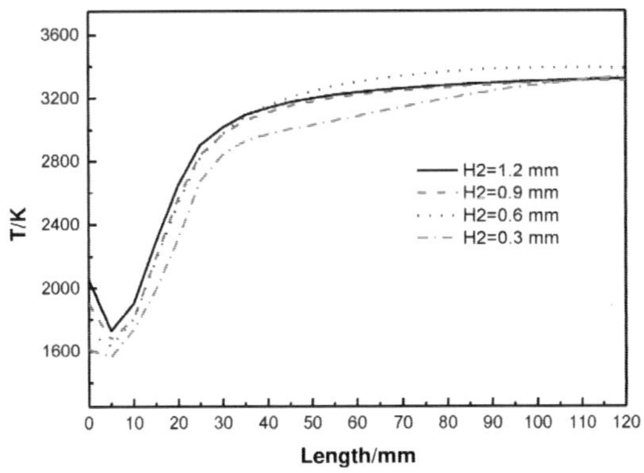

Figure 7. Axial distribution of the chamber wall temperature

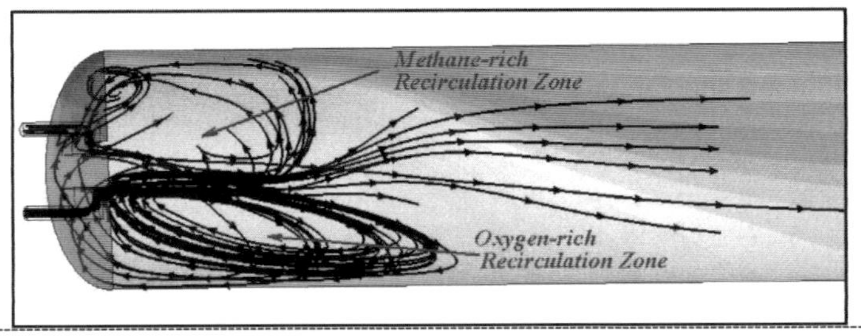

Figure 8. A 3D view of the streamline near faceplate with H2=0.3mm

When studying the effect of design parameters of the platelet injector on the combustion characteristic, the main results include temperature field, component distribution and combustion efficiency. The walls were modeled with no slip adiabatic wall boundary condition in this simulation, so are unlikely to represent the true value of wall temperature, but the methodology adopted in quantitative analysis can be helpful for this research.

The temperature contours for different middle plate thickness are shown in Figure 6, which serve as a reasonably marker for the GO_2/GCH_4 combust reaction temperature, the result shows that the maximum temperature is about 3400K, very close to the theoretical value, and the temperature contours for all cases indicate an apparent similarity. Non-uniformity combustion is found in the recirculation zones, which comes mainly from the non-uniformity of the velocity distribution. The average axial distribution of the chamber wall temperature are shown in Figure 7, the impinging plane operates a lower temperature than the faceplate area because of incomplete burning, and there is a slight difference between the four cases. When $H2 = 0.6mm$, which operate a higher temperature compared to other cases.

On the other hand, two big scale oxygen-rich and methane-rich recirculation zones are produced near the injector face (see Figure 8), these recirculation zones enhanced mixing and combustion, but can cause a heat flux peak on the faceplate. However, it also serves as a flame holder. These works will be studied in the future.

4.3 Composition distribution

When the excess oxidizer coefficient was set as $\alpha = 0.8$, and the specific products only take CO_2, CO, H_2O, O, H, OH, H_2. By analysing the thermal calculation, the percentage of mole fraction of CO_2, CO,

H_2O was 10.96%, 19.23%, 43.82% respectively, the closer the values approximate to thermal calculation, the more complete combustion has been done. The influence of middle-plate thickness $H2$ on percentage of mole fraction of CO_2, CO, H_2O are shown in Figure 9. It can be seen that with the value of $H2$ increases, the percentage of the mole fraction of main products increase first and then decrease and finally reach a balance at the position about 40mm away from the faceplate. The concentration of oxygen near the oxidizer post is higher than the fuel post because of different velocity ratio, which can induce to incomplete combustion and efficiency decline.

For methane combustion, the proportion of water was larger than other products, so we can reveal the main combustion area and combustion length by analysing the mole fraction of H_2O. Here we define the combustion length L^* when the mole friction of water was 90% of the theoretical values. Figure 9 (a) shows the mole fraction of H_2O and the combustion length. This result indicates that larger $H2$ is beneficial to promote the combustion performance and reduce combustion length, although its influence is not obvious. In general, for $H2 = 0.6mm$, it has better combustion performance, shorter combustion length.

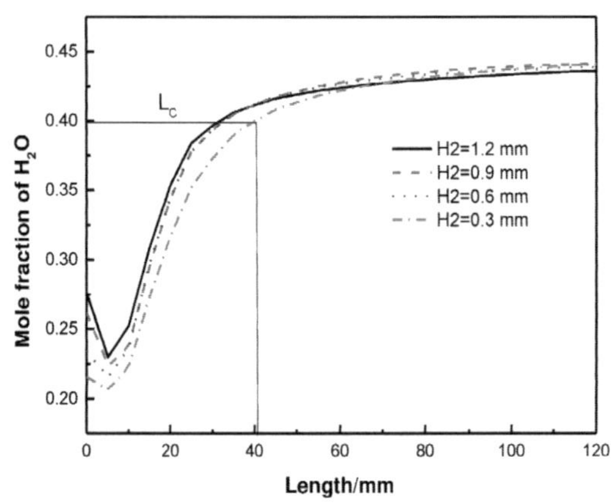

(a) H_2O

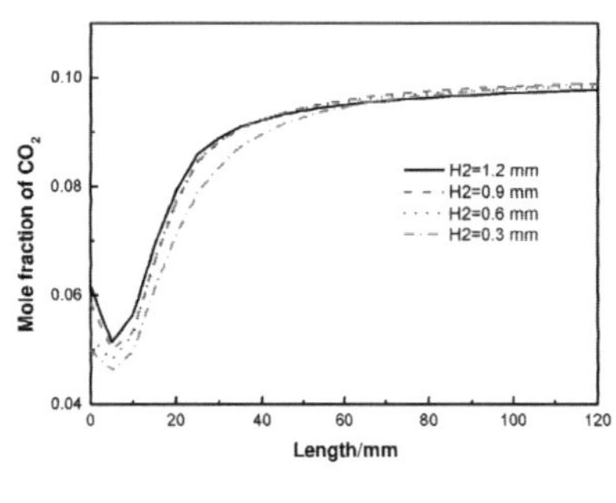

(b) CO_2

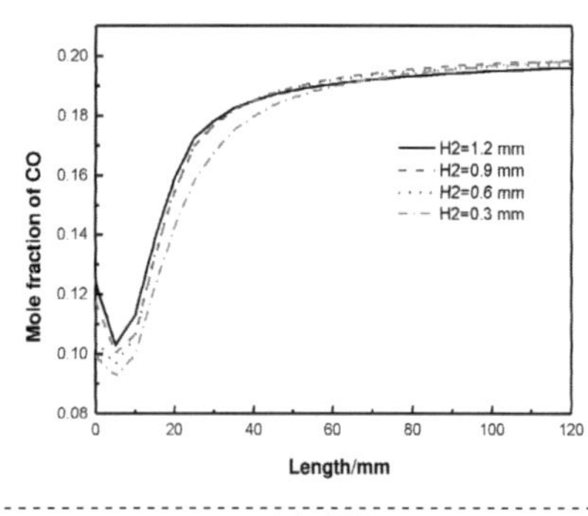

(c) CO

Figure 9. Axial mole fraction distribution of the main products

5. Conclusions

The combustion characteristic of single-element platelet injector with GO_2/GCH_4 as propellant is discussed by numerical method, middle-plate thickness and characteristic chamber length is calculated for improving combustion efficiency. The following conclusions can be obtained: 1) the combustion distance can be shortened effectively by using platelet injector. Good combustion can be achieved for reasonable characteristic chamber length L^*. Characteristic velocity efficiency can reach above 99% when $L^*=600mm$. 2) the influence of *H2* on temperature, composition distribution and combustion efficiency of splash injector is not very noticeable. When $H2=0.6mm$, the combustion performance can reach the best. In general, this configuration has a better combustion performance, shorter combustion length. 3) these efforts can aid in the optimization of the splash platelet injector in order to achieve the highest mixing characteristic and combustion efficiency. For the later, other key parameters including expending angle, relative thickness, nozzle space and the heat load of the injector faceplate et al. will be studied, and detailed chemical reaction model should be adopted. This works are still going on.

6. Acknowledgments

The authors gratefully acknowledge the financial support of National Science Foundation of China (Grant NO.90916018) and National Science Foundation of Hunan (Grant NO.13JJ2002)

7. References

[1] Robbers B A, Anderson B J, Hayes W A. et al. Platelet devices-limited only by one' s imagination, A1AA 2006-4542, 42nd AIAA/ASME/SAE/ASEE Joint Propulsion Conference & Exhibit , July9-12, 2006, Sacramento, California.
[2] Philip J. Robinson, Eric M. Veith, Eric A. Hurlbert, et al, 445N (100-lbf) LO2/LCH4 reaction control engine technology development for future space vehicles, Acta Astronautica, 66(2010) 836-843.
[3] Todd Neill, Donald Judd, Eric Veith, Donald Rousar, Practical uses of liquid methane in rocket engine applications, Acta Astronautica, 65(2009) 696-705.
[4] O. Schwarz, P. Q. Duong, G. Schafer, R. Schomacker, Development of a microstructured reactor for heterogeneously catalyzed gas phase reactions: Part I. Reactor fabrication and catalytic coatings, Chemical Engineering Journal, 145(2009)420-428.
[5] Liu Hongpeng, Liu Weiqiang, A numerical model for the platelet heat-pipe-cooled leading edge of hypersonic vehicle, Acta Astronautica,118(2016) 210-217.
[6] Charles E. Hickox. Characterzation of typical platelet injector flow configurations, NASA Report N75-30468.
[7] H. H Mueggengurg. Platelet injector design and development history. AIAA/SAE/ASME/ASEE 27th Joint Propulsion Cong, 1991.
[8] Li Junhai, Zeng Peng, Yu Nanjia, Cai Guobiao, Design and experimental research of a gas oxygen/propane small thrust engine, Journal of Aerospace Power, 26(10): 2352-2357, 2011.
[9] Bassham L B, Labotz R J, Michel P W. Space Shuttle Orbital Maneuvering Engine Platelet Injector Program, NASA Report N76-18231.
[10] KahL R C, Labotz R J, Bassham L B, Platelet Injector for Space Shuttle Orbital Maneuvering Engine, AIAA/SAE 10th Propulsion Cong, 1974.
[11] Lei Fanpei, Zhang Zhenpeng, Yao Mingming et al. Experimental investigation on flow rate and mixing characteristics of splash platelet injector element, Journal of Aerospace Power, 24(6): 1402-1406, 2009.
[12] Guo Zhihui, Cao Yong, Huang Yong, Spray characteristics of side unlike impinging platelet injector, Journal of Aerospace Power, 20(2): 202-207, 2005.
[13] Sozer E, Vaidyanathan A, Segal C, et al, Computational assessment of gaseous reacting flows in single element injector , AIAA Paper No.2009-449.
[14] Lin J, West J S, Williamst R W, CFD code validation of wall heat fluxes for a GO2/GH2 single element combustor , AIAA Paper No. 2005-4524.
[15] Wang Xiaowei, Cai Guobiao, Jin Ping, Scaling of the flowfield in a combustion chamber with a gas–gas injector, 2010 Chinese. Phys. B, 19 019401.

[16] Conley A, Vaidyanathan A, Segal C, Heat flux measurements for a GO2/GH2 single-element, shear injector , Journal of Spacecraft and Rockets, 44(3): 633-639, 2007.
[17] Gao Yushan, Du Zhenggang, Jin Ping, Cai Guobiao, Numerical simulation on the combustion characteristic of shear coaxial GO2/GCH4 injector, Journal of rocket propulsion, 35(5): 18-23, 2009.
[18] Botz R J L, Development of the platelet micro-orifice injector, 35th IAF Cong, 1984.

IOP Conf. Series: Earth and Environmental Science **63** (2017) 012022 doi:10.1088/1755-1315/63/1/012022

Optimization of the Working Cycle for an Underwater Propulsion System Based on Aluminium-Water Combustion

Xianhe Chen, Zhixun Xia, Liya Huang and Jianxin Hu

Space Science and Engineering College, National University of Defense Technology, Changsha, China

Zhixun Xia: zxxia@nudt.edu.cn

Abstract. The working cycle of a novel underwater propulsion system based on aluminium combustion with water is researched in order to evaluate the best performance. The system exploits the exothermic reaction between aluminium and water which will produce high temperature, pressure steam and hydrogen mixture that can be used to drive turbine to generate power. Several new system configurations corresponding to different working cycles are investigated, and their performance parameters in terms of net power, energy density and global efficiency are discussed. The results of the system simulation show that using the recirculation steam rather than hydrogen as the carrier gas, the system net power, energy density and efficiency of the system are greatly increased compared, however the system performance is close either using adiabatic compression or isothermal compression. And if an evaporator component is added into system in order to take full use of the solid product heat, the system performance will be improved.

1. Introduction

As the world pays more attention to the ocean, unmanned underwater vehicle power technology has become an important field of scientific research [1,2]. To a certain extent, the traditional power battery and the lithium battery has been hardly to meet the future demand for power system, and the development of new fuel cell technology is not mature enough, especially in the transport and storage of hydrogen which is a big problem in the power system [3-7]. Especially in recent years, scholars have put forward the underwater propulsion system based on aluminium-water combustion such as HAC(Hybrid Aluminium Combustor) system[8-12]and HAC-SOFC(Hybrid Aluminium Combustor - Solid Oxide Fuel Cell) [13-14]. At the same time, the application of aluminium-water in power generation and hydrogen production are discussed [15-17]. All of these similarly system exploits the exothermic reaction between aluminium and steam to produces thermal power to driven turbine and hydrogen as a by-product of the reaction. According to the results of the performance [13], this system has the advantages of high energy density that can meet the demand of power system to a certain extent.

The present paper is focused on the development of some new underwater propulsion system based on the combustion of aluminium and water described in Ref. [13]. Because these system consists of fuel seeder system, combustor, compressor and turbine et al, so we can improve the system performance by increasing some other components or change the system component layout. However, there is no research reported about this section by now. Consequently, several kinds of new systems are investigated and the system performance such as system net power output, energy density and the efficiency are researched respectively by means of the establishment of mathematical model of each system.

2. Propulsion System Concept

These six systems are named with System A, B, C, D, E, and F respectively. Figure 1 shows a simplified scheme of the System A which is more extensively described in Ref. [13]. It includes eight main elements: combustor, fuel feeder system, separator, compressor, turbine, recuperator, condenser and pump. The system consists of a fluidized bed fuel feeder in which aluminium powder is suspended by hydrogen. The Al and H_2 are injected into combustor where they will be mix and react exothermically with the recirculation steam to form Al_2O_3 and additional H_2. The second water (quenching water) injected along the combustor walls prevents molten alumina particles from attaching to the combustor walls and fouling the system. This quenching water also evaporates to produce large amounts of high-temperature steam. The combustion products pass through a cyclonic separator to remove the solid Al_2O_3. Two streams of steam/hydrogen mixture are separated from the splitter, in which the small one is mixed with quenching water and moved into the combustor through compressor to form recirculation steam for the sustain the reaction of the combustor, and the most rest steam/hydrogen mixture is used for driving the turbine work. The steam/hydrogen mixture moves into exchanger through the drive turbine for preheating the seawater. The steam is fully condensed into liquid water downstream from the heat exchanger, separated from H_2 and returns back into the intake system of seawater. A part of the H_2 is compressed into the fuel feeder system through the compressor to be reused as the carrier gas, and the rest with the excessive for compression storage or discharged.

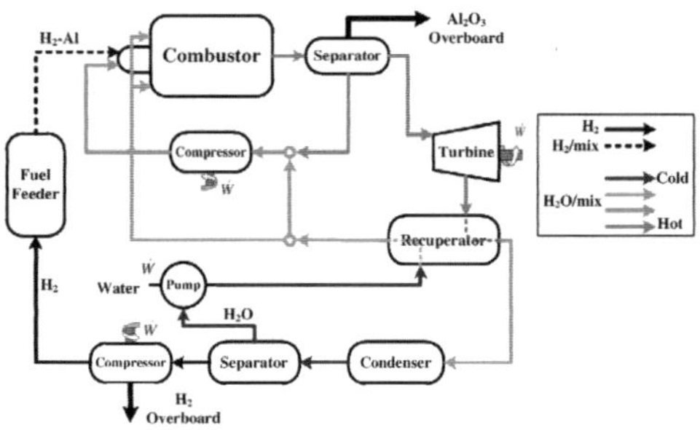

Figure 1. The diagram of System A.

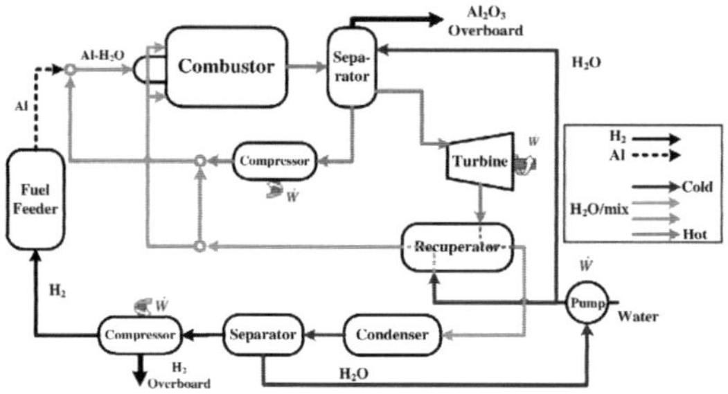

Figure 2. The diagram of System B.

Figure 2 shows a simplified scheme of the System B which is modified on the basis of System A. First, the amount of water used to separate solid particles is increased. Second, considering that the power are mainly consumed by the hydrogen compressing in System A, so the hydrogen as a carrier gas is replaced by high-temperature steam, thereby the compression power is reduced. Last, the recirculation steam will be compressed first before mixed. A method of fluidized bed [8] is used in fuel seeder traditionally and it's possible to achieve with high-temperature steam as the carrier gas in principle. However, on one hand, after use steam as carrier gas, aluminium and water are premixed which means that it could accelerate combustion, on the other hand, the duct thermal insulation and safety issues should be considered. There is no mention about the fuel seeder design in this paper.

Figure 3, Figure 4 shows System C and D diagram respectively. As we all know that the combustion of aluminium and water is difficult, especially with liquid water. So high-temperature steam is used in all of these systems. In order to ensure the efficiency of aluminium-water combustion, increasing the temperature of recirculation steam is considered in System A and B to obtain System C and D. The main improvement is that it will lead out high-temperature steam from the separator to enter into combustor and react as recirculation steam with aluminium under the compressor driving.

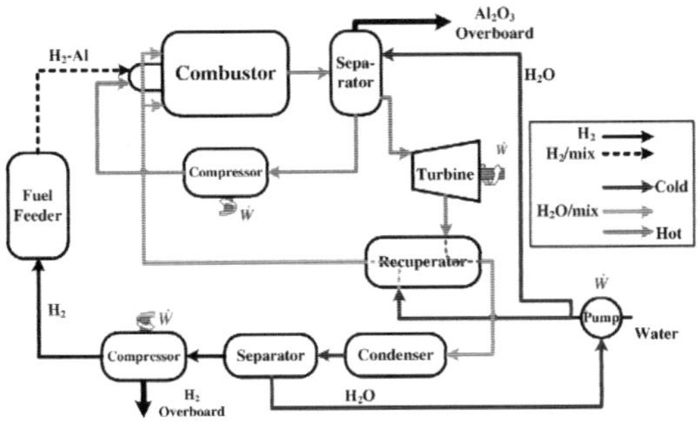

Figure 3. The diagram of System C.

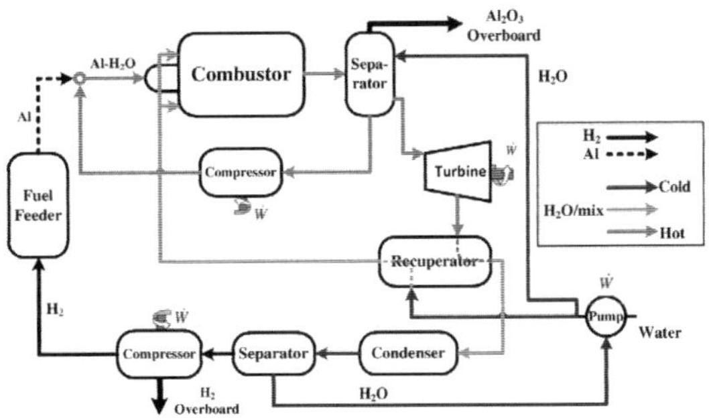

Figure 4. The diagram of System D.

According to the results of reference, the mass fraction of the steam/hydrogen mixture return back to combustor from the separator is about 20%. On one hand, the power output of the turbine is reduced, on the other hand, a large amount of energy is consumed in order to compress the steam/hydrogen mixture back to the combustor. The performance of the system is reduced. Thus, an improved System E and F are put forward and respectively showed in Figure 5 and Figure 6. Compared to the previous System A, an evaporator and a solid heat exchanger component are added into these systems, and the high-temperature steam compressor is eliminated. The evaporator aims to cool water evaporated to slight superheated steam. The purpose of adding solid heat exchanger is using the high-temperature solid-phase products to turn the slightly superheated steam into the recirculation steam. However, combustion temperature needs to be improved in these systems.

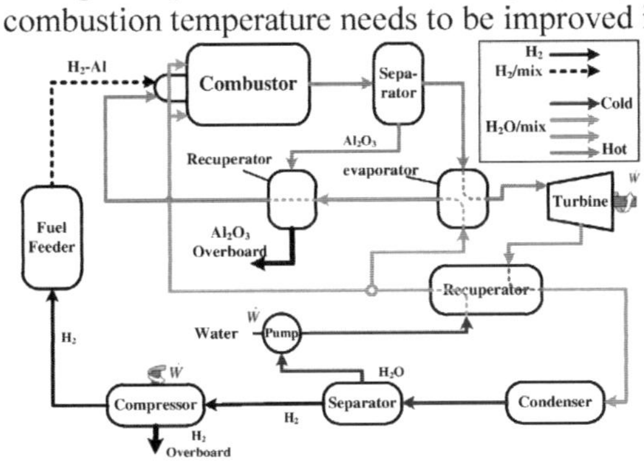

Figure 5. The diagram of System E.

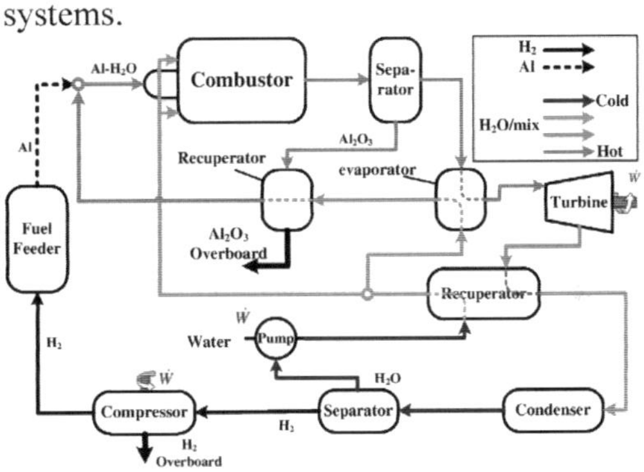

Figure 6. The diagram of System F.

3. Numerical Analysis

The performance calculation model of these six systems at present can be found in Ref. [13]. The global numerical model of each system is able to relate the system performance to the operating parameters, such as the aluminium mass flow, the water temperature, the water mass flow, the combustor pressure and temperature and the component efficiency. The thermo-fluid dynamics properties of hydrogen and steam are introduced in the numerical model according to the NIST database by means of lookup tables. An important feature of this model is that it can account for the phase change of the species as a function of the pressure and temperature variations. Because the combustor temperature is about 1200K and the combustor pressure is about 3MPa, the thermodynamic calculation software CEA result shows that the product after the reaction are alumina and hydrogen, other components are trace and negligible. Thus no chemical kinetics or surface reactions are included in the numerical model, and the heat release is calculated only in terms of the global aluminium-water reaction [15]:

$$2Al(s)+3H_2O(g)\rightarrow Al_2O_3(s)+2H_2(g)+15152kJ/kg(T=298.15;P=1.0atm) \tag{1}$$

Suppose that the combustion chamber with external insulation, total enthalpy value before and after reaction is equal.

$$\sum_{i=1}^{n}\dot{m}_{i,r}(h_f^{T_s}+\int_{T_s}^{T}c_p dT)_i=\sum_{j=1}^{m}\dot{m}_{j,p}(h_f^{T_s}+\int_{T_s}^{T}c_p dT)_j \tag{2}$$

$\dot{m}_{i,r}$ is the mass flow of ith-reactant, $\dot{m}_{j,p}$ is the mass flow of jth-product, $h_f^{T_s}$ is the Standard enthalpy, c_p is specific heat capacity, T is the combustion temperature and T_s is the reference temperature.

Suppose that there is no external heat exchange in the evaporator, and a mathematical model is

$$\dot{m}_{cold}\left(h_{out,cold}-h_{in,cold}\right)=\dot{m}_{hot}\left(h_{in,hot}-h_{out,hot}\right) \tag{3}$$

$\dot{m}_{cold}$ is the mass flow of cold side, $\dot{m}_{hot}$ is the mass flow of hot side, $h_{out,cold}$ is the outlet specific enthalpy of cold side, $h_{in,cold}$ is the inlet specific enthalpy of cold side. $h_{in,hot}$ is the inlet specific enthalpy of hot side and $h_{out,hot}$ is the outlet specific enthalpy of hot side.

Other components of the mathematical model and the calculation method of the component volume and mass can be found in Ref. [13]. The volume and mass of the evaporator refers to heat exchanger results.

In order to estimate the overall performance of the six systems, the net power can be expressed as:

$$\dot{W}_{sys}=\dot{W}_{turb}-\dot{W}_{comp}-\dot{W}_{pump} \tag{4}$$

$$\dot{W}_{turb}=\dot{m}C_pT_1\eta_T\left[1-PR^{\frac{\gamma-1}{\gamma}}\right] \tag{5}$$

$\dot{W}_{sys}$ is the system net power, $\dot{W}_{turb}$ is the turbine power, $\dot{W}_{comp}$ is the compression power, $\dot{W}_{pump}$ is the pump power, $\dot{m}$ is the mass flow rate of turbine, C_p is the hydrogen and steam mixture's specific heat capacity, η_T is the turbine efficiency, PR is the turbine pressure ratio and γ is the mixture's ratio of specific heats . Because the hydrogen is overboard, hydrogen utilization is not considered here. The energy density ED_V is defined as:

$$ED_V=\dot{W}_{sys}\frac{\rho_{Al}}{\dot{m}_{Al}}\frac{V_{Al}}{V_{sys}} \tag{6}$$

ρ_{Al} is the aluminium density, $\dot{m}_{Al}$ is the aluminium mass flow, V_{Al} is the aluminium fuel total volume, V_{sys} is the system total volume ($V_{sys}=1000$L see Ref.[13]), the aluminium fuel total volume: $V_{Al}=V_{sys}-V_C$, V_C is the component volume. On the condition of neutral buoyancy:

$$V_{Al}=\frac{\rho_{water}V_{sys}-M_C}{\rho_{Al}} \tag{7}$$

M_C is the component mass.

The overall system efficiency:

$$\eta_{sys}=\frac{net\ power\ output}{chemical\ energy\ input}=\frac{\dot{W}_{sys}}{\dot{m}_{Al}\Delta H_{reac}} \tag{8}$$

ΔH_{reac} is the volumetric energy content of reactants.

4. Results and discussion

The results are compared with the values of Ref. [13] and the results showed in Figure 7. The turbine power is slightly larger than the reference value, and the system net power is slightly less than the value in the reference. Under the condition of adiabatic compression, system net power slightly less than the reference value and under the condition of isothermal compression is greater than reference values. Combustion chamber temperatures were taken from the reference value, so the temperature value is basically the same. Mass flow calculation results are slightly lower than the reference value. The overall results are in good agreement with the reference results prove the accuracy of the model calculation.

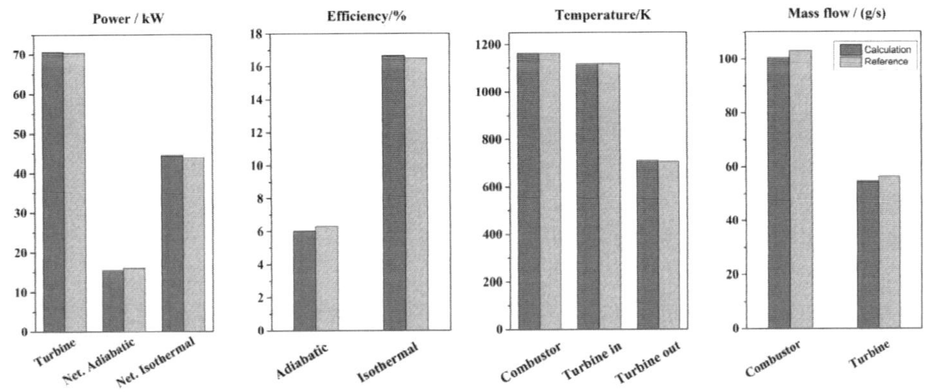

Figure 7. Calculation and reference results.

Following, a detailed analysis of these six different systems would be introduced. According to the results from the reference, low pressure turbine ratio can improve the energy density of the system. Therefore, this paper assume turbine pressure ratio PR equal to 20. According to the new system using high temperature steam as the carrier gas, the hydrogen fuel feeder system will become less; therefore, this paper selects the seeding ratio with 10 when hydrogen is used as the carrier gas, consistent with the reference. Non-idealities in the combustion process are estimated by considering the combustion efficiency equal to 95% [15], which represents the effective oxidized amount of aluminium.

Table 1 shows the initial calculation conditions of the program [13] , and Figure 8 to Figure 14 show the main results of each system.

Table 1. System operating condition

Operating parameters	Value	Units
Combustor pressure	2.76	MPa
Pump pressure	4.0	Mpa
Hydrogen store pressure	4.8	Mpa
Combustor temperature	1150	K
Recirculation steam temperature	755	K
Sea water temperature	298.15	K
Aluminum fuel	17.6	g/s
Recirculation steam	17.6	g/s
Carrier hydrogen	1.76	g/s
Turbine efficiency	0.65	
Adiabatic efficiency	0.7	
Isothermal efficiency	0.7	

Recuperator efficiency	0.5
Condenser pressure ratio	0.9
Recuperator pressure ratio	0.9
Turbine pressure ratio	20

Table 1 shows the initial calculation conditions of the programs[13], and Figure 8 to Figure 14 show the main results of each system.

Figure 8 shows that the turbine mass flow of System B, D and F are greater than System A, C and E. This result is due to System B, D and F using the recirculation steam replace hydrogen as the carrier gas which means much more quenching water could be injected into the combustor.

Figure 9 shows that the distribution of the turbine power is C>A>E>D>B>F, and the largest difference is 8kW. Moreover the turbine power of System A, C, E is much bigger than system B, D, F but the power of System A, C and E are substantially equal, so as System B, D and F. According to the formula eq(5), the turbine power is related to the mass flow, turbine efficiency and the mixture property. Since System B, D and F using the recirculation steam replace hydrogen as the carrier gas, the mass fraction of hydrogen in turbine entrance is decreased. Because hydrogen's specific heat capacity is several times higher than steam's, relatively small changes in the hydrogen gas mass fraction can have a significant impact on *Cp* of the mixture, so the mixture specific heat capacity is decreased. Although the mass flow is increased (shown in Figure 8), the turbine power is decreased still. Because the aluminum mass flow is equal(shown in table 1) and turbine inlet temperature is basically consistent(about 1100K), the turbine power of System B, D and F are substantially equal to each other, and similarly the same result for System A, C and E.

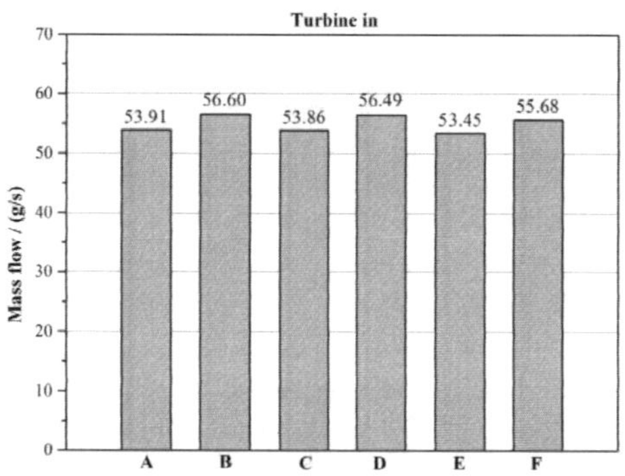

Figure 8. Mass flow of turbine in for each system.

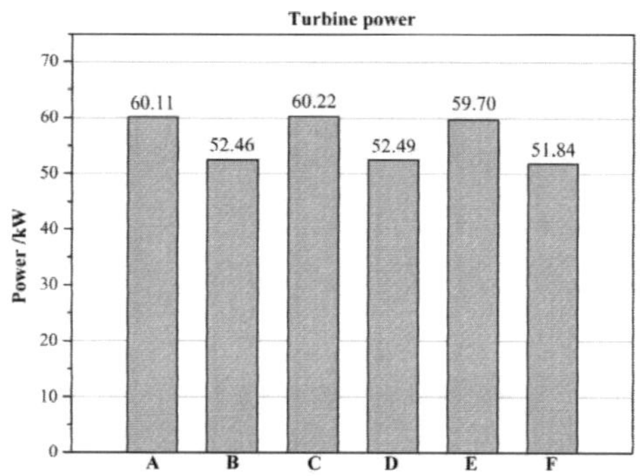

Figure 9. Turbine power for each system.

Table 2. Hydrogen parameter distribution

	A	B	C	D	E	F
Carrier gas(g/s)	1.76	0	1.76	0	1.76	0
Back to fuel feeder(g/s)	1.7663	0.0063	1.7663	0.0063	1.7663	0.0063
Overboard(g/s)	1.8514	1.8514	1.8514	1.8514	1.8514	1.8514
Adiabatic compression power(kW)	20.3	0.8	20.3	0.8	20.3	0.8

Figure 10 shows the net power of adiabatic compression and isothermal compression conditions of each system. Obviously, the net power of isothermal compression conditions is greater than adiabatic compression which is consistent with the results of Ref.[13]. under the condition of adiabatic compression, the net power distribution is F>B>D>E>A>C. The net power of System F is the maximum because it makes full use of the alumina heat and the net power of System C is the minimum because the consumption of compression power to compress the high-temperature circulation steam is large despite the fact that the turbine power is the largest (shown in Figure 9). It is interesting to notice that the system net power is greatly increased while using the recirculation steam to replace hydrogen as the carrier gas (such as System A and B, an increase of 42% approximately): this result is due to the mass flow of hydrogen which need to compress sharp decrease when using the recirculation steam as the carrier gas (shown in Table 2). Under the condition of isothermal compression, the net power distribution is also F>B>D>E>A>C and the difference of each system is significantly reduced compared to adiabatic compression: the reason is that the consumption power of isothermal compression is lower than adiabatic compression.

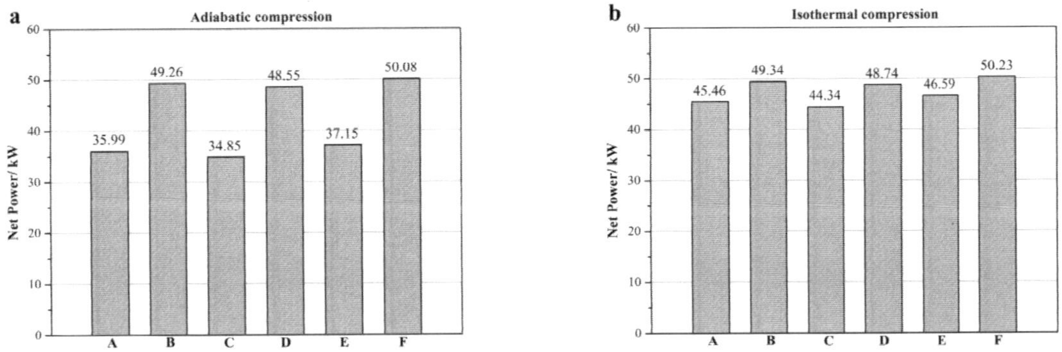

Figure 10. Net power for each system.

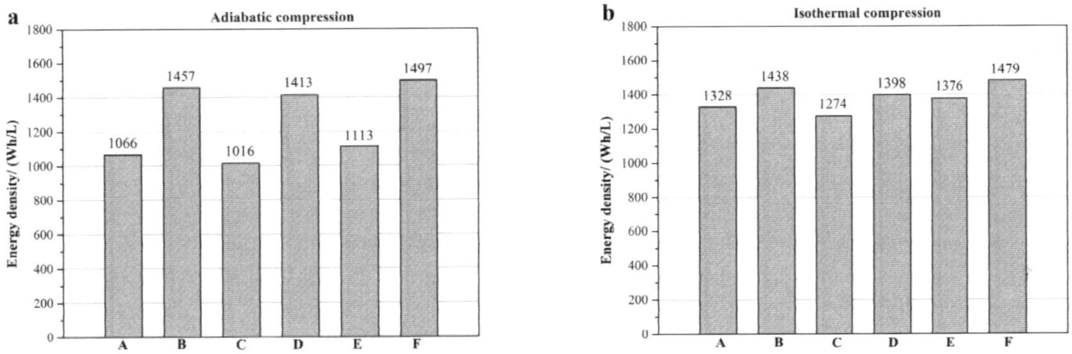

Figure 11. Energy density for each system.

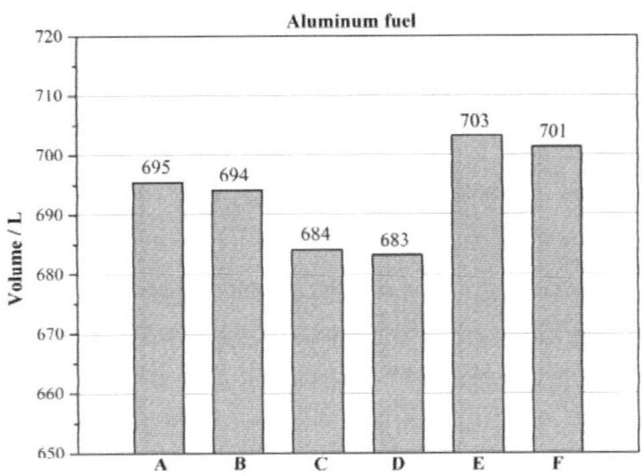

Figure 12. Aluminium fuel volume for each system in adiabatic compression.

According to the formula eq(6), the energy density is both related to the net power and aluminium fuel volume. Figure 10 illustrates the negatively buoyant energy density for each system. Under the condition showed in the figure, the energy density distribution is F>B>D>E>A>C. Such variation can be obtained from the net power (shown in Figure 10) and aluminium fuel volume (shown in Figure 12,

IOP Conf. Series: Earth and Environmental Science **63** (2017) 012022 doi:10.1088/1755-1315/63/1/012022

and decrease 10L respectively when using isothermal compression) of each system. Besides, the energy density of isothermal compression conditions is below adiabatic compression in System B, D and F. This result is due to an additional 10L volume of compressor for isothermal compression with the almost equal net power [13].

Figure 13 shows the total efficiency for each system during the present work. In the case of the constant aluminium mass flow, overall efficiency is proportional to the net power of system. The highest efficiency is System F (about 18.84%), the lowest is System C (about 13.07%), under two compression conditions.

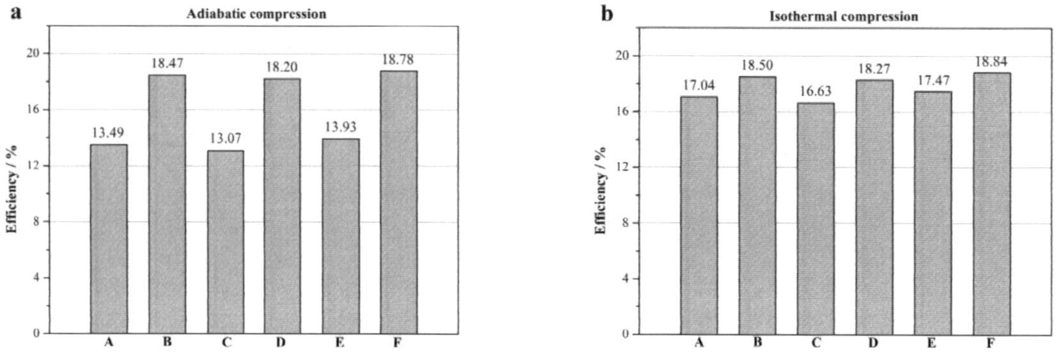

Figure 13. Total efficiency for each system.

Figure 14 compares predicted the neutral-buoyancy energy densities for each system under the condition of 15kW net power. The solid portion of the bars shows the performance with adiabatic compression. The top of the cross-hatched bars shows the performance that is possible with an isothermal compressor that adds negligible mass and volume compared to an adiabatic compressor.

As depicted in Figure 14, System B, D and F is obtained a relatively high energy density, indicating that after using the recirculation steam replace hydrogen as the carrier gas, the system performance can be increased significantly. The highest energy density is up to 824Wh/L and the maximum difference is up to 2.5 times. An evaporator component is added into System E and F, thus the total quality of system increases, which leads to the volume of aluminium fuel is relatively small, so the energy density is relatively low. Using hydrogen as the carrier gas, the energy density distribution is: A>C>E. However, the difference between System A and C is small, thus, it can be concluded that with the increase of the circulation steam temperature, on one hand, the combustion of aluminium particles can be promoted, on the other hand, a lower impact on the system performance can be obtained.

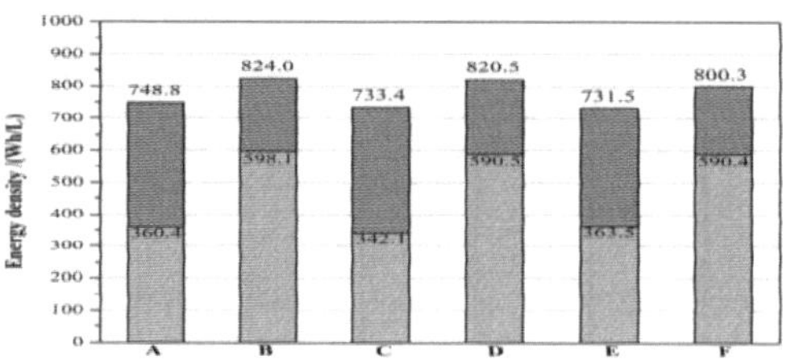

Figure 14. Comparison of the neutrally-buoyant energy densities for each system.

5. Conclusions

The performance of these six underwater propulsion system based on aluminium combustion with water has been researched. The results demonstrated that it is possible to improve significantly the energy density and efficiency of the proposed system by modifying the working cycle.

The net power of isothermal compression conditions is greater than that of adiabatic compression. In addition, the difference of each system in isothermal compression is significantly reduced compared to adiabatic compression, and the energy density of isothermal compression conditions is below adiabatic compression in System B, D and F.

In particular, according to the neutral-buoyancy energy densities, the energy density distribution is A>C>E when using hydrogen as carrier gas; system B, D and F are obtained a relatively high energy density after using the recirculation steam replace hydrogen as a carrier gas, and the highest energy density is achieved to 824Wh/L.

The results of present work implied that System B, D, E and F can improve the performance of the system compared with the original System A. However, the circulation steam temperature is increased in System C, and it's conducive to the aluminium water combustion efficiency meanwhile the system performance decline slightly.

Although so far, there have not been related application examples or related experiments are reported, so the research of this paper is highly important. According to the calculation results, we are trying to solve a series of problems and complete the development of the system finally.

6. References

[1] The Navy Unmanned Undersea Vehicle (UUV) Master Plan. Nov. 9, 2004

[2] Long endurance undersea vehicle propulsion. *Office of Naval Research BAA Announcement* #11-016, 2011

[3] Chalk SG, Miller JF. Key challenges and recent progress in batteries, fuel cells, and hydrogen storage for clean energy systems. J. Journal of Power Sources 159(2006) 73-80

[4] Vasiliev LL, Kanonchik LE, Kulakov AG, Mishkinis DA, Safonova AM, Luneva NK. New sorbent materials for the hydrogen storage and transportation. J. International Journal of Hydrogen Energy 32(2007) 5015-5025

[5] Kim J, Moon I. Strategic design of hydrogen infrastructure considering cost and safety using multiobjective optimization. J. International Journal of Hydrogen Energy 33(2008) 5887-896

[6] Zheng J, Liu X, Xu P, Liu P, Zhao Y, Yang J. Development of high pressure gaseous hydrogen storage technologies. J. International Journal of Hydrogen Energy 37(2012)1048-1057

[7] Farshid Zabihian, Alan S. Fung. Performance analysis of hybrid solid oxide fuel cell and gas turbine cycle (part I): Effects of fuel composition on output power. Journal of the Energy Institute 87(2014) 18-27

[8] T.F.Miller, J.L.Walter, D.H.Kiely. A next-generation AUV energy system based on aluminum-seawater combustion. Proceedings of 2002 Workshop on Autonomous Underwater Vehicles. San Antonio. 2002

[9] W. Ethan Eagle. Modeling of a high energy density propulsion system based on the combustion of aluminum and steam. MS Thesis. University of Maryland, College Park. 2007

[10] W. Ethan Eagle, Daniel F. Waters, Christopher P. Cadou. System Modeling of a Novel Aluminum Fueled UUV Power. *AIAA* 2012-1135

[11] Daniel F. W. , Christopher P.C. , W. Ethan Eagle. Quantifying Unmanned Undersea Vehicle Range Improvement Enabled by Aluminum-Water Power System. J. JOURNAL OF PROPULSION AND POWER 29(2013) 675-685

[12] Daniel F. W. ,Christopher P.C.. Modeling a Hybrid Rankine-Cycle/SOFC UUV Propulsion System Powered by Aluminum-Water Combustion. *AIAA* 2012-1134

[13] Daniel F. W. ,Christopher P. C. . Modeling a hybrid Rankine-cycle/fuel-cell underwater propulsion system based on aluminum water combustion. J. Journal of Power Sources 221(2013) 272-283

[14] Daniel F. W. ,Christopher P. C.. Estimating the neutrally buoyant energy density of a Rankine-cycle/fuel-cell underwater propulsion system. J. Journal of Power Sources 248(2014) 714-720

[15] Franzoni F, Mercati S, Milani M, Montorsi L. . Operating maps of a combined hydrogen production and power generation system based on aluminum combustion with water. J. International Journal of Hydrogen Energy 36(2011) 2803-2816

[16] Mercati S, Milani M, Montorsi L, Paltrinieri F.. Design of the steam generator in an energy conversion system based on the aluminum combustion with water. J. Applied Energy 97(2012) 686-694

[17] Stefano Mercati, Massimo Milani, Luca Montorsi, Fabrizio Paltrinieri. Optimization of the working cycle for a hydrogenproduction and power generation plant based on aluminum combustion with water. J. international journal of hydrogen energy 38(2013) 7209-7217

Theoretical Study of Mercury Species Adsorption on MgO(001) Surface

Chong Zhang*, Yu Liu, Li Zhao, Min-shu Cui andQiang Lu

National Engineering Laboratory for Biomass Power Generation Equipment, North China Electric Power University, Beijing 102206, PR China

jl-zc@163.com

Abstract. Traditional thermal power generation releases mercury, which will cause important environmental issues. As one of the main components of fly ash produced by coal-fired power generation, MgO can adsorb mercury in HCl atmosphere. Aiming to investigate the interaction mechanism of MgO with mercury in coal-fired flue gas in HCl atmosphere, density functional theory method is used to study the adsorption of mercury species on the fly ash component. The MgO(001) surface is built to study the stable adsorption configurations of Hg^0, HCl, HgCl and $HgCl_2$ on all possible adsorption sites. Results show that Hg^0, HgCl and $HgCl_2$ are all chemisorbed on MgO(001) surface weakly, while HCl is dissociatively adsorbed. HgCl cannot exist stably on the surface, so it has no obvious effect on the adsorption and oxidation of Hg^0.

1. Introduction

Mercury is an important pollutant in coal-fired flu gas, and its main existing forms are elemental mercury (Hg^0), oxidized mercury (Hg^{2+}) and particulate mercury (Hg^P) [1,2]. The fly ash produced by coal-fired power generation can absorb Hg^0 which cannot dissolve in water and is difficult to be removed by existing pollution control devices, to a certain extent [3-5]. Since experimental research is difficult to explain the adsorption and oxidation of Hg^0 on the surface of fly ash component in micro scale, the quantum chemical Density Functional Theory (DFT) calculation is used to create the conditions for revealing the adsorption and oxidation mechanism of Hg^0. On the basis of experimental research, many scholars have conducted deep studies using DFT method on the adsorption and oxidation mechanism of Hg^0 on the surface of fly ash components including Fe_2O_3[6,7], CuO[8,9] and MnO_2[10,11]. However, theoretical studies on the adsorption mechanism of Hg^0 by MgO are rare.

In this paper, the detailed mechanism of mercury species adsorption and dissociative adsorption of HClon MgO(001) surface are studied by using first-principles calculations based on the density functional theory and the periodic slab models.

2. Computational details

2.1. Model

A four-layer MgO(001) slab is used with a 2 × 2 (8.5088Å×8.5088Å) unit cell, as is shown in Figure 1. The slabs are separated by a vacuum space of 12 Åto prevent interaction between periodic images. The positions of all atoms are allowed to relax with the exception of the bottom 2 layer kept fixed in their bulk positions throughout the calculations.

IOP Conf. Series: Earth and Environmental Science **63** (2017) 012023 doi:10.1088/1755-1315/63/1/012023

2.2. Computational methods

Structural optimizations are performed using DMol^3 package in Materials Studio software in this study. The Perdew–Burke–Enzerhoff generalized-gradient approximation (PBE, GGA) is used for the exchange-correlation functional. The double numerical basis set with polarization functions (DNP) and atomic orbital cutoff of 5.5 Å is used. The (3×3×1) k-points Monkhorst–Pack grid is used for bulk lattice optimization.

For the adsorption of mercury species on MgO(001) surface, the adsorption energy E_{ad} is defined as:

$$E_{ad} - E_{sys} - E_{ads} - E_{sur} \quad (1)$$

WhereE_{sys} isthe system total electronic energy after adsorption; E_{ads} is theelectronic energy of adsorbate; E_{sur} istheelectronic energy of the surface.

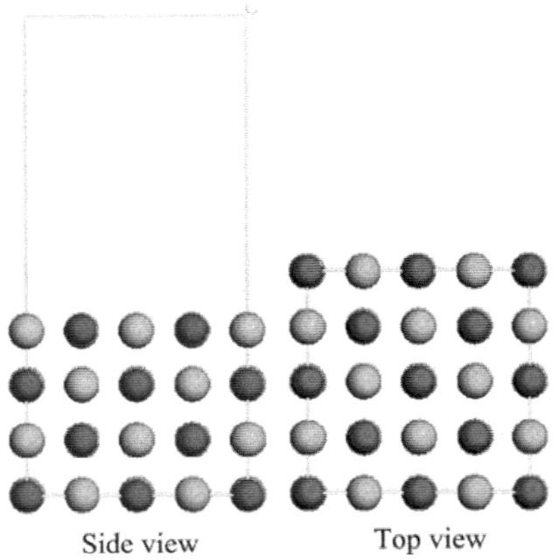

Figure 1. MgO(001) surface. Green and red balls represent Mg and O atoms, respectively.

3. Results and discussion

3.1. Hg^0 adsorption on MgO(100) surface

There are four possible adsorption sites for Hg^0 on MgO(001) surface, Mg site (1A), O site (1B), bridge site (1C) and hollow site (1D), respectively. The optimized configurations ofHg^0 adsorption on each site are shown in Figure 2. The adsorption energies and geometric parameters are given in Table 1.The adsorption energies results show that the stability of Hg^0 adsorption is in the order of 1A < 1D < 1C < 1B. All of the adsorption energy values are negativewith less absolute value, which indicates the Hg^0 adsorption is weak chemisorption. The adsorption energy of Hg^0 adsorption on Mg site is relatively lower, and Hg^0 is closer to the surface with a distance of 3.296Å, indicating that Hg^0 is more likely to be bound to the surface Mg atoms. The above results show that Hg^0 cannot be absorbed directly on MgO(001) surface.

IOP Conf. Series: Earth and Environmental Science **63** (2017) 012023 doi:10.1088/1755-1315/63/1/012023

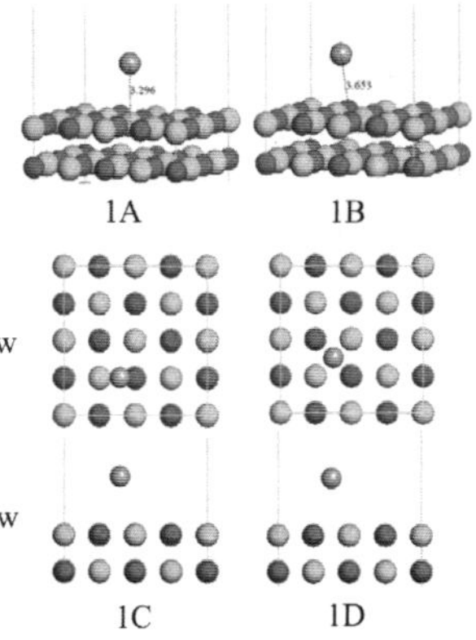

Figure 2. Optimized configurations of Hg^0adsorption on MgO(001) surface. Grey balls represent Hg atoms.

Table 1. The adsorption energies and geometric parameters of Hg^0 adsorption on MgO(001) surface.

No.	Hg-O（Å）	Hg-Mg（Å）	Eads(kJ/mol)
1A	-	3.296	-11.087
1B	3.653	-	-8.690
1C	3.446	3.591	-9.266
1D	3.616	3.649	-9.438

3.2. HCl adsorption on MgO(001) surface

HCl in coal-fired flu gas can act as the oxidative agent of the Hg^0 catalytic oxidation reaction, while the mechanism of HCl affecting the Hg^0 oxidation on the fly ash composition is still not clear. All possible adsorption sites and initial configuration of HCl are considered. The optimized configurations of HCl adsorption on each site are shown in Figure 3. The adsorption energies and geometric parameters for HCl adsorption on the surface are given in Table 2. The adsorption energy results show that the stability of HCl adsorption is in the order of 2E < 2C < 2D < 2B < 2F < 2A. All of the adsorption energy values are negative, among which the absolute value of adsorption energy of 2A, 2B, 2D and 2F are relatively low, indicating they are weak chemisorption, while that of 2C and 2E are relatively high. In 2C, the adsorption energy is -94.749 kJ/mol, the distance of H-Cl elongates from 1.293 Å to 1.715Å. H atom of HCl interacts with surface O and the length of new-formed H-O bond is 1.073Å. 2E is strong chemisorption with an adsorption energy of -94.749 kJ/mol, in which dissociation adsorption of HCl take place and the H-Cl bond breaks, the distance of H-Cl elongates to 1.923Å, H atom interacts with surface O and the length of H-O bond is 1.022Å. After the dissociative adsorption of HCl, the Cl radical doesn't adsorb on the surface but releases to the gas phase, which shows that Cl radical may combine with Hg in the gas phase to form HgCl, an intermediate of oxidation reaction.

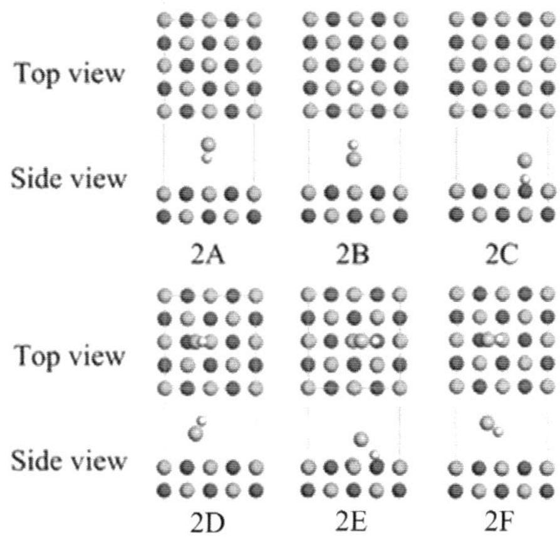

Figure 3. Optimized configurations of HCl adsorption on MgO(001) surface. White and grey-green balls represent H and Cl atoms respectively.

Table 2. The adsorption energies and geometric parameters of HCl adsorption on MgO(001) surface.

No.	H-O (Å)	H-Mg (Å)	Cl-O (Å)	Cl-Mg (Å)	H-Cl (Å)	E_{ads} (kJ/mol)
2A	-	3.353	-	-	1.293	-10.125
2B	-	-	3.021	-	1.295	-14.496
2C	1.073	-	-	-	1.715	-94.749
2D	-	-	3.144	3.366	1.295	-15.267
2E	1.022	-	-	2.51	1.923	-118.092
2F	3.415	-	4.2	-	1.291	-10.540

3.3. HgCl adsorption on MgO(001) surface

HgCl is a possible intermediate of oxidation reaction of Hg^0. So, it is necessary to study the adsorption of HgCl onMgO(001) surface. All possible adsorption sites and initial configuration of HgCl are considered. The optimized configurations of HgCl adsorption on each site are shown in Figure 4. The adsorption energies and geometric parameters for HgCl adsorption on the surface are given in Table 3. The adsorption energy results show that the stability of HgCl adsorption is in the order of 3E < 3G < 3A < 3C < 3D < 3F < 3B. All of the adsorption energy values are negative, but their absolute values are all less than 80 kJ/mol, showing they are relatively weak chemisorption. In addition, the difference of adsorption energy between various configurations is not large. Except for 3E, the absolute values of adsorption energies of other configurations are less than 50 kJ/mol, and all of them are vertical adsorption. The most stable structure for HgCl adsorption on the surface is 3E, in which HgCl adsorbs on the surface horizontally, and Hg atom is close to the surface O atom, Cl atom is close to the adjacent surface Mg atom, the distance between Hg and O is 2.768 Å, the distance between Cl and Mg is 2.614 Å, and Hg-Cl bond elongates slightly from 2.495 Å to 2.700 Å. All these results show that HgCl cannot be absorbed stably on MgO(001) surface, i.e., the intermediate of Hg^0-Hg^{2+} oxidation cannot be formed on the surface.

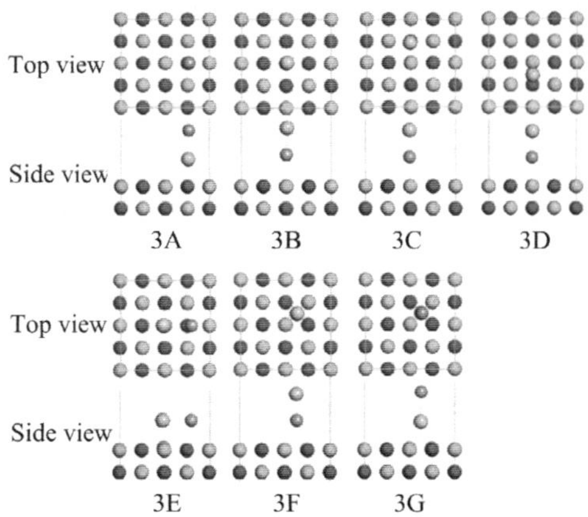

Figure 4. Optimized configurations of HgCl adsorption on MgO(001) surface.

Table 3. The adsorption energies and geometric parameters of HgCl adsorption on MgO(001) surface.

No.	Hg-O (Å)	Hg-Mg (Å)	Cl-O (Å)	Cl-Mg (Å)	Hg-Cl (Å)	E_{ads} (kJ/mol)
3A	-	-	2.701	-	2.756	-49.927
3B	-	3.009	-	-	2.528	-43.035
3C	2.709	-	-	-	2.526	-45.938
3D	2.935	3.014	-	-	2.510	-45.277
3E	2.768	-	-	2.614	2.700	-72.103
3F	3.145	3.201	-	-	2.519	-44.963
3G	-	-	3.064	2.963	2.822	-51.748

3.4. $HgCl_2$ adsorption on MgO(001) surface

In the Hg^0 oxidation reaction with HCl, $HgCl_2$ is the final product of the reaction. All possible adsorption sites and initial configurations of $HgCl_2$ are considered. The optimized configurations of $HgCl_2$ adsorption on each site are shown in Figure 5. The adsorption energies and geometric parameters for $HgCl_2$ adsorption on MgO(001) surface are given in Table 4. The adsorption energies results show that the stability of $HgCl_2$ adsorption is in the order of 4B < 4A < 4C < 4D. The most stable structure for $HgCl_2$ adsorption on the surface is 4B, in which $HgCl_2$ adsorbs on the surface horizontally, and Hg atom is close to the surface O atom, Cl atoms are close to 2 adjacent surface O atoms, The Hg-Cl bond has no significant elongation, but the Cl-Hg-Cl angle is reduced to 163.399°. Similar to HgCl, the adsorption of $HgCl_2$ on the surface is also a very weak chemisorption. Based on the adsorption of HgCl on MgO(001) surface, it can be inferred that MgO(001) surface has no obvious effect on the adsorption and oxidation of Hg^0.

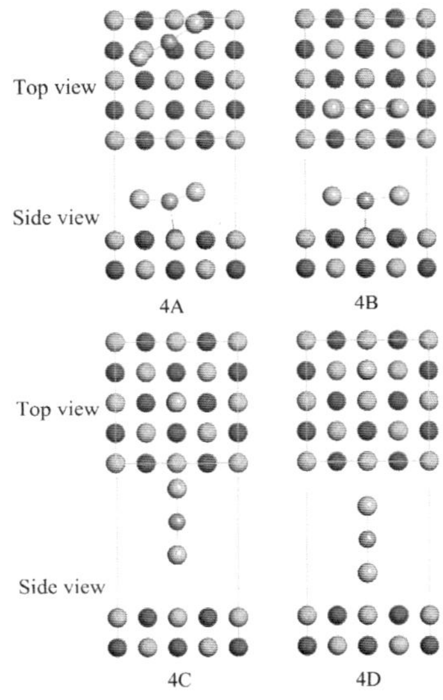

Figure 5. Optimized configurations of $HgCl_2$ adsorption on MgO(001) surface.

Table 4. The adsorption energies and geometric parameters of $HgCl_2$ adsorption on MgO(001) surface.

No.	Hg-O (Å)	Hg-Mg (Å)	Cl-O (Å)	Cl-Mg (Å)	Hg-Cl (Å)	Cl-Hg-Cl (°)	E_{ads} (kJ/mol)
4A	2.629	3.889	-	-	2.379	157.442	-31.195
4B	2.502	-	-	-	2.377	163.399	-51.625
4C	-	-	-	4.455	2.339	179.809	-12.955
4D	-		2.933	-	2.342	179.953	-12.923

4. Conclusions

The detailed adsorption mechanism of Hg on MgO(001) surface is investigated by a systematic density functional theory study and the periodic slab model. Hg^0, HgCl and $HgCl_2$ are all chemically absorbed on MgO(001) surface weakly, while HCl is dissociatively absorbed. Since HgCl cannot exist stably on the surface, the intermediate of Hg^0-Hg^{2+} oxidation cannot be formed. As a result, on MgO(001) surface, the adsorption and oxidation of Hg^0 cannot be obviously promoted.

5. Acknowledgments

The authors thank the Beijing Natural Science Foundation (3172030), Beijing Nova Program, and Fundamental Research Funds for the Central Universities (2016YQ05, 2016MS55) for financial support.

6. References

[1] Lu W, Tian Y, Zhan J and Xie M 2012 *Journal of Environment and Health* **29** 761
[2] Ren J, Zhou J, Luo Z and Cen K 2002 *Acta Scientiae Circumstantiae* **22** 289
[3] Meng S, Duan Y and Huang Z 2009 *Proceedings of the CSEE* **20** 66
[4] Wu C, Duan Y and Zhao C 2008 *Proceedings of the CSEE* **14** 55
[5] Zhao Y, Liu S and Ma X 2008 *Proceedings of the CSEE* **28** 55
[6] Guo P, Guo X and Zheng C 2010 *Applied Surface Science* **256** 6991
[7] Guo P, Guo X and Zheng C 2011 *Fuel* **90** 1840
[8] Sun S, Zhang D and Li C 2014 *Chemical Engineering Journal* **258** 128
[9] Xiang W, Liu J and Chang M 2012 *Chemical Engineering Journal* **200** 91
[10] Zhang B, Liu J and Yang Y 2015 *Chemical Engineering Journal* **280** 354
[11] Zhang B, Liu J and Zheng C 2014 *Chemical Engineering Journal* **256** 93

Geochemical Characteristics of Mercury in Oil and Gas

Qituan Yan, Zhongxi Han and Shuying Wang

Lang fang Branch, Research Institute of Petroleum Exploration and Development, Langfang 065007, China

729927878@qq.com.

Abstract. As an important element in oil and gas, the process of mercury formation, migration and enrichment, has an important practical significance to determine the nature and preservation status of deep oil-gas reservoirs, which can provide reliable geochemical basis for oil-gas exploration departments. Therefore, basing on the study of mercury formation, migration and enrichment, the geochemical anomaly patterns of mercury are determined, and the geochemical characteristics in oil and gas are discussed. Study shows that the mercury have strong affinity with oil and gas, and it will migrate and accumulate with the maturation of kerogen; Because of the abyssal faults, magmatic activity and deep geological process, in some petroliferous basins, the oil type gas have higher mercury content; On the basis of geochemical anomaly characteristics and morphological combination of mercury in surface, the anomaly patterns in oil and gas geochemical exploration are summarized to guide the anomaly interpretation and evaluation of testing zone, so as to provide the reliable geochemical basis for oil and gas exploration department.

1. Introduction

Because of the high dispersion of mercury in the lithosphere, the natural gas has the low mercury phenomenon. However, due to the secondary enrichment caused by chelation and cation exchange of coal, humus and sulfide to mercury, making the natural gas in many mercury enrichment areas are rich in mercury[1]. Form the migration, transformation and enrichment of mercury in the ore source layer, to a series of physical and chemical reactions of mercury from the ore source layer into the natural gas, they will be controlled by various geological factors, and which will determine the origin, content and distribution of mercury in natural gas [2]. Therefore, based on the origin and associated effect of mercury in petroleum and natural gas, the mercury anomaly pattern and formation mechanism are summarized, so as to provide a reference for oil and gas exploration.

The abnormal enrichment of mercury is a new geochemical prospecting method, which has the advantages of good stability and high abnormal recurrence rate [3]. In the early 80s, the adsorption phase mercury technique was applied to oil and gas geochemical exploration, and proved by practice that it has good oil and gas exploration effect, and is an effective index for oil and gas geochemical exploration. But at present, we lack the systematic understanding of mercury anomaly formation processes, morphological types or patterns, and methods for recognizing anomalies and techniques for evaluating anomalies [4]. Therefore, it is urgent to summarize the abnormal patterns of mercury above oil and gas reservoirs, analysis the formation mechanism, and innovate the effective methods, to guide oil and gas exploration and play the role of oil and gas geochemical exploration.

2. Sources of mercury in oil and gas

More and more evidences prove that the mercury has strong affinity with oil and gas, the mercury in the nature circle migration, circulation and accumulation constantly [5]. For the different mercury in different spheres, their origin and occurrence are also different. So, in the process of migration and transformation, the different geological background and environmental conditions (oxidation and reduction) make the distribution of mercury in coal and oil are different [6].

2.1. Mercury in source rocks

The formation of oil and gas is a process of thermal evolution, and with this process, the mercury in natural gas will be enriched from the source bed to the oil and gas continuously, and the different ability to absorb mercury, will inevitably lead to different mercury content from the different source rocks of parent materials [7]. The mercury is relative rich in terrigenous organic matter, and the content of mercury in oil type gas which is produced by mixed humic organic matter is usually lower than produced by humic type organic matter, so the difference of mercury content can be used as an important index to identify coal formed gas and oil type gas effectively.

In general, the rock and coal are not the main reasons that the mercury enrichment in natural gas, and it is probably produced by the magmatic hydrothermal activity. Coal has strong absorption capacity and can be used as the "middle storage station" of mercury. In the coalification, due to mercury in coal with the lowest of melting point, boiling point and heat of vaporization in the metal, so it is easy to deepen gasification with burial depth, and it will migrate with the gas channel that generated by coalification, into the trap. Therefore, when the gas-forming of coal is formed, the mercury content is high. So in the analysis of mercury content in natural gas, on the one hand from the main source of mercury-hydrothermal activity, on the other hand to study mercury enrichment and preservation conditions, this may be the reason why the mercury in gas reservoir is higher than that in coal field gas. (Table 1).

Table 1. Mercury content in natural gas bearing basins in China

Basins	Natural gas		Coal-formed gas		Oil-type gas	
	Scope (ng/m^3)	Arithmetic mean	Scope (ng/m^3)	Arithmetic mean	Scope (ng/m^3)	Arithmetic mean
Ying Qiong basin	430000-450000	440000	430000-450000	440000	—	—
Liu He basin	14400-1930000	216100	25200-52500	36900	14400-19300	298800
Huang Hua basin	170-7000	1190	—	—	170-7900	1510
Ji Zhong basin	170-256000	64640	610-256000	108600	170-142000	20600
Pu Yang basin	130-140000	12540	1440-140000	48640	130-1100	510
Si Chuan basin	40-125100	7830	250-125000	36400	40-400	690
Ordos basin	40-48100	7780	390-48000	10830	40-230	150
Junnar Basin	200-5450	2010	5450	5450	270-320	290
Traim Basin	170-8780	—	3530-8780	5680	170-2340	890
Qaidam Basin	760-9848	3970	4510-9840	5330	760-1840	1300

2.2. Inorganic mercury

The mercury from the deep earth is meaningful, according to the phenomenon that the mercury in hydrothermal origin mercury and crude oil near San Joaquin Valley oilfield in California, USA, Bailey, etc. proposed that mercury in oil and gas comes from ore-forming fluid in the deep of the earth.

In space, mercury deposits, oil and gas field and deep fracture zone is often three-in-one. For example, the mercury content is generally high in California, USA, while the mercury content of oil are mostly from San Joaquin Valley, which Cymric oilfield produced the highest mercury content of oil, up to 3×104ng/m^3. Cymric San Joaquin Valley oilfield is located in the west, and associated with the mercury deposits of New Almaden, New Idria (Fig.1). The above oil and mercury deposits are

IOP Conf. Series: Earth and Environmental Science **63** (2017) 012024 doi:10.1088/1755-1315/63/1/012024

controlled by the San Andreas great fault, which be explained that the oil and gas fields with high mercury content have the same tectonic background, that is, they are distributed in the compound and transfer sites of the giant tectonics, which is related to the great faults and deep faults zone.

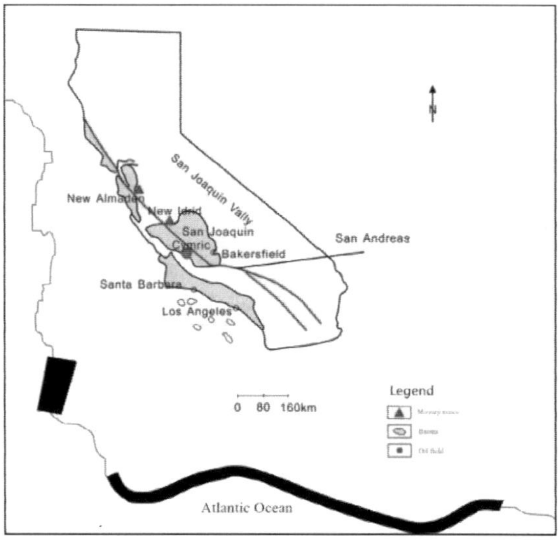

Figure 1. The location of Cymric San Joaquin Valley oilfield

At present, many scholars believe that the mercury is related to the tectonic activity and deep geological action of the basin [8]. Under normal circumstances, the faults are exists in the basin, due to the presence of the earth's deep source of mercury, the mercury content in oil and gas is high. For example, the mercury content is highest in extensional active basin in eastern China, while the middle cratonic stable basin is relatively low, and followed by the Northwest Asia stable Basin (Fig. 2).

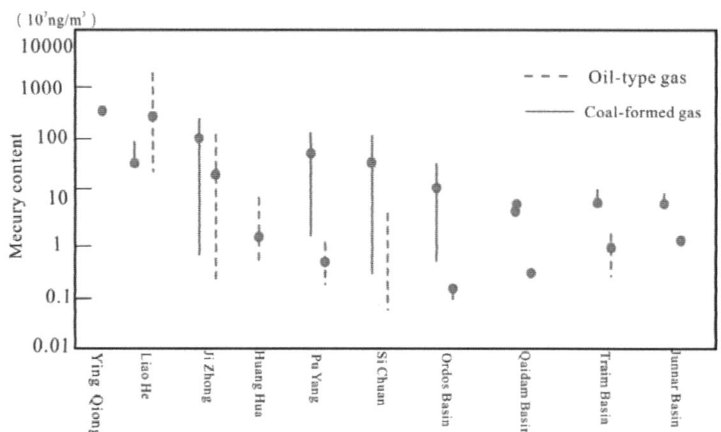

Figure 2. The distribution of mercury in oil-formed gas and coal-formed gas

3. Mercury migration

Due to the large ionization potential of mercury, mercury can be reduced to Hg0 in various forms, while Hg0 is highly volatile, and the volume is smaller than water, so that when the mercury diffusion and migration, it will has a great ability to penetration [9]. The mercury in oil-gas reservoirs, in order to maintain a certain physical and chemical equilibrium with the outside world, it will through the pores, peripheral faults and fractures, and accompanied with the cyclic action of groundwater, penetrating through thick rock upward diffusion, migration to the surface, and the formation of mercury anomalies in the surface soil layer.

Mercury can not only migrate in gaseous, but also in the form of complexes with organic matter. In recent years, the geochemical effects of Hg in oil-gas exploration have been paid more and more attention, and the inspiration has been drawn from the similarity of oil and gas fields and mercury deposit formation conditions [10]. The global mercury ore belt distribution along with the convergent plate boundary with compressional properties, in the space, adjacent to the convergent sedimentary basin of huge oil and gas fields in Alaska, California, Indonesia and some parts of the Mediterranean Sea (Fig.3). About 90% of the Hg ore bodies occur at the top of the anticline, about 86.4% of oil and 88.4% of natural gas are distributed in anticline type reservoirs.

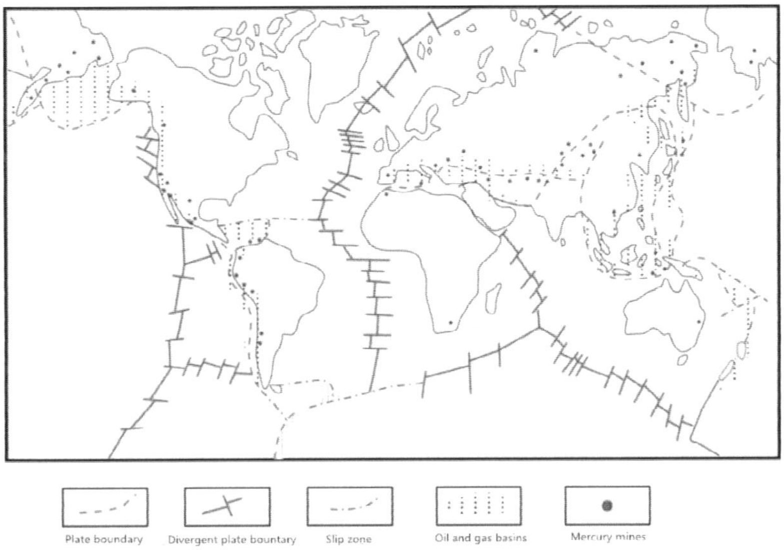

Figure 3. The major mercury mine and oil and gas basin in the world

The high porosity rocks under the weak permeable layer are both favorable positions for Hg and oil and gas reservoir occurrence. 75% of natural gas and 42% of oil are distributed in sandstone; terrigenous rocks (clastic rock) in the occurrence of Hg reserves of 75%, about 20% in carbonate rocks. Both of them have similar formation temperature and pressure conditions. Oil generation is mainly in the 1.5~6 km depth, and the temperature is range from 60 to 150Mpa, while the temperature and pressure conditions of the Hg deposit are from 70 to 170 ℃ and from 60 to 100Mpa accordingly. Mercury and oil and gas are similar in metallogenic epoch, both of them are mainly formed in Mesozoic Cenozoic Era.

4. Geochemical significance of mercury in oil-gas

The physical, chemical and geochemical properties of mercury are special, creating the complexity of mercury distribution in natural environment and oil and natural gas. Oil and gas contain different amounts of mercury more or less, which is directly related to the migration of mercury. But practice shows that because of different types of reservoirs, the hydrocarbon and associated group structure, characteristics of reservoir and cap rock, the migration loss method of components are different, the corresponding mercury enrichment patterns are also different, showing that mercury anomaly patterns are not the same.

4.1. Diversification characteristics of mercury anomaly

4.1.1. Mercury anomaly in annular. This abnormal index not only showed annular, half annular, embayment distribution in gas reservoir of edge, but also above the gas reservoir is the massive and banded, but its abnormal strength is slightly lower than the abnormal side gas reservoir (Fig.4). But at the specific natural gas reservoir, the abnormal distribution of different indicators of annular may be having a certain displacement or the horizontal zoning law has some differences. However, most of the indicators, especially the natural gas reservoir hydrocarbon anomaly of ring are basically consistence to each other. Such as the Yangxin sag area stores found the heat release mercury anomaly, which the anomaly were distribution of irregular large annular with strong north and weak south, and the anomaly of hydrocarbon forms a high concentration of half annular anomaly in the inner ring of mercury anomaly. It is inferred that the anomaly may be related to oil and gas reservoirs, and probably is the gas reservoirs. After verification by Y-16 well, the $418\times104\ m^3$ of natural gas and nearly 30m thick oil reservoir were obtained under the well depth of 1500m.

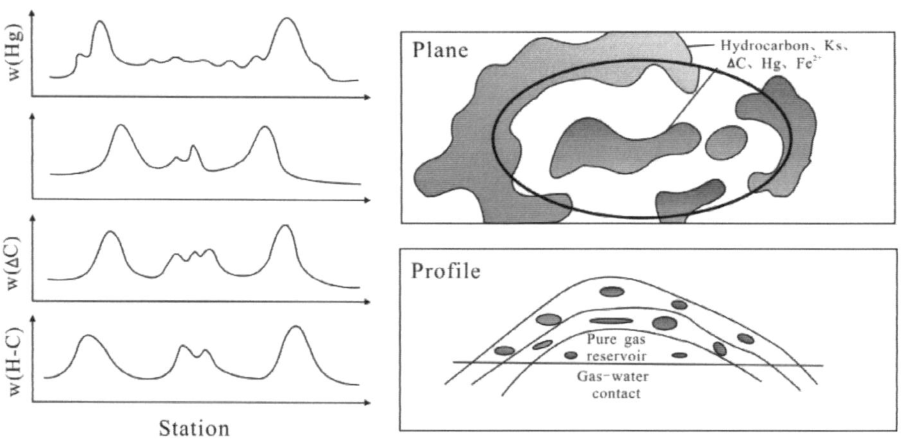

Figure 4. Geochemical anomaly pattern in the upper part of the gas reservoir

4.1.2. Mercury anomaly in massive. The main indicators of the model are regular or irregular planar or massive in shape and distributed in the upper part of the natural gas reservoir, that is often called the "top of the halo". This is mainly due to when the cap layer sealing property is poor, the hydrocarbons and mercury are transported vertically to the surface from the oil-gas reservoirs in various ways, such as percolation, diffusion and micro bubble, and showing a variety of irregular massive mercury anomaly. The mercurometric survey showing, the southwest of Wan'an Basin found a large mercury anomaly area, including the two anomalies of No. 6 and No. 11 (Fig.5). The No. 6 is the medium anomaly of the sea, the morphology is regular and with a large scale. The anomalies located at the sedimentary center of north and south basin, the world famous Daxiong oil field is in the abnormal range, the area is about $50km^2$, the recoverable oil is estimated at 7-8 billion barrels ($1.1\sim13m^3$), and with a large number of unassociated gas.

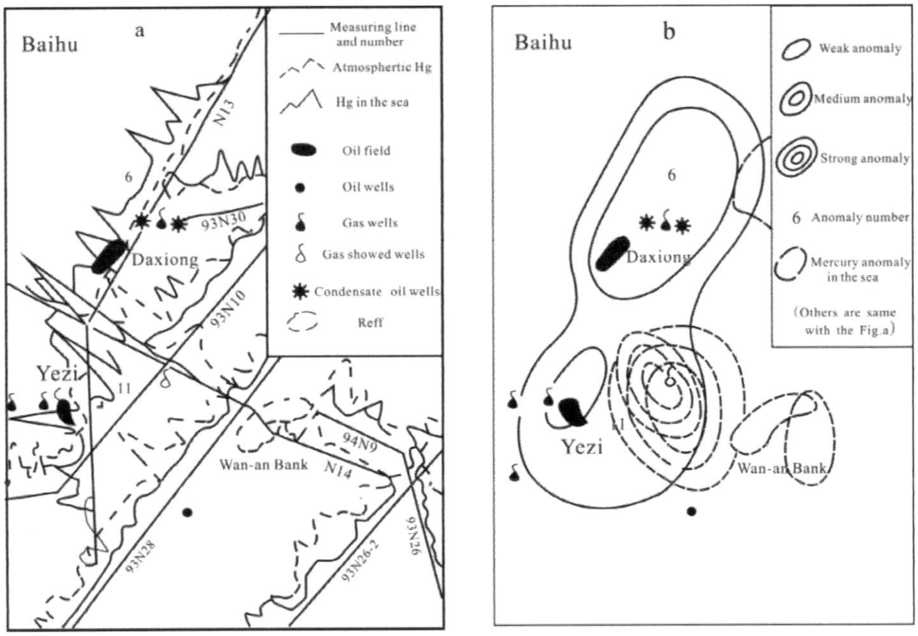

Figure 5. The anomaly distribution and content of mercury in Wan'an Basin

The 11 anomaly is located in the southern slope of the NO.6. It is a complex anomaly which is composed by the atmospheric mercury and mercury in the sea. The anomaly has a large scale and with good continuity, the Yezi oil field and a lot of wells are in the range.

4.2. Duality of mercury sources

The results show that, the anomaly of adsorption phase Hg and adsorption hydrocarbon often showed great differences, which absorbed mercury exhibits good ring halo or ring top halo anomaly, but the adsorption of hydrocarbon anomaly is very weak, scattered or not even, At the same time, the mercury in the adsorbed phase is not a reflection of oil and gas reservoirs, but an empty trap structure in the deep underground. For example, in Southwest Sichuan Z1 gas reservoir, the combination of the adsorbed phase mercury and the adsorbed hydrocarbon anomaly is very different, and the C1 anomaly of the Z1 gas reservoir is characterized by the top halo, and the abnormal body is in the middle of the abnormal prospect area; While the Hg is a "halo feature top" loop, the abnormal C1 abnormal distribution around the outer edge of the subject (Fig.6).

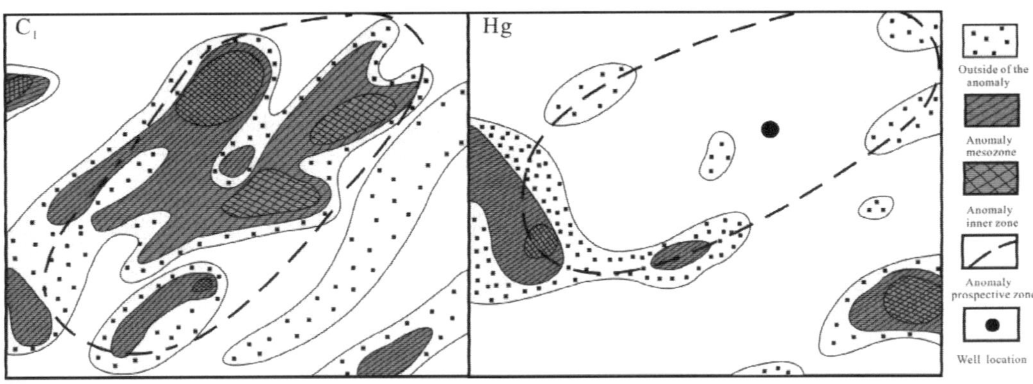

Figure 6. Duality of mercury sources

It can be seen the Hg from the deep underground oil and gas has two kinds of source, one is released in the process of oil and gas generation is released in the oil and gas generation process of Hg; And the other is from the deep of the earth to the oil and gas production layer through the fault channel. Because of the two character of the Hg source, leading to the double indication of the adsorption phase Hg anomaly.

4.3. The mechanism of oil-gas geochemical anomaly model

In the past 40 years, a large amount of dates obtained from the practice of oil and gas geochemical exploration have proved that the formation of geochemical anomaly is not controlled by a single factor[11-12]. The dispersion and migration of oil and gas is complex with each other. In addition, the complex effects of physical, chemical, physical, chemical and biochemical factors will occur in the process of percolation, that is to say, the fugitive migration is a dynamic system, its initial or boundary conditions with the change of geological environment and dynamic change, this is an important point of the main oil-bearing basins of the world. The conceptual model of Figure 7 is mainly based on the main factors of migration mechanism, dynamics, channel, phase, rate, intensity, microbial action and mineral alteration, and the formation mechanism of oil and gas geochemical anomalies were displayed by the system.

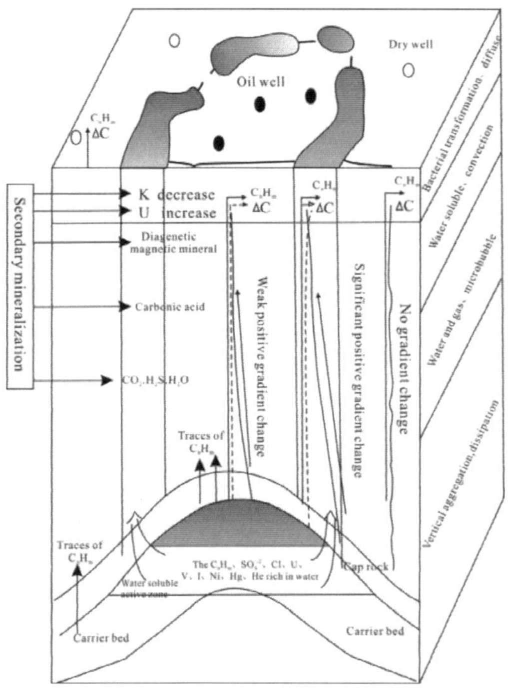

Figure 7. Basic mechanism of oil and gas geochemical exploration

5. Conclusion

1) The formations of mercury in natural gas are organic and inorganic. The organic origin refers to the mercury in natural gas from hydrocarbon source rocks, and the gas content is influenced by different types of parent material source, in general, the mercury content in coal type gas is higher than the oil type gas; The inorganic origin refers to the mercury in natural gas, which is mainly comes from the deep of the earth's ore-forming fluid, and the mercury content in natural gas is higher than that the oil bearing basin that with the basement fault structure and magmatic activity.

2) Both of the oil and gas are contain of the mercury composition, physical and chemical properties of mercury element can not only determine it can migrate as the form of gaseous, forming a certain

abnormal pattern on the surface, but also can be transported in the form of complexes with organic matter.

3) In the geochemical exploration of oil and gas, the characteristics of surface geochemical anomalies and morphology can determine the nature and preservation status of deep oil and gas reservoir. According to the above rules, the geochemical anomaly of oil and gas can be comprehensively interpreted and evaluated, and get the more accurate results, so as to make an objective and scientific prediction for the prospective area of oil and gas exploration and the location of underground oil and gas reservoirs, and provide reliable geochemical exploration for oil and gas exploration departments.

6. References

[1]He, W., Tang, D.Z,Yan, Q.T.,Xu, H., 2011. Distribution of mercury in natural gas and its origin mechanism. Resources and Industries, 13(6): 110-117.

[2] Li, G.Z., Yuan, Z. Y., Zhuang, Y., Jiang, H., 2008., Geological significance of mercury element for petroleum exploration. Geophysical and exploration, 32(2): 143-147.

[3] Li, S.J., 1998. The features of mercury and their application in oil-gas geochemical field mineral. Resources and geology, 12(66): 281-286.

[4] Yao, Z.G., Gao, P., 2008. Anomaly model for oil and gas geochemical exploration. Geophysical an exploration, 38(6): 604-610.

[5] Gustkiewicz, M. S., Kwiecinska, B., 2001. Organic matter in the upper Silesian (Mississippi Valley– type) Zn- Pb deposits, Poland. Econ. Geol, 94(7): 981-992.

[6] Han, Z.X., Li, J., Yan, Q. T., Gou, Y.X., Wang, S.Y., Wang, C.Y., 2013. Discussion of natural gas mercury content as an identification index of coal type gas and oil type gas. Natural gas geosciences, 24(1): 129-134.

[7] Tu, X.Y., 1980. Existence of mercury in gases from Biyang basin, Henan province. Oil and gas geology, 1980, 1(3) : 241-247.

[8] Dong, W.L., Huang, B.J., 2000. Identification and source identification of coal type gas in Ying Qiong basin, South China Sea. Natural Gas Industry, 20(1) : 23-27.

[9] Krupp R.Physicochemical., 1998. Aspects of mercury metallogenesis. Chemical Geology,69(3-4): 345-356.

[10] Dreyer R. M., 1940. Geochemistry of quick silver mineralization. Econ. Geol, V. 35: 17-48 and 140-157.

[11] Cheng, J.T., Tang, Y.P., 1999. Near surface display and geochemical exploration of hydrocarbon migration, Petroleum Industry Press, Beijing.

[12] Zhou, Q.M., Jia, X.G., Zhao, Y.F., 2000. Characteristic of supergene geochemical anomaly in different development period of oil-gas field. Mineral resources and geology, 14(2): 33-37.

A Review on Advanced Treatment of Pharmaceutical Wastewater

Y Guo, P S Qi and Y Z Liu

State Key Laboratory of Urban Water Resource and Environment, Harbin Institute of Technology, Heilongjiang, China

qipeishi@163.com

Abstract. The composition of pharmaceutical wastewater is complex, which is high concentration of organic matter, microbial toxicity, high salt, and difficult to biodegrade. After secondary treatment, there are still trace amounts of suspended solids and dissolved organic matter. To improve the quality of pharmaceutical wastewater effluent, advanced treatment is essential. In this paper, the classification of the pharmaceutical technology was introduced, and the characteristics of pharmaceutical wastewater effluent quality were summarized. The methods of advanced treatment of pharmaceutical wastewater were reviewed afterwards, which included coagulation and sedimentation, flotation, activated carbon adsorption, membrane separation, advanced oxidation processes, membrane separation and biological treatment. Meanwhile, the characteristics of each process were described.

1. Introduction

In China, pharmaceutical technology can be classified by productive methods. First is biopharmaceutical, which contains fermentation engineering, cell engineering, enzyme engineering and genetic engineering. The most popular is fermentation engineering. It depends on activities of microorganisms to ferment, filter and extract from organic raw materials, such as antibiotics, vitamins, amino acids. Next is chemical pharmacy, which is synthesis using multilevel chemical reactions of organic matter or inorganic matter. In addition, there is another kind of method by extracting from plants and animals or directly, called Chinese patent medicine, also known as natural medicine.

In recent years, the fast development of pharmaceutical industry has become an important part of China's national economy. In 2014, the output of chemical original medicine had reached 3.034 million tons, rose by 15.2 %, compared to the same period in 2013. There are some characteristics of the pharmaceutical industry in China, such as a multitude of enterprises and products with small scale, scattered overall arrangement. By the end of 2014, the pharmaceutical enterprises have reached 7108 in China, which was 8.9 % higher than 2013. Only 1609 among them are large or medium-sized enterprises.

However, with the development of pharmaceutical industry, the environmental pollution is becoming more and more serious. Due to the variety of the pharmaceutical industry products, and the difference of production scale and process, there are many kinds of pharmaceutical wastewater. Biopharmaceutical wastewater is mainly generated by high-concentrated antibiotic wastewater, which is characterized as strong fluctuation in quantity, low C/N, high SS concentration, high sulfate concentration, complicated composition, biological toxicity and high chroma. The composition of chemical pharmacy is single, leading to lack of nutrition. It also has high concentration and salt content. And it is hard to biodegrade and toxicity to microbiology [1].

Along with the strict government standard, traditional treatment methods of pharmaceutical wastewater are difficult to satisfy the demand. Therefore, advanced treatment of pharmaceutical wastewater is essential.

The review aims to introduce the fundamental advanced treatment of pharmaceutical wastewater.

2. Characteristics of pharmaceutical wastewater

2.1. Characteristics of pharmaceutical wastewater

In general, the composition of pharmaceutical wastewater is complex, which has high concentration of organic matter, microbial toxicity, high salt, and it's hard to biodegrade [2,3]. In addition, most of pharmaceutical factories are batch process, and there are different raw materials and production process, which causes huge varieties in different wastewater [4]. Table1 provided a summary of the characteristics of pharmaceutical wastewater [5].

Table 1. Common pharmaceutical wastewater physicochemical properties

COD (mg/L)	BOD_5 (mg/L)	TN (mg/L)	TP (mg/L)	SS (mg/L)	Chromaticity (times)	Temp. (℃)	pH
1000-10000	500-2500	500-1500	50-250	200-500	500-1000	25-80	1-8

Otherwise, different kinds of pharmaceutical wastewater has different characteristics. Biopharmaceutical wastewater is characterized as strong fluctuation in quantity, low C/N, high SS concentration, high sulfate concentration, complicated composition, biological toxicity and high chroma. Chemical pharmacy is lack of nutrition, hard to biodegrade and toxicity to microbiology, and it also has high salt content. The characteristics of the wastewater of Chinese patent medicine is containing sugar, glycosides, organic pigment, anthraquinone, tannins, Alkali content, cellulose, lignin and other organic matter [6].

2.2. Characteristic of secondary effluent

Biological treatment is mainstream for pharmaceutical wastewater at home and abroad, which is the most economical way to remove organic pollutants. Therefore, organic matter are the main pollutants in pharmaceutical wastewater. Biological treatment technologies can be generally divided into aerobic process, anaerobic process and combination of anaerobic-aerobic process [7]. After secondary treatment, almost every kind of pollutants concentration are decline.

However, due to their own complex features of pharmaceutical wastewater, and discharge standard is increasingly strict, the current methods are hard to satisfy the demand.

3. Methods of Advanced treatment of pharmaceutical wastewater

In recent years, the main emphasis of the scientific research and engineering application has shifted to advanced treatment of pharmaceutical wastewater, which main method is physicochemical technology [7]. It means that wastewater is treated by physical or chemical methods, like coagulation and sedimentation, flotation, activated carbon adsorption, advanced oxidation processes, membrane separation.

3.1. Coagulation and sedimentation

Coagulation is adding chemical agents to wastewater, dispersing by rapid mixing, then making stable pollutants into unstable and precipitable matters. The mechanism of coagulating is complex. For advanced treatment of pharmaceutical wastewater, the key is how to squeeze and remove bound water round hydrophilic colloid. So the character of flocculent is important, which related to the effect of coagulation. Inorganic metal salts and polymers are frequently used as flocculent. This method can remove SS, chromaticity and toxic organic matter [8]. Meanwhile, it can improve the biodegradability of pharmaceutical wastewater.

Sedimentation is the most common method after coagulation. Under the gravity, pollutants can be separated, which has greater density than wastewater.

Coagulation and sedimentation have some advantages, such as easy operation and mature technology, but it is hard to remove dissolved organic matter.

3.2. Flotation

Except for sedimentation, flotation can also remove suspended solids of secondary effluent. The technology characteristic is producing a large number of tiny bubbles by injecting air into wastewater, forming floating floc with smaller density than wastewater. And it can float to the surface of wastewater to separate.

3.3. Activated carbon adsorption

Activated carbon, as a kind of adsorbent, has many advantages. It has large specific surface area, multilevel pore structure, high adsorption capacity and stable chemical property. Therefore, it is widely used as adsorbent or catalyst carrier to remove pollutants [9,10]. In industrial effluents treatment, activated carbon is used for effluent, which is toxic and hard to achieve discharge standard. It is an important method of advanced treatment of pharmaceutical wastewater as well.

Activated carbon adsorption can be classified as physical adsorption and chemical adsorption. Physical adsorption is reversible, and no selectivity to adsorbate. When activated carbon saturated by adsorbates, it is easy to desorb. To the contrary, chemical adsorption adsorbs only one or several specific adsorbates, which is irreversible and hard to desorption. For cyclic utilization, saturation of activated carbon restores its adsorption property by regeneration.

This method is widely used for advanced treatment, because it can be recycled, its better treatment effect and wide suitability. But there are some disadvantages, such as high costs relatively, low efficiency of regeneration and complex operation, which limit application.

3.4. Advanced oxidation processes

Advanced oxidation processes (AOPs), which can oxidize pollutants by forming free radicals. Those kinds of pollutants cannot be degraded by common oxidizing agent. There are many kinds of AOPs, such as wet air oxidation, supercritical water oxidation, Fenton reagent, photocatalytic oxidation, ultrasound oxidation, electrochemical oxidation and ozonation.

3.4.1. Wet air oxidation(WAO). WAO has been put forward by F. J. Zimmer Mann in 1958, which was used for papermaking black liquid treatment. By using of air or oxygen as the oxidant, this method decomposes organic matter into inorganic or small molecules at high temperature (150-350 ℃) and high pressure (0.5-20 Mpa). WAO is generally used in pretreatment of wastewater advanced treatment.

This method has wide range of applications, high efficiency of COD removal, which can even reaches more than 90 % under appropriate conditions, low energy consumption, less secondary pollution, and it is easy management.

3.4.2. Supercritical water oxidation(SCWO). SCWO is chemical reaction between dissolved oxygen and organic pollutants in supercritical water. Organic matter, air, and supercritical water were complete mixed at 24 Mpa pressure and 400 ℃ temperature, becoming homogeneous phase. Under these conditions, organic compounds spontaneously initiate the oxidation reaction. With the increase of the reaction temperature, 99.9 % or more of the organic matter is rapidly oxidized into simple non-toxic small molecules in a period of time, achieving the purpose of removing pollutants.

SCWO has high oxidation efficiency, will not cause secondary pollution, organic can be oxidized completely. However, this method has some shortcomings, such as it requires high operating conditions and high cost.

3.4.3. Fenton reagent. Fenton system was initially used in organic synthesis, with the deepening of people's understanding of it, the system was gradually applied to industrial wastewater treatment.

Fenton reaction can be carried out under normal temperature and pressure, and less damage to the environment [11]. It is an advanced oxidation method, which is simple, mild reaction conditions, and high oxidation efficiency. However, the shortcomings of the Fenton reaction cannot be ignored. One is due to free radical scavenging and decomposition of H_2O_2 itself, leading the loss of oxidant. The other is the formation of iron mud in the neutral conditions, making it hard to the subsequent processing [12].

3.4.4. Photocatalytic oxidation. Photochemical oxidation, also known as ultraviolet photocatalytic oxidation, is a method combination of UV radiation and oxidant. In the excitation of ultraviolet light, oxidative decomposition of oxidants produces stronger oxidative capacity of free radicals, which can oxidize more difficult to decomposition of refractory organic pollutants with oxidants only.

According to the types of oxidants, photochemical oxidation can be divided into UV / O_3, UV / H_2O, UV / H_2O_2 / O_3 and so on. At present, the steps of the reaction process still can't be described. However, the effect of this method on the treatment of refractory organic matter is undeniable.

3.4.5. Ultrasound oxidation. Ultrasonic degradation of organic matter in wastewater is a physical and chemical degradation process, mainly based on ultrasonic cavitation effect and the resulting physical and chemical changes, which includes the following three ways: free radical oxidation, pyrolysis and supercritical water oxidation.

There are many factors that affect the effect of ultrasonic degradation of organic pollutants in wastewater. The main factors are ultrasound frequency, sound power, ultrasonic time, solution temperature and pH, and the nature of organic matter.

3.4.6. Electrochemical oxidation. The use of electrochemical reaction to remove toxic and harmful pollutants in the water, is known as the electrochemical method. Electrochemistry is a new method in water treatment field. The principle of the method is as follow: in the electrochemical reaction process, the reactant will lost electrons and be oxidized in the anode. Conversely, the reactant in the cathode will lose electrons and be reduced. In general, the removal of refractory organic matter, mainly due to the oxidation of the anode. Conventional electrochemical wastewater treatment processes includes: electrolytic recovery, electrochemical oxidation, electrolytic air flotation, electrodialysis and micro-electrolysis.

Electrochemical method is also known as "Environmentally Friendly" process, has a great advantage compared with other methods. For example, the electrochemical method is generally carried out under normal temperature and pressure and has high efficiency, can be used alone or in combination with other processes, and it covers a small area and no secondary pollution, has relatively high degree of automation. The main research direction of the electrochemical reaction in the future is the research of the anode and electrochemical reactor [13].

3.4.7. Ozonation. For a long time, ozone is considered to be a very effective oxidant and disinfectant. In acidic conditions, ozone is primarily an oxidant. But in neutral and alkaline conditions, it mainly relys on free radical reactions. Table 2 shows the oxidation potential of various oxidants. It can be seen that the oxidation capacity of ozone is significantly higher than other conventional oxidants.

Table 2. The oxidation potential of various oxidants

Oxidant	Half-reaction	Oxidation potential(V)
• OH	• $OH + H^+ + e \rightarrow H_2O$	3.06
O_3	$O_3 + 2H^+ + 2e \rightarrow O_2 + H_2O$	2.07
H_2O_2	$H_2O_2 + 2H^+ + 2e \rightarrow 2H_2O$	1.77
HClO	$2HClO + 2H^+ + 2e \rightarrow 2Cl^- + 2H_2O$	1.63
Cl_2	$Cl_2 + 2H^+ + 2e \rightarrow 2Cl^-$	1.36

Ozone can quickly oxidize and decompose most of the organic matter in the water, which can effectively remove pollutants in water. However, it is difficult to directly mineralize. Instead, a product which is easily biodegradable can be produced. At the same time, it can also effectively remove turbidity and pathogens in wastewater.

Ozone can be combined with other wastewater treatment technologies, forming ozone advanced oxidation technology. This method has stronger oxidation ability and lower selectivity to reactants. Such as O_3 / H_2O_2, O_3 / UV, etc.

3.5. Membrane separation

Under certain driving force across the membrane, a component in water selectively permeates it by using permselective membrane separating media, which is called membrane separation. In this way, achieve the separation, purification, concentration of the target substance from the mixture. There are several membrane separation techniques in wastewater treatment. Such as microfiltration, ultrafiltration, reverse osmosis and electrodialysis.

3.5.1. Microfiltration(MF). MF is based on static pressure as the driving force, and the separation process is performed by the action of the sieve separation of the membrane, which principle is similar to traditional filtration. The slight difference is that the pore size of MF is smaller. This method can effectively remove SS and microorganisms in wastewater.

3.5.2. Ultrafiltration(UF). UF driving force is the pressure difference between the membranes on both sides, the filter medium is the ultrafiltration membrane. Under certain pressure, when water passes through the membrane surface, water, inorganic salts and small molecules penetrate, other macromolecules are trapped.

This method is mainly used for the removal of macromolecules and colloids in wastewater. In the application of this method, it should be ensured that the membrane has adequate membrane flux and is easily disassembled, replaced, cleaned.

3.5.3. Reverse osmosis(RO). There are two categories of RO membrane, cellulose ester and aromatic polyamide. Its component form includes tube, plate and frame, roll and hollow fiber type.

RO process can remove a wide range of impurities, dissolved inorganic salts and a variety of organic matter. Meanwhile, it has a high efficiency of salt removal and water reuse rate. However, this method requires a high pretreatment of the feed water. With the development of RO, nanofiltration (NF) has been proposed as a new method.

3.5.4. Electrodialysis(ED). ED is a combination of electrolytic and dialysis diffusion process. Under the action of the DC electric field, anions and cations of the dissolved salts in the wastewater are moved to the anode and the cathode respectively. In this way, the concentration of anions and cations in the intermediate compartment is gradually reduced, and the separation and recovery are achieved.

This method has many advantages, such as less energy and pharmaceutical consumption, less environmental pollution, easy to operate and automate. But it can only remove the salt in water, and desalination efficiency is lower than RO.

3.6. Biological treatment

In theory, after secondary treatment, pharmaceutical wastewater should not be treated by biological methods, because of its poor biodegradability. However, we cannot ignore the advantages of biological treatment, such as low cost, stable treatment effect. It can be used as a way of pretreatment in advanced treatment.

4. Conclusion

Due to the complexity of pharmaceutical processes, pharmaceutical wastewater has some characteristics, such as poor biodegradability and high concentration. From these characteristics, advanced treatment of pharmaceutical wastewater is very necessary. There are many kinds of advanced treatment, each method has its own features. Through rational utilization of various methods, can effectively improve the quality of pharmaceutical wastewater effluent.

5. Acknowledgments

During the writing of this paper, Prof. P S Qi and Mrs. Y Z Liu gave me great support and encouragement, I would like to express my sincere thanks to them. And thanks to my classmates, B Liu and L L Liu.

6. References

[1] Rui D 2013 Study on Advanced Treatment of Pharmaceutical Wastewater (Guangzhou: South China University of Technology) pp 1-3
[2] Shubin X and Liren R 2009 *J. Fine. Specialty. Chemicals.* **3/4** 16-8
[3] Huiqiang X, Liangjun L and Yuting H 2005 *J. Environ. Sci. Techno.* **28** 92-3
[4] Wenxin M, Weizhong C, Jianjun R, Zhangqian W and Yueming L 2001 *J. Environ. Pollution. Control.* **2** 87-9
[5] Xin L and Guoyi L 2015 *A Review: Pharmaceutical Wastewater Treatment Technology and Research in China* (Zhuhai: Asia-Pacific Energy Equipment Engineering Research Conference) pp 345-7
[6] Yu Y 2013 *Experimental on Pharmaceutical Tail Water before Biochemical Pretreatment* (Changchun: Jilin University) pp 12-9
[7] Liang W, Qi X and Wenfei Z 2015 *J. Technology. Innovation. Application.* **6** 1-2
[8] Yiping G and Yu B 2010 *Advanced Treatment and Recycling Technology of Wastewater Treatment Plant* (China Architecture Press) pp 198-206
[9] Bingyu M, Yubin T, Fangyan C and Tao W 2015 *J. Chinese. J. Environ. Eng.* **4** 1863-4
[10] Sandoval R, Cooper A M and Aymar K 2011 *J. J. Hazard. Mater.* **20** 296-303
[11] Joseph J, Pignatello, Esther O and Allison M 2006 *J. Crit. Rev. Env. Sci. Tec.* 1-84
[12] Bokare A D and Choi W. 2012 *J. J. Hazard. Mater.* **275** 122-4
[13] Hui W and JianLong W. 1999 *J. Chin. J. Envir. Sci.* **19** 441-4

Mechanism of Electro-Coagulation with Al/Fe Periodically Reversing Treating Berberine Pharmaceutical Wastewater

Zhaonan Sun[1,*], Zheng Liu[2] and Xiaomin Hu[3]

[1]Department of Chemical and Material Engineering, Yingkou Institute of Technology, Yingk, China;
[2]State Grid Liaoning Maintenance Company, Yingkou, China;
[3]College of resources and civil engineering, Northeastern University, Shenyang, China.

280728183@qq.com

Abstract. The method of treating pharmaceutical wastewater by electro-coagulation with Al/Fe periodically reversing (ECPR) was proposed based on traditional electrochemical method. The principle of ECPR was generalized. Mechanism of ECPR to treat berberine pharmaceutical wastewater was investigated. Treatability and mechanism studies were conducted under laboratory conditions. For berberine wastewater (800 mg/L), decolourization efficiency and COD removal efficiency were highest to 98% and 95% respectively when voltage was 8V, reaction time was 60 min, alternating period was 10 S electrolyte concentration was 0.015 mol/L, stirring speed was 750 rpm, pH value was 3-10 and distance between two plates was 0.6 cm. For removal berberine, flocculation ,floatation and oxidation provided 73%, 8% and 18% removal efficiency, which can be inferred by analysing UV-visible absorption spectrum, acidification experiment, EDTA shielding experiment, structure-activity relationship, oxidation and floatation. Meanwhile decolourization and COD removal conformed to apparent pseudo-first order and zero-order kinetics for 200mg/L and 400-1000 mg/L berberine wastewater respectively.

1. Introduction

Berberine is an important alkaloid, and traditional Chinese medicine applied long in China, which has a significant inhibitory effect. Its structure is shown in Table 1. Berberineis producted using natural plant extracts and chemical synthesis, through relatively mature technology of sulfuric acid and lime method [1].The mother liquor of berberine contains a high concentration of berberine which has strong antibacterial. Duo to it can't be directly treated by biochemical methods, some pretreatment methods are needed like coagulation, membrane filtration or chemical before biochemical treatment. Electro-coagulation has been used to treat dyeing wastewater [2-3], papermaking wastewater [4], the coking wastewater [5], pharmaceutical wastewater [6], oily wastewater [7-8], heavy metal wastewater [9] and other wastewater. But traditional electro-coagulation technology uses of unidirectional power and one kind anode, which easily lead to extreme concentration and polarization plate passivation. And then treatment efficiency is reduced, power consumption will be increased. With regards to these, power supply was improved on the basis of traditional electrochemical, berberine wastewater was treated using homemade periodically reversing power [10] and aluminium and iron as electrodes. The mechanism of ECPR treating wastewater under the optimal conditions was investigated. Flocculation,

IOP Conf. Series: Earth and Environmental Science **63** (2017) 012026 doi:10.1088/1755-1315/63/1/012026

electric floatation and oxidation for berberine removal were quantified, which had guided significance to improve treating efficiency and reduce power consumption through adjusting reaction factors.

Table 1. The characteristics of berberine

Name	Structure	Molecular weight (g/mol)	Molecular formulae	(nm)
Berberine	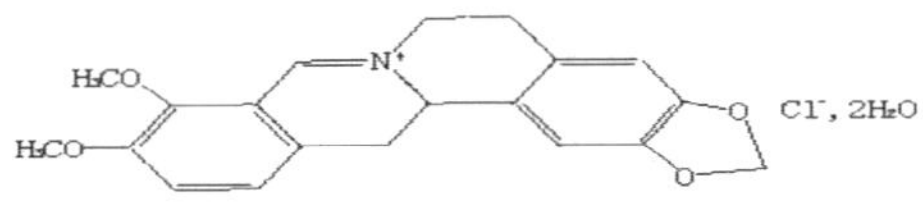	373	C20H18ClN O4	344

2. Materials and procedure

2.1. Principle of ECPR

In the process of treating organic wastewater by Fe/Al ECPR, Fe and Al plates as anode in turn dissolved and released Fe^{2+}、Fe^{3+} and Al^{3+} under commutation current function. And then these ionic are formation of various hydroxyl complexes, multi-core hydroxy complex and even hydroxides via a series of hydrolysis, polymerization and ferrous oxidation process, which could separate suspended and colloidal impurities by cohesion precipitation. Meanwhile charged pollutant particles moving in an electric field, could be destabilized and removed though part of charges neutralized. Moreover pollutant can directly be oxidized and reduced like ·OH and Fe^{2+} on surface of anode and cathode in ECPR process. And the principle can be shown in Figure 1.

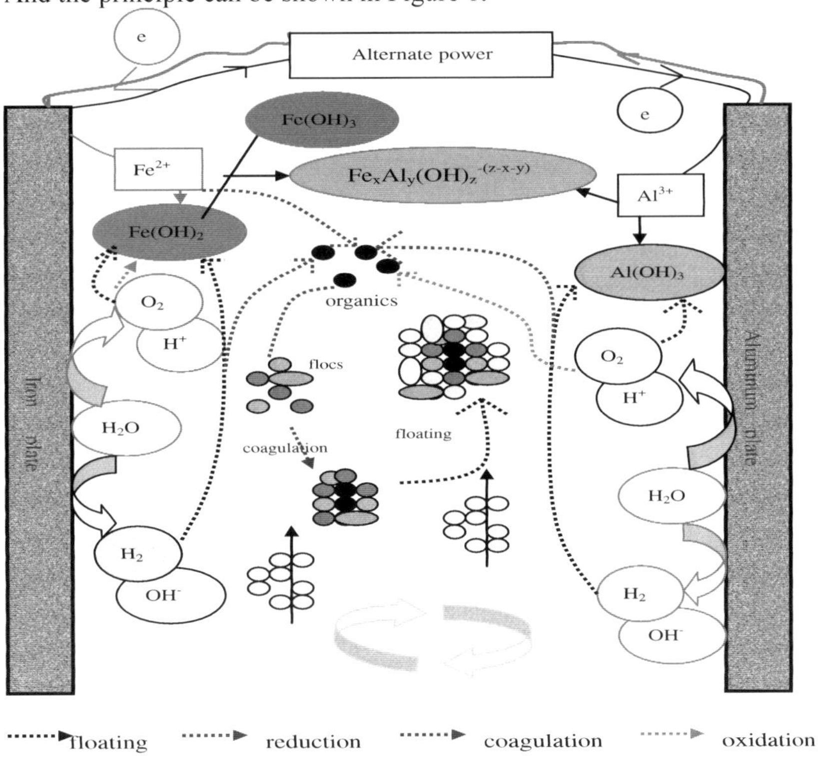

Figure 1. Basic principle of ECPR with Al/Fe

2.2. Materials
The berberine was obtained from a pharmaceutical factory in northeast China. The main characteristics of this organic substance and its structure are shown in Table 1. The simulated concentration of the berberine wastewater was 800 mg/L. The other agents were of analytical grade.

2.3. Procedure
Self-designed batch-type electrolytic reactor [10] was applied to the experiment, which was shown in Figure 2. Reactor is made of Plexiglas column with the effective volume of 400 mL. Aluminium and iron plates (80 mm × 60 mm × 2 mm, adjustable distance between the electrode plates) were used as electrodes in turn. And the solution was mixed by a heating magnetic stirrer to ensure a constant concentration of ions. The simulated wastewater was prepared with pure berberine (800 mg/L), and Na_2SO_4 was added as the supporting electrolyte.

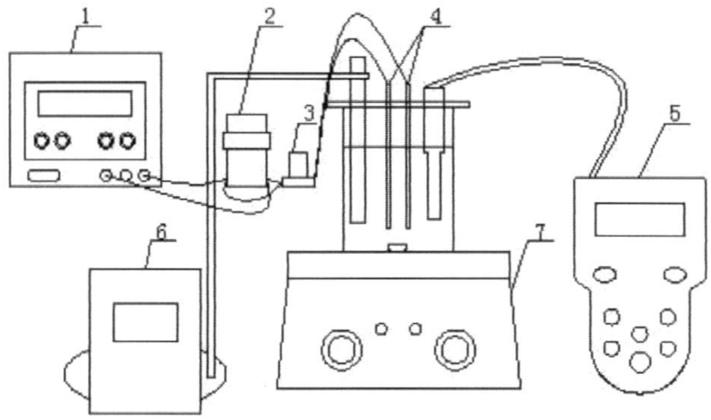

1 D.C.Power Supply
2 Clock relays
3 Relay
4 Electrode (Al, Fe)
5 Conductivity meter
6 pH meter
7 Magnetic stirring controller

Figure 2. Experimental apparatus

3. Results and discussion
For berberine wastewater (800 mg/L), decolourization efficiency and COD removal efficiency were highest to 98% and 95% respectively when voltage was 8V, reaction time was 60 min, alternating period was 10 S electrolyte concentration was 0.015 mol/L, stirring speed was 750 rpm, pH value was 3-10 and distance between two plates was 0.6 cm. Validity and stability of treating berberine wastewater by ECPR under optimal conditions had been verified [11]. The mechanism of ECPR treating berberine wastewater was investigated in next experiments.

3.1. Analysis of UV-visible absorption spectrum
Changes process of berberine molecules were researched through analysing UV – visible absorption spectrums after treating at different reaction time. Figure 3 shown that characteristic absorption peaks of berberine (228 nm, 263 nm and 344 nm) were significantly reduced as the reaction proceeding, the absorption peaks disappeared at 60 min, and there weren' t new absorption peaks appeared in the whole reaction process. Yellow bubble layer was found on the uppermost layer of the solution, which due to the gas adsorbent berberine molecules. Yellow-green bubble layer was under yellow bubble layer, which formed because of divalent iron ion complex and aluminium ionic flocculants berberine molecules. Therefore it can be inferred that most berberine molecules were removed directly by floatation of gas and flocculation of forming inorganic salt flocculants with aluminium ions and metal ions. Through further analyzing, the reason may be that berberine molecules dissolved in water and dissociated into colloid charging positively and chloride ions, and the hydroxide flocculants forming of the iron ion and aluminium ion usually charge negatively. Two oppositely charged colloidal occurs electroluminescence, so most berberine molecules were directly flocculating removed. Meanwhile Chloride ions can be formed to Cl_2 near the anode, and further generating strong oxidizing ClO^- and ClO_4^- can oxidize berberine molecules, but the oxidation product needed to further proof.

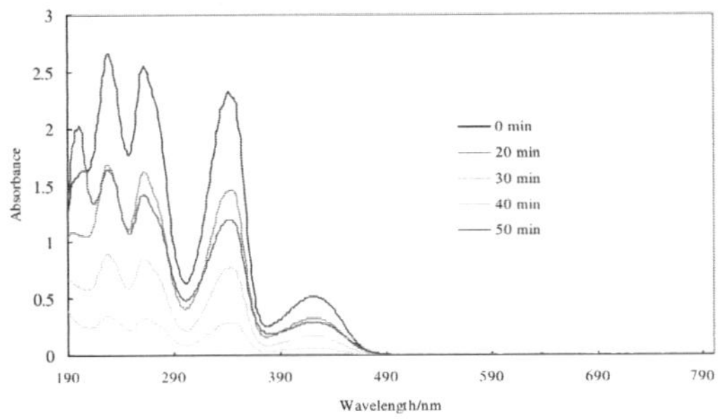

Figure 3. Scanning spectrum of UV - Vis

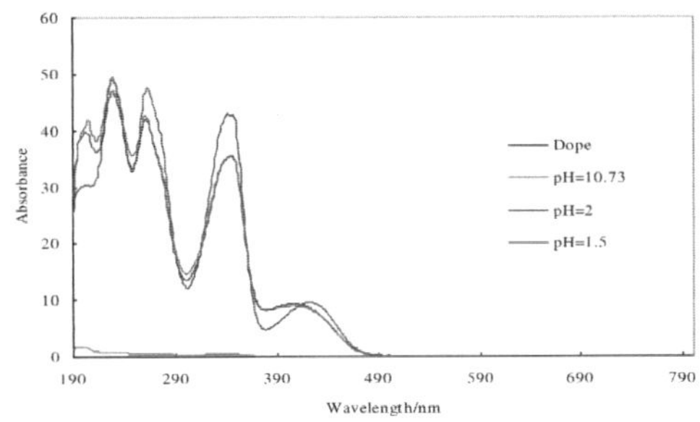

Figure 4. The effect of different pH value on scanning spectrum of UV – Vis

3.2. Analysis of flocculation of mental ions by acidification

The process of treating barberine wastewater by ECPR, may include flocculation of Fe $(OH)_3$, Fe $(OH)_2$ and Al $(OH)_3$ and adsorption of metal hydroxy complex. The pH value of solution had important impact on precipitation of Fe^{2+}, Fe^{3+}, Al^{3+} and formation of complex. Fe^{3+} began to precipitate at pH value of 2 and Fe $(OH)_3$ formed. Fe^{2+} and Al^{3+} successively precipitate with the pH value increasing. If the pH value of solution was adjusted to 2 or less, then metal ion flocculants dissolved, which will course flocculated berberine molecules re-released back to solution. The following experiment was designed to further study flocculation of metal ions in the process of ECPR.

Table 2. The effect of different pH value on COD value of berberine solution

	COD of dope (mg/L)	Reaction 60 min	Acidification after reaction 60 min	
		pH=10.73	pH=2	pH=1.5
Solution of berberine	1023.8	58.6	1085.6	1098.2
Solution of Na_2SO_4	0	20.5	79.6	87.7
Difference	1023.8	38.1	1006	1010.5

Figure 4 shown that the decolourization rate was up to 99% when wastewater (800mg/L) reacted 60 min under optimal conditions. When the pH value of the solution was reduced to 2 or less, flocculation and flotation were eliminated, and then the decolourization rate dropped to 18%. As seen from Table 2, when the pH value was 1.5, the COD value of the solution was close to the COD value of the berberine dope. It can be inferred that 19% berberine molecules were bleaching via oxidation or reduction, but formed intermediate molecules still remained in solution. And 80% berberine molecules redissolved with the dissolution of the metal flocculants, which indicated that flocculating and floatation of the metal ions was dominant in the process of treating berberine wastewater by ECPR.

3.3. Analysis of flocculation by EDTA shielding experiment

Flocculation in the process of ECPR was further analyzed through shielding flocculation utilized EDTA. The results were shown in Table 3. Decolorization rate gradually decreased with addition of EDTA increasing, indicating EDTA could constant shield flocculation. Decolorization rate descended to constant 25% when addition of EDTA was up to 5000 mg/L or more. Decolourization rate difference (about 73%) can be considered contribution rate of flocculation.

Table 3. The effect of EDTA addition

Addition of EDTA（mg/L）	1000	2000	3000	4000	5000	6000	7000
Decolorization rate（%）	90.7	72.3	60.2	38.66	25.41	23.8	26.8

3.4. Analysis of structure-activity relationship

Metal ions of solution exist in the form of structural states polyhydroxy complexes in the process of ECPR. Figure 5 showed coordination number of Polyhydroxy complexes was higher, the complex ion charging negatively electricity tariff was higher, and the adsorption ability of the outer attracting opposite charged ions was stronger. So berberine dissolved in solution could be occur electrical hop reaction with polyhydroxy complexes of metal ions. Otherwise berberine molecule can replace the hydroxyin polyhydroxy complexes and coordinate with metal ions by N atom in the molecule structure of berberine, which made berberine firmly adsorbed with complexes of the metal ions.

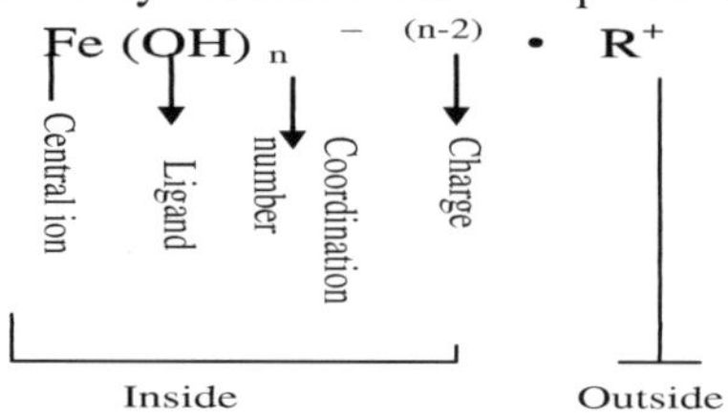

Figure 5. Diagram of ferrous hydroxyl complex

3.5. Analysis of electric floatation

As the principle of ECPR known, O_2 generated on the anode and H_2 generate on cathode when the reaction was carried out. Electric floatation cannot be ignored in the electro-coagulation process. The following two groups were carried out under optimal conditions to research floatation to remove berberine. The first group was contrast: absorbance and COD value of filtrate were directly measured after reaction ending. And the second group: the power was turned off and the plates were removed, but stirring was continued. Stirring was stopped when the bubble layer on solution surface. And then absorbance and COD value of filtrate were measured. The results were shown in Table 4.

Table 4. The effect of floating on treating efficiency

	Decolourization rate（%）	Removal rate of COD（%）
First group	99.21	94.5
Second group	91.34	88.4
Difference	7.87	6.1

In the second group, decolourization rate and removal rate of COD declined, indicating berberine molecules adsorbed on yellow bubble layer re-dissolved to solution with bubble layer disappearing. Therefore difference (about 8%) between two groups was roughly understood as floating occupied proportion. But re-dissolved berberine molecules may be removed through the flocculation of metal ions complex, so practical proportion of floating should be higher than experimental date (8%).

3.6. Analysis of oxidation

Structure of berberine molecule showed Cl^- existed in berberine solution, and Cl^- converted into Cl_2 by anodic chlorine evolution. Cl_2 converted into reactive oxidants HClOvia hydrolysis. While HClO participated in sires of main and side reactions and generated Cl^-. Therefore recycling mechanism of Cl^- existed, which was favorable to HOCl playing oxidation.

3.7. Analysis of reaction kinetics

Berberine pharmaceutical wastewaters (200--1000mg/L) were treated under optimal conditions, and removal kinetics of berberine molecules were researched through investigating the changes of decolourization rate and COD removal rate. The results showed that decolourization and COD removal rate conformed to apparent pseudo-first order kinetics while 200 mg/L berberine wastewater was treated by ECPR. And decolourization and COD removal rate conformed to apparent zero-order kinetics while 400--1000 mg/L berberine wastewater was treated by ECPR.

4. Conclusions

The principle of ECPR with Fe/Al was generalized. And through a series of designed experiments, flocculation, floatation and oxidation were quantified in process of treating berberine wastewater by ECPR, which provided 73%, 8% and 18% removal efficiency. Meanwhile decolorization and COD removal conformed to apparent pseudo-first order and zero-order kinetics for 200mg/L and400-1000 mg/L berberine wastewater respectively.

5. Acknowledgments

This research was financially supported by the National Natural Science Foundation of China (no: 51178088), and the Doctoral Fund of Ministry of Education of China (no: 20120042110025).

6. References

[1]QIN Weiwei, XIAO Shuhu, Song Yonghui, et al. Treatment of Berberine Pharmaceutical Wastewater by O_3/UV Synergetic Oxidation[J]. Research of Environmental Sciences, 2010, 23（07）:877-881.

[2]Tak-Hyun K., Chulhwan P., Eung-Bai S., et al. Decolorization of disperse and reactive dyes by continuous electrocoagulation process [J], Desalination,2002,150:165-175

[3]Mahmut B., Mehmet K., Orhan T. C., et al. Operating costanalysis of electrocoagulation of textile dye wastewater [J], Separation and Purification Technology, 2004, 37:117-125.

[4] K.S. ParamaKalyani， N. Balasubramanian, C. Srinivasakannan. Decolorization and COD reduction of paper industrial effluent usingelectro-coagulation[J].Chemical Engineering Journal，2009，1-3（151）：97-104.

[5]ZHANG Xuan, WEN Yibo, CHEN Jinsong. Advanced treatment of coking wastewater with electro-coagulation [J]. Shanxi Architecture, 2009, 35（09）: 192-193.

[6]REN Meijie, SONG Yonghui, ZENG Ping, et al. Treatment of Berberine Hydrochloride Pharmaceutical Wastewater by Pulse Electro-coagulation[J]. Research of Environmental Sciences, 2010, 23（07）:892-896.

[7]Long, warrenP. Uniflox EM dash a machine employing the method of electroflocculation for enhance soap separation. TAPPIAikaline Pulping Conference Preprint Nov 7-10,1977.347-354.

[8]Svetashov, Dobrevskji. Effeteleetric current Pulse shape on efficiency of electrochemical treatment of water containing Petroleum Products. Khimiyal Technology1992, 856-859.

[9]Dikuscar, Zajdman.Anodic dissolution during electrochemical water Purification from heavy metal ions. Electronnaya Obrabotka Mterialovn, 1993,2227～31.

[10]Hu Xiaomin, Fu Zhongtian, Sun Zhaonan, et al. A method and apparatus of treating dye wastewater by electro-coagulation with periodically reversing [P]. Patent number: 200910012599 .X，Authorized announcement date: 2011-7-20.

[11] Sun Zhaonan, Hu Xiaomin, Dong Chang-e. The study on treatment of simulate pharmaceutical wastewater with berberineby electro-coagulation with periodically alternate [J]. Journal of Safety and Environment, 2011，11（6）：60-64.

The Substituent Effects on π-type Pnicogen Bond Interaction

Jian-Qing Zhu[1], Sheng-Wei Cao[2], Wei Wang[3], Xiao-Lu Xu[4] and Hui-Ying Xu[4]*

[1] Department of Basic Courses, Zhejiang Shuren University, Hangzhou 310015, China
[2] ZhejiangLinix Motor CO., LTD, Dongyang 322118, China
[3] Zhejiang Surveying Institute of Estuary and Coast, Hangzhou 310008, China
[4] College of Biology & Environment Engineering,ZhejiangShuren University, Hangzhou 310015, China

xuhy65@163.com

Abstract. Intermolecular interactions between PH_2Cl and Ar-*R*（*R*=H, OH, NH_2, CH_3, Br, Cl, F, CN, NO_2）were calculated by using MP2/aug-cc-pVDZ quantum chemical method. It has been shown from our calculations that the aromatic rings with electron-withdrawing groups represent much weaker binding affinities than those with electron-donating groups. The charge-transfer interaction between PH_2Cl and Ar-*R* plays an important role in the formation of pnicogen bond complexes, as revealed by NBO analysis. The π-type halogen bond was also calculated and comparison of these two π-type interactions was made. It has been revealed that the π-type pnicogen bond systems are more stable than the halogen bond ones.

1. Introduction

Molecular interaction, closely correlated to various physicochemical properties, environmental governance and material structures and so on, has been of great concern in chemistry subject, environmental science, material science and some other fields[1-2]. The hydrogen bond interaction is of typical molecular interaction. Then, the halogen bond [3], lithium bond [4] and other weak interaction were found. In 2009, Hey-Hawkins, et al [5] proved the existence of P···P non-bond interaction via $^{13}C\ \{^1H,\ ^{31}P\}$ NMR experiment. Scheiner[6] theoretically calculated the geometrical structure and interaction energy of a series of pnicogen bonds systems including P···P and N···P, etc. Furthermore, the research on pnicogen bond interaction attracted the attention from theoretical and experimental chemists [7,8]. PH_3 is the simplest donor molecule in the pnicogen bond with a weak pnicogen bond interaction. When one of the H atoms is replaced by Cl atom (PH_2Cl), the P···N pnicogen bonds interaction formed between PH_2Cl and the representative electron donor NH_3 even exceeds the hydrogen bond interaction between water molecules [9], so PH_2Cl is often used as the donor model molecule in the research on pnicogen bond interaction [10]. In this paper, the interactions between PH_2Cl and the substituted benzene (Ar-*R*, *R* = H, OH, NH_2, CH_3, Br, Cl, F, CN, NO_2) are calculated by the quantum chemistry method, and to explore the geometrical structure, electronic structure and interaction energy of π-type pnicogen bond between PH_2Cl and the aromatic compound. Furthermore, the effect of the substituting group in the aromatic ring to the stability of π-type pnicogen bond complex are considered, in order to provide a theoretical basis for recognizing the nature of π-type pnicogen bonds, the application of the environmental governance and material structures.

2. Computational Methods

The structure optimizations and energy calculations of all molecules are performed by Gaussian 09[11] program package. In consideration of the importance of electron correlation for accurate interaction energy, the structure optimizations of all monomers and complexes are performed by MP2 method with aug-cc-pVDZ basis set, and no symmetry limit is applied to the optimization. References also show this method is applicable for the research on pnicogen bond interaction [12]. The interaction energy ($\Delta \boldsymbol{E}_{\text{int}}$), obtained from the complex energy deducting two monomer energies, is corrected to interaction energy ($\Delta E_{\text{int}}^{CP}$) after the BSSE (Basis set superposition error) is corrected by CP (counterpoise)[13]. To further investigate the geometric and electronic properties of the complex systems, the second-order perturbation stabilization energy ($\Delta \boldsymbol{E}^2$) and the charge transfer quantity are calculated by the NBO program in the Gaussian 09 program package.

3. Results and Discussion

3.1. Geometrical structure and interaction energy

The stable structures of the complexes constructed between PH_2Cl and the substituted benzene (Ar-*R*, *R* = H, OH, NH_2) are obtained at level MP2/ aug-cc- pVDZ, and the geometrical structures are shown as figure 1. As can be seen from figure 1, the pnicogen bond interaction is generated between the P atom in PH_2Cl and the π-electron in the aromatic ring. According to our aforesaid definitions [14], the α angle is the included angle between the vector from the P atom to the hetero ring centroid and the vector at P-Cl bond direction (see figure 2), and the values of angle α for complexes **1**~**9** are shown in table 1. The angle α values in table 1 show that the angle α of 8 complexes is almost 180° except complex **3**, i.e. the three points of Cl, P, and the centroid are almost at the same straight line. The optimization result for complex **3** shows that one of the H atoms (H17) in PH_2Cl is also close to the aromatic ring center, i.e. a certain π-type hydrogen bond interaction is formed between the atom H in PH_2Cl and the π-electron in the aromatic ring, so the decreasing of angle α is caused by the interference of interaction P15-H17······π.

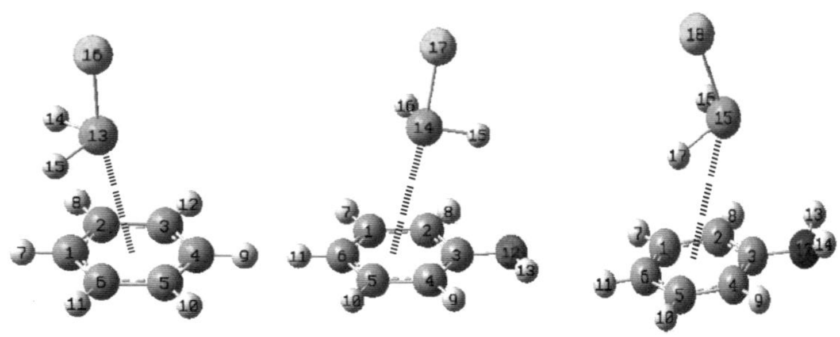

Complex **1** Complex **2** Complex **3**

Figure 1. The optimized geometries of complexes**1-3** at MP2/ aug-cc-pVDZ level.

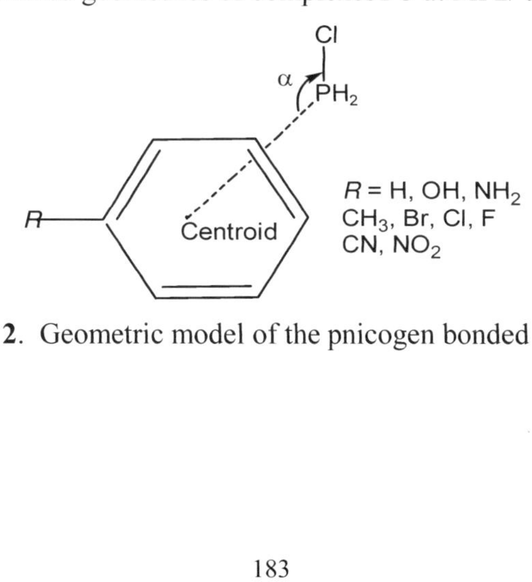

Figure 2. Geometric model of the pnicogen bonded angle α.

The uncorrected interaction energies ($\Delta \boldsymbol{E}_{\text{int}}$) and the corrected interaction energies ($\Delta E_{\text{int}}^{CP}$) of the complexes obtained at MP2/ aug-cc-pVDZ level are listed in table 1. Compared the $\Delta \boldsymbol{E}_{\text{int}}$with $\Delta E_{\text{int}}^{CP}$, it has been showed that the differences between the $\Delta \boldsymbol{E}_{\text{int}}$ and $\Delta E_{\text{int}}^{CP}$ are 2.7 kcal·mol^{-1} to 3.5 kcal·mol^{-1}, so BSSE correction is necessary for the π-type pnicogen bond system. Complex **1** is the one formed by PH_2Cl and the non-substituted benzene, and which corrected interaction energy is -3.81 kcal·mol^{-1}. The corrected interaction energies of complex **2** and complex **3** are respectively -7.56 kcal·mol^{-1} and -9.65 kcal·mol^{-1}, which are much larger than the that of complex **1**. Because the substituted groups, -OH and -NH_2, are the electron donor groups , the electron density of π-electron system in the benzene ring is increased, the interaction between PH_2Cl and π-electron is stronger, and the interaction energy of the complex is larger and the complex is more stable. Similarly, the electron donating capacity of the group -CH_3 is weaker than the above two groups, so the interaction energy of complex **4** is smaller than that of complexes **2** and **3**. The corrected interaction energies of complexes **7**~**9** are respectively -3.55 kcal·mol^{-1}, -3.44 kcal·mol^{-1}, and -3.45 kcal·mol^{-1} which are smaller than the interaction energy of complex **1**, i.e. the complexes **7**~**9** are less stable than complex **1**. This is because the electron-withdrawing groups, -F、-CN and -NO_2, reduce the π-electron density in the benzene ring, and result in the weakening of the interaction between P and π-electron. It is remarkable that the substituting groups -Br and -Cl in complex **5** and complex **6** are electron-withdrawing groups, but their interaction energies are a little larger than complex **1**. The reason may be the weak hydrogen bonds interaction between the hydrogen atom in PH_2Cl and halogen atom of the substituting group. It can be clearly seen in the geometrical structure of figure 1, and also proven by the following NBO analysis.

Table 1. The geometric parameters and interaction energies of complexes at MP2/aug-cc-pVDZ level [a]

Pnicogen bonding Complexes	$\Delta \boldsymbol{E}_{\text{int}}$	$\Delta E_{\text{int}}^{CP}$	α/(°)	Halogen bonding Complexes	ΔE_{XB}^{CP}
Complex **1** ($H_2ClP\cdots$Ar-H)	-6.54	-3.81	169.1	$PH_2Cl\cdots$Ar-H	-1.58
Complex **2** ($H_2ClP\cdots$Ar-OH)	-10.66	-7.56	169.3	$PH_2Cl\cdots$Ar-OH	-4.99
Complex **3** ($H_2ClP\cdots$Ar-NH_2)	-13.09	-9.65	145.6	$PH_2Cl\cdots$Ar-NH_2	-6.00
Complex **4** ($H_2ClP\cdots$Ar-CH_3)	-7.19	-4.28	167.1	$PH_2Cl\cdots$Ar-CH_3	-2.09
Complex **5** ($H_2ClP\cdots$Ar-Br)	-7.57	-4.12	171.5	$PH_2Cl\cdots$Ar-Br	-2.05
Complex **6** ($H_2ClP\cdots$Ar-Cl)	-7.22	-4.05	170.1	$PH_2Cl\cdots$Ar-Cl	-1.97
Complex **7** ($H_2ClP\cdots$Ar-F)	-6.49	-3.55	171.8	$PH_2Cl\cdots$Ar-F	-1.82
Complex **8** ($H_2ClP\cdots$Ar-CN)	-6.43	-3.44	168.8	$PH_2Cl\cdots$Ar-CN	-2.35
Complex **9** ($H_2ClP\cdots$Ar-NO_2)	-6.63	-3.45	170.1	$PH_2Cl\cdots$Ar-NO_2	-2.74

[a] interaction energies are given in kcal·mol^{-1}

The halogen bond is the interaction between the halogen atom and the electron donor in the halogenous molecule. It was found earlier than the pnicogen bond, and the research on halogen bonds is more mature than the pnicogen bond. In this research system, when the π-type pnicogen bond interaction can be generated between the PH_2Cl molecule and the aromatic compound, a π-type halogen bond complex can be generated between π electron of the aromatic ring and atom Cl in the PH_2Cl molecule via the π-type halogen bond interaction. The molecular structures of nine typical halogen bond complexes are shown in Figure 3. As a comparison, their structures are optimized at MP2/aug-cc-pVDZ level, and then their corrected interaction energies (ΔE_{XB}^{CP}) which are corrected through the BSSE are shown in Table 1. Comparing the corrected interaction energy $\Delta E_{\text{int}}^{CP}$ of the pnicogen bond system and the corrected interaction energy (ΔE_{XB}^{CP}) of the corresponding halogen bond system, it has been found that the π-type pnicogen bond system is more stable than the π-type halogen bond system. It is caused by the unbalanced distribution of positive electrostatic potential in the PH_2Cl. As can be seen in Figure 3, in the same aromatic π electron donor, the positive electrostatic potential at the top of atom P is larger than that at the top of atom Cl.

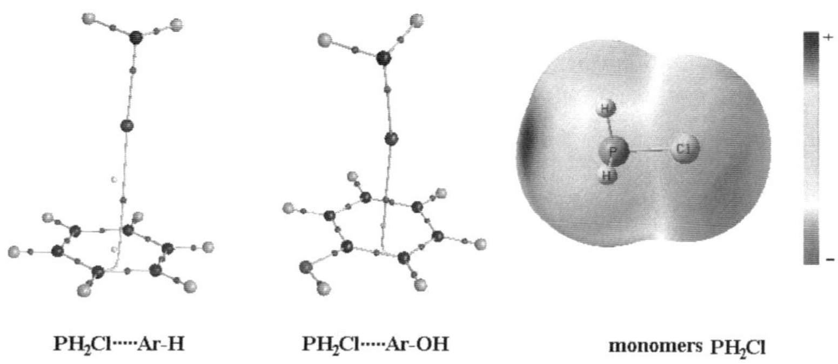

Figure 3. Molecular graph of PH_2Cl.....Ar-H and PH_2Cl.....Ar-OH, and electrostatic potential surface of monomers PH_2Cl.

3.2. NBO analysis

Table 2 shows the donor-acceptor orbital, the second-order perturbation stabilization energy (ΔE^2), and the charge transfer quantum (Q_{CT}) of 9 π-type pnicogen bond complexes at level MP2/aug-cc-pVTZ.

Table 2. Natural Bond Orbital Analysis of complexes **1-9** at the MP2/ aug-cc-pVDZ level

Complexes	Donor NBOs	Acceptor NBOs	ΔE^2/ kcal.mol^{-1}	Q_{CT}/ e
Complex **1** (H_2ClP···Ar-H)	BD (2) C1-C6 BD (2) C2-C3	BD* (1) P13- Cl16	3.05 0.56	0.01115
Complex **2** (H_2ClP···Ar-OH)	BD (2) C1-C6 BD (2) C2-C4	BD* (1) P14- Cl17	3.50 0.62	0.01536
Complex **3** (H_2ClP···Ar-NH_2)	BD (2) C2-C3 BD (2) C1-C6	BD* (1) P15- Cl18 BD* (1) H17- P15	4.47 1.02	0.01614
Complex **4** (H_2ClP···Ar-CH_3)	BD (2) C1-C6 BD (2) C2-C3	BD* (1) P16- Cl19	2.30 1.10	0.01007
Complex **5** (H_2ClP···Ar-Br)	BD (2) C2-C3 Lp (3)Br	BD* (1) P13- Cl16 BD* (1) H14- P13	2.85 0.81	0.01250
Complex **6** (H_2ClP···Ar-Cl)	BD (2) C2-C3 Lp (3)Cl	BD* (1) P13- Cl16 BD* (1) H14- P13	2.91 0.49	0.01179
Complex **7** (H_2ClP···Ar-F)	BD (2) C2-C6 BD (2) C1-C6	BD* (1) P13- Cl16	3.06 0.43	0.01077

Complex **8** ($H_2ClP\cdots Ar$-CN)	BD (2) C6-C7	BD* (1) P1- Cl 4	2.78	0.00643
Complex **9** ($H_2ClP\cdots Ar$-NO_2)	BD (2) C5-C6 BD (2) C7-C8	BD* (1) P1- Cl 4	2.30 0.57	0.00953

The second-order perturbation stabilization energy (ΔE^2) can be obtained from the following equation (1):

$$\Delta E^2 = \Delta E_{ij} = q_i \frac{F(i,j)^2}{\varepsilon_j - \varepsilon_i} \tag{1}$$

Where, q_i is the donor orbital occupancy, ε_i and ε_j are the diagonal elements (orbital energies), and F is the NBO Fock matrix element. It has been shown from the donor-acceptor orbital for complexes **1-9** in table 2, that the pnicogen bonds interaction is mainly the interaction between the C-C π bonding orbital and the σ anti-bonding orbital of P-Cl in the monomer PH_2Cl. The three complexes **3**, **5** and **6** are further involved with the hydrogen bonds interaction in addition to the pnicogen bond interaction. The second-order perturbation stabilization energy values indicate that the pnicogen bond interaction, followed by the hydrogen bond interaction, plays a key role in the complex stability. The existence of the supportive hydrogen bond interaction is the fact that the interaction energies of complexes **5** and **6** are higher than that of complex **1** in the above analysis, and why the angle α is smaller in the complex **3**.

The molecular interaction is always accompanied by charge transfer, and the charge transfer quantities in complexes **1-9** are shown in table 2. As can be seen in table 2, the charge quantities transferred from the aromatic compound to PH_2Cl molecule are from 6me to 16me. The charge quantity transferred is not so much, and it complies with the soft acid – soft alkali model (PH_2Cl molecule as the soft acid and the aromatic compound as the soft alkali). The plot of the charge transfer quantity and the corrected interaction energy is shown in Figure 4. The curvilinear equation (2) is as below:

$$\Delta E_{\text{int}}^{CP} = -11.215 + 1852.819\, Q_{CT} - 106824.298\, (Q_{CT})^2 \tag{2}$$

$$R = 0.954 \qquad SD = 0.540 \qquad N = 9$$

Figure 4 shows that the charge transfer quantities of the 9 complexes decrease with their interaction energies decreasing, and this result indicates that the charge transfer plays an important role in the complex stability.

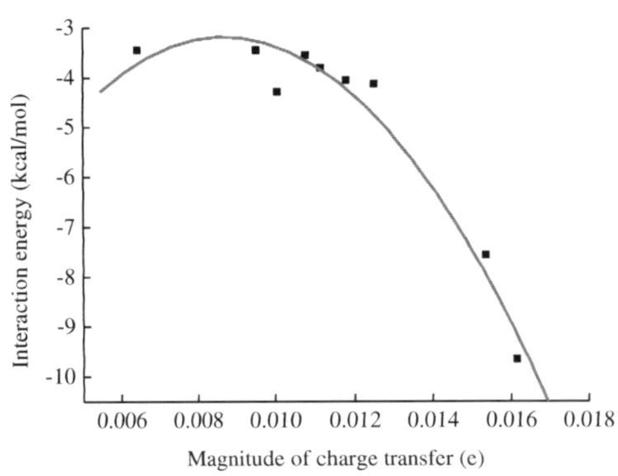

Figure 4. The relationship between charge transfer and interaction energy.

4. Conclusion

The geometrical structural optimization, energy calculation and topological and graphic analyses for the various pnicogen bond system of PH_2Cl and Ar-*R* (*R*=H, OH, NH_2, CH_3, Br, Cl, F, CN, NO_2) have been made at level MP2/ aug-cc-pVDZ. The results have showed, the complex pnicogen bond interaction will be strengthened when the substituting group in the benzene ring is the electron donating group, and the complex pnicogen bond interaction will be weakened when the substituting group is the electron withdrawing group. To compare the interaction energy of the π-type pnicogen bond system with the halogen bonds system, the interaction energy of π-type halogen bond system is also calculated, and it is showed that the π-type pnicogen bond system is more stable. NBO theory is used to analyze the correlation between the second-order perturbation stabilization energy, the charge transfer quantum and the interaction energy, and the result shows that the charge transfer plays an important role in the stability of the pnicogen bond complex.

5. Acknowledgments

This study was financially supported by the Science and Technology Project of Zhejiang Province (2016C33039).

6. References

[1] Gillet N, Chaudret R, Contreras-García J, Yang W, Silvi B and Piquemal J 2012 *J. Chem. Theory Comput.* **8**, 3993
[2] Mahadevi A S and Sastry G N 2013 *Chem. Rev.* **113**, 2100
[3] Wash P L, Ma S, Obst U and Rebek J J 1999 *J. Am. Chem. Soc.* **121**, 7973
[4] Ault B S 1978 *J. Am. Chem. Soc.* **100**, 2426
[5] Bauer S, Tschirschwitz S, Lönnecke P, Frank R, Kirchner B, Clarke M L and Hey-Hawkins E 2009 *Eur. J. Inorg. Chem.* 2776.
[6] Scheiner S 2011 *J. Chem. Phys.* **134**, 164313
[7] Del Bene J E, Alkorta I, Sánchez-Sanz G and Elguero J 2012 *Chem. Phys. Lett.* **538**, 14
[8] Alkorta I, Sánchez-Sanz G, Elguero J and Del Bene J E 2013 *J. Phys. Chem. A* **117**, 183
[9] Scheiner S 2011 *J. Phys. Chem. A* **115**, 11202
[10] Adhikari U and Scheiner S 2012 *Chem. Phys. Lett.* **532**, 31
[11] Frisch M J 2009 *Gaussian 09 (Revision A.01)*, Gaussian, Inc., Wallingford CT
[12] Adhikari U and Scheiner S, 2012 *Chem. Phys. Lett.* **536**, 30
[13] Boys S F and Bernardi F, 1970 *Mol. Phys.* **19**,553
[14] Xu H Y, Wang W and Zou J W 2013 *Acta Chim. Sinica* **71**, 1175

IOP Conf. Series: Earth and Environmental Science **63** (2017) 012028 doi:10.1088/1755-1315/63/1/012028

Geological Features of Iron Formations and Associate Rock in Bulunkuole Group, West Kunlun, Xinjiang, China

Chaoyang Huang[1,a], He Wang[2,b], Guangli Ren[3,c] and Wenying Wu[1,d]

[1] Chongqing Institution of Engineering, 402260, Chongqing, China;
[2] Guangzhou Institute of geochemistry; Chinese Academy of Sciences, 510640, Guangzhou, China
[3] Xi'an Center of China Geological Survey, 710054, Xi'an China.

a:345245972@qq.com, b:wanghe@gig.ac.cn, c:155219321@qq.com, d:569111860@qq.com

Abstract. West Kunlun is located at northwest part of the Tibetan plateau, and it's divided into North Kunlun terrane, South Kunlun terrane and Tianshuihai terrane. Bulunkuole Group is located in Tianshuihai terrane, and Taaxi, Yelike, Laobing, Zankan, Mokaer and Jiertiekegou iron deposit was found in this terrane. Those iron deposits are set to the important iron ore mineralization belt of Xinjiang and even China. By investigation the geological features, ore distributions, and mineralization characters; and analyze the electron microprobe of magnetite and pyrite, stable isotope of pyrite, magnetite and anhydrite to discussion the tectonic setting and its genesis analysis iron ore deposit in Blunkuole Group, and establish the metallogenic model in this area. By contrastive analyzing of Taaxi, Yelike Zankan-Mokar iron ore deposits in Bulunkuole Group, it was found that all the iron ore deposit share the similar sedimentary environment. Those deposits have 3 ~ 4 ore bodies, each get a meter to dozens of meters in width, extend to thousands of meters. The wall rocks are biotite quartz schist and plagioclase amphibole schist, and the two rocks interbed usually. It should be distinguish in the stratigraphic correlation of Bulunkuole Group for the influenced by the magma intrusion, such as granite and felsite porphyry.

1. Introduction

Precambrian is one of the most important period for iron deposition, and it's proved about 65.6% iron ore resource of the total in China[1]. In China, Banded Iron Formations(BIFs) are mainly distributed in the North China Craton [2], and the distribution of iron-formation in the west kunlun has been reported widely since[3]. The Taxkorgan is one of the most important iron-ore mining districts in the west Kunlun Orogenic belt, in the province of Xinjiang, China [4] (Fig. 1).

Since the discovery of the iron deposit in 2005, the reconnaissance studies and mineralization studies have been done gradually. Based on the analyzed of the prospecting exploration data of the depsosit geology, ore bodies and ore-structure characteristics, it is concluded that the Taxkorgan magnetite iron deposit belongs to sedimentary-metamorphic type, or the Algoma-type BIF [5]. By tested the metamorphic zircon age of the ore deposit (532±3.9Ma) in biotite-quartz-schist of Bulunkuole Group, and the combination of a anhydrite-gypsum-iron formation, Changhai think it's a new type of iron deposit named Pamir-type [6].

The depositional time frame of the Taxkorgan BIF in Bulunkuole Fomation, is not as precisely

IOP Conf. Series: Earth and Environmental Science **63** (2017) 012028 doi:10.1088/1755-1315/63/1/012028

known. It is wildly believed that the Blunkuole Group is a part of early Paleozoic component by comparing with the metamorphic grade and isotopic data in southwest Pamir region as well as the magneite-quartz combination in it. By test the detrital zircon from the Taxkorgan region and got the ages of 540-220Ma, suggested that the deposition time of this region should be later than the Neoproterozoic [7]. Ji got the insitu ziron U-Pb LAICPMS date on rhyolite zircons yielded an age of 2481±14Ma and interpreted the as the eruption age of the Bulunkuole Group volcanic rocks [8]. Yan C.H. obtained a metamorphic age early Paleozoic period age from zircon in the biotite-quartz schist in Laobing iron mine.

The purpose of this study is to describe the geology characters of these mines in Taxkorgan, and study on the mineralogy and geochemistry elements of the iron ores of the Bulunkuole Group and in aims at discussing the origin of the Taxkorgan iron formations in the West Kunlun.

2. Geological settings

The West Kunlun orogenic belt is located in the northwest part of the Qinghai-Tibet Plateau, southwest of Tarim Basin, Western China (Fig. 1). The two main structural faults, Kangxiwa Fault and Karakunlun Fault are divided the west Kunlun into the North Kunlun, South Kunlun, and Karakunlun domains, and the Paleoproterozoic Bulunkuole Group is between the two faults. And the two regional faults not only control the tectonic framework of this region, but also to monitor the development of secondary faults, magmatism, volcanism and sedimentary formations.

Bulunkuole Group is one of the oldest sections of Paleoproterozoic terrane in west Kunlun Orogen, with low-grade greenstone successions and comprises BIF, mafic-felsic volcanic rocks, ultramafic rocks, chert, shale and minor carbonates. The majority of the iron ore bodies in the Taxkorgan are hosted in biotite-quartz schist and plagioclase-hornblende schist of the Bulunkuole Group (Figure 1).

Based on their mineralogical composition, the Blulunkuole Group successions can be subdivided into four lithostratigraphic units, namely, BIF, metavolcanic rocks, garnet-rich biotite quartz schist, and marble successions.

The west part of the Bulunkuole Group is BIF-bearing section is biotite-quartz schist and plagioclase-hornblende schist. In this succession, three important iron deposits are founded including Taaxi, Yelike-Laobing, Zankan-Mokar. The metavolcanic succession is in the middle of the Bulunkuole Group, with plagioclase-hornblende schist and biotite-quartz schist. In this succession, it is regarded volcanic eruption features as dark rock with light-colored rock rhythm recurring.

The garnet-rich biotite-quartz schist is located in the middle-east of the Bulunkuole Group. This succession comprises Garnet-biotite-quartz schist with minor sillimanite-garnet-plagioclase-biotite-quartz schist, and the garnet and sillimanite content of up to 20-30%.

In the east part of Bulunkuole Group is marble succession consist of marble, tremolite marble with biotite-quartz schist layers in it. Silurian sedimentary successions unconformably overlie the Paleoproterozoic Bulunkuole Group successions, which consists of predominantly clastic sediments with minor metapyroclastic rocks interbedded with minor lava sheets.

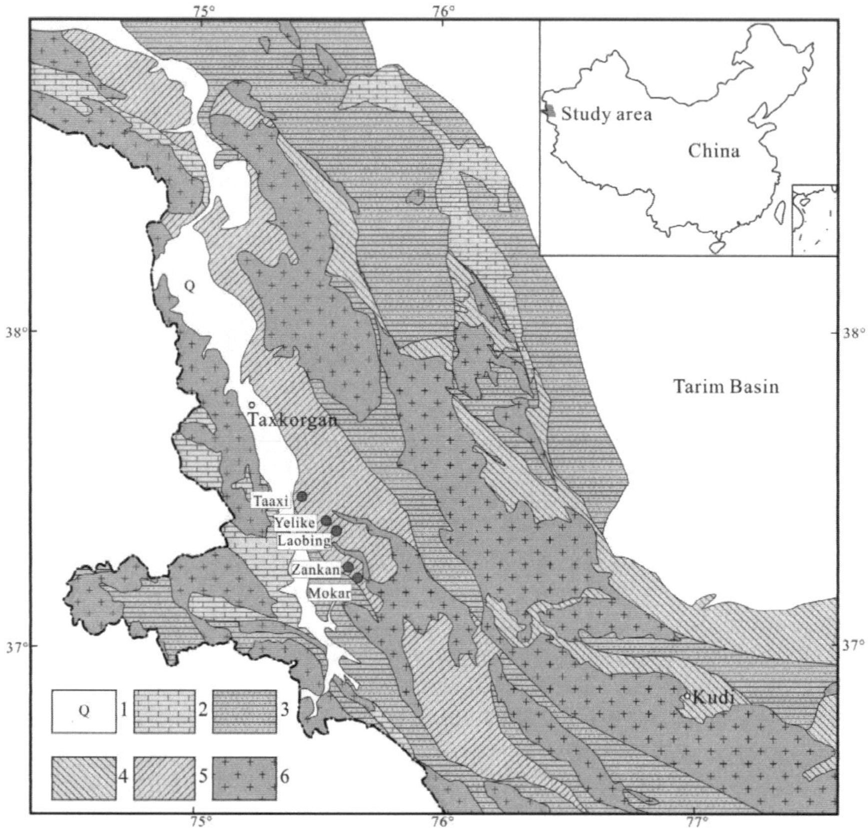

Figure 1. Geology and mineral resources in Taxkorgan area [5-7]

3. Local geology and petrography

The Taaxi, Yelike-Laobing, and Zankan-Mokar iron ore deposits are located in the southeast of the Taxkorgan, West Xinjiang, China, trending northwest-southeast, about 120 kilometers long.

3.1. Taaxi iron deposit point

The Taaxi iron ore point is located in the western edge of Bulunkuole Group, trending northeast in the eastern part of the mine and turn west-east trending in west part of this iron deposit. The iron deposit is composed mainly of three parallel ore bodies, are about 2200m long and 0.5–6m wide, with TFe2O3 40%wt in average, trending of 270, and an angle of 50-75.

The Taaxi iron deposit is hosted in metamorphic rocks of Paleoproterozoic Bulunkuole Group consists of iron formations, plagioclase-hornblende schist, biotite-quartz schist and marble. From the geological cross-section map in the Taaxi iron deposit, three ore bodies are interrelated in plagioclase-hornblende schist, biotite-quartz schist, and interbedded marble in the northern part, strike NW at the east part of the mine and become strike EW at west. Marble was just found at the northern part of this district, with only 2-3m in width, laminated in biotite-quartz schist. The cross-section can be divided into three groups based on the rock assemblage characters. The southern part of the cross-section comprises plagioclase-hornblende schist interlayer minor biotite-quartz schist, and some of the rock units were covered by quaternary sediments.

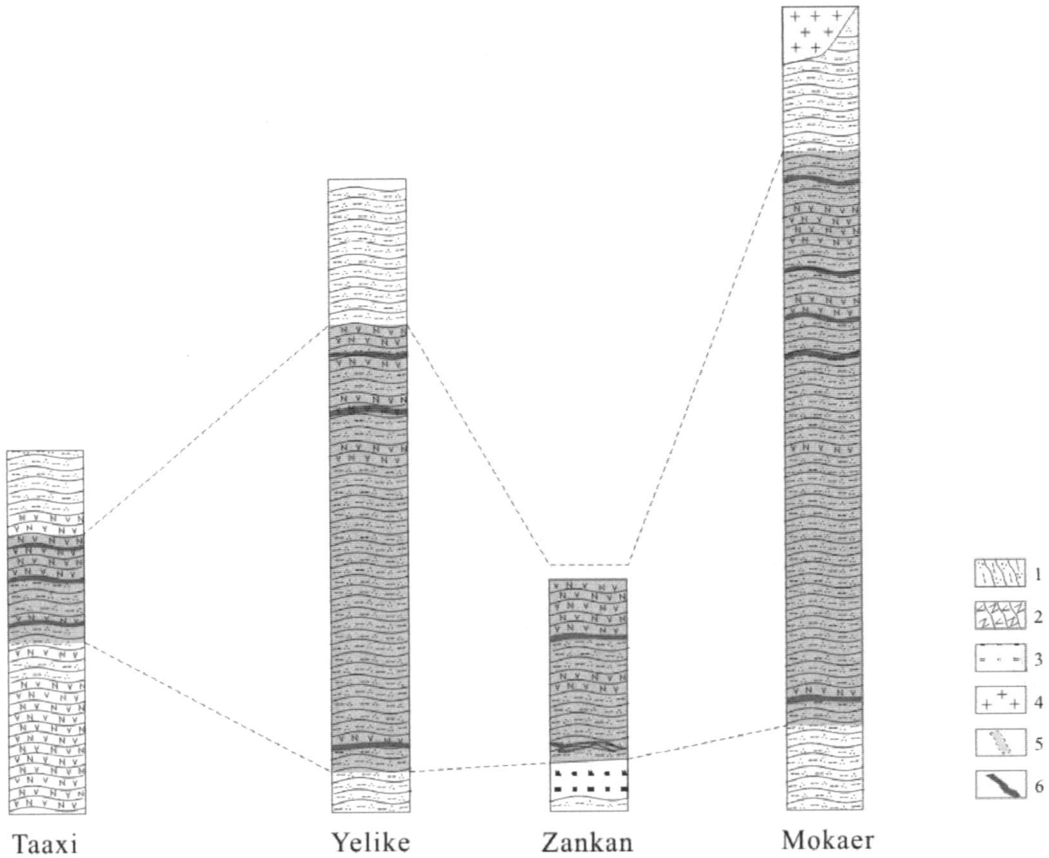

Figure 2. Columnar sections of the iron formations in the West Kunlun orogenic

3.2. The Yelike-Laobing Iron Ore Deposit

The Yelike-Laobing iron ore deposit is one of the most important ore districts in Taxkorgan area. The ore deposit is about 5000m in strike length and composed 3 to 5 parallel ore bodies, which can be divided into two mines by the different mining company: Yelike and Laobing. The two iron ore deposits are of the same geological characters. The 3-5 parallel ore bodies are interlayered with plagioclase-hornblende gneiss and biotite-quartz schist, strike EW at the east part of the mine and turn NW at west (Fig. 2). Almost all massive ores have a high Fe grade of about 45wt. %.

In the east part of this mine (Laobing), some ore bodies are deformed slightly, and one Syncline was found at the east, trend NWW. While in the western part of the mine (Yelike), most of the ore bodies are interlayered with plagioclase–hornblende schist and biotite-quartz schist. In the iron ore, the major minerals are magnetite, quartz, biotite and plagioclase. Under the microscope, iron ore consist of magnetite and quartz, and they exhibit a weak linear fabric, and have a granular texture with sharp boundaries in some mineralization wall rock. The banded iron content of magnetite is between 20% and 40%, whereas that of quartz is between 50% and 70%.

3.3 Zankan-Mokar iron ore deposit

The Zankan-Mokar iron ore deposit hosted total premining resources about 150 millionmetric tons (Mt), with an average grade of 41.1%, is located in the southern part of the iron ore deposit district in Taxkorgan. The Zankan-Mokar iron ore deposit consists of more than 10 ore bodies, which can be regarded as 4 parallel ore bodies from the geology map. The mineralogy and geological characters are similar to those of Taaxi and Yelike-Laobing. Biotite-quartz schist and plagioclase-hornblende schist are the most important wall rock in this deposit. In this iron ore deposit, the ore bodies are interlayer in the biotite-quartz schist and plagioclase-hornblende schist and ore bodies are conformable contact with their wallrocks, especially the contact zone between the two rock units.

The ore body at the southern part of the cross-section is about 40m in width, and most of the ores are banded, while some of them are massive. Biotite-quartz schist and plagioclase-hornblende schist are the most rock units, and conformable contact with the ore bodies.

4. Mineralogy and petrography

4.1. Iron formations
Most of the iron formations in Taxkorgan area are mesobands to massive structure. The petrology and mineralogy of the Bulunkuole Iron Formations are representative of most Precambrian banded iron-formations over China. Microphotographs of various iron ore samples from the West Kunlun shows there are 3 types of the iron formations, laminated magnetite quartzite, coarse laminated magnetite quartzite and fine laminated magnetite quartzite.

In sample ZK10-110, mineral abundances are quartz (40–50%), magnetite (25–30%), amphibole (10–15%), biotite (10–15%), chlorite and epidote (1–5%) and accessory plagioclase (<2.5%).

4.2. Biotite-quartz schist
Biotite-quartz schist is one of the most common rock type in this district. These schists contain quartz (50–60%), biotite (40–50%). The quartz grains are inequant and showing both strained properties, and biotite often forms continuously elongated flakes in those samples.

4.3 Plagioclase-hornblende schist
Plagioclase-hornblende schist in Bulunkuole Group show that it contains quartz (30–40%), biotite (10–20%), sericite (10–20%), muscovite (5–10%), The amphiboles are grained and some strained to , show pronounced pleochroism, and contain inclusions of allanite.

In West Kunlun area, the ore bodies are mostly occurring between the plagioclase-hornblende schist and biotite-quartz schist.

5. Summary
1. Iron ore deposits in West Kunlun have 3 to 4 ore bodies, each get one meter to dozens of meters in width, and extend to thousands of meters.
2. The plagioclase-hornblende schist and biotite-quartz schist are the most important wall rock in this area.
3. By analyzing of ore body and stratum from wild, it shows that the deposit in Bulunkuole Group has the characteristics of sedimentary deposit.

6. Acknowledgments
This study was supported by National Science & Technology Pillar Program during the Twelfth Five-year Plan Period (2015BAB05B00) and Chongqing Vaccination Institute of Engineering Commission (KJB201513).

7. References
[1]Santosh, M., 2010. Assembling North China Craton within the Columbia supercontinent: the role of double-sided subduction. Precambrian Research 178, 149-167

[2]Yuan, C., Sun, M., Zhou, M. F., Zhou, H., Xiao, W. J., Li, J. L. 2002. Tectonic evolution of the West Kunlun: Geochronologic and geochemical constraints from Kudi granitoids. International Geology Review, 44(7): 653-669.

[3]Xiao, W. J., Windley, B. F., Fang, A. M., Zhou, H., Yuan, C., Wang, Z. H., Li, J. L. 2001. Palaeozoic-Early Mesozoic Accretionary Tectonics of the Western Kunlun Range, NW China. Gondwana Research, 4(4): 826-827.

[4]Jiang, Y. H., Jia, R. Y., Liu, Z., Liao, S. Y., Zhao, P., Zhou, Q. 2013. Origin of Middle Triassic high-K calc-alkaline granitoids and their potassic microgranular enclaves from the western Kunlun orogen, northwest China: A record of the closure of Paleo-Tethys. Lithos, 156, 13-30.

[5]Feng Changrong,Wu Haicai and ChenYong. 2011.Geological characteristics and genesis of the Zankan iron deposit in Taxkorgan, Xinjiang. Geotectonica et Metallogenia, 35(3): 407-409.

[6]Yan Changhai. 2012. Pamir Type Iron Ore. Beijing: Geological Publishing House. 1-256.

[7]Ji Wenhua, Li Rongshe, Chen Shoujian et al. 2011. The discovery of Palaeoproterozoic volcanic rocks in the Bulunkuole Group from the Tianshuihai massif in Xinjiang of Northwest China and its geological significance. Science China Earth Science, 54(1): 61-72.

[8]Zhang Lianchang, Zhai Mingguo, Wangyusheng et al. 2012. Study of the Precambrian BIF-iron deposts in North China Craton: Progresses and questions. Acta Perologica Sinica, 28(11): 3431-3445.

Comparative Research to Surface Aeration and Blasting Aeration System Based on LCC Theory

Liai CHEN [1,2], Hongxun HOU [2,*], Weibiao FEI [2] and Eryan ZHAO [2]

1. College of Mechanical & Electrical Engineering, Anhui Jianzhu University, Anhui Hefei,China;
2.Anhui Guozhen Environmental Protection Science and Technology Limited Liability Company, Anhui Hefei 230088, China)

best_2002@163.com

Abstract. It is difficult to select the suitable aeration system for the designers of wastewater treatment plant (WWTP). In this paper, taking two WWTPs with surface aeration systems and blasting aeration respectively for an example, *LCC* theory was adapted to analysis the cost of consumption and the environmental impact, which caused by the different aeration system. Research results showed that: (1) In the 20-year life cycle, the *LCC* mainly depended on the cost of energy consumption whatever blasting aeration system or surface aeration, while the LCC of blasting aeration system affected by the equipment maintenance cost, maintenance cost, economic losses caused by wastewater loss and environmental load in maintenance period. (2) The LCC of blasting aeration system was lower than the surface aeration in general, on the premise of the standard discharge; (3) the blasting aeration system estimated a saving of 60,0000RMB annually in costs compared with the surface aeration.

1. Introduction

The energy consumption of secondary treatment unit was more than 50% compared with that of the total wastewater treatment plant (WWTP) [1-3], the research on the choose of aeration devices has been widely attracted attention [4, 5].

In one WWTP using typical A^2/O process, the energy consumption distributions of various units and equipment in the entire wastewater treatment process were analyzed [5]. The methods and solutions of energy saving potential could be put forward, and the importance of the optimal aeration system was also presented. By comparing surface aeration process and blasting aeration process in oxidation ditch WWTP, L G Shao [6] analyzed the treatment effect and energy consumption in 2010(one year) , and the results showed that: surface aeration system paid more cost than the blasting aeration system, that was about 1,300,000RMB one year in 10Wm^3/d per day. Another typical example was illustrated by P yang [7], the same results coincided that surface aeration need cost more energy consumption than blasting aeration at the similar circumstance. In addition, Y Liang [8] carried on the research that the variation of energy consumption changed with influence of oxygenation capacity of fine bubble aerator. To reduce energy consumption of WWTP, B Kumar[9], A Thapinta [10], I Lee [11], M Jolly [12] and so on were respectively studied the oxygenation capacity of surface aeration or blasting aeration equipment.

Based on the above analysis, much research had been devoted to the energy consumption of aeration equipment and obtained certain research results. However, the conclusions weren't considered the other economic costs and environmental negative benefits. In other words, they didn't conclude the costs of surface or blasting aeration in the whole life cycle. In particular, the operation and maintenance of blasting aeration system were difficult, and the average length of 10 years should replace the blasting aeration disc, such as the plug, tear, corrosion, aging and so on, which might bring out more economic consumption and environmental loads than the surface aeration. The principal purpose of this paper was confirmed to concern with this problem.

2. Theory of LCC analysis

The LCC was a methodology for calculating the whole cost of a system or product from inception to disposal. The systems or the products were applicable to long series of industry or production, which could be a building, a equipment, a factory or a power station. So LCC could found its place in calculate the cost of aeration system in WWTP.

Generally, the entire life-cycle of system could be divided into the follows: construction stage(planning and research, investment), operation stage(use or operation, maintenance) and the dismantling stage(disposal), while the total cost also provided a variety in different ways depending on the application filed, the specific content could be changed based on the requirement of the users. For this paper, the purpose of LCC analysis didn't only make a decision of the total cost, but also compare costs of different alternative system plans between the surface aeration and blasting aeration system. So we could check out which choice would be the most cost efficient, and what caused difference in cost, these would be one of the main benefits of the study and be shown as data. Taking into account the operation stage cost and energy consumption were more than 60% in LCC analysis, so we paid more attention and gave a detailed analysis, so that the conclusion was more in line with the actual situation. The implementation of LCC for aeration system was divided into three parts. First, a specific and representative WWTP was illustrated, which had two treatment systems and distinguished from each other only by the aeration system. Second, the LCC analysis asset list of different aeration system was developed and calculated including the process form inception to disposal. Third, the total cost was calculated and discussed the several alternatives of the efforts to improve efficiency.

3. WWTP introductions: a case study

The investigation was applied to the two WWTPs (defined A and B) located in Guangdong province which were considered a relatively consistent except the aeration system [6]. The capacity for A was 100,000 m^3/d operational with A^2/O oxidation ditch process, and the aeration system adopted inverted umbrella surface aerators. The costs and energy consumption of A could represent that generated by the surface aerator equipment. The capacity for B was also 100,000 m^3/d operational with A^2/O oxidation ditch process, and the aeration system adopted rubber membrane middle and fine bubble diffusers as the source of oxygen. At present, there was a large proportion of the WWTP using the aeration of rubber membrane, so the aeration plate material used in B factory could have certain representativeness in china. A and B plant layout plan were shown in figure 1.

Plant A: There were three anaerobic tanks and one blenders installed per tank for favoring P removal, two anoxic tanks per tank were employed and two anoxic blenders installed using for nitrate supplied by the internal recirculation. There were 6 sets of inverted umbrella surface aerator installed in aerobic tank and each 3 units (dual-use option). Each surface aerator impeller adopted turbo type with 8 blades and actual rotational velocity was 35/26rpm, the power of electric machine was 150kW and rotational velocity was 900/740rpm.

Plant B: The layout of the plant B were exactly similar with A, except 4 sets of underwater impellers and 3 units high-speed single-stage centrifugal blasting aerator (two in service one standby).The power of blasting aerator was 250kW and maximizes air volume was 10,200 m^3/h. The diameter of micro-porous aeration diffuser disc was 230mm and total amounts were 8160.

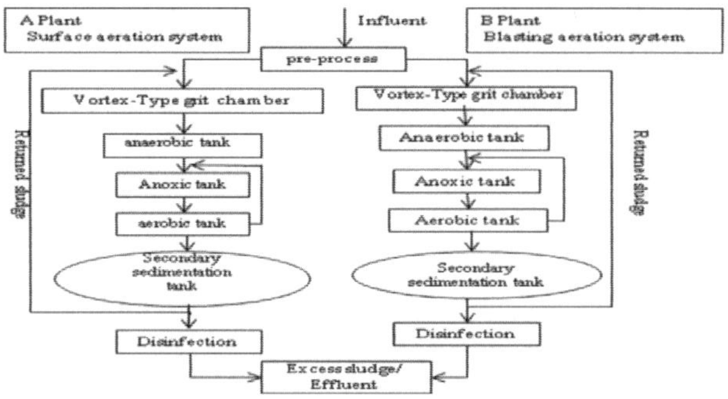

Figure1. Process flow diagram

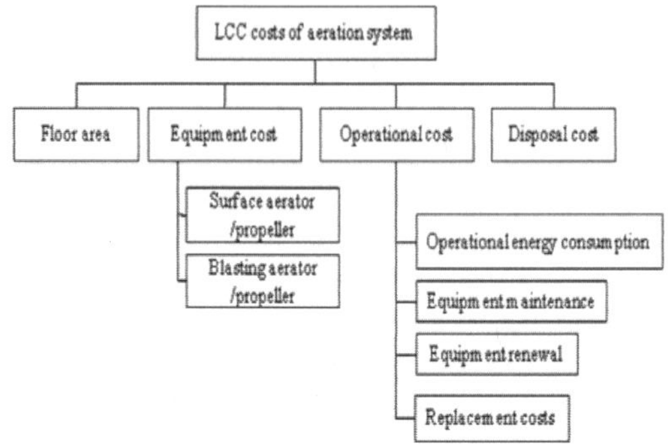

Figure2. Scope definition of aeration system LCC

4. LCC analysis of aeration system

4.1. LCC-oriented range definition of aeration system

Minimum LCC of aeration system was the most straightforward and easy-to-interpret measure defined from the user's point of view, so life-cycle range was focus on equipment purchases, installed, operational, removal costs and environmental load brought in the process. The consumption of various raw materials and the environment load caused by mining, processing and removal were not taken into account.

Based on the life-cycle range of aeration system, the required data in the LCC list were determined and operational cost was paid more attention because of its high proportion in total cost. The details were shown in figure 2.

4.2. LCC analysis of aeration system

4.2.1. Covers area analysis. The surface aeration system (A) and the blasting aeration system (B) resembled one another in the scale, including the influent quality, the effluent quality, the process with modified A^2/O, and also had many features similar in volume of secondary sedimentation tank, the grit clarifier as well as other process units. The significant difference was aeration equipment and came with the covers area of oxidation ditch which induced by different aeration system. Therefore, it was reasonable that only design parameters of biological treatment stage made a statistics and analysis in oxidation ditch. The details about the parameters of oxidation ditch were shown in table 1.

Table1. Design parameters of biological treatment stage in oxidation ditch

Items(unit/m^3)	A	B
Affective volume of anaerobic zone	3177×2	3330×2
Affective volume of anoxic zone	5678×2	5210×2
Affective volume of aerobic zone	16500×2	14600×2
Available water depth	4.9	5.25

Generally, the immersion depth of surface aeration equipment was from 10 to100mm under the water surface in operating, principle of operation were as followed: firstly, high-speed-working impellers splash the water into the air: secondly, these splashing water mixed the oxygen in the air drops into the water surface, then the content of dissolved oxygen in wastewater would be increased. The oxygen transfer mainly occurred at the air-water surface, which limited the depth of oxidation ditch to achieve the effluent quality standards. The depth of oxidation ditch usually was 3 to 4 meters; the deepest was less than 5 meters, and more deeper more energy consumption. Blasting aeration utilized a bottom-up method for continuous creation of foams comprising of air micro-bubbles, that was, the oxygen transfer not only came on the air-water surface but also in the depths of wastewater. Therefore, it could save a certain construction area because the oxidation ditch with blasting aeration was deeper than surface aeration's. According to the volume and the depth shown in table 1, the construction areas of the aeration system were approximate calculated, as shown in figure 3.

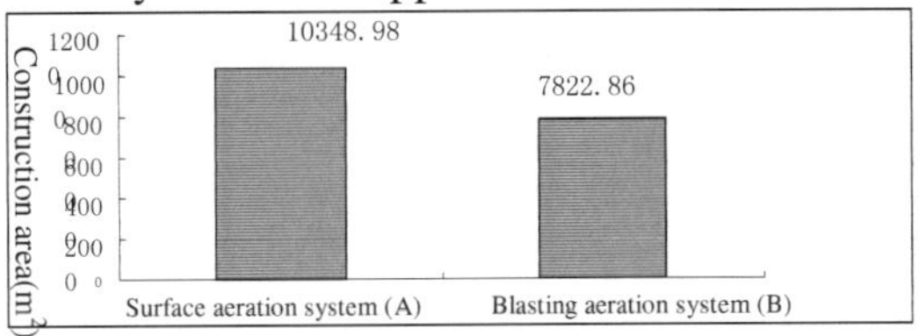

Figure 3. Construction area of different aeration system

As shown by the data in figure 3, the construction areas of surface aeration system and blasting aeration system were about 10348.98m^2 and 7822 m^2, respectively. This brought about 2500m^2 construction areas economized if blasting aeration system was applied. However, it should be noted that some auxiliary facilities such as blower room for blasting aeration also needed to be built, so the construction area was less than 2500m^2.The costs were calculated at current price 500,000 RMB /mu, there was a difference about 1870,000RMB between surface aeration system and blasting aeration system.

4.2.2. Analysis of equipment list in aeration stage. The equipments for surface aeration system (A) and the blasting aeration system (B) in oxidation ditch were confirmed and formed a statistical list, which was showed in table 2.

Table 2. The number of equipment and the power of biological treatment stage

Surface aeration system (A)			Total Power
Equipment	quantity	power(kW)	
Blender for anaerobic(T1)	6	11	1086kW
Blender for anoxic zone(T2)	4	30	
Surface aerator (T3)	6	150	
Blasting aeration system(B)			Total Power
Equipment	quantity	power(kW)	
Propeller for anaerobic(T4)	4	2.3	828kW
Propeller for anoxic(T5)	4	4.3	
Propeller for aerobic(T6)	12	4.3	
Blasting aerator(T7)	3	250	

Due to the different selling price for various brands, as a representative, the equipments with the medium-price were chosen for a substitute. The quantity and the unit-price of aeration system were shown in table 3.

Table 3. Price of major equipments

surface aeration system (A)			Blasting aeration system(B)		
Equipment	quantity	unit-price(RMB)	Equipment	quantity	unit-price(RMB)
T1	6	53000	T4	4	53000
T2	4	93000	T5	4	58000
T3	6	500000	T6	12	58000
			T7	3	720000
			T8		500000
total	14	37,020,000	Total	23	3,800,000

T8 (in table 3) referred to the assistive devices such as diffuser disc, gas pipeline and so on, which were necessary for blasting aeration system, but surface aeration system were not. As shown by the data in table 3, total cost of surface aeration was 3702,000 RMB and blasting aeration was 3800,000 RMB, the difference between two aeration systems was about 98,000 RMB.

4.3. Inventory and cost analysis in operation stage

4.3.1. Inventory analysis. Operation experience based on the WWTP proved that most of the WWTPs need be conducted technical modification or equipment revamping for 15 to 20 years guaranteeing to satisfy the development of wastewater treatment technology, the increasing amount of wastewater and the improvement of the water discharge requirements, etc.

In the paper, the energy consumption was evaluated in terms of 20-year life cycles. Factoring the costs of operation and maintenance stage was in closing of 60% to total costs, the following sections explained the costs of these in more detail. Item lists in operation and maintenance stage were shown in table 4:

Table 4. Item list in operation stage

Number	Item list
1	Equipment maintenance
2	Update of aerator
3	Economic losses
4	Environmental load
5	Energy consumption(Electricity)

The costs of day upkeep and repair projects for the blasting and surface aeration system were roughly similar every year, thus daily maintenance costs were ignored and not covered in item lists.

4.3.2. LCC analysis in operational stage. Based on the practical experience of the WWT and the data in table 4, the consumption in the operational stage was divided into three parts:

(1) Maintenance and repair list during operations. Basically, it was not necessary for surface aerators to require massive overhaul in 20 years of operation, involved only routine maintenance. As there was no significant difference between surface aeration and diffused aeration in terms of routine maintenance costs, so this part of cost could be neglected. Currently, combined with practical engineering, rubber was mainly used for gas distribution membrane in micro-porous aeration system. In contrast to micro-porous aeration with the titanium or ceramic plate, they could effectively avoid the blockage. However, rubber had a certain degree of flexibility; it was prone to causing breakage, tearing, corrosion or decline of gas distribution membrane. These damages on the aeration tray might greatly reduce the oxygenation efficiency of the diffused aeration system, and might affect the treatment effect of wastewater treatment plants. Therefore, LCC of the diffused aeration system was closely related with the service life of the aeration tray. With the increase of operation time, it could cause a significant reduction in oxygenation power efficiency, due to the lack of timely replacement and maintenance. Therefore, it was necessary to pay close attention to the changes of effluent quality and conduct timely contrast with the historical data. On the premise of the influent quality unchanged and blasting aerator fault excluded, only by extending the aeration time could meet the discharge standard, it would be considered that there was something wrong with diffusers. In this situation, they had to be repaired or replaced. To obtain better economic costing, the diffusers should be replaced by a certain year. The optimal option was about 10 years. There was a large price volatile because of different manufacturers, to make the results representative we assumed that the price of aeration diffusers was taken as 150RMB and the life-cycle was 10 years, it arrived at the conclusion that the diffusers should be replaced twice within a 20-year life cycle, ignored the labor costs and ancillary expenses because of a small proportion.

(2) Maintenance and repair based on sensitivity analysis. As a comparison, the present value method (formula 1, 2) was adopted to convert the cost into the base year, due to the replacement time of the diffusers different.

$$PCOST = COST \times PW \tag{1}$$

$$PW = (1+i)^n \tag{2}$$

Where PCOST was the present value convert to base year(RMB),COST referred to cost present value (RMB), PW was the present value factor, I was the bank interest rates, I=0.05.

Based on the formula (1) and (2)for calculation, the replacement cost of the aeration plate in the air blasting aeration system during the operation period of 20 years was 751,000 RMB (P1) (in the 10th) and 461,000 RMB P2(in the 20th) respectively, therefore the total costs were P1+P2=1,212,000 RMB.

(3) Economic loss cost in the repair process. In the replacement of diffusers of blasting aeration system, two ways was adopted to dispose wastewater : (1) supposed that other WWTP or backup systems could have a capacity to disposal the wastewater ,the government can carry out the deployment of the pumping station to accept the wastewater during the diffusers replacement days,(2) Supposed that no other WWTP or backup systems could receive a large volume of waste water, with approval of authority, the WWTP could only declare parking maintenance and the wastewater could be directly discharged without treatment. The economic loss caused by first method was wastewater quantity loss. There were two classes of loss caused by the second method: wastewater quantity loss of WWTP and serious damage to the environment which might cause a series of chain reactions. Therefore, wastewater quantity loss and environmental load were taken into consideration during calculation.

1) Cost of wastewater quantity loss. The replacement of 8160 aeration disc generally required 3 days under full preparations. During this period, wastewater wasn't disposal, thus causing cost of wastewater quantity loss than they otherwise would have. At current prices for wastewater (0.9 RMB /ton) in the domestic market, the wastewater loss in 3 days was $10\text{Wm}^3 \times 3\text{day} \times$water price. The entire aeration disc should be replaced 2 times in the 20-year lifecycle. The cost caused by replaced of aeration disc calculated with the present value method, that was166,000 RMB (P3)、102,000 RMB (P4) the 10th year and the 20th year respectively. So the total cost brought out by replacement of disc was P3+P4=268,000 RMB.

2) Environmental load cost. If the wastewater was discharged to other WWTP or backup systems under the coordination of the government, loads would not excessively generated to the environment during the replacement of diffusers. The environmental load cost under this circumstance (denoted as T1) was negligible. If the wastewater was directly discharged into the natural environment (denoted as T2), pollutants containing in the untreated wastewater entered into the environment, which caused serious damage to the natural environment such as rivers and lakes and so on. The environmental load cost caused by the equipment replacement was calculated according to COD emission reduction. The actual annual influent COD concentration of plant B was 157mg/L, and the annual effluent COD concentration was 23mg/L. The environmental load generated from COD replacement was W= $(157\text{-}23)\text{g/m}^3 \times 3\text{days} \times 10^5\text{m}^3$. Calculated as the COD emission trading price of 5,034,000RMB /ton provided in Literature [11], the environmental load cost generated in the 10th and the 20th year converted to the base year was 1,240,000 RMB (P5) and 763,000 RMB (P6) respectively. Therefore, within the life cycle of 20 years, the economic loss converted by environmental load generated from the COD emission reduction was P5+P6=2,003,000 RMB.

3) Energy efficiency costs. The oxygenation capacity of the micro porous aeration device for the blasting aeration system would be gradually decreased with the accumulation of being operated years. In order to meet the wastewater discharge standard, the WWTP increased the air volume of blasting aeratio system so as to achieve the required treatment effect. Therefore, the operating cost of the blasting aeration system increased year by year, and the actual energy efficiency costs. In Literature [12], the oxygen transfer efficiency and the increased power of blasting system of a new aeration and an old one operating for years had been contrasted and examined. The power of new one was 969.438kW, and the old one was about 1077.15kW. Total power of the aeration system increased by 11% in ten years. Based on this, the power of the blasting aeration system in plant B was 828kW at the

IOP Conf. Series: Earth and Environmental Science **63** (2017) 012029 doi:10.1088/1755-1315/63/1/012029

beginning, the power of the blower would change into 919.1kW during the replacement of the new aeration plate at the tenth years, under the circumstance of normal discharge standards. In order to simplify the calculation, the average value of 873.5kW was taken as the total installed power of blasting aeration system. Considering the cost sensitivity, the energy cost of the blasting system converted to the base year according to formula (3) and (4) was 65,225,000 RMB

$$PCOST = TP \times UPC \times 24 \times 365 \times PW \tag{3}$$

$$PW = \frac{(1+i)^n - 1}{i(1+i)^n} \tag{4}$$

Where PCOST was the present value of total operating cost (RMB). TP was the power of blasting aeration system (kW). PW was the present value coefficient. UPC was the unit electric charge (RMB / kWh). n was operation time of aeration equipment (year). I was interest rate, which was taken as 0.05. Refer to the national general industrial electricity prices, and set UPC as 0.684 RMB /kWh.

The energy costs for blasting aeration system were 65,225,000 RMB based on Formula (3) and (4).For the surface aeration system; the power of the whole life cycle was basically unchanged. According to Formula (3) and (4), the total costs of energy consumption converted to the base year were 81,093,000 RMB.

4.3.3. The stage of demolition. The costs of aeration system in demolition stage were largely depended on the considered elements. However, WWTP as an important public infrastructure would not be dismantled after ending the lifecycle, but another beginning of the WWTP lifetime, so the costs in demolition stage were not a consideration.

5. Comprehensive compare of LCC

5.1. Economic consumption comparison within life-cycle

In dismantling stage, the economic consumptions of A and B were basically the same, thus the comparison calculation wasn't carried out in total cost. Normal maintenance and maintenance wouldn't significantly different for the two aeration systems. Therefore, the analysis hadn't concentrated on normal maintenance cost of the two aeration systems. The comprehensive comparison results of surface and blasting aeration systems energy consumptions in various stages were shown as table 5.

Table 5. The total costs of surface aeration and blasting aeration system

Item(unit/thousand RMB)	A	B (T1)	B (T2)
Energy consumption	81093	65225	65225
Major equipments	3702	3800	3800
Replacement of equipments	0	1212	1212
Wastewater quantity loss	0	260	260
Environmental load cost	0	0	2005
Total	84795	70497	72502
Area cover(m^2)	10349.0	7822.9	7822.9

Shown as table 5, the *LCC* mainly depends on the cost of energy consumption whatever blasting aeration system or surface aeration. The WWTP adopting blasting aeration system could save 60,000 RMB per year and floor space of 2500m^2 ,comparing with the surface aeration system within a 20-year life cycle comparing with surface aeration system, which was also the reason why some WWTPs carried out remolding project or select blasting aeration equipment. Therefore, the blast aeration system should be the priority selection in wastewater plant upgrading and reconstruction to reduce its life cycle cost.

6. Conclusion and suggestion

On the premise of the standard discharge, the LCC comparison research result of surface aeration system and blasting aeration system shows that:

(1) The energy consumption of the blast aeration system was lower than that of the surface aeration system.

(2) The life cycle costs of blasting aeration system were also better than that of surface aeration system, which could save about 600,000RMB per year and effectively reduce the occupied area. Besides, the annual costs saving were related to the service life and equipment cost of air blasting aerator.

(3) Considering the cost of environmental load, and there was no other WWTP or available backup system within the life cycle, which exercised a great influence on the total costs of blasting aeration system. During the operation maintenance period, the unavoidable discharge into the water environment was proposed to be arranged in the rainy season, in order to minimize the adverse effects of wastewater on the environment through the rain attenuation.

(4) On the premise of the standard discharge, the NH_4^+-N and TN concentration of A^2/O wastewater plant adopting surface aeration system was superior to that adopting blasting aeration system.

7. Acknowledgements

Research for this paper was partially supported by National water pollution control and governance of science and technology major project (2015ZX07218001).

8. Reference

[1] Wang M H, Wang L K 1979 *Journal of Environment science* **22** (4):23-26
[2] Chang J, Yang A M, Gan Y P, Meng C L , Peng Y Z and Zhang S J 2011 *China Water & Wastewater* **27**(4):33-36
[3] Xia W H, Liu F, Zhou B 2009 *Water & Wastewater Engineering* **35**(1):121-125
[4] Lee I, Lim H, Jung B, Kim H 2014 *Chemosphere* **140**(3):72-78
[5] Verrecht B, Judd S, Guglielmi G, Brepols C and Mulder JW 2008 *Water Research* **42**(19):4761-70
[6] Huang H H, Zhang J, Wen X H 2009 *Chinese ournal of Environmental Engineering* **3**(1):35-38
[7] Shao L G, Yang P and Xiong F 2012 *Industrial Safety & Environmental Protection* **38**(9):18-20
[8] Yang P 2012 Wuhan: *Wuhan University of Science and Technology* 12-24
[9] Liang Y, Wang J, Li J, Ying Q F,Lu **C** S ,Xie J L,Hu J 2011 *Water & Wastewater Engineering* **37**(1):42-45
[10] Kumar B and Rao A R 2010 *Korean Journal of Chemical Engineering* **27**(6):1796-00
[11] Thapinta A 2015 *Proceedings of International Academic Conferences: International Institute of Social and Economic Sci*ences(IAC, 2015)
[12] Lee I, Lim H, Jung B, Colosimo *MF and Kim H 2014 Chemosphere* **140**:72-78.

IOP Conf. Series: Earth and Environmental Science **63** (2017) 012030 doi:10.1088/1755-1315/63/1/012030

The Study and Application of Hydrometallurgical Gold Leaching in the Analysis of Refractory Precious Metals

M Yang[1, 2], X Geng[2], Y L Wang[1, 2] and D X Li [2*]

[1]Shanghai Cooperative Center for WEEE Recycling, Shanghai Second Polytechnic University, Shanghai 201209, China
[2]School of Environmental Science and Engineering, Donghua University, 2999 North Renmin Road, Shanghai 201620, China.

lidengxin@dhu.edu.cn

Abstract. Three orthogonal tests are separately designed for each hydrometallurgical gold leaching process to finding the optimum reaction conditions of melting gold and palladium in each process. Under the optimum condition, the determination amount of gold and palladium in aqua regia—hydrofluoric acid, Sodium thiosulfate, and potassium iodide reaches 2.87g/kg and 8.34 g/kg, 2.39g/kg and 8.12 g/kg, 2.51g/kg and 7.84g/kg. From the result, the content of gold and palladium using the leaching process of combining Aqua regia, hydrofluoric acid and hydrogen peroxide is relatively higher than the other processes. In addition, the experiment procedure of aqua regia digestion operates easily, using less equipment, and its period is short.

1. Introduction

In recent years, with the development of technology and electronic industries, the pace of upgrading electronic devices is accelerating, and the electronic wastes from it are also increasing [1]. Printed circuit boards is an important part of every electronic devices, generally containing gold, silver, palladium and other precious metals, whose recycling value is high but quantity is low[2]. So the pretreatment method is very important for the recycling of waste circuit board.

There are many literature reports of hydrometallurgical treatment process [3-5]. But the report of the design of orthogonal tests on the process is few. In this article, three orthogonal tests are designed for each hydrometallurgical gold leaching process to finding the optimum reaction conditions of melting gold and palladium in each process and determined the content.

2. Experimental

2.1. Orthogonal tests on aqua regia system

2.1.1. Ingredients. The ingredients of the experiment are the powder of circuit board below 100 mesh.

2.1.2. Materials, chemicals and apparatus. Nitrate, hydrochloric acid, hydrofluoric acid, hydrogen peroxide (All the reagent is AR), 50ml Teflon crucible, separating funnel, heating plate, ICP-AES, electronic analytical balance.

2.1.3. Experimental procedure. The sample (0.5000g) was taken from the crushed circuit board and put into Teflon crucible. A little deionized water was added. New aqua regia was made (V_{HCl}: V_{HNO3}=3:1) and slowly added. When the yellow smoke gradually disappeared, heated until the solution became transparently green, with stable level and no yellow smoke (otherwise aqua regia should be added). Cooled to room temperature, hydrofluoric acid was added and heated again. The lid was sealed when white smoke appeared. Cooled down when about 0.5ml solution remained in it. Hydrogen peroxide was added and heated until the solution is almost dry out, then nitrate (15ml, 10%) was added. The digested solution was filtered by separating funnel, diluted to volume by 50ml volumetric flask, and mixed to prepare for measurement.

2.1.4. Result and analysis. The amount of aqua regia, hydrofluoric acid and hydrogen peroxide, the reaction time and the reaction temperature are chosen as the factors. Each factor has four levels. A orthogonal test, shown in Table 1.

Table 1. The factors of orthogonal tests and their levels.

Factors / Levels	Aqua regia /ml	HF /ml	H_2O_2 /ml	Time /h	Temperature /℃
1	10	5	3	1.5	100
2	15	7	5	2	120
3	20	9	7	2.5	140
4	25	11	9	3	160

The value K and the range R gotten from analyzing the result are shown in the Figure 1. to Figure4.

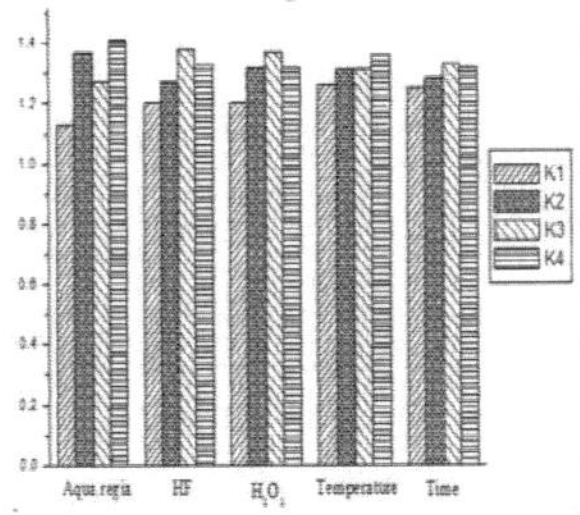

Figure 1. The analysis on the value K of leaching gold in aqua regia system.

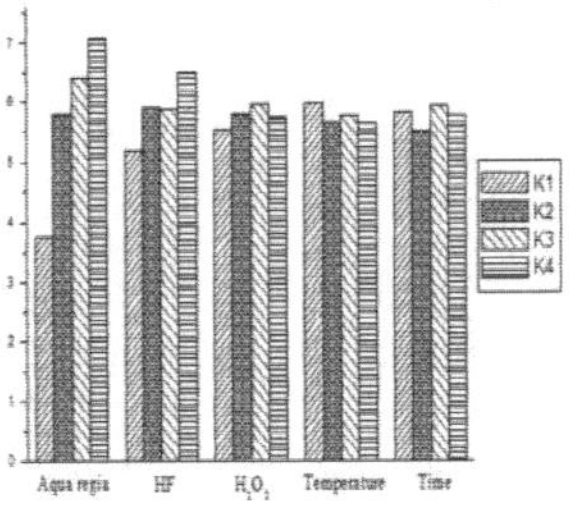

Figure 2. The analysis on the value K of leaching palladium in aqua regia system.

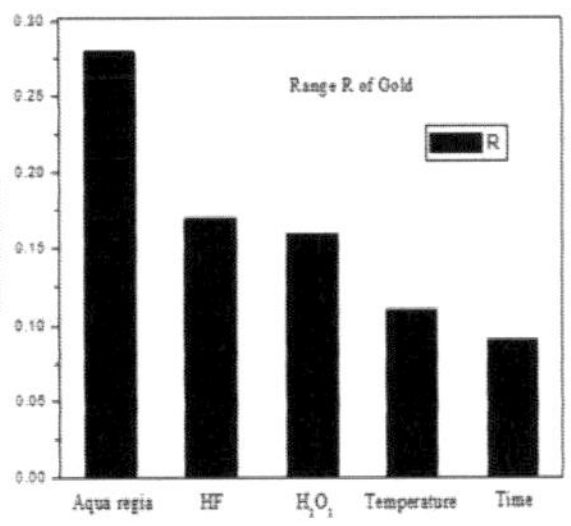

Figure 3. The analysis on the range R of leaching gold in aqua regia system.

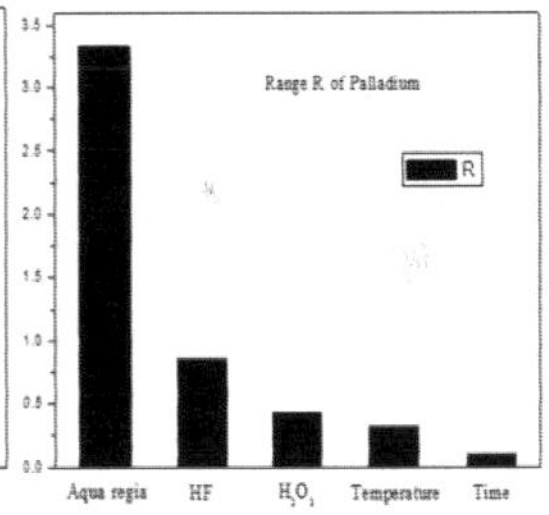

Figure 4. The analysis on the range R of leaching palladium in aqua regia system.

2.1.5. The choice of best experimental condition. From the analysis of the result, the value K and the range R of gold and palladium can be calculated. It can be concluded that the amount of aqua regia has the most significant influence on the result of leaching, followed by hydrofluoric acid; the reaction time has the smallest influence.

2.1.6. The situation of gold and palladium leaching under the optimum condition. Based on the orthogonal experiment, the optimum condition and the result is shown in the Table 2.

Table 2. The situation of gold and palladium leaching in aqua regia system.

Factors	Best condition of Au	Result	Best condition of Pd	Result
Aqua regia/ml	25		25	
HF/ ml	9		11	
H_2O_2/ ml	7	2.87g/kg	7	8.34 g/kg
Temperature/℃	160		160	
Time/h	2.5		2.5	

Under this reaction condition, the content of gold and palladium dissolved reached 2.87g/kg and 8.34g/kg. It can indicate that the dissolve effect is great under such conditions and the result is reliable.

2.2. Orthogonal tests on sodium thiosulfate system

2.2.1. Ingredient. The ingredients of the experiment are the powder of circuit board below 100 mesh.

2.2.2. Materials, chemicals and apparatus. The reagent: sodium thiosulfate, sodium chloride, copper sulfate, ammonia (all AR). The apparatus used in this experiment was Magnetic stirrer, Electronic balance, Thermostatic bath (HH-1), ICP-AES.

2.2.3. Experimental procedures. The powdered sample was weighed and then put into 250ml beaker. Some sodium thiosulfate was slowly transferred into beaker by pipette. Then some copper sulfate and ammonia was transferred into beaker. Sodium thiosulfate concentration, copper sulfate concentration, ammonia concentration, reaction time and temperature were chosen as the factors of the experiment. Each factor has four levels. This orthogonal experiment is shown in the Table 3.

Table 3. The factors and levels of orthogonal tests on sodium thiosulfate system.

Factors / Levels	$S_2O_3^{2-}$ /mol/L	$CuSO_4$ /mol/L	Ammonia /mol/L	Time /h	Temperature /℃
1	0.2	0.1	0.2	1.5	20
2	0.3	0.2	0.4	2	40
3	0.4	0.3	0.6	3	60
4	0.5	0.4	0.8	4	80

2.2.4. Result and analysis. The value K and the range R gotten from analyzing the result are shown in the Figure 5. to Figure 8.

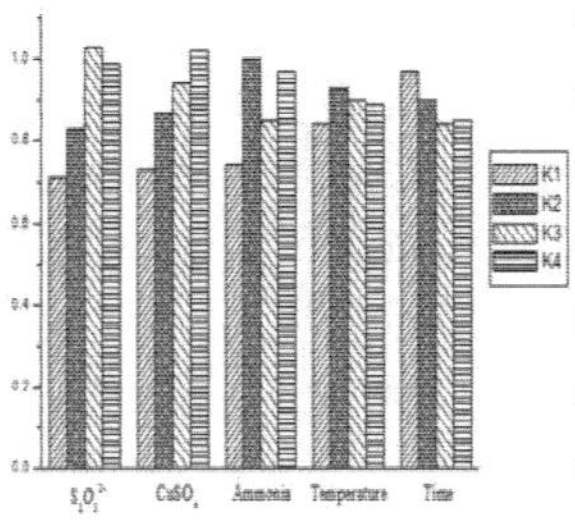

Figure 5. The analysis on the value K of leaching gold in sodium thiosulfate system.

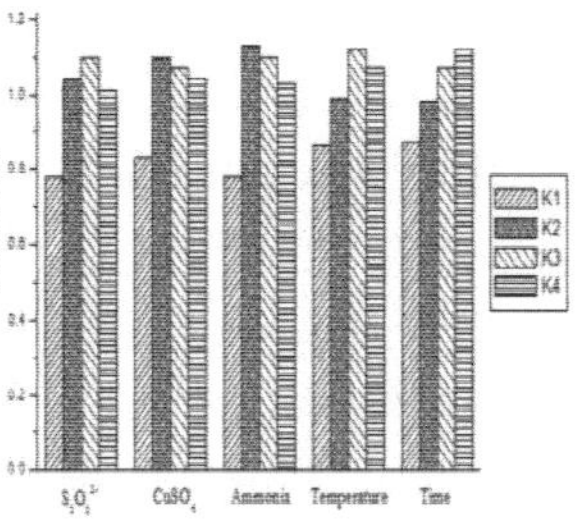

Figure 6. The analysis on the value K of leaching palladium in sodium thiosulfate system.

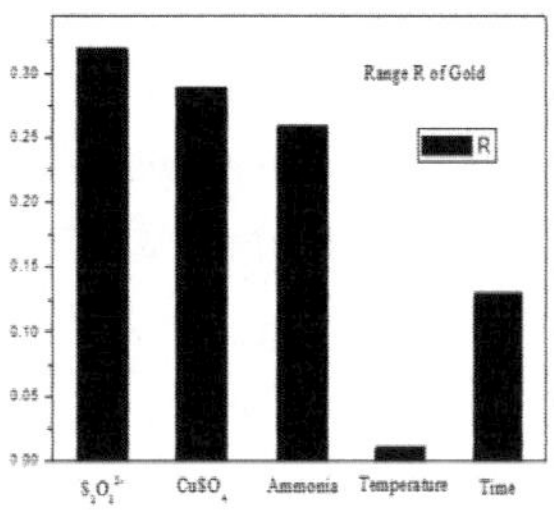

Figure 7. The analysis on the range R of leaching gold in sodium thiosulfate system.

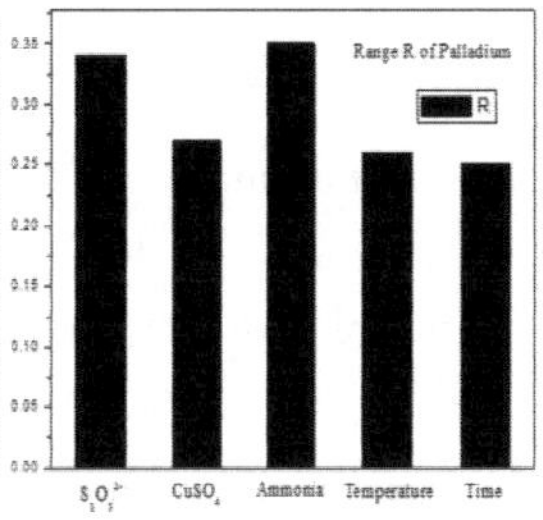

Figure 8. The analysis on the range R of leaching palladium in sodium thiosulfate system.

2.2.5. The choice of best experimental condition. From the value K and the range R, it can be shown that the amount of sodium thiosulfate, copper sulfate and ammonia has a significant influence on the result. When the concentration of $S_2O_3^{2-}$ is increased, the compound $[Cu(S_2O_3)_3]^{5-}$ will be synthesized in the solution, resulting the decrease of the concentration of $S_4O_6^{2-}$. $S_2O_3^{2-}$ can be oxidized to $S_4O_6^{2-}$ by Cu^{2+}. The higher concentration of Cu^{2+} leads to the larger consumption of $S_2O_3^{2-}$. The final gold leaching rate will be low. So these three factors should be strictly controlled. The influence of temperature and reaction time is small.

2.2.6. The situation of gold and palladium leaching under the optimum condition. Under the premise of sodium thiosulfate leaching experiments, the optimum condition and the result is shown in the Table 4.

Table 4. The situation of gold and palladium leaching in sodium thiosulfate system.

Factors	Best condition of Au	Result	Best condition of Pd	Result
$S_2O_3^{2-}$/ mol/L	0.4		0.4	
$CuSO_4$/ mol/L	0.2		0.2	
Ammonia/ mol/L	0.2	2.39g/kg	0.2	8.12 g/kg
Temperature /℃	80		60	
Time/h	4		4	

Under the best dissolution condition, their content reaches 2.39g/kg and 8.12 g/kg. It shows that under this condition the dissolution effect of gold and palladium is well and the result is reliable.

2.3. Orthogonal tests on KI system

2.3.1. Ingredients. The ingredients of the experiment is the powder of circuit board below 100 mesh.

2.3.2. Materials, chemicals and apparatus. KI (AR), Sodium hypochlorite (effective chlorine content ≥ 12%), deionized water. Digital homothermal magnetic stirrer, homothermal water bath kettle, standard sieve set, ICP-AES.

2.3.3. Experimental procedures. The powdered sample was weighed and then put into 250 ml beaker. Some KI was slowly transferred into beaker by pipette. The orthogonal experiment is designed as what is shown in the Table 5.

Table 5. The factors and levels of KI system.

Factors / Levels	Solid-liquid ratios	KI /mol/L	NaClO /mol/L	Time /h	Temperature/ ℃
1	1:6	0.05	6	2.5	25
2	1:8	0.1	8	3	30
3	1:10	0.2	10	4	40
4	1:12	0.3	12	5	50

2.3.4. Result and analysis. The value K and the range R gotten from analyzing the result are shown in the Figure 9. to Figure 12.

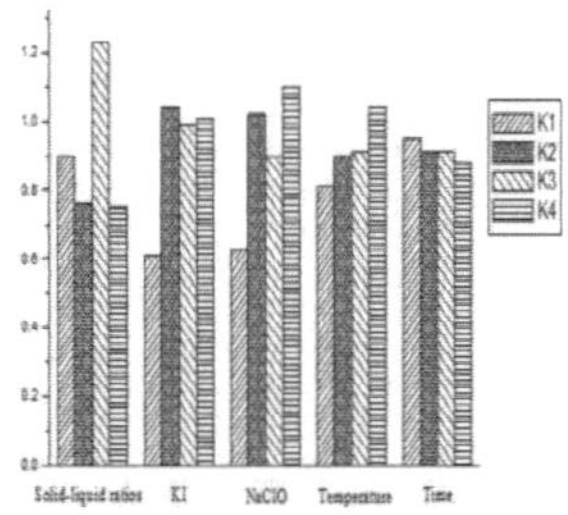

Figure 9. The analysis on the value K of leaching gold in KI system.

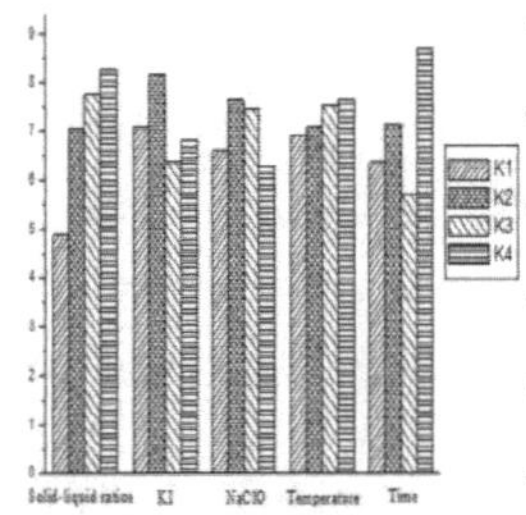

Figure 10. The analysis on the value K of leaching gold in KI system.

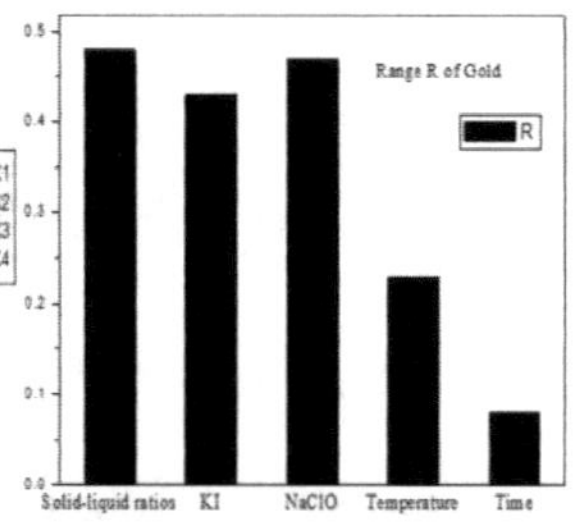

Figure 11. The analysis on the value K of leaching gold in KI system.

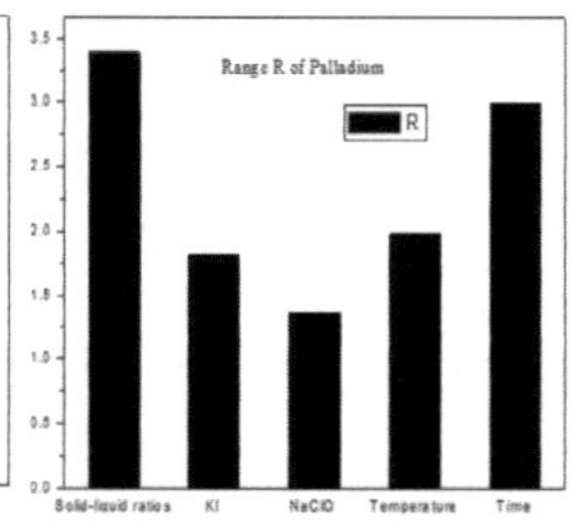

Figure 12. The analysis on the value K of leaching gold in KI system.

2.3.5. The choice of best experimental condition. The experiment of using KI method to leach precious metal from circuit board indicates that sodium hypochlorite, as an additional oxidant, can effectively increase the leaching rate of precious metal. In addition, the solid-liquid ratio and the concentration of KI both have significant influence on the leaching rate, so these two factors should be strictly controlled in practice application.

2.3.6. The situation of gold and palladium leaching under the optimum condition. Under the premise of KI leaching experiments, circuit boards are dissolved in the best reaction condition. Then the digestion solution is determined. The result is shown in Table 6.

Table 6. The situation of gold and palladium leaching in KI system.

Factors	Best condition of Au	Result	Best condition of Pd	Result
Solid-liquid ratio	0.1		0.08	
KI/ mol/L	0.1		0.1	

NaClO/ mol/L	12	2.51g/kg	8	7.84 g/kg
Temperature/℃	50		50	
Time/h	2.5		3	

Under the best dissolution condition, their content reaches 2.51g/kg and 7.84 g/kg. It shows that under this condition the dissolution effect of gold and palladium is well and the result is reliable.

3. Conclusion

Three orthogonal tests are separately designed for each hydrometallurgical gold leaching process to finding the optimum reaction conditions of melting gold and palladium. The result shows that: aqua regia—hydrofluoric acid, Sodium thiosulfate, and potassium iodide, under the optimum condition, the determination amount of gold and palladium reaches 2.87g/kg and 8.34 g/kg, 2.39g/kg and 8.12 g/kg, 2.51g/kg and 7.84g/kg respective. From the result, we can see that the content of gold and palladium determined in the digestion solution from digesting circuit board using the leaching process of combining Aaqua regia—hydrofluoric acid is relatively higher than the others. In addition, the experiment procedure of aqua regia digestion operates easily, using less equipment, and its period is short. So there is reliable to choose aqua regia gold leaching process as the pretreatment method.

4. Acknowledgments

This project was funded by Shanghai Cooperative Center for WEEE Recycling. And the authors are thankful to the grant support of Shanghai Cooperative Center for WEEE Recycling.

5. References

[1] Long E, Kokke S, Lundie D, Shaw N, Ijomah W and Kao C C 2016 Technical solutions to improve global sustainable management of waste electrical and electronic equipment (WEEE) in the EU and China *J. Remanuf.* **6** pp 1-27

[2] Yamane L H, Moraes V T D, Espinosa D C R and Tenório J A S 2011 Recycling of weee: characterization of spent printed circuit boards from mobile phones and computers *Waste Manage..* **31** pp 2553-8.

[3] Petter P M, Veit H M and Bernardes A M 2014 Evaluation of gold and silver leaching from printed circuit board of cellphones *Waste Manage.* **34** pp 475-82.

[4] Liu Y. The test research of precious metal recycling technology of circuit board from mobile phones 2013 *Southwest Jiaotong University*

[5] Flandinet L, Tedjar F, Ghetta V and Fouletier J 2012 Metals recovering from waste printed circuit boards (wpcbs) using molten salts *J. Hazard. Mater.* **213–14** pp 485-90.

The Effect of AOT and Octanoic Acid on the Formation of Stable Water-in-diesel Microemulsion

Yue Zhang [1], Misni Bin Misran [2], Zhicheng Wang [1] and Yu Zhang [1]

[1] Energy and Environmental Research Institute of Heilongjiang Province, Heilongjiang Academy of Sciences, Harbin, Heilongjiang 150027, China
[2] Department of Chemistry, Faculty of Science, University of Malaya, 50603 Kuala Lumpur, Malaysia

hnh_wzc@126.com (Zhicheng Wang); zy_nyy@163.com (Yue Zhang)

Abstract. Sodium bis(2-ethylhexyl)sulfosuccinate (AOT) and octanoic acid (OA) were used as surfactants to prepare water-in-diesel microemulsion. The effect of mixed surfactants ratio on the phase behavior of water-in-diesel microemulsion was investigated. The R_0-T plot phase diagrams for the diesel/AOT and OA/water system with different surfactant ratios were constructed at 30-80 °C. The results indicate that the largest single phase region could be obtained when OA to AOT molar ratio was 1. The temperature had a significant influence on phase transformation behavior. The single phase separated into two immiscible phases with the increase of temperature when R_0 value was above 10. Compared with applying AOT alone, mixing AOT with appropriate amount of OA is benefit to form smaller nanosized W/O droplets. The determination of particle size was performed to verify the phase transformation behavior, and the results were consistent with the phase diagrams.

1. Introduction

The increasing demand for fuel energy, continuing consumption of reserved petroleum and environmental pollution problems such as the emission of NO_x, particulate matter (PM) and CO_2 due to the combustion of fuels have challenged the existing fuel energy [1-3]. Recently, more attention has been paid to the development of cleaner, alternative and sustainable fuels in order to meet higher emission standards and to reduce the dependence on pure fossil fuels.

The addition of water into diesel to form water-in-diesel microemulsion fuel exhibits several benefits. First of all, it has been proved that the vaporization of water and the mixing process can effectively reduce soot, PM and NO_x emissions [4-6]. Another benefit is that water in the form of microstructure droplets can exert some positive influences on the combustion process of fuels. When microemulsion fuel is heated, the vaporization of water will cause continuous hydrocarbon phase to ‘‘explode’’ [7]. This phenomenon helps in improving fuel combustion process and combustion efficiency [8]. Furthermore, since water droplets are wrapped by diesel fuel, there is no direct contact between water and engine cylinder, which can eliminate the negative effect of water on lubricating oil contamination and engine wear [9]. There is no doubt that water-in-diesel microemulsion fuel is energy saving and environmental protection.

Sharon et al. investigated the co-surfactant chain length, alcohol isomerism, solution salinity, CTAB/alcohol ratio and the effect of combining various alcohols were studied [10]. Ochoterena et al. investigated physical properties, spray behavior and combustion characteristics of water-in-diesel emulsion, water-in-diesel microemulsion and conventional diesel fuel [11]. Spray behavior studies demonstrated droplets of water-containing fuels penetrated further than droplets of regular diesel fuel. Combustion studies showed that water-in-diesel fuels yielded to flames with lower temperature and lowered soot concentrations than diesel fuel. Kannan and Anand reported that the emission characteristics like carbon monoxide (CO), carbon dioxide (CO_2), nitric oxide (NO) and smoke emissions for biodiesel and microemulsion fuels were lower than diesel fuel[12]. Bulent and Mudhafar investigated the effects of water concentration in a biodiesel nanoemulsion fuel on engine performance and exhaust emissions of a 4-cylinder diesel engine [13]. Biodiesel nanoemulsions containing 5%, 10% and 15% water were used for the engine tests. They concluded that increasing water concentrations in biodiesel nanoemulsions increased the engine brake specific fuel consumption and CO emissions. The rate of NO_x reduction was greater than the rate of CO increase when the water concentration in biodiesel nanoemulsions increased from 10% to 15%.

However, papers have been reported were mainly concerning combustion properties, spray behavior and reduction of emissions. In our previous studies, several kinds of nonionic surfactants were used to form water-in-diesel emulsions, whereas the stability of such emulsions was not satisfactory. Hence, we embarked on the study of phase behavior of water-in-diesel microemulsion. The objective of this research is to experimentally study the effect of AOT and octanoic acid on the formation of water-in-diesel microemulsion. AOT is a popular surfactant used for the formation of water-in-oil microemulsion droplets containing large amount of water under wide range of conditions (such as water content, temperature, solvent, electrolyte type and concentrations) [14-16]. OA is used as co-surfactant and its role is to enhance AOT surfactant property and compatibility with diesel at the interface.

2. Materials and methods

2.1. Materials

Sodium bis(2-ethylhexyl)sulfosuccinate (AOT) with 98% purity was purchased from Fluka. Octanoic acid (OA) with 98% purity was purchased from Sigma-Aldrich. Double distilled de-ionized water was used in sample preparations. Diesel was purchased from local Petronas petrol station (national oil company of Malaysia).

2.2. Preparation of microemulsion and phase diagram

The microemulsions were prepared by mixing given concentration of AOT and OA solution with diesel oil. Then different volumes of water were added into the mixture. The samples were mixed by using Minishaker for 10 s. The phase diagram was constructed by calculating ratios of water concentration ([H_2O]) to surfactant concentration ([Surf]) [17] (formula 1). By plotting different R_0 values with respect to temperatures, the phase diagram was obtained. For all the samples, the total surfactant concentration was kept constant as 0.222 mol/L (the total surfactant concentration = [AOT] + [OA]). The phase behaviors of samples were observed with eyes after pulled out 30 s, and then the phase numbers were recorded.

$$R_0 = \frac{[H_2O]}{[Surf]} \quad (1)$$

2.3. Instrumentation

The water-in-diesel microemulsion samples were prepared in 3.5 mL vials sealed with caps. The phase behaviors of samples under different temperatures were investigated by changing temperature in the range of 30-80 °C in a Thermo Haake water bath, which accuracy is ±1.0 °C. Water content was measured by Karl Fisher Titrator (Mettler Toledo DL38). Dynamic Light scattering (DLS) experiments were performed to obtain the mean diameter of the water-in-diesel microemulsion droplets. The measurements were carried out by using a Zeta Sizer Nano Series (Malvern). All measurements were performed at the temperature of 40°C. Each sample was monitored for 3 times.

3. Results and discussion

3.1. Water content and R_0 values

In order to confirm actual water content in the samples, Karl Fisher Titrator was employed to determine the water quantity, and then the water concentration could be calculated. After the water concentrations were figured out, R_0 values could be calculated according to formula 1. Results of R_0 values for all the samples with different surfactant ratios were listed in Table 1.

Table 1. R_0 values of samples with different surfactant ratios.

Water volume of addition (μL)	R_0						
	AOT alone	$n_{OA} : n_{AOT}$ = 1:4	$n_{OA} : n_{AOT}$ = 1:2	$n_{OA} : n_{AOT}$ = 1:1	$n_{OA} : n_{AOT}$ = 4:3	$n_{OA} : n_{AOT}$ = 2:1	$n_{OA} : n_{AOT}$ = 4:1
50	5.23	5.38	4.05	4.66	5.17	4.15	4.39
100	7.66	9.00	9.18	9.00	9.05	8.97	8.87
150	10.17	14.50	13.27	13.11	13.90	13.99	14.99
200	15.94	17.35	19.14	17.34	17.81	19.43	18.15
250	22.36	23.54	23.36	20.50	21.87	23.09	23.19
300	25.78	28.38	27.05	27.91	26.33	28.52	27.54
350	32.04	33.44	34.69	32.03	31.94	32.68	31.64
400	38.13	36.27	38.96	39.96	37.34	36.08	38.29

3.2. Construction of phase diagrams

The molar ratio of surfactant OA and AOT is an important factor that influences the phase behavior of microemulsion. The phase behavior was divided into several categories as shown in figure 1. The single phase seemed to be clear and transparent, while the phase behavior of two-phase or three-phase altered with variation of conditions, such as surfactant ratio, water content and temperature. To investigate the effect of surfactant ratios on the transformation of W/O microemulsion domain, R_0-T plot phase diagrams of diesel/surfactants/water system for different surfactant molar ratios at temperature range of 30-80 °C were presented in figure 2.

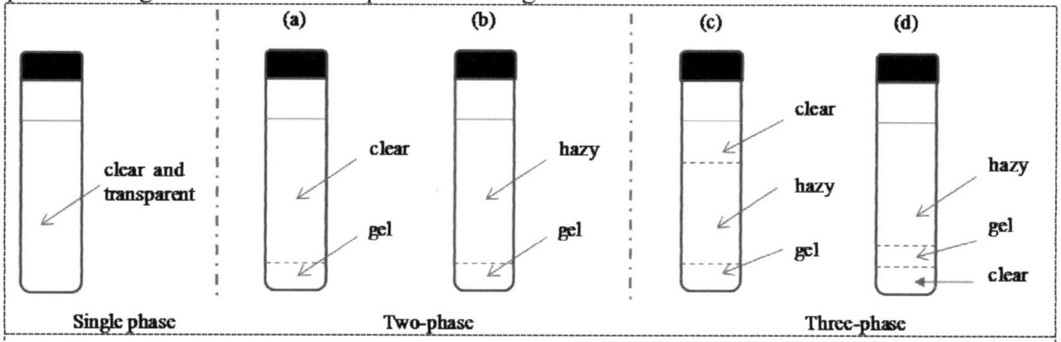

Figure 1. Schematic diagrams of phase behavior. (a) Phase behavior at lower water content; (b) Phase behavior at higher water content; (c) Phase behavior for samples with $n_{OA} : n_{AOT}$= 1:1; (d) Phase behavior for samples with $n_{OA} : n_{AOT}$= 4:1.

Figure 2 (a) reveals that single phase separated into two immiscible phases with the increase of temperature for all cases. The highest R_0 value for single phase boundary was 20.50 when n_{OA}/n_{AOT} was 1 at 30 °C, and the corresponding water content was 10.8 %. Figure 2 (b) reveals the three-phase region (inside the boundary) transformation process with the increase of OA amount. For n_{OA}/n_{AOT} below 1, the three-phase region transformed from lower R_0 value to higher R_0 value. When n_{OA}/n_{AOT} was increased to 2, the phase diagram appeared the minimum three-phase region area. However, the three-phase region not only transformed to higher R_0 value but also expanded again when n_{OA}/n_{AOT}

was 4. Furthermore, the single phase region disappeared at this ratio. The results illustrate that it is not suitable to form stable water-in-diesel microemulsion by applying AOT alone, and the addition of appropriate amount of OA is benefit to increase water content in water-in-diesel microemulsion. This is due to almost 70 % of the components in diesel are saturated linear alkanes with C_8-C_{16} [18-19], while the carbon number of AOT hydrophobic tail is six and the shape of AOT molecule is wedge. This may lead to an unfavorable interface between AOT and diesel oil. The presence of OA will effectively reduce the interfacial energy of water-oil surface and stabilize the water-oil surface [20]. Consequently, the optimum molar ratio of OA to AOT was 1 in this experiment.

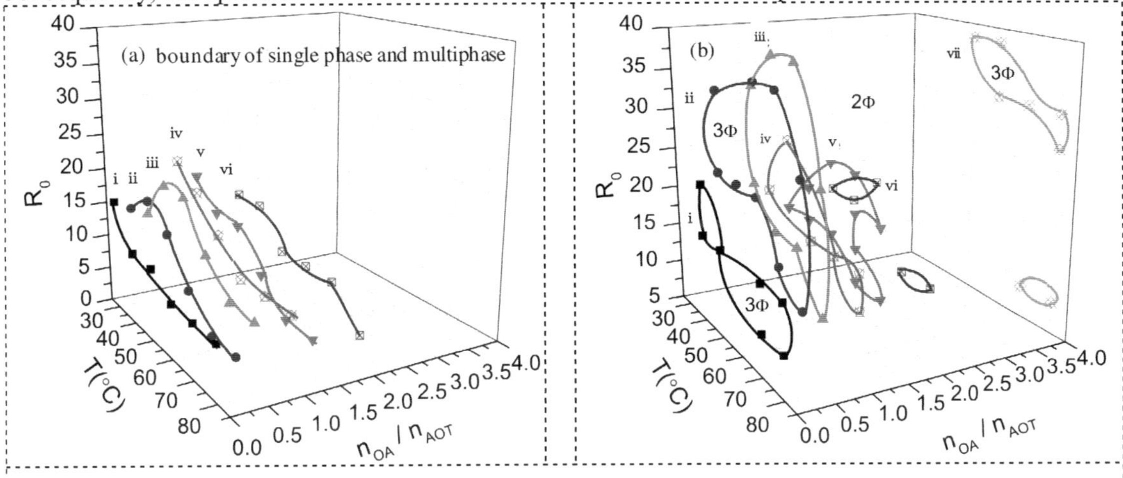

Figure 2. R_0-T plot phase diagram of diesel/surfactants/water system with different OA/AOT molar ratios: (i) AOT alone; (ii) n_{OA} : n_{AOT}= 1:4; (iii) n_{OA} : n_{AOT}= 1:2; (iv) n_{OA} : n_{AOT}= 1:1; (v) n_{OA} : n_{AOT}= 4:3; (vi) n_{OA} : n_{AOT}= 2:1; (vii) n_{OA} : n_{AOT}= 4:1. (b) boundary of two-phase and three-phase.

3.3. Particle size and polydispersity index

The surfactant ratio has an important influence on the particle size, and then the particle size of the droplet determines the phase behavior of the samples. In order to identify the relation between particle size and phase behavior, the particle size of clear water-in-oil (W/O) region in samples has been measured. As shown in figure 3, the curves for different surfactant ratios were fluctuated with the increase of added water volume. The variation of particle size was consistent with phase behavior, the phase boundary of single phase and multiphase could be deduced. For example, the particle size at point 4 and point 5 in the case of n_{OA} : n_{AOT} = 1:1, corresponding to R_0 value of 17.34 and 20.50, were not linear increase. This tendency exactly indicates a phase transition, which also can be seen in figure 2. Point 4 was the phase boundary of single phase and multiphase, and point 5 was the phase boundary of two-phase and three-phase. The particle size at point 5 decreased to 21.7 nm, which was mainly due to the formation of a clear phase on the top of the sample (see figure 1 c). Figure 3 also reveals that the addition of excessive OA is more feasible to form larger size droplets. This may be due to AOT is interfacial active to increase the interfacial energy, which favors to form negative curvature. However, OA is not so interfacial active as AOT that it will lower the interfacial energy. With an overdose of OA is added to the mixture, there is an increase in the magnitude of the spontaneous curvature of the surfactant monolayer, and then the droplets swell in size [21].

In an emulsion-based solution, polydispersity index (PDI) is one of the key characteristics, as PDI contributes to physical stability and rheological properties of the solution [22]. PDI values of the samples with different surfactant molar ratios were listed in Table 2. The PDI value provides a measure of the narrowness of particle size distribution, with values ≤ 0.1 indicating a very narrow distribution [23]. That means it is more possible to obtain stable and monodisperse samples at lower PDI value. As shown in Table 2, the stability of the samples decreased with the increase of OA amount. Moreover, the lowest PDI values for all the cases were exactly corresponded to particle size transition points in figure 3.

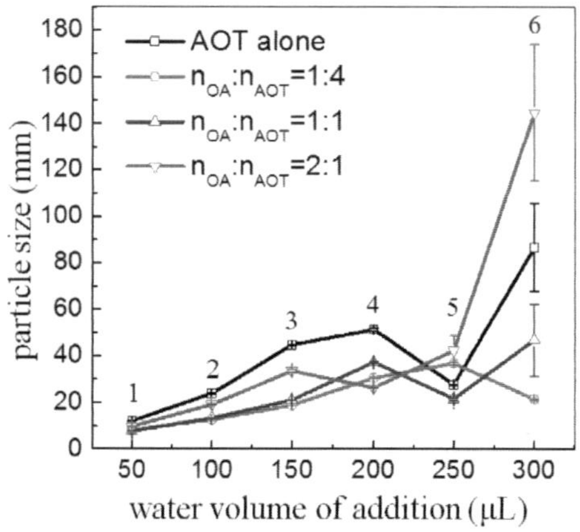

Figure 3. Particle size of droplets in water-in-oil region as a function of water volume of addition. The samples were soaked in water bath with temperature of 40°C, and the measurement of particle size was done immediately after the sample was taken out.

Table 2. PDI values of samples with different surfactant ratios.

Water volume of addition (μL)	PDI±SD			
	AOT alone	$n_{OA} : n_{AOT}$ = 1:4	$n_{OA} : n_{AOT}$ = 1:1	$n_{OA} : n_{AOT}$ = 2:1
50	0.195±0.007	0.213±0.095	0.162±0.002	0.102±0.005
100	0.070±0.010	0.078±0.029	0.049±0.012	0.139±0.050
150	0.033±0.007	0.026±0.010	0.051±0.013	0.053±0.017
200	0.062±0.016	0.025±0.012	0.036±0.005	0.182±0.043
250	0.099±0.020	0.026±0.003	0.058±0.014	0.349±0.059
300	0.483±0.149	0.068±0.030	0.451±0.240	0.519±0.154

4. Conclusions

In this work, the effect of mixed molar ratio of surfactant AOT and OA on the phase behavior of water-in-diesel microemulsion was tested. The results confirmed that nanosized water-in-diesel droplets formed by using the mixture of AOT and OA. The largest single phase region could be obtained when the molar ratio of OA to AOT was 1. The single phase separated into two immiscible phases with the increase of temperature when the R_0 value was above 10. It is not suitable to form homogeneous and stable water-in-diesel microemulsion by applying AOT alone. The addition of appropriate amount of OA is benefit to form clear and transparent microemulsion with smaller nanosized droplets. The measurement of particle size and PDI was performed to verify the phase transformation behavior, and the results were consistent with the phase diagrams.

5. Acknowledgments

Heilongjiang Academy of Sciences in China supported this work. We also gratefully acknowledge the guidance of Prof. Misni Misran in Malaya.

6. References

[1] Sheehan J, Cambreco V, Duffield J, Garboski M and Shapouri H 1998 *A report by US Department of Agriculture and Energy* 1–35

[2] Salvi B, Subramanian K and Panwar N. Alternative fuels for transportation vehicles: A technical review 2013 *Renewable & Sustainable Energy Reviews* 25 404–419
[3] Zhang Z, Cheung C, Chan T and Yao C Experimental investigation of regulated and unregulated emissions from a diesel engine fueled with Euro V diesel fuel and fumigation methanol 2010 *Atmos. Environ.* 44 1054–1061
[4] Lin C and Wang K The fuel properties of three-phase emulsions as an alternative fuel for diesel engines 2003 *Fuel* 82 1367–1375
[5] Subramanian K A comparison of water–diesel emulsion and timed injection of water into the intake manifold of a diesel engine for simultaneous control of NO and smoke emissions 2011 *Energy Convers. Manage.* 52 849–857
[6] Armas O, Ballesteros R, Martos F and Agudelo J Characterization of light duty Diesel engine pollutant emissions using water-emulsified fuel 2005 *Fuel* 84 1011–1018.
[7] Anna L and Krister H Water-in-diesel emulsions and related systems 2006 *Adv. Colloid Interface Sci.* 123–126
[8] Kadota T and Yamasaki H Recent advances in the combustion of water fuel emulsion 2002 *Prog. Energy Combust. Sci.* 28 385–404
[9] Yang W, An H, Chou S, Chua K, Mohan B, Sivasankaralingam V, Raman V, Maghbouli A and Li J Impact of emulsion fuel with nano-organic additives on the performance of diesel engine 2013 *Appl. Energy*
[10] Sharon L, Deane T and Simon B The formation of water-in-oil microemulsions using a concentrated saline aqueous phase 1998 *Colloid Surf. A-Physicochem. Eng. Asp* 137 25–33
[11] Raúl O, Anna L, Magnus N, Sven A and Ingemar D Optical studies of spray development and combustion of water-in-diesel emulsion and microemulsion fuels 2010 *Fuel* 89 122–132
[12] Kannan G and Anand R Experimental investigation on diesel engine with diestrol-water micro emulsions 2011 *Energy* 36 1680–1687
[13] Bulent K and Mudhafar A Performance and NO_x emissions of a diesel engine fueled with biodiesel-diesel-water nanoemulsions 2013 *Fuel Processing Tech.* 109 70–77
[14] Sujan C, Rajib K, Bidyut K and Subhash C Interface of AOT/Brij mixed reverse micellar systems: Conductometric and spectrophotometric investigations 2006 *J Colloid Interf. Sci.* 298 935–941
[15] Li Q, Li T and Wu J Water solubilization capacity and conductance behaviors of AOT and NaDEHP systems in the presence of additives 2002 *Colloid Surf. A-Physicochem. Eng. Asp.* 197 101–109
[16] García R, Mejuto J and Pérez L Ester aminolysis by morpholine in AOT-based water-in-oil microemulsions 2006 *J Colloid Interf. Sci.* 301 624–630
[17] Alireza S, Julian E, Kevin J and Rico F Tuning aggregation of microemulsion droplets and silica nanoparticles using solvent mixtures 2008 *J Colloid Interf. Sci* 318 244–251
[18] Shinya S, Yoshikazu S, Kinya S, Ikuo S and Sok Y Diesel quality and molecular structure of bitumen-derived middle distillates 2004 *Fuel* 83 1915–1927
[19] John Holladay Thermochemical Conversion Processes to Aviation Fuels 2012 *A report by US Department of Energy* 1–28
[20] Catherine E, Sharon J and Matthew J Unique crystal morphologies of glycine grown from octanoic acid-in-water emulsions 2006 *J Am Chem Soc* 128 7718-7719
[21] Catherine M, Justin W and Stephanie R Effect of added α-lactalbumin protein on the phase behavior of AOT–brine–isooctane systems 2003 *J Colloid Interf. Sci* 261514–523
[22] Hamed M, Chin P, Nazimah S and Salmah Y Optimization of the contents of Arabic gum, xanthan gum and orange oil affecting turbidity, average particle size, polydispersity index and density in orange beverage emulsion 2008 *Food Hydrocolloids* 22 1212–1223
[23] Amir H, Yuan F and David J Fabrication of vitamin E-enriched nanoemulsions: Factors affecting particle size using spontaneous emulsification 2013 *J Colloid Interf. Sci* 391 95–102

IOP Conf. Series: Earth and Environmental Science **63** (2017) 012032 doi:10.1088/1755-1315/63/1/012032

Study on the Configuration and Capacity of the Lateral Runway Based on the Airport Green Operation

Xiong Li[1*], Feng Li[1] and Xiaoqing Chen[2]

[1] Planning and Design Institute, China Airport Construction Group Corporation of CAAC, Beijing 100101, China
[2] Aviation Industry Development Research Center of China, Beijing 100101, China

lixiong_cacc@163.com

Abstract. Airport as an important transportation infrastructure, its efficient operation, energy saving and emission reduction are of great significance for the sustainable development of the city. The planning of the lateral runway is gradually being applied in the construction of large airports. First, the different utilization modes of the lateral runway and adjacent parallel runway were analyzed. After which four operational correlations and intervals were discussed, i.e.: independent operation, take-off & take-off dependent operation, take-off & landing dependent operation and fully dependent operation. Taking Chengdu new airport as an example, the airfield simulation models were established to evaluate the efficiency of the lateral runway. The simulation result suggests: A group of longitudinal and transverse runways can meet the hourly take-off and landing movements of about 75-77 in the case of independent operation. The take-off & landing dependent operation has relatively large impact on capacity. Under the conditions of fully dependent operation, the hourly movements are only 65, lower by about 16%, 10% and 6% compared with other three modes.

1. Introduction

The continuous growth of social economy drives China air transport on the fast-growing track, during which China civil aviation has become the second largest air transport system in the world. The growing demand for air transport leads to the increasing scale of airport system in a response to the higher requirement. As an important transportation infrastructure, the airport is of great significance to the sustainable development of the city. To build a green airport, we should reduce the impact on the environment while utilizing the resources at the same time; realize the harmony between man and nature, development and environment. However, the key to achieving green airport is the airport green operation, that is, to improve the efficiency of the airport ground operation, reasonable planning runway configuration to enhance the airport capacity.

In conventional airport planning and design, parallel runway configuration is usually the first choice for large airports planning multi-runway system. But too many parallel runways will reduce the efficiency of the ground operation, bypass taxiways and runway-crossing taxiways are to be planned as support. Considering the factors such as airspace, terrain, noise and terminal layout, the lateral runway has been applied in new airports, such as Beijing New Airport and Chengdu New Airport. Therefore, it is of great significance to research the capacity of airport multi-runways system, particularly of the lateral runway and adjacent parallel runway (LR&APR) system, which will be used to instruct the construction of China airports in the future.

Three methods are usually employed to evaluate the capacity and operational efficiency of airport runways, i.e.: statistical data analysis, mathematical analysis and computer simulation. (1) Statistical data analysis is used to estimate the airport capacity based on the capacity envelope diagram drawn against the statistical flow data of existing airports, a model usually performed in the early stage of capacity analysis, but the disadvantage is that it can only evaluate the capacity of existing airport runways system[1]. (2) Mathematical analysis is used to obtain the capacity of runways by establishing the mathematical equation set of airport and airspace operational parameters through the appropriate hypothesis, which is mainly used for macro-level capacity evaluation [2,3]. (3) Computer simulation is an alternative in the most extensive application to evaluate the airport capacity and operational efficiency, which turns out the capacity level and efficiency index through the analysis of simulated operational data [4-6]. Simmod (Airport and Airspace Simulation Model) and TAAM (Total Airspace & Airport Modeller) are the most popular simulation software for the evaluation of airport capacity [7,8].

However, the existing airport capacity studies mostly focus on single runway or parallel runway configurations, lacking in-depth analysis of lateral runway operation mode and capacity. This paper will present in-depth study on the use patterns and operating efficiency of LR&APR. The simulation model of the runway was established by using Simmod software. Efficiencies of two runway operation modes based on different operation intervals, namely take-off & landing isolated operation and mixed operation, were considered separately. Then, evaluations on runway capacity were given, by which theoretical references was provided for the planning and construction of lateral runways.

2. Application of lateral runway

The lateral runway is relative to the main parallel runways of the airport. A lateral runway is normally located on one side of the parallel runway, with their center line extensions intersecting. In the US, runway configurations of Denver International Airport and Dallas International Airport are parallel runways + lateral runways, as shown in Figure 1.

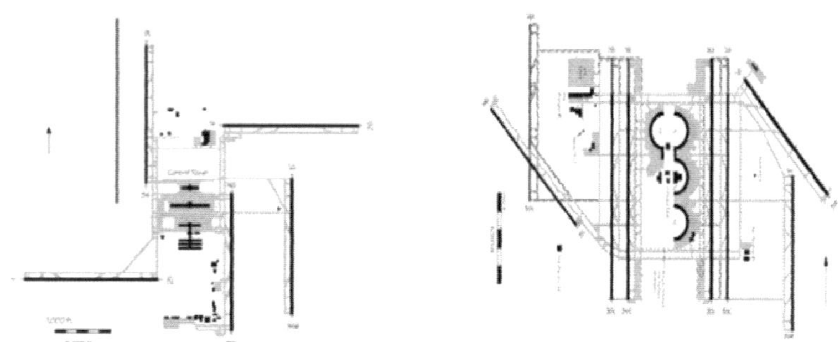

Denver International Airport   Dallas International Airport

Figure 1. Airport with lateral runway in the US.

Multi-runway configurations of China's existing airports are parallel runways. Beijing and Chengdu are both planning the construction of a new airport that equipped with lateral runways. The runway configuration in current construction program is shown in Figure 2. Beijing new airport is planned to build four runways (N1, E1, W1 and W2). Chengdu new airport is planned to build three (N1, E1, W1).

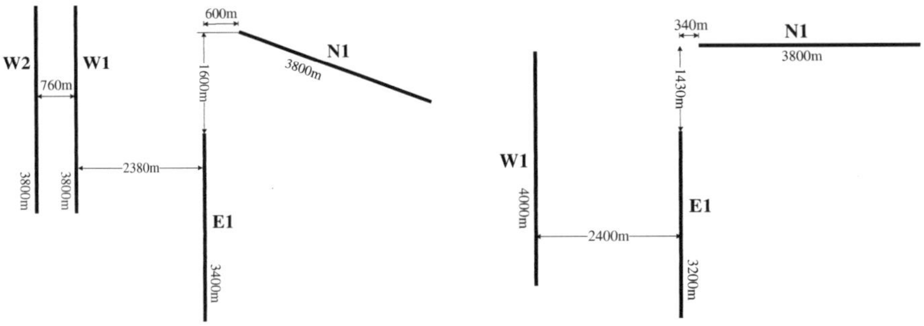

Beijing New Airport Chengdu New Airport

Figure 2. Airport with lateral runway in China.

Taking Chengdu new airport as an example, when a group of parallel + lateral runways runs northward from the airport, there are mainly two modes of operation: (1) isolated operation, N1 for take-off, and E1 for landing; (2) Semi-mixed operation, N1 for take-off, E1 mainly used for landing and auxiliary take-off. When the airport is operating southward, the runway use pattern is similar.

3. Operational Correlations and Intervals

As to the correlations between lateral runway and adjacent parallel runway, as well as the requirements on operational intervals, there is no clear stipulation given by the International Civil Aviation Organization (ICAO) or the Civil Aviation Administration of China (CAAC) [9]. Therefore, according to different runway operation modes, this paper analyses four operational correlations and take-off & landing intervals.

(1) Independent operation (IO). The aircraft taking off on the lateral runway (N1) does not interfere with the aircraft taking off or landing on the parallel runway (E1), taking no account of the take-off & landing coordination.

(2) Take-off & take-off dependent operation (TTDO). The aircraft taking off on the lateral runway (N1) runs in relation to the aircraft taking off on parallel runway (E1). Suppose that the departure protection area (DPA) boundary of runway E1 and the center line of runway N1 intersect in the point A, on runway E1 a flight flies over point A with t_z seconds, and a take-off flight on runway N1 taxies to point A with t_h seconds. Then, the departure clearance of a flight on runway E1 is t_h seconds after the take-off flight on runway N1 starting to taxi; departure clearance of a flight on runway N1 is t_z seconds after the take-off flight on runway E1 starting to taxi.

(3) Take-off & landing dependent operation (TLDO). The aircraft taking off from the lateral runway (N1) runs in relation to the aircraft landing on the parallel runway (E1). This mode takes into account mainly the impact of the runway E1 touch-and-go flights on the N1 departure flights. Draw a circle arc with the runway threshold of N1 as center and 3Nm as radius. The arc intersects with the extension of runway E1 at point C, as shown in the Figure 3. The distance between end of runway E1 and point C is d_c. Suppose that the miss-approach plane boundary of a landing flight on runway E1 and center line of runway N1 intersect at point B, and that take-off flights on runway N1 taxi to B point in t_m seconds. The duration from pilot hearing the take-off command to the aircraft starting to taxi is t_n seconds. The distance of an approach flight on runway E1 flies within t_m+t_n seconds is d_y. Then, the release time of take-off flight on runway N1 must meet the condition: landing flight on runway E1 having a distance from the runway end greater than d_c+d_y.

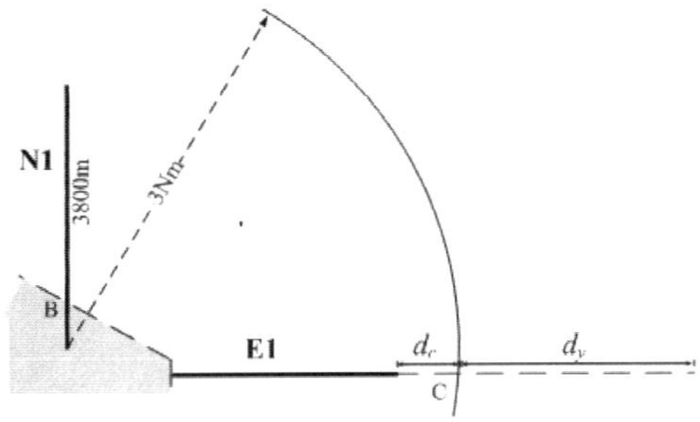

Figure 3. Take-off & landing dependent operation intervals.

(4) Fully dependent operation (FDO). The operational interval standards in (2) and (3) above are applicable for aircraft take-off on the lateral runway (N1) and those take-off or landing on parallel runways (E1).

4. Simulation modeling

This paper operates Simmod plus 7.6 to make a simulation analysis of the operational efficiency in different operational patterns of LR&APR. Simmod, one dynamic and comprehensive airport simulation micro-software firstly proposed by FAA in 1978, through the continuous upgrading and perfection, has become one of the most applied airport and airspace simulation software [7,8].

Given that this paper focally analyses the capacity and efficiency of airport runways system, the simulation modelling will put aside the restrictive factors of airspace. Airfield layout is given with Chengdu new airport as a reference, runway use modes are considered in two situations: (1) E1-N1 isolated operation, (2) E1-N1 semi-mixed operation. The take-off & landing proportion is allocation as shown in Figure 4. On this basis, simulation models are established for four kinds of operational correlations and take-off & landing intervals.

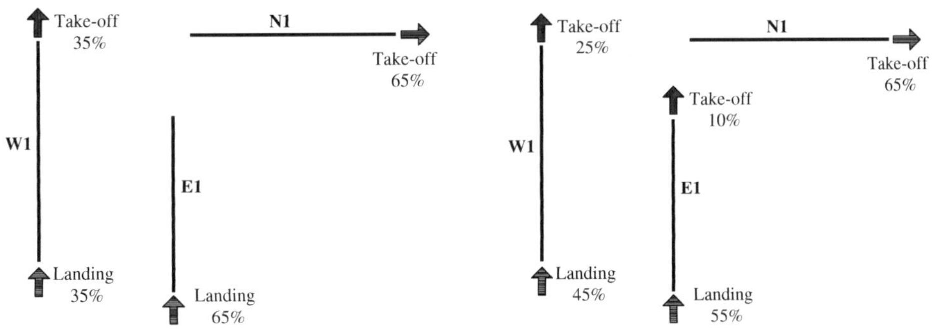

Figure 4. Runway operation modes.

Landing interval is controlled mainly based on wake vortex separation, as shown in Table 1. In addition, extra safe redundancy is taken into account. According to differences between aircrafts, departure separation of each runway is arranged at 90~130 seconds. When one single runway is in alternate operation and distance from landing flight to runway is less than 8km, departure flights are forbidden to taxi out from waiting point. This paper assumes that the ratio of heavy and middle aircraft is p_h=18%, p_m=82%, respectively.

Table 1. Wake vortex separation of landing flights.

Latter aircraft	Former aircraft		
	A380	Heavy	Medium

A380	8km	6km	6km
Heavy	12km	8km	6km
Medium	13km	10km	6km

For operation of N1 and E1, t_h=25 seconds, t_z=70 seconds, d_c+d_y=4.63km (2.5Nm).Taking the airfield model, under northward operation, as example, the simulated operation demonstration is shown in Figure 5.

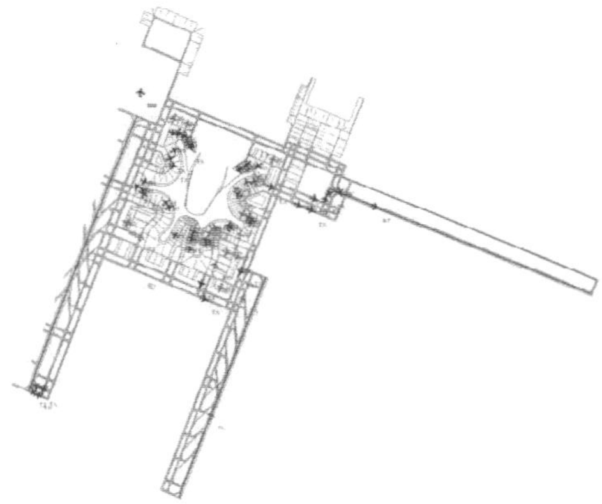

Figure 5. Airfield simulated operation demonstration.

5. Simulation results and analyses

Under the conditions of E1-N1 isolated or semi-mixed operations, four correlations and intervals are discussed, i.e.: IO, TTDO, TLDO and FDO. The peak hour take-off and landing capacity of the lateral runway and adjacent parallel runway are shown in Table 2.

Table 2. Peak hour take-off and landing capacity.

Simulation model number	Runway operation mode	Operational correlations	Take-off flights	Landing flights	Total movements
1	Isolated	IO	40	35	75
2	Isolated	TLDO	34	33	67
3	Semi-mixed	IO	48	29	77
4	Semi-mixed	TTDO	39	33	72
5	Semi-mixed	TLDO	43	26	69
6	Semi-mixed	FDO	36	29	65

In the case of IO, the semi-mixed mode, compared to the isolated mode, is more superior by about 3% in flight movements. In the semi-mixed operation mode, the FDO is with the lowest capacity, hourly movements are only 65, lower by about 16%, 10% and 6% compared with the other three modes.

Movements of LR&APR, under above mentioned use patterns, are as shown in Table 3 and Figure 6. In the case of semi-mixed operation mode, TTDO has the highest impact on the runway E1 take-off flights, with the take-off flights only 13% of IO; TLDO has the biggest impact on the take-off flights on runway N1, compared to IO, take-off flights decreased by about 10%.

Table 3. LR&APR peak hour take-off and landing flights.

Simulation model number	Runway operation mode	Operational correlations	E1 movements		N1 movements
			Landing	Take-off	Take-off
1	Isolated	IO	35	0	40
2	Isolated	TLDO	33	0	34
3	Semi-mixed	IO	29	8	40

4	Semi-mixed	TTDO	33	1	38
5	Semi-mixed	TLDO	26	7	36
6	Semi-mixed	FDO	29	1	35

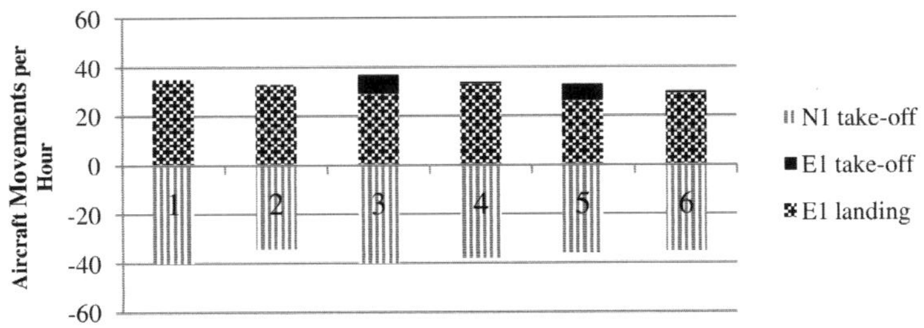

Figure 6. LR&APR movements per hour.

6. Conclusions

In this paper, Simmod simulation software is used to simulate the lateral runway and adjacent parallel runway under different operating modes and correlations. Runway capacity and the efficiency of airport green operation are evaluated. The conclusions are as follows:

(1) The planning of lateral runway and flight procedures should be designed to ensure that it operates independently with the adjacent parallel runway. In the case of independent operation, a group of longitudinal and transverse runways can meet the hourly take-off and landing movements of about 75-77.

(2) Lateral runway is mainly used for take-off, and parallel runway mainly for landing. In this way, the take-off & landing dependent operation has relatively large impact on capacity. In this case, the two runways have an hourly take-off and landing movements about 67-69.

(3) In the semi-mixed operation mode, if take-off and landing are fully dependent operation, the air traffic control and coordination would be more difficult, with the capacity only 65 per hour. Hence in this case, an isolated operating mode is recommended.

The construction and application of lateral runway are still in the initial stage in China. The collision risk analysis on take-off and landing flights, and the interval usage standards of dependent operation, are still there to be further studied.

7. Acknowledgments

This work is supported by two research projects:

(1) China civil aviation security ability building funds--Research on application of simulation technology in civil airport planning and design;

(2) National Key Technologies R&D Program (2014BAJ04B02).

8. References

[1] Gilbo E P 1993 Airport capacity: representation, estimation, optimization, *IEEE Transactions on Control Systems Technology* vol. 1 no. 3 p 144-154

[2] Neufville R D and Odoni A R 2002 *Airport systems planning, designs, and management* (New York: McGraw-Hill) p 1-28

[3] Chen Y, Cao Y and Zhou Y 2005 Estimation of Beijing international airport capacity, *Flight Dynamics* vol. 23 no. 4 p 86-89

[4] Wang C, Zhang X Y and Xu X H 2008 Simulation study on airfield system capacity analysis using SIMMOD, *2008 Int. Symposium on Computational Intelligence and Design* Wuhan China

[5] Gao W and Jiang S Y 2010 Simulation study on closely spaced parallel runway analysis using SIMMOD plus, *Int. Conf. on Intelligent Computation Technology and Automation* Los Alamitos CA, USA

[6] Gao W and Huang C W 2010 Research on capacity evaluation for Chongqing TMA, *Journal of Civil Aviation Univ. of China* vol. 28 no. 5 p 1-4

[7] Li X, Wei D X, Li D B and Chen X Q 2015 Utilization pattern of closely spaced parallel runways and simulation of operational efficiency, *Proceedings of 2015 IEEE International Conference on Progress in Informatics and Computing* Nanjing China, p 158-162

[8] Wu H N, Guo Y C, Mu T and Li X 2016 Apron taxiway operation mode and simulation evaluation in Beijing new airport, *Journal of Transportation Systems Engineering and Information Technology* vol. 16 no. 3 p 214-220

[9] Yan Y, Zhu D W and Zhang Q 2016 Research on the wake interval of cross runway, *China Science and Technology Information* vol. 7 p 43-47

Hindrances to Green Building Developments in Nigeria's Built Environment: "The Project Professionals' Perspectives"

[1]S G Dalibi, [2]J C Feng, [3]Liu Shuangqin, [4]Abubakar Sadiq, [5]B. S. Bello & [6]I.I. Danja

[1, 2, 3, 4,5]Business School, Hohai University, Nanjing, Jiangsu Province, China.
[6,] MSc. candidate at Architecture Department, South East University, Nanjing, Jiangsu Province, China.

[1] salgidos@yahoo.com

Abstract. Developing countries like Nigeria are faced with the growth of residential housing sector accompanied by huge power, water, and material consumptions etc., which were due to the population growth, increased households and the increased urbanization. The construction industry is guilty of many practices because its activities have adversely affected the environment negatively. It responded with new initiative called Eco / green / sustainable buildings to ensure environmental sustainability. Despite all these glaring challenges green building developments and sustainable practices are embraced very slowly and practiced at slow pace in the Nigeria's construction industry. This is worrisome and is due to some factors hindering such pace. The aim of this paper is to identify, examine and assess the factors that are hindering green building developments based on the project professionals' perceptions. Reviewed literatures were used to identify the various factors that hindered the adoption of green measures and practices within the Nigeria's built environment. A questionnaire survey was conducted within the industry. The results showed the major factors hindering green building developments and practices in the Nigeria's built environment.

1. Introduction

1.1. Background to the study

The world's population continued growth has led to the implementation of resource-efficient measures in all areas of human activities especially in the built environment (BE) which has a significant impact on all resources while also affecting the air quality and transportation patterns of communities; both the present and the future generations. [1-3]. The friendliest way to handle the environment is not to build. However, without construction, life can be miserable and threatening [4], [5]. What is needed is a dynamic equilibrium without any form of threat especially to the environment [6]. The combination of these challenges gave birth to a new concept in design, construction/renovation, operation and maintenance of buildings in conformity with "sustainable practices for buildings" known as Green Buildings [7].

Green building (GB) is the foundation of sustainable construction and refers to a structure that is resource-efficient in terms of economy, utility, durability, and comfort [8], [9]; GBs are those sensitive to the "Environment, Resource and energy consumption, Impact on people, Financial impact and the world at large [10]; GBs are environmentally friendly which encompasses energy use, water use, and storm-water and wastewater re-use [11]; GBs combine energy and water efficiency systems, Day Lighting strategies, Indoor Environmental Quality (IEQ) systems and efficient Building Envelope system to provide comfort and positive impact to the occupants and the environment[3]. Sustainability in construction is all about following suitable practices in terms of choice of materials, their sources, construction methodologies, as well as design philosophy so as to be able to improve performance, decrease the environmental burden of the project, minimize waste and be ecologically friendlier [12]. It enhances the environment against the negative side effect of construction activities [5]; a clear answer to health, economic and environmental challenges [13]; maximize the resource savings, protect the environment, reduce pollution, provide people with healthy, comfortable and high efficient space, and exist harmoniously with nature" [14-16]. The term "green building", or "more sustainable building", does not have an exact definition, but, nevertheless, these terms have been used frequently [17].

These clearly indicate that GB is a major response to ensure environmental sustainability globally within the construction industry. However, countries like China, USA, UK, Germany, Japan, Korea and other developed nations have embraced the concept fully; while other countries are prioritizing which feature to incorporate in to their buildings; whereas other countries are ignoring or slowly embracing the concept due to some reasons.

1.2. The Research Problem

Developing countries like Nigeria are faced with the growth of residential housing sector accompanied by huge power, water, natural and processed material consumptions due to increased urbanization. This increasing urbanization is associated with loss of arable land, material and water crisis, and serious environmental problems like air pollution, noise pollution and waste generated from buildings [18], [19]. The construction industry is guilty of many practices in this regard and it responded with a concept called GBs to ensure and promote environmental sustainability.

Despite all these glaring challenges and drastic measures, GB developments and sustainable practices are embraced very slowly and practiced at slow pace in the Nigeria's construction industry. This is worrisome and may be due to some factors affecting the sustainable practices within its built environment. Such factors may also be attributed to project stakeholders or the concept itself etc.

1.3. Research Aim

The aim of this paper is to identify, examine and assess the factors that are hindering green building developments based on the project professionals' perceptions with a view of exposing the impacts of such hindrances on the Nigeria's BE.

1.4. Research Hypotheses

The following hypotheses were formulated for this research; T-test statistical tool was used in testing these formulated hypotheses:

- *Null hypothesis* (H_O); Green building developments ***are not*** hindered significantly by some factors in the Nigeria's Built Environment.
- *Alternative hypothesis* (H_A); Green building developments ***are*** hindered significantly by some factors in the Nigeria's Built Environment.

1.5. Research Scope and Limitations

This research work is limited to the perception of various practicing professionals working in the Nigeria's Built Environment. Such professionals include the Architects, Quantity Surveyors, Civil Engineers, Building Engineers, Mechanical and Electrical Engineers, Project Managers, Construction Managers, Land surveyors, Town planners and Estate Surveyors.

IOP Conf. Series: Earth and Environmental Science **63** (2017) 012033 doi:10.1088/1755-1315/63/1/012033

2. Review of Literature

2.1. Green Building: Developments & Hindrances

Construction of GB entails tailoring a building and the site to the local climate, site conditions, cultural and community in order to reduce resource consumption, augment resource supply, and enhance the quality and diversity of life, while optimizing all these in an integrated design [5]. In other words, it is a total quality management approach to building so as to ensure the achievement of synergistic design through interdisciplinary teamwork [5]; it is neither, an assemblage of 14 environmental' components nor a piecemeal modification of an already designed standard building, rather, it is a building philosophy in which natural and resource efficient features are incorporated in a building [13]. To achieve synergy in GB, all the professionals that would be involved in the planning, design and construction of such building must be brought early (at the design stage), for professional input in the design of GB from the beginning to undertakes Site Analysis and Environmental Impact Assessments etc. [5].

Every project or development comes with its unique benefits, challenges and factors that hindered its success; GB developments in Nigeria are not an exception. "To be sustainable, buildings should usefully last for many generations. This requires some knowledge of the future climate and the resources available to maintain the operations, in particular the energy consumption, of buildings" [20]. This practice is a two-three decades old trend, but with insufficient data about the costs and absence of measured building performance data from currently operating sustainable designed buildings [21], [22], [23]. Industry professionals, in both the design and construction disciplines, are generally slow to change, tend to be risk-adverse, lack sound knowledge, experience, and understanding of how to apply ecology to construction design; moreover, environmental or economic benefit of some green building approaches has not been scientifically quantified [24].

Lisa and Morris (*2004*); opined that the first question often asked about sustainable design is: *what does "green" cost? Typically meaning does it cost more?* This raises the question: *More than what? More than comparable buildings? More than the available funds?* Or *more than the building would have cost without the sustainable design features?* The answers to these questions have been thus far elusive, due to the lack of hard data [25]. Morris (*2007*); argued that "The most common reasons cited in studies for not incorporating green elements into building designs is the increase in first cost" [26]. While a Davis Langdon Report in 2007 stated that "*there is no one size that fits all answer to the question of the cost of green"* [27].

However, the appreciation of the significance of non-technical issues (soft issues) has grown, giving recognition to Economic and Social sustainability concerns as well as Cultural heritage of the BE as being equally important and provide further challenges[12]. Many housing estates developments in the Abuja FCT, Nigeria do not reflect the desired housing needs of the end-users, the developers'/ clients' lack of comprehensive data about the financial obligations with regards to incorporating green features into renovation or proposed projects which has impact on the total development cost which in turn affects end-users / occupants in terms of Rental value, Sales value, Envisaged savings due to green elements, Future asset value of the green building etc. [28].

Incompatibility of interests amongst stakeholders caused conflicts and disputes in construction [29]. Incorporating the various interests of stakeholders should be extremely important for the preparation of green specifications, construction and maintenance [30]. GB projects design and construction is new in Nigeria and is characterised by the problem of lack of shared perception and agreement on the objectives and success/failure of the green building projects by stakeholders [31]; also Different Set of Criteria for success/failure for the project [32] etc. As such, each stakeholder perceives the success according to a hierarchy of dimensions, which comply with their personal agenda.

The initial emphasis of sustainability was on Technical issues such as materials, building components, construction methodologies and energy related design concept [8], [9]. However, recognition of (soft issues) Economic and Social sustainability concerns as well as Cultural heritage of the built environment as being equally important [12]; Sustainable world progress is dependent upon continued Economic, Social, Cultural, and Technological progress [33]. These four main factors; Economic, Social, Cultural and Technological factor, each of which is found to have a significant effect towards adopting Green Building Technology [34].

The awareness of green building by the general public will form the market-driven power for such developments especially in the urban area. However, the difficulties include: Lack of basic data of using GB assessment system, Lack of professionals, Lack of interest from real estate developers and Difficulty of having a unified GB assessment standard etc. [35].

Building materials have been playing an important role in the construction industry, no field of engineering is conceivable without their use [36], [37]. The cost of building materials poses a significant threat to both the construction industry and people aspiring to own houses [38], [39], [40], [41]. The setback witnessed in housing efforts in Nigeria was hinged on the high cost of imported materials used for construction [42], [43], [44]. A report of UNCHS (1993) found that building materials remain the most significant input in project development; with GBs, most of the material components are not locally produced / manufactured in Nigeria e.g. solar panels, Switchable glazing, water conserving appliances and grey water systems etc. [45], [46].

2.2. The Green Building Hindrances.

The hindrances to GB developments identified from the reviewed literature above were listed in the table 1.

Table 1. Hindrances to GB developments

S/N	HINDRANCES TO GB DEVELOPMENTS	SOURCES
1	GB Technical Know How (GB requisite knowledge among the Built environment Professionals & the scarcity of GB certified professionals)	[5], [13], [20], [24], [25].
2	Lack of GB cost data and other performance related Data	[21], [22], [23], [25], [26], [27], [28].
3	The Perception of GB as Expensive Concept (Perceived Increased cost for incorporating GB features etc.)	[21], [25], [26].
4	GB as a new change (which comes with its associated risks)	[24]
5	Cultural, Economic, Social and Technological (CEST) barriers.	[8], [9], [12], [33], [34].
6	Divergent interests and views of success factors and success criteria of GB developments among stakeholders	[28], [29], [30], [31], [32], [35]
7	GBs awareness	[35]
8	Lack data for using GB assessment systems	[35]
9	Lack of Locally or a single unified/standard GB assessment system	[35]
10	In-availability of local GB material and other components and High cost of Imported GB materials	[8], [9], [36], [37], [38], [39], [40], [41], [42], [43], [44], [45], [46].

Source: Authors' review of literature.

3. The Research Methodology

Secondary sources of data such as journals, conference / seminar / workshop papers, textbooks, newspapers, magazines and internet sources etc. were used to review literatures on the GB field, which helps in identifying and narrowing the various factors that hindered GB developments. These identified hindrances form the main body of the Questionnaire which was manually distributed to the various professionals working in Nigeria's Built Environment; selected randomly. A ***5-point Likert scale*** Questionnaire format (Strongly Agree =5, Agree=4, Neutral/Undecided=3, Disagree=2, strongly Disagree=1) was used to obtained the various perceptions of the professionals working in the Nigeria's Built Environment. Frequency and percentage count tables, Mean item score and T-test statistics were used for data analyses.

4. Data Presentation, Analysis and Results

4.1 Results from the Administered Questionnaires

The Primary data for this research work was obtained through manually distributed questionnaires to professionals working in the Nigeria's Built Environment and the responses obtained were shown in the table 2.

Table 2. The Questionnaires (Qns) Responses by Professional Disciplines

S/N	PROFESSIONAL DISCIPLINES	Nr. Of Qns Distributed	Nr. Of Qns Returned	Percentage (%) of Total Qns returned per discipline
1	Architects	80	51	10.63%
2	Quantity Surveyors	40	31	6.46%
3	Civil Engineers	40	33	6.88%
4	Building Engineers	40	21	4.38%
5	Mechanical Engineers	40	20	4.17%
6	Electrical Engineers	40	27	5.63%
7	Project Managers	40	29	6.04%
8	Construction Managers	40	22	4.58%
9	Land surveyors	40	14	2.92%
10	Town planners	40	24	5.00%
11	Estate Surveyors	40	30	6.25%
	TOTAL	**480**	**302**	**62.92%**

Source: Authors' Field work

From the table 2 above, the following observations and deductions were made:

i. 480 numbers of questionnaires were distributed randomly to various professionals in the Nigeria's built environment. 302 numbers of questionnaires were returned while 172 were not returned. This puts the response rate at 62.92% which is good.

ii. All the professionals were administered equal number of questionnaires (40nr. each) with the exception of the Architects (80nr.). This is because as the Main design professionals they have much more relevance in GB design, specification and construction.

iii. The Architects have the highest total responses (10.63%), followed by Civil engineers (6.88%), Quantity surveyors (6.46%), Estate surveyors (6.25%) and Project managers (6.04%); while land surveyors (2.92%) have the least.

The table below shows the assessment of the various GB Hindrances by the professionals based on the structured 5-point Likert scale.

Table 3. An Assessment of the Hindrances to GB developments by the Respondents

S/N	HINDRANCES TO GB DEVELOPMENTS	Strongly Agreed	Agreed	Neutral	Disagreed	Strongly Disagreed	TOTAL	Mean Item Score	Rank
		5	4	3	2	1			
1	GB Technical Know How (GB requisite knowledge among the Built environment Professionals & the scarcity of GB certified professionals)	53	27	192	19	11	302	**3.3**	**7th**
2	Lack of GB cost data and other performance related Data	106	78	92	24	2	302	**3.87**	**4th**
3	The Perception of GB as Expensive Concept (Perceived Increased cost for incorporating GB features etc.)	187	104	11	0	0	302	**4.58**	**1st**
4	GB as a new change (which comes with its associated risks)	88	67	44	102	1	302	**3.46**	**6th**
5	Cultural, Economic, Social and Technological (CEST) barriers.	39	111	33	90	29	302	**3.14**	**8th**
6	Divergent interests and views of success factors and success criteria of GB developments among stakeholders	188	12	77	17	8	302	**4.18**	**3rd**
7	GBs Awareness	25	87	101	66	23	302	**3.08**	**9th**
8	Lack data for using GB assessment systems	10	173	79	38	2	302	**3.5**	**5th**
9	Lack of Locally or a single unified/standard GB assessment system	3	37	177	77	8	302	**2.83**	**10th**
10	In-availability of local GB materials and other components and High cost of Imported GB materials	203	67	5	6	21	302	**4.41**	**2nd**

Source: Authors' statistical computations

From the table 3 above, the following observations and deductions were made:

i. The respondents strongly agreed with "The Perception of GB as Expensive Concept" as a major Hindrance to GB developments; ranked 1st.

ii. In-availability of local GB materials and other components and High cost of Imported GB materials (2nd), Divergent interests and views of success factors and success criteria of GB developments among stakeholders (3rd), Lack of GB cost data and other performance related Data and Lack data for using GB assessment systems (4th) were also agreed as GB development Hindrances by the respondents.

iii. Lack of locally or a single unified/standard GB assessment system (10th), GBs Awareness (9th) and Cultural, Economic, Social and Technological (CEST) barriers (8th) were the GB Hindrances with the least impact on GB Developments in Nigeria.

4.2. Testing of Hypotheses

The hypotheses formulated for this research work was tested using T-statistics. The values of the mean item scores in table 3 above were used as the data for the statistical computations with the result shown in the table 4.

Table 4. T-Test statistical computation result

Professional Perceptions	MEAN	Standard Deviation	Standard Error	N	DF	Alpha (Significance level)	P value	Tcal	$Ttab_{0.05,\ 9}$
GB Hindrances	3.6351	0.5957	0.1884	10	9	5%	0.0001	6.3514	-1.8331

With 9 degrees of freedom (DF) and 5% level of significance, the T-test calculated (Tcal = 6.3514) is greater than T-test tabulated ($Ttab_{0.05,\ 9}$ = 1.8331) the As such, the Alternative hypothesis was accepted; which states that "*Green building developments* ***are*** *hindered significantly by some factors in the Nigeria's Built Environment*".

5. Conclusions

This study identified ten hindrances to GB developments in the Nigeria's Built Environment, out of which: The Perception of GB as Expensive Concept is the major Hindrance to GB developments; which ranked 1st; followed by In-availability of local GB materials and other components and High

IOP Conf. Series: Earth and Environmental Science **63** (2017) 012033 doi:10.1088/1755-1315/63/1/012033

cost of Imported GB materials (2nd), Divergent interests and views of success factors and success criteria of GB developments among stakeholders (3rd). Whereas, Lack of locally or a single unified/standard GB assessment system (10th), GBs Awareness (9th) and Cultural, Economic, Social and Technological (CEST) barriers (8th) were the GB Hindrances with the least impact on GB Developments in Nigeria. The impact of these GB hindrances is significant which is further attested by the T-test statistical test of the research hypotheses by accepting the Alternative hypothesis; which states that "*Green building developments **are** hindered significantly by some factors in the Nigeria's Built Environment*".

6. References

[1] EUROSTAT (2011): "Consumption of energy" -*Statistics Explained* .Available at http://epp.eurostat.ec.europa.eu/statistics_ explained/index.php/Consumption_of _energy *Accessed in Jan. 2012.*

[2] Michael Bauer, Peter Mosle, Michael Schwarz (2007): "Green Building: *Guide book For Sustainable Architecture".* by Springer©

[3] Salisu Gidado Dalibi (2012): Cost Impact Assessment of Green Buildings in China *(A Case Study of Few Selected Green Building Projects In Shanghai, China). Msc Thesis submitted to* Hohai University Nanjing – Jiangsu Province of China.

[4] Kolawole, J. and Anigbogu (2005) "Impact of Construction Activities on the Environment." A Paper Presented at the National Conference Towards a Sustainable Built Environment. Ahmadu Bello University, Zaria – Nigeria. Sept. 21st – 23rd

[5] D Dahiru, K Bala and AD Abdul'Azeez (2013); *Professionals' Perception on the Prospect of Green Building Practice In Nigeria.* SBE 13: Creating a Resilient and Regenerative Built Environment. 15-16 October 2013, Cape Town, South Africa.

[6] Zubairu, S. (2012) The Importance of Evaluation and Sustainability in the Built Environment In: Laryea, S., Agyeponng,S., leiringer, R. and Hughes, W. (Eds) Procs 4th West Africa Built Environment Research (WABER) Conference, 24 – 26 July Abuja, Nigeria, 9 – 13.

[7] Salisu Gidado Dalibi (2014): *Green Buildings: The Clients' & The End Users' Common Ground in Environmental Sustainability.* Nigerian Institute of Quantity Surveyors (NIQS) National Training Workshop; in Uyo, Akwa Ibom State of Nigeria. 9th – 11th October, 2014.

[8] U. S. Green Building Council, How to achieve certification, Accessed: February 21 2012, Available: http://www.usgbc.org/Display-Page.aspx?CMSPageID=1991, 2012.

[9] Environmental Protection Agency (EPA), GREEN BUILDING, available at http://www.epa.gov/greenbuilding/pubsabout.htm.

[10] Greg Kats (2003): "*The cost and financial benefits of green building"* – A report to California's Sustainable building task force.

[11] Zane Satterfield, P. E(2009): Tech Brief On Green Building, Published By: The National Environmental Services Centre (Winter 2009), Vol. 8, Issue 4 Tech Briefs.

[12] Abolore, A. A. (2013). Comparative study of environmental sustainability in Building Construction in Nigeria and Malaysia, journal of Emerging trends in Economics and Management Sciences, 3 (6):951-961.

[13] Karolides, A. (2002) Project Planning & Cost Estimating, In: Green Building Approaches in Green Building RS Means Co Ltd; USA Pp 1-22.

[14] MOC 2006. Evaluation standard for green building (GB/T 50378-2006). 1 ed.: MoC.

[15] LI, Y. & CURRIE, J. (2011). Green Buildings in China:Conception, Codes and Certification. Available: http://www.institutebe.com/Institute BE/media/Library /Resources/ Green%20 Buildings /Issue_Brief_Green_Buildings_in_China.pdf [Accessed Feb 11, 2012].

[16] Yujun Liu (2012): *Green Building Development in China: A Policy-Oriented Research With a Case Study of Shanghai.* Master's degree Thesis in Environmental Studies and Sustainability Science submitted in partial fulfillment of the requirements of Lund University International Master's Program in Environmental Studies and Sustainability Science.

[17] Berardi, U., 2013. Clarifying the new interpretations of the concept of sustainable building. Sustainable Cities and Society 8, 72–78WU, Z. 2010. Current Status and Future Trend of Green Building in China. 2010 Shanghai EXPO.

[18] WU, Z. 2010. Current Status and Future Trend of Green Building in China. 2010 Shanghai EXPO.

[19] CHINESE SOCIETY FOR URBAN STUDIES 2008. Green Building 2008, Beijing, China Architecture & Building Press.

[20] Byrd, H and Leardini, P (2011) Green buildings: issues for New Zealand. "Procedia Engineering", 21, 481-488.

[21] Building Design and Construction (BD&C), editors. 2003. *White Paper on Sustainability: A Report on the Green Building Movement.* Building Design and Construction Supplement.

[22] ENSAR Group (2003). *Sustainable Federal Facilities Task 2.1 Business Case.* November 2002 workshop notes. Boulder, Co.

[23] Carmona Andreu, Isabel and Tadj Oreszczyn. (2004). Architects Need Feedback on Environmental Performance. *Jul-Aug 2004, Vol. 32 (4). Building Research and Information.*

[24] Zhiyong Wang (2013): New cost structure approach in green buildings: Cost-benefit analysis for widespread acceptance and long-term practice. submitted to the System Design and Management Program in partial fulfillment of the requirements for the degree of Master of Science in Engineering and Management at the Massachusetts Institute of Technology January

2013.
[25] Hankinson, M., & Breytenbach, A. (2012). Barriers that impact on the implementation of sustainable design. http:// cumulushelsinki2012.org/cumulushelsinki2012.org/wp-content/uploads/2012/05/ Barriers-that-impact-on-the-implementation-of-sustainable-design.pdf. Accessed Nov 2013.
[26] Lisa Fay Mathiessen and Peter Morris (2004): Costing green: "*A comprehensive database and budgeting methodology"*
[27] Peter Morris (2007). "What does Green Really Cost?" *PREA QUATERLY*
[28] Salisu Gidado Dalibi, Hadiza Balarabe and Jamilu Bala MaiAuduga (2016): *Green Buildings: A Concept aligning the interests of Stakeholders (Developers / Clients and End-users) in Estate Development Projects in Abuja - F.C.T (Federal Capital Territory), Nigeria.* SBE16 Hamburg: Strategies – Stakeholders – Success factors. International Conference on Sustainable Built Environment in Hamburg, Germany. 7-11th March, 2016.
[29] Fenn, P., Lowe, D., Speck, C. (1997). Conflict and dispute in construction. Construction Management and Economic, 15 (6), 513–528. http://dx.doi.org/10.1080/014461997372719
[30] Berke, P.R., 2002. Does sustainable development offer a new direction for planning? Challenges for the twenty-first century? Journal of Planning Literature 17, 21–36.
[31] Salisu Gidado Dalibi and Hassan Ali Kumo (2016): *Success Criteria for Green Building Projects in the Nigeria's Construction industry: "The Stakeholders' perception"*. SBE16 Hamburg: Strategies – Stakeholders – Success factors. International Conference on Sustainable Built Environment in Hamburg, Germany. 7-11th March 2016.
[32] Kumo, Hassan Ali (2012): Perception of stakeholders on success criteria for world bank water supply projects in kaduna state – Nigeria. MSc thesis submitted to Department Of Building Faculty of Environmental Design Ahmadu Bello University, Zaria, Kaduna State –Nigeria.
[33] Shelbourn A., Bouchlaghem, D. M., Anumba, C. J., Carillo, P. M., Khalfan, M. M.K., Glass, J., (2006): ITcon Vol. 11(2006), available at http://itcon.org/2006/4/ pg. 57
[34] Nuruddeen Usman and Usman Mohammed Gidado (2015): An Assessment of the Factors Affecting Green Building Technology (GBT) Adoption. WASET 18th IC Jeddah Saudi Arabia Jan 26-27, 2015, 13 (01) Part XIII
[35] Qian Shi (2008): *Strategies of Implementing a Green Building Assessment System in Mainland China.* Journal of Sustainable Development Vol. 1, No. 2 July, 2008. Available at www.ccsenet.org /journal.html.
[36] Akanni, P. O. (2006). Small scale building material production in the context of the informal economy. The Professional Builders, pp. 13-18.
[37] Udosen, J. U., & Akanni, P. O. (2010). A factorial analysis of building material wastage associated with construction projects. Journal of Civil and Environmental Systems Engineering, 11(2), 81-90.
[38] Anosike, P. (2009, April 6). Nigerian groans under high cost of building material. The Daily Sun, pp. 38-39.
[39] Mekson, J. (2008, August). Prices change of building materials in developing communities in Nigeria. The Professional Builders, pp. 21-27.
[40] Mohammed, H. Y. (2008, December 25). Nigeria: Builders groan on rising cost of building materials. Daily Trust, p. 29.
[41] Njoku, J. (2007, April 9). Grappling with escalating cost of construction materials. The Vanguard, pp. 36-37
[42] National Housing Policy (1991); *A National Housing Policy (NHP) of the Federal Republic of Nigeria.* Federal Government Press, Lagos, 1991.
[43] B.U. Iwuagwu, N. Eme-anele (2012); Earth construction technology and design: A positive solution to mass housing in Africa, International journal of scientific innovations and sustainable development, 2(2012) 89-92.
[44] Oloruntoba Kayode and Ayodele E. Olusegun (2013): Local Building Materials: a Tool Towards Effective Low-Income Housing in Nigeria. Middle-East Journal of Scientific Research 18 (4): 492-497, 2013. ISSN 1990-9233
[45] Ibn-Homaid, N. T. (2002). A comparative evaluation of construction and manufacturing materials management. International Journal of Project Management, 20, 263-270.
[46] United Nations Centre for Human Settlement. (1993). Building materials for housing: Appropriate intermediate, cost effective building materials, technology and transfer mechanism for housing delivery. Retrieved from http://ww2.unhabitat.org/ programmes/housingpolicy/ documents/ HS.C.14.7.htm

IOP Conf. Series: Earth and Environmental Science **63** (2017) 012034 doi:10.1088/1755-1315/63/1/012034

Vernacular Architecture of Northern Nigeria in the Light of Sustainability

Isa Ibrahim Danja[1], Xue Li [2] and S.G. Dalibi[3]

[1]MSc. Architecture candidate at Department of Architecture, South East University Nanjing, Jiangsu Province, China.

[2] Asst. Professor in the Department of Architecture, South East University Nanjing, Jiangsu Province, China

[3] PhD candidate at Business School, Hohai University Nanjing, Jiangsu Province, China.

[1]danjaisah@yahoo.com

Abstract. Nigeria, a country located in the Western part of Africa has been the site of numerous kingdoms and tribal states over the millennia and is divided in to Northern and Southern parts with the Hausa-Fulani's, a diverse but culturally homogeneous people living in the northern part of Nigeria. The vernacular architecture of the Northern Nigeria (VANN) is perhaps one of the least known but most beautiful of the medieval age, over the centuries a tradition of fine architecture has flourished in the area, the variety and quality of buildings are bound to generate much delight and enthusiasm in anyone who is interested in buildings. Many of their early buildings such as mosques and palaces are bright and colorful. This architectural style is known as *Tubali* which means the traditional architecture in the Hausa language and globally known as Vernacular Architecture (VA). VANN has many unique features which distinguish it from any other architecture. All buildings, be it modern or old has an impact on the environment; hence the need for sustainable measures and practices regarding such buildings. The aim of this research paper is to examine the features of VANN in the light of sustainability. Reviewed literatures were used to identify the various features of the Vernacular Architecture in the Northern Nigeria which were critically examined in the light of sustainability and its concept. The result showcased the various features of VANN in the light of sustainability and also discussed the factors hindering the sustainable endeavors of VANN with respect to such features.

1. Introduction

1.1. Background to the study

Moving towards sustainability is also a social challenge that entails international and national law, urban planning and transport, local and individual lifestyles and ethical consumerism. Ways of living more sustainably can take many forms from reorganizing living conditions (e.g., eco-villages, eco-municipalities and sustainable cities), reappraising economic sectors (permaculture, green building, sustainable agriculture), or work practices (sustainable architecture), using science to develop new technologies (green technologies, renewable energy and sustainable fission and fusion power), or designing systems in a flexible and reversible manner,[1][2] and adjusting individual lifestyles that conserve natural resources.[3]

"The term 'sustainability' should be viewed as humanity's target goal of human-ecosystem equilibrium (homeostasis), while 'sustainable development' refers to the holistic approach and temporal processes that lead us to the end point of sustainability"[4]. Human attempts to achieve environmental sustainability has been, and continues to be, questioned—in light of environmental degradation, climate change, overconsumption, population growth and societies' pursuit of indefinite economic growth in a closed system.[5],[6]

The term sustainability has divergent view because it involves all disciplines. Each discipline views it from its own perspective. However, it is defined as: the endurance of systems and processes covering various domains such as ecology, economics, politics and culture [7]; the study of sustainable development and environmental science [8]; the development that meets the needs of the present without compromising the ability of future generations to meet their own needs [9]; the development that does not compromise the ability of future generations to meet their own needs [10]; Sustainable development consists of balancing local and global efforts to meet basic human needs without destroying or degrading the natural environment [11],[12],[13]; something that improves "the quality of human life while living within the carrying capacity of supporting eco-systems"[14]; presumes that resources are finite, and should be used conservatively and wisely with a view to long-term priorities and consequences of the ways in which resources are used." [15].

Therefore, the emergence of sustainability (in the last century ago) transcends disciplinary or industry boundaries as attested from the few definitions above especially making sure that present actions do not compromise the ability of future generations to meet their own needs. This entails conserving and preserving traditional buildings based on sustainable practices within the Vernacular Architectural (VA) field; which will enable the future generations to embrace or have a link with their past histories, traditions, cultures and heritages based on the preserved ancient buildings and structures.

The term “Vernacular Architecture” (VA) is not an easy term to define. Many professionals define it in many different ways as; the ‘idea and technology’ of a particular group of people[16]; Adhorcist’s approach to architecture [17]; A true reflection of how generality of people want to build [18]; A “native science of building”[19]; As architecture that is the outcome of anonymous design period [20]; design and construction techniques using locally available resources based on the environmental, cultural and historical background of people [21], [22],[23]; VA is an architectural style reflects local traditions [24]; the unconscious “realization” and “embodiment” of the culture of the society with the requirements of the people in nature [25],[26].

From the above definitions, this research work simply defines VA as *"a building/structure or a constructed shelter of a group of people according to their ethnicity, culture, traditions, religion and environment which is constrained by their climate, locally available materials and their unique features"*. The particular area of interest for this research work will be limited to VA practices covering Traditional / Historical buildings of the Hausa-Fulani people in the Northern part of Nigeria. The Hausa-Fulani are diverse but culturally homogeneous people living in the northern part of Nigeria with a civilization dating back to a millennium; which was evident from their cultural heritages and ancient buildings/ structures.

1.2. Research Problem

Vernacular Architecture in Northern Nigeria (VANN) has suffered a great deal, be it from human negligence, socio-economic conditions, weather and climatic factors or modernization. According to Oneh and Ati (2010), the present day traditional architecture of a place usually results from solutions to climatic problems through reasoning, countless experiments, experiences, and accidents as well. This usually takes the effort and generation of builders who make use of whatever works for them while discarding what did not [27]. Fatty (2006), also states that in the tropics, only scientific evaluation of new ideas will save its traditional architecture [28]. In Nigeria, the relics of earth buildings are seen in our traditional city centers as reminiscences of history, the old earth buildings are being replaced by modern structures [29]. Mud buildings always suffer from common defects such as surface erosion, partial crumbling, humidity, and hallowed bases [30]. It is apparent that the state of quality control for earth construction hangs in critical balance with very limited tolerance for satisfactory performance; that is why people use sand-cement wall system because they have been very useful in term of durability and wider tolerance in tropical environment and performs satisfactorily [29]. The combination of these challenges further compound the sustainable practices within the VA especially that of Northern Nigeria which may gradually erode and diminish conservational practices of the historical / ancient buildings or structures. Hence, the need to examine sustainable measures and practices regarding such buildings.

1.3. Research Aim

The aim of this research paper is to examine and discuss Vernacular Architecture of Northern Nigeria in the light of sustainability based on its few selected features.

1.4. Research Methodology

The main sources of data were journals, conference / seminar / workshop papers, text books, newspapers, magazines and the internet sources etc., which were used to review literatures in the VA field and help in identifying and narrowing some features of the VANN. These features were examined and the factors impeding their sustainable application within the VANN where identified and discussed in relation to the global context.

2. Literature

2.1. Vernacular Architecture of Northern Nigeria

The architecture of the Hausa/Fulani is perhaps one of the least known but most beautiful of the medieval age. Many of their early mosques and palaces are bright and colorful, including intricate engraving or elaborate symbols designed into the façade [31]. Traditional buildings of Northern Nigeria have a special aura around them and they are of various shapes and sizes. Some of which are seen over the centuries. A tradition of fine architecture has flourished in the area.

The variety and quality of buildings in Northern Nigeria are bound to generate much delight and enthusiasm in anyone who is interested in buildings. Indeed, the manner in which the buildings are constructed, and the sheer artistry that is deployed in the effort, combines to reveal aspects of a people's development, their history and culture, and assists in projecting the precise conditions of a people's soul [32]. Hence, the need to sustain and preserve these sheer artistry, history and culture for future generations especially in the Northern Nigeria.

2.2. Features of Vernacular architecture of Northern Nigeria

Vernacular architecture of northern Nigeria has many unique features which distinguish it from any other architecture. Some of the notable features are the engravings on the façade of the building walls, use of building materials such as mud, reeds, stones, and timber within the structures like foundations, walls, columns, slab, beams, doors, windows, plastering, and process of renovations. These features form an intricate, diverse and colorful form of architecture which can only be found in northern Nigeria. As such this research will discuss the above features of VANN in the light of sustainability.

2.2.1. Engravings. Engraving is the practice of incising a design onto a hard, usually flat surface, by cutting grooves into it. The result may be a decorated object in itself. This feature dates back to hundreds of millennia BC [24], [33] In northern Nigeria, the wall engravings are designed by traditional builders, professional artisans and highly experienced hand engravers who are able to draw out minimal outlines directly on the wall surface just prior to engraving. Hausa traditional architectural decoration is categorized in to three groups, namely surface design, calligraphy and ornamental [34]. Hence the façade decoration usually demonstrates wealth and social prestige [35]. Unfortunately, these earth buildings are considered as relics for they are being replaced by modern structures. Figure 1 below shows engraving symbols carved and or designed into the building facades in Northern Nigeria's VA.

Figure 1. Engraving works (as a unique feature of Vernacular Architecture of Hausa-Fulani in Northern Nigeria).

Source: [36]

2.2.2. Building materials. The materials used in prominently in the building traditions of Africans are; stone, straw, and earth which have been independently and jointly used and skillfully applied [37]; in Hausa land, the four major building materials are: earth, timber, reeds/grasses, and stones. Though in most cities, stones are less used, apart from this mud is more tolerant to climate because of its poor conductivity. [38], [39], [40] None of the materials used by traditional architecture was company manufactured, processed or fabricated [41]; Examples of local materials used to achieve outstanding architectural monuments are: Bight of Benin (stabilized mud brick and plaster, thatch, timber, shingles); The Kano wall (mud and vegetable mat); Centenary Hall, Ake, Abeokuta (stone, mud, timber) [42].

2.2.3. Earth as a building material. Earth as a building material date back to 12000 BC [43]; the basic house walls in most of traditional architecture in Nigeria were built of 'earth' in simple low-cost and self-help construction arrangement [29]; Very little of adobe / earth / mud / brick architecture have lasted, apart from some of the monuments, temples and mosques. Nevertheless, the enduring cultural practices of rural people indicates that adobe or mud surely has been one of the most common and abundantly obtainable material [37]; the generally employed method was wattle-and-daub earth technology [37] [39] [44]; As soon as it is plastered and properly covered with overhanging roofs, these earth buildings were structurally firm, environmentally sound and could exist for years as long as the day to day maintenance was adhered to [37]; the building earth greatly differs in quality from the excellent brown clay to the blackish type [27]; the development of the adobe brick, a performed modular masonry unit of sun dried mud, came in to existence with higher civilization levels [45]. Unfortunately, modernization affects the sustainability of these earth buildings as they are being replaced with new ones made of modern building material like sand-cement, concrete and glass etc. Some of the earth buildings are abandoned and therefore crumbled due to lack of proper preservation and conservation.

2.2.4. Stone as a building material. Stones as a building material in Northern Nigeria has been discussed in many studies, some of which are; the basic form of the use of stones/rock for dwellings had begun with the habitation of naturally occurring caves about which walls and roofs are in mud or thatch, bricks and straws were erected converting them in to livable places [37]; The deliberately planned and built shelters began around the same time around the world. Even those times, similar to the present scenario, development involves many levels based on the social evolutions stage [45]; the main materials in traditional houses are stones, wood, bricks and metals. Stones are the most utilized materials; some of them prevent moisture and humidity which are used in building foundations etc. [46] [26]; in some parts of Northern Nigeria, stones are used in building foundations because of its moisture proofing property. In some areas with heavy rainfall, layers of stones or un-coursed rubbles are used in other to reduce the amount of the moisture movement from the foundation upwards [27]. With the emergence of modern materials such as sand-cement block, and reinforced concrete which can be used conveniently, local artisans and building industries have abandon the use of stones as the main building material and therefore the use of stone as a building material is receding greatly which affects the sustainable use of this material in VANN.

2.2.5. Thatch and Grass as building materials. Grass is a building material with a wider usage in VANN. The grass and earth forms a composite material. While the grass is the reinforcement, the earth serves as a matrix or binder as it surrounds the straw or thatch. The thatch/ straw possess a tensile strength while the earth has compressive strength [47]; thatch is one of the oldest of building materials known; grass is a good insulator and easily harvested. Many African tribes have lived in homes made completely of grasses and sand year-round [24]; thatch construction uses matted or baled straw from wheat, oats, barley, rye, rice and others as wall or covered b earthen or lime stucco [37]; thatching methods have traditionally been passed down from generation to generation, and numerous descriptions of the materials and methods used over the past three centuries. Thatch is still employed by builders in developing countries, usually with low-cost, local vegetation [24]. Sadly, with the emergence of zinc, aluminum roofing sheets, and cement (used as a binder in building practices), the use of thatch as a building material was greatly affected and diminishes its use in the VANN.

2.2.6. Timber as building materials. Timber as a building material has a wider usage also in VANN. The best timber used in Hausa traditional buildings are obtained from the trunks of male palm tree (Daleb or Giginya). The timbers are commonly called "Azara" beams. They are rigid and heavy, resistant to termite attack, and very durable because they took several decades if not a century without signs of decay. They serve as wooden reinforcement to strengthen the structures of the wall and pillars [27]; the "Azara" beams are also used to make frames constructions, beams, brackets and corbels as elements for carrying flat and domed roofs. The ashes of the timber are often used as an insulating layer when spread on top of flat roofs, treated with in fusions from pods or roots to water proof the top of flat roofs [27] [39].

Hausa people usually harvest their timber from the locally available tree trunks mostly male palm trees (Daleb or Giginya) and use them in construction as beams and columns. They also use the ashes for insulation and water proofing on flat roofs. Due to the availability of modern construction materials such as steel rods and other steel sections with concrete, the use of Azara as building material has vanished completely; invariably affecting the VANN sustainable practices.

2.3. Factors Hindering Sustainability of VANN

Various architectural concepts have certain factors hindering their sustainability and vernacular architecture of Northern Nigeria is no exception, some of the factors are; human neglect such as neglect of heritage buildings, forgetting and ignoring of traditional architectural element [48], [49]; socio-economic conditions such as financial wherewithal, social class factor, and high cost of maintenance of traditional buildings [29], [37]; modernization such as replacement of the old buildings with new ones using modern building materials, poor social acceptability of traditional buildings, and societal advancement somehow had downgraded these practices of using traditional building materials in favoring the machine intensive [29], [50]; discontinuity such as lack of qualified artisans and master builders, disappearance of traditional buildings in the cities, and many of the traditionally significant buildings of earth have weathered badly and are partially derelict [29], [48]; weather and climatic conditions such as surface erosion, partial crumbling, unhealthy conditions due to constant humidity and hollowed bases and wear and tear of the buildings, constant humidity, and moisture from underground [29], [51] . The main factors Hindering the sustainability of VANN were shown in Table 1 below.

Table1: Factors hindering sustainability of VAAN and their Impacts

S/N	Factors hindering sustainability of VAAN	Impacts	Sources
1	Human neglect	Neglect of heritage buildings; Forgetting and ignoring of traditional architectural element.	Osasona et.,al(2009), Ibrahim (2015)
2	Socio-economic conditions	Financial wherewithal; Social class factor; and high cost of maintenance.	Egenti et.,al (2014), Ejiga et.al,(2012)
3	Modernization	Replacement of the old buildings with new modern ones; Poor social acceptability of traditional buildings.	Egenti et.,al (2014), Khalil (1999)
4	Discontinuity	Lack of qualified artisans and master builders presently; Disappearance of traditional buildings in the cities; Traditionally significant buildings of earth have weathered badly and are partially derelict.	Egenti et.,al (2014), Osasona et.,al(2009)
5	Weather and climatic conditions	Surface erosion, partial crumbling, unhealthy conditions due to constant humidity and hollowed bases; Wear and tear of the buildings; Constant humidity, and moisture from underground.	Egenti et.,al (2014), UNCHSHabitat (1986).

Source: Authors, 2017; from literature review

3. Discussion, summary and conclusion

This study discusses the VANN in the light of sustainability; some desirable features and characteristics were identified and the factors hindering the sustainability of VANN with respect to such features were identified (in table 1 above) and discussed. Thus, the following conclusions were made:

- The main features of VANN are identified as: engravings on the building facades and use of building materials such as earth, grass, stones and timbers etc., which are unique to the VANN.
- Earth is the prominent building material used in VANN because it is easy to use; it is the most abundantly obtainable building material and therefore became the most utilized building material. It is used as a binder when mixed with grass to form a composite material and also used to make the best possible use of other materials.
- The main factors hindering "the sustaining of VANN" were identified as: Human neglect, Socio-economic conditions, Modernization, Discontinuity and Weather and climatic conditions.
- The emergence of modern building materials such as Cement (used in concretes and plasters), steel (used as reinforcement rods or in structural steel section), Zinc and Aluminum roof covering sheets, glass and other composite materials etc., really impede and hampered the continuity of the concepts and construction processes used in VANN while also diminishing its conservation, preservation and socio-cultural relevance within the Hausa-Fulani tribal communities in Northern Nigeria.

4. References

1. Fawcett, William; Hughes, Martin; Krieg, Hannes; Albrecht, Stefan; Vennström, Anders (2012). *"Flexible strategies for long-term sustainability under uncertainty"*. Building Research. **40** (5): 545–557. *doi:10.1080/09613218.2012.702565.*
2. Zhang, S.X.; V. Babovic (2012). *"A real options approach to the design and architecture of water supply systems using innovative water technologies under uncertainty"* (PDF). Journal of Hydroinformatics.
3. Black, Iain R.; Cherrier, Helene (2010). *"Anti-consumption as part of living a sustainable lifestyle: Daily practices, contextual motivations and subjective values"* (PDF). Journal of Consumer Behaviour. **9** (6): 437. *doi:10.1002/cb.337.*
4. Shaker, R.R. (2015). The spatial distribution of development in Europe and its underlying sustainability correlations. Applied Geography, 63, 304-314. doi.org/10.1016/j.apgeog.2015.07.009.
5. State of the World 2013: Is Sustainability Still Possible? worldwatch.org
6. Strong sustainable consumption governance — precondition for a degrowth path? degrowth.org
7. *James, Paul*; Magee, Liam; Scerri, Andy; Steger, Manfred B. (2015). *Urban Sustainability in Theory and Practice:*. London: Routledge.; Liam Magee; Andy Scerri; Paul James; Jaes A. Thom; Lin Padgham; Sarah Hickmott; Hepu Deng; Felicity Cahill (2013). "Reframing social sustainability reporting: Towards an engaged approach". Environment, Development and Sustainability. Springer.
8. Lynn R. Kahle, Eda Gurel-Atay, Eds (2014). Communicating Sustainability for the Green Economy. New York: M.E. Sharpe. ISBN 978-0-7656-3680-5.
9. A. A. Bartlett, "Reflections on Sustainability, Population Growth and the Environment," in Marco Keiner (ed.) The Future of Sustainability, (Springer, The Netherlands, 2006).
10. A. A. Bartlett, "Reflections on Sustainability, Population Growth and the Environment," in Marco Keiner (ed.) The Future of Sustainability, (Springer, The Netherlands, 2006).
11. Kates, R., Parris, T. & Leiserowitz, A. Harvard (2005). "What is Sustainable Development? Goals, Indicators, Values, and practice" Environment 47(3): 8–21.
12. International Institute for Sustainable Development (2009). What is Sustainable Development?. Retrieved on: 2009-02-18.
13. EurActiv (2004). "Sustainable Development: Introduction." Retrieved on: 2009-02-24.
14. IUCN/UNEP/WWF (1991). "Caring for the Earth: A Strategy for Sustainable Living." Gland, Switzerland. Retrieved on: 2009-03-29.
15. URL:https://www.sustain.ucla.edu/about-us/what-is-sustainability/ accessed 28/11/2016
16. Porphyrios, D., 2006. Classicism is not a style. In : Jencks, C. and K. Kroopf, (Eds.), Theories and Manifestoes of Contemporary Architecture. 2nd Edn., John Wiley and Sons Ltd., Sussex, pp: 179-180.
17. Attoe, W.O., 1979, Theory, Criticism and History of Architecture. In: Snyder, J.C. and A.J Catanese, Introduction to Architecture. McGraw Hill Book Company, New York, pp:30.
18. Osasona, C.O., The concept of the "traditional" in African architecture (Chapter 2). Principles of Traditional Culture, ed. M. Okediji, BARD Books: Ibadan, pp:18-25, 1992
19. Oliver P., 2006, Built to Meet Needs Cultural Issues in Vernacular Architecture. Elsevier Ltd., Oxford, UK. 2006.

20. Kagan Gunce,& Zafer Ertuk & Sevic Ertuk., Questioning the "*prototype dwellings*" in the framework of Cyprus traditional architecture, available online at www.Sciencedirect.com., Elsevier Ltd. 2007.
21. Oliver P. Encyclopedia of Vernacular Architecture of the World. Cambridge University Press,; 1997 pp 14.
22. S.S Chandel, V. Sharma & B.M. Marwah., Review of energy efficient features in Vernacular Architecture for improving thermal comfort conditions, available online at www.Sciencedirect.com.
23. Oliver P. Earth as a building material today. 5. Oxford Art J Arch,; 1983. Pp 2.
24. URL: Wikipedia.org accessed 10/29/16
25. Glassie, H. (1990), Architects, Vernacular Traditions and Society, *Traditional Dwellings and Settlement Review* Vol, 1, No 2 (spring), 9-21. Retrieved in August 17 from; http//www.jstor.org/stable/23566248.
26. Kirbas and Hizli, 2016., *Learning from Vernacular Architecture: Ecological solutions in traditional Erzurum Houses*., available online at www.Sciencedirect.com., Elsevier Ltd. 2016.
27. Anselm E.O, &O.F Ati., 2010,. *The influence of rainfall on Hausa traditional architecture.* Research journal of applied science, engineering and technology. Maxwell Scientific organization 2010.
28. Fatty H., 2006. Natural energy and Vernacular Architecture. In: Jencks, C. and K. Kroopf, (Eds.), Theories and Manifestoes of Contemporary Architecture. 2nd Edn., John Wiley and Sons Ltd., Sussex, pp: 144-145.
29. C. Egenti, J.M. Khatib, & D. Oloke., Conceptualization and Pilot Study of Shelled Compressed Earth Block for Sustainable Housing in Nigeria, 2014. Publication source, 2212-6090/c 2014 the gulf Organization for research and development.
30. Heathcote, K.A., 1995. Durability of earth wall buildings. Constr, Build. Mater. 9 (3), 185-189
31. kano chronicle, 1970-72 accessed 10/29/16.
32. URL: http://allafrica.com/stories/200202200154.html accessed on 31/10/2016.
33. World's oldest engraving discovered, Australian Geographic, 4 December 2014.
34. Adamu, M.S.T (2005): 'interpretation of significant and messages in Hausa traditional architecture. Case of ''zaure" entrance hall. Journal of association of architectural educators in nigeria. Vol4, No 1. January –March, 2005. Pp10-21'.
35. O.P. Agboola & M.S. Zango. (2014) Development of traditional architecture of Northern Nigeria; A Case Study of Hausa house form. International Journal of African Society Culture and Traditions. Vol.1, No 1, pp 61-74, June 2014.
36. URL: www.pinterest.com/engravings/Nigeria
37. O. Ejiga, O. Paul, O.O. Cordelia. Sustainability in traditional African architecture: a springboard for sustainable urban cities. June 2012. Sustainable futures: architecture and urbanism in global south Kampala, Uganda 27-30 June 2012.
38. Moughtin, J.C., 1985. Hausa Architecture. Ethnographica Ltd., London, pp: 1-123.
39. Dmochowski, Z.R., 1990. An introduction to Nigerian architecture-Northern Nigeria. Ethnographica Ltd., London. pp: 1.4,-1.10, 1.2, 1.39, 1.45, 4.20.

40. Anselm E.O, &O.F Ati., 2010,. The influence of rainfall on Hausa traditional architecture. Research journal of applied science, engineering and technology. Maxwell Scientific organization 2010.
41. Adeyemi, E. A. (2008): Meaning and Relevance in Nigerian Traditional Architecture: the Dialectics of Growth and Change. Public Lecture Series one. (21), pp 1-33.
42. T.O. Odeyale &T.O. Adekunle. Innovative and sustainable local material in traditionalAfrican architecture – Socio cultural dimension. Structural Analysis of Historic Construction – D'Ayala & Fodde (eds) © 2008Taylor & Francis Group, London, ISBN 978-0-415-46872-5.
43. Pacheco-Torgal, F., Jalili S., 2012. Earth construction: lessons from the past for future eco-efficient construction. Constr. Build. Mater. 29, 512-519.
44. Fathy, H. (1973). Architecture for the poor; an experiment in rural Egypt. The university of chicago press, Chicago. ISBN:0-226-23916-0.
45. Hamed. N., M.F.M. Zain., M. Jamil., S. Niroumad. Earth architecture from ancient until today. 2nd Cyprus international conference on educational research 2013.
46. Karpuz, H. (1984a), Turk Islam Meskin Mimarisinde Ezerum Evleri, Ankara: ministry of culture and tourism publication.
47. Eneh, A.E.O., 2006. Composite materials and their uses: A review, An unpublished paper,pp2.
48. C.O Osasona,. F.O. Ewemade. Upgrading Ille-Ife's vernacular architecture heritage. WIT Transactions on the Built environment, V109, 2009 WIT Press, ISSN 1743-3509.
49. H.G.A. Ibrahim. Regeneration of sustainability in contemporary architecture: approach based on native function and activities strengthen identity. Urban planning and architecture design for sustainable development, UPADSD 14-16 October 2015.
50. Khalil, N. (1999) Ceramic houses and earth architecture, how to build your own, california: calEarth press.
51. UNCHS Habitat (1986). Earth construction technology: Manual on surface protection. Nairobi: UNCHS (Habitat).

Overview of Modular Design Strategy of the Shipping Container Architecture in Cold Regions

Zexin Sun, Hongyuan Mei and Ruixian Ni

School of Architecture, Harbin Institute of Technology, Room 325, No .66Xidazhi St, Harbin, Heilongjiang, China

736087330@qq.com

Abstract. In recent years, the shipping container architecture, as an emerging green building mode, has grown in popularity. In this article, we analyse the advantages of applying shipping container architecture in cold regions, such as shortening the construction cycle, saving construction materials, reducing energy consumption and adapting to diverse sites. Then we analyze the modular design strategy and the modular design strategy in architecture. In the end we introduce the modular design strategy of shipping container architecture in cold regions with the technology roadmap, classification and combination.

1. Introduction

In recent years, the shipping container architecture has grown in popularity. The shipping container architecture refers to a type of architecture transformed from the steel intermodal containers (Figure1). With the crises of resources and environment, sustainable and green are becoming the consensus of architecture gradually. As an emerging type of green building, the shipping container architecture has shown great potential in construction efficiency, economy, mobility and sustainability [1]. As a transportation product, shipping containers have standard sizes and high structural strength so they are very suitable for standardized construction. Besides, the shipping containers is aneco-friendly building material which can be easily recycled and reused and also the shipping containers are with a relatively low expense and have a wide range of sources, there are more than 1 million TEU (Twenty-Feet Equivalent Units) of shipping container seliminated every year, provides sufficient materials for the construction[2]. Architecture in cold regions have the problems of short construction cycle, high building energy consumption and high construction material consumption, the introduction of the shipping container architecture provides a new opportunity to solve these problems and develops a new territory of architecture in cold regions.

2. The advantages of shipping container architecture in cold regions

The cold regions are at high latitudes and have relatively longer winter (Figure2). Due to the limitation of region and environment, architecture in cold regions faces a great shackle on greenization [3]. Under such circumstances, there are many advantages of applying shipping container architecture in cold regions as follows.

2.1. Prefabrication, shorten the construction duration

In cold regions, it is very difficult to carry out construction in the long cold winter, which leads to a poor economic performance due to the short construction cycle and long-duration of the project. However, the shipping container architecture is more adapted to the industrialization since it is easier to be prefabricated in the factory with a more stable construction process and a higher degree of standardization. The construction duration will be shortened by using standardized prefabricated modules and the construction efficiency will be improved greatly as well. Compare to the traditional reinforced concrete structure buildings, the construction duration will be shortened to 1/3, and the useful life will be increased by 40% [4].

Figure 1. Stacked shipping containers form temporary pavilion in Shanxi, China, by People's Architecture Office

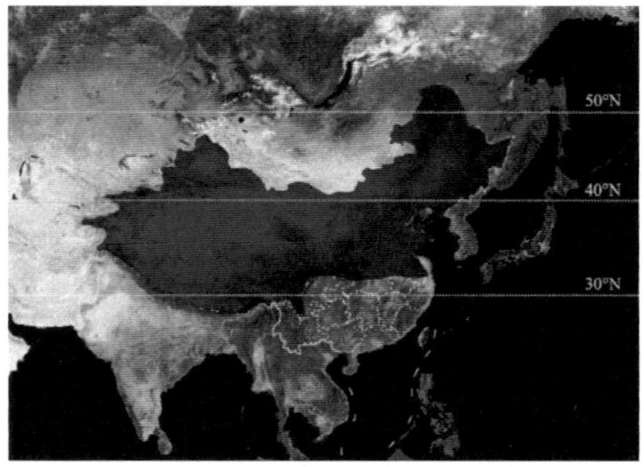

Figure 2. Cold regions in China (the blue part) according to the “Thermal Design Code for Civil Buildings”.

2.2. Recycling, save the building material

The building envelope is thick and heavy in cold regions due to the requirement of insulation performance, and the components of therein forced concrete structure buildings are in large size, both of them lead to a low effective area coefficient of the buildings. Besides, the traditional buildings cannot ensure the maximum performances of the materials and also the construction materials are difficult to recycle and reuse, it will produce a lot of construction waste during the construction process, leading to a huge waste. Shipping container architecture applies the standard container, which is an ideal building material with good structural strength and long durability as the basic module, lead to a relatively high effective area coefficient of the building. At the same time, shipping container is easily to recycle. The introduction of shipping container architecture in cold regions will bring more than90% of reducing, reusing, and recycling the building materials.

2.3. Standardization, reduce building energy consumption

The climate of cold regions can be characterized by low temperature, little sunshine and long winter, therefore, buildings should take full account of the demand of adaptive with the cold climate. As an industrial product, the shipping container is easier to standardized prefabricated in factory. The performance of the building will be improved through a systematic design and construction process from the inside to outside, through using the integrated heat preservation and heat storage technology as well as the BIM (Building information modeling) platform; it will effectively improve the thermal performance and reduce the energy consumption of the building. Compare to the traditional construction methods, the shipping container architecture can achieve 40% of water saving, 90% of energy saving, 20% oflandsavingand80%of construction waste reducing [5].

2.4. Removable, adapt to diverse sites

Cold regions are with varied landscapes, such as snow, ice, lakes, forests, grasslands, wetlands, farmlands and so on; the diverse sites have more requirements for the buildings. Besides, the buildings are significant seasonal, the frequent demolition and construction of seasonal service buildings will cause a huge waste of resources. While the shipping container architecture has the characteristic of moveable and convenient to assemble or disassemble, it cannot only be used to build mobile homes for various sites, but also can be applied in the seasonal service facilities and shelters, so that it will greatly expand the scope of buildings in cold regions.

3. Analysis of the modular design

In the perspectives of globalization and industrialization, the concept of modular design was formed based on the idea of standardization. The modular design aims at improving the degree of versatility, reducing the variety and numbers of components in order to use less cost to get more series of productions. The concept of modularity was applied to manufacturing industry first, and then it was widely applied in architecture gradually.

3.1. Modular design

The modularity is based on the idea of standardization with an advanced form. The so-called standardization is to develop common and repeated service regulations for specific or potential problems in order to get best orders. The essence of standardization is achieving unity through customizing, distributing and implementing the standards. And the modularity is aimed at simplifying the design process in order to use the manpower and resources effectively, shorten the construction cycle and improve the quality of products, so that it will reduce the cost and maximize economic benefits in the end.

Modular design doesn't have a unified and precise definition (Table1), it usually refers to break up a large system into several subsystems within a series of relations, and each subsystem can be replaced or combined with each other. Masahiko Aoki indicated that each module is a half self-discipline subsystem, the modules are connecting with each other in certain rules in order to interrelate and constitute a more complex system, the modularity contains the two behaviors of division and concentration of the modules[6].

Table 1. Narrow sense and broad sense of modular design

	Connotation	Denotation
Narrow sense	1. The system has a clear multistage modular hierarchies 2. Modules have compatibility on function and size	Modular production composed by different modules.
Broad Sense	1. Project is with a clear hierarchy 2. Modules are with generality and representativeness	Everything composed by typical generic units

The modular design strategy is of great significance for contemporary manufacturing. The practice of modular design started long before, in 1964, IBM designers creatively developed the IBM360 system through using modular design strategies, achieved the important landmark in the history of computer innovation. Then, the modular design strategy was widely applied in the industry of automotive, furniture, integrated circuits, manufacturing and so on. In recent years, the modular design strategy has been adapted to the new industrial revolution with the emerging computer technology, become the important force to promote the development of manufacturing industry, which is on a transition from the standardized production to mass customization. Professor Baldwin and Clark from the Harvard Business School indicated that the contemporary is the modular time, pointed out that the modularity has significant impacts on the upgrade of the industrial structure and the manufacturing industry has entered a time of modularization design, modularization production and modularization consumption[7]. Masahiko Aoki even believed the essence of the new industrial is modularity.

The service object of the modular design is system, the unit is the module. Modular design is based on the system and it realizes combination and utilization of the different modules with the greatest extent. Modular design strategy is not only a kind of design and technical method, but also a complicated system engineering and a systematic process.

3.2. Modular design in architecture

The purpose of applying the modular design strategy in architecture is to improve the efficiency of construction. In the development of architecture, the modular design concept was always implied in the architectural design process. As early as back to the days of the Song Dynasty in China (960–1279), the book "Yingzao Fashi"(Treatise on Architectural Methods or State Building Standards) recorded the utilization of the modular strategy in the traditional Chinese architecture. The modular system was adopted to construct different components such as Dougong, which is a unique structural element of interlocking wooden brackets. This system stipulated different modules such as "cai", "fen" and "qi". The height of "cai" is 15 "fen" and the thickness of "cai" is 10 "fen", the distance between the layers of "Dougong" is 6 "fen", which is the "qi". Almost all of the wood components used "cai", "fen" and "qi" to determine the size, making a unified standard between the components [8] (Figure 3).

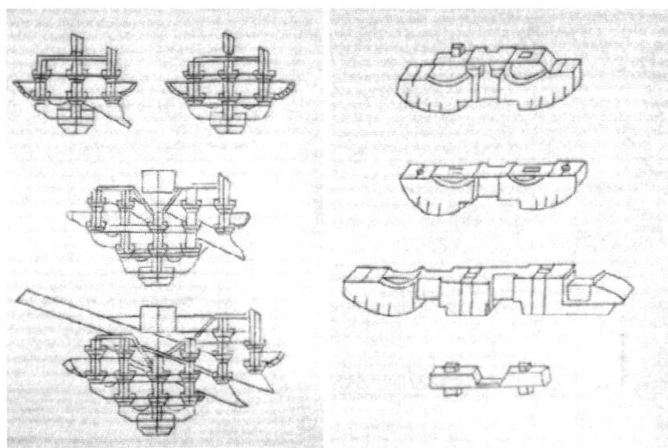

Figure 3. "Yingzao Fashi" by Li Jie indicated the modular wood components of dougong.

After the Renaissance, the architectural design and construction were gradually separated, the design process is operating in the architectural work space based on the orthogonal projection system and applies the architectural drawings to represent. The concept of modularity is throughout the process of design, such as using the modulus and axis, dividing the building into components and so on. In modern architecture, the modularity idea was widely applied by the pioneer modernism architects, for example Le Corbusier adopted the standardized and modular strategy not only as his starting point of architectural aesthetics, but also an important designing tool in his projects. His design process was based on division and combination of the building components such as foundations, walls, columns, floors, stairs, roofs and so on, reached a high degree of unity of the logic and aesthetics[9].

In the 1960s, the concept of modular design gradually raised with the high-tech architecture movement and utopian ideas. In the 1967 International and Universal Exposition, Safdie Moshe finished his first project Habitat 67 with the idea of providing affordable residences for the low-income groups, 158 houses were constructed from354 modular units. Each box type residential unit was prefabricated and then scattered piled up together (Figure4). At the same time in 1960s, Japanese architects brought the Metabolism which also applied the modular design strategy; the difference from Safdie is that the modules were plugged into the structural and service core. One of the most famous Metabolism projects is the Nakagin Capsule Tower by Kurokawa, he set up are cycling strategy that the modules could be easily extracted and updated if needed (Figure 5).

Figure 4. Habitat 67.

Figure 5. Nakagin Capsule Tower.

In recent years, the contemporary architects combine the modular design strategy with the green and ecological concept, create the "modular homes" which are sectional prefabricated buildings. The module sections are prefabricated at an offsite facility and then delivered to the site for construction. The "modular homes" realize more variability of configurations and styles in the building lay out through the systematic combination of the function modules and building envelop modules [10]. Now, the concept of modular design grows more and more popular with the rise of mass customization and the rapid development of computer technology [11].

4. Modular design strategy of shipping container architecture in cold regions

In the design and construction process of the shipping container architecture in cold regions, the modular design strategy refers to using the shipping container as a standard unit to carry out the process, the shipping container is not only the unit module of construction and the basis of the design operation, but also the structural components due to its high structural strength and durability. The design work is not limited on paper but rather applying the BIM technology and operating the design in three dimensional, the modular design strategy of shipping container architecture in cold regions has significance as follows:

1. The mechanism of modular design simplifies the design process; architects are able to solve specific problems with modular logic by applying the serialized modules.
2. The modularity is an advanced form of standardization with higher performance; the shipping container architecture in cold regions will be more integrated and diversified by applying multi-functional composite modules.
3. The application of modular design strategy in shipping container architecture is not only during the design process, but also throughout the whole process of construction and operation, such as modular management and modular construction, and thus it will significantly improve the construction efficient in cold regions.

4.1. Classification of the modular design strategy

The utilization of the modular design strategy is the key to promote the development of architecture industrialization that can be characterized as standard design, industrialized production, prefabrication and informatization management, and thus will have revolutionary impacts on architecture in cold regions. The modular design strategy of shipping container architecture in cold regions is based on the existing adaptive design techniques and different demands of cold regions aimed at energy saving, sustainable and movable. And the modular design strategy is an integrated system of operation method and technology strategy with the subsystems of building envelope modules, equipment modules, transportation and construction modules. In this system, the subsystem of building envelope modules include the roofs, panels, windows, floors and so on; focus on adaptive design, insulation technology, air tightness of the joints and so on. The equipment modules include solar dynamic power system, thermal storage battery system, water treatment system, auto leveling system and so on. The transportation and construction modules include the prefabrication and transportation system, construction management system and so on; form the self-sufficient passive high-performance shipping container architecture design and construction strategy as well as the municipal pipe network interface modules of the shipping container architecture. The system is targeting at greenization and based on the strategy of serialization, recycle and integration. All in all the modular design strategy is not only a design method of the shipping container architecture in cold regions, but also the technology road map that indicates the implementation of the operations (Figure 6).

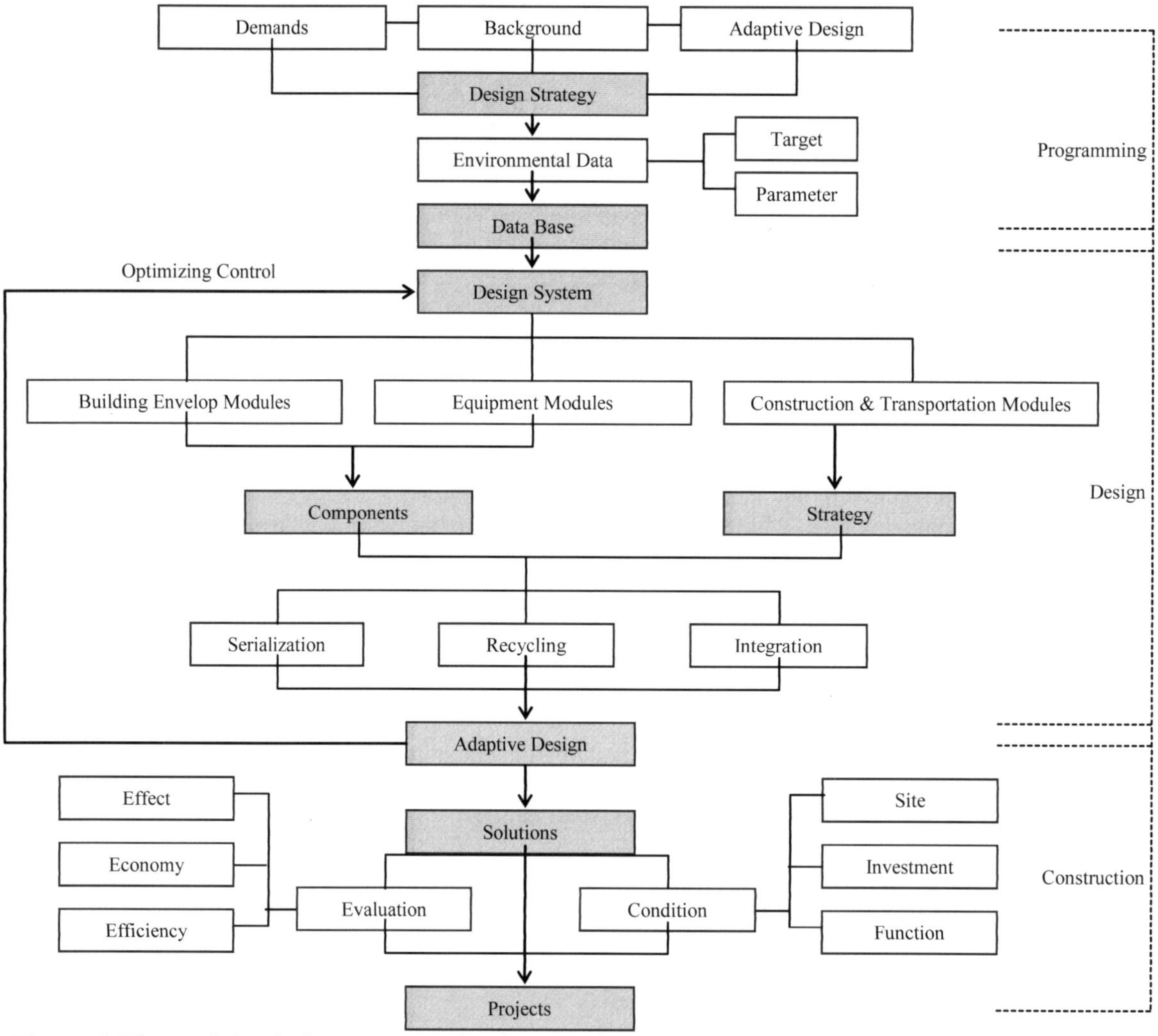

Figure 6. The modular design system technology road-map of shipping container architecture in cold regions.

4.2. Combinations of the modular design strategy

The combinations of the modular design strategy of the shipping container architecture in cold regions can be characterized with the equation as:

- Project = Universal Modules + Special Modules + Functional Modules

In this equation, the universal modules refer to the invariant parts which are the structural components and the standard shipping container units; the special modules refer to the variable parts which are the building envelope modules and equipment modules mentioned before; the functional modules are the adaptive parts which are the construction and transportation modules. The combinations of the modular design strategy are not only the different combination of modules, but also the adaptive operation methods. On the other hand, the combinations are also indicating the form design strategy of the shipping container architecture; usually refer to the combinations of the shipping container units. The shipping container unit is a box-type element with characters of simple and high strength performance, it is easier for the architects to carry out the adaptive design with simple logics through different combinations of the units, such as side-by-side style, converging style, unit style, stacked style, mixed style and so on.

As mentioned before, the modularity is a systemic operation method of the shipping container architecture in cold regions. The modularity put forward an application strategy of standardized construction which is adapted to the mass customization of the contemporary architecture industrialization, so that it will achieve intelligent construction and lean production of the shipping container architecture. By applying different combinations of the modules and shipping container units according to the functions and sites, so that it will realize the variability of configurations and styles of the buildings.

5. Conclusion

As an emerging green building mode, the shipping container architecture will provide an effective way of construction in cold regions. The application of the modular design strategy of the shipping container architecture is not only the way to optimize the design process, but also an implementation strategy of the whole industry. Develop the modular design system with the subsystems of building envelope modules, equipment modules, transportation and construction modules based on the strategy of serialization, recycle and integration, thus establishing an innovative operation method of the shipping container architecture in cold regions.

6. References

[1] Buchmeier, M., et al. "Container Atlas: A Practical Guide to Container Architecture." Berlin: Gestalten (2010).
[2] Slawik, H., et al. Container atlas. Die Gestalten Verlag, 2016.
[3] Han Pei;Mei Hongyuan."A Studyon Dynamic Design Strategy of Cold Regions Architecture Based on Photokinetic Mechanism of Plant"[J]. New Architecture,2016,03:64-68,ISSN: 1000-3959
[4] Giriunas, Kevin, Halil Sezen, and Rebecca B. Dupaix. "Evaluation, modeling, and analysis of shipping container building structures." Engineering Structures 43 (2012): 48-57.
[5] Martinez-Garcia, Mary. "Shipping Container Home." (2014).
[6] Aoki, Masahiko. "Institutional complementarities between organizational architecture and corporate governance." RIETI (Research Institute of Economy, Trade and Industry) Conference on Corporate Governance. 2003.
[7] Baldwin, Carliss Y., and Kim B. Clark. "Managing in an age of modularity." Managing in the modular age: Architectures, networks, and organizations 149 (2003).
[8] Feng, Jiren. Chinese Architecture and Metaphor. University of Hawai'i Press, 2012.
[9] Corbusier, Le. The Modulor and Modulor 2. Vol. 1. Springer Science & Business Media, 2004.
[10] Brisson, Dale. "Modular homes." U.S. Patent No. 7,028,440. 18 Apr. 2006.
[11] Smith, Ryan E. Prefab architecture: A guide to modular design and construction. John Wiley & Sons, 2011.

IOP Conf. Series: Earth and Environmental Science **63** (2017) 012036 doi:10.1088/1755-1315/63/1/012036

Study on the "3F-in-1" Sustainable Reconstruction of Rural Architecture from Placeality Perspective--A Case Study of Caiyuan Village in Jingmen City, Hubei Province

Fu Fangyu and Cao Yu

PhD; School of Architecture and Urban Planning, Huazhong University of Science and Technology;
Lecturer; School of Architecture and Urban Planning, Huazhong University of Science and Technology

27481733@qq.com

Abstract. This paper takes Caiyuan Village in Jingmen City of Hubei Province as the research object, analyzes the production, life and ecological functions of rural buildings and the "3F-in-1" inherent mechanism from the local perspective. Based on the concept analysis of placeality and "3F-in-1", this paper clarifies the relationship among the value of life function, production function, ecological function so as to analyze the "3F-in-1" mode of rural architecture with placeality. On this basis, this thesis puts forward the strategy of sustainable spatial transformation (1) preserve the traditional overall spatial structure of villages, (2) improve the adaptability and function of rural architecture, (3) extend the rural social culture, (4) pay attention to local perception, with a view to explore an organic system design method for the exhibition of placeality and sustainable development of beautiful countryside.

1. Introduction

With the rapid development of China's economy and the acceleration of urbanization, the reconstruction of rural buildings and the construction of ecological human settlements will be the focus of future development[1]. The village is facing spatial reconstruction and functional transformation [2]. The village production, lifestyle and family structure have undergone various changes and face with new demands, while the renovation and reconstruction of rural buildings is one of the core content.

Caiyuan village is under the jurisdiction of Macheng town, Duodao District, Jingmen City, Hubei Province Knife, it is located 18 kilometers southeast of Jingmen City, in the Golden Triangle of Hubei Province, borders upon Jiaozhi Railway, 207 State Road to the west, stays close to Han River, JingQian provincial highway, Xiang Jing, Jing-Yi Highway to the east. Right now, there are 11 villager groups, 332 households, 1358 people, 7231 acres of arable land in this village. The case studied in this paper is one of the construction projects of Jingmu Yangzhuo ecological civilization and beautiful village demonstration areas. After the implementation of this project, Caiyuan village was nominated as the "Model village for the new rural construction in Hubei Province".

The survey found that a large number of rural construction and renovation design just simply learn from foreign villa design and reconstruct the rural architecture as unified country "villa" or rows of barracks building regardless of the urban and rural differences, ignoring the geographical characteristics of the rural architecture. However, in the transferring course from "township" to "city" and from "villager" to "citizen", the material production structure, spatial settlement form, resident social structure, urban and rural cultural form has witnessed a dramatic change[3]. After the thorough transformation, the original material form of the village has disappeared and lifestyle has changed. In face of silent dramatic change, we can feel the spectacular of change and the lamentable loss of rural architecture landscape and traditional local favor which resulting in rural local recession. In recent years, this phenomenon has attracted the attention of experts and scholars, who have done quite a lot researches on the new rural planning, the protection of rural landscape and traditional houses, green building and ecological construction, with researches methods changing from single disciplinary to multi-disciplinary [4]. In the study of rural building renovation, Zhao Rui et al. (2016) analyzed the symbiosis theory of KISHO KUROKAWA so as to intending to explore ways to realize the coexistence of old and new architecture style; Song Chengnuo (2016) investigated the significance of regional culture in rural architecture reconstruction, and put forward the core concept of adjusting measures to local conditions; Xue Kaiqiang et al. (2016) researched the rural construction in the central region, and put forward the concept of "renewal of history", "preservation of life", "adaptation of change". On the whole, the reconstruction of rural buildings is both a focus and a difficult issue. Rural architecture has its own characteristic, its spatial elements are the basis of internal links of rural production and life, and it is a constant reproduction, adjustment and evolution of a basic unit from bottom to top. Therefore, for the transformation of rural construction and the construction of beautiful village, the top-down policy approach need to be more cautiously implemented so as to avoid the unacclimatized development. In this paper, the author tries to discuss the approaches of countryside building reconstruction from placeability perspective.

2. Concept Analysis

2.1. Placeability

In the 1970s, geographers rethought the logical positivist method of measurement [5] and scholars of humanistic geography represented by Duan Yifu started to take "placeality" as an important research topic and the essence description of local culture. The nature of the description at the same time gets the attention of scholars. Starting from the subjectivity of human beings, humanism geography believes that place is space with meaning given by human. The spatial experience of human is an important way of composing and interpreting the place, and placeality is not only a geographical phenomenon, but also a rich human experience. From the perspective of humanism, some scholars pointed out that the placeality is the spirit or trait given to the composition of the place in the long-term interaction between the human and the natural environment, it is the surface humanistic phenomenon produced and developed through the "creative activity" of human being and a regional place with local characteristics [6].

2.2. 3F-in-1

Because of the different nature of the occupation, rural residents are essentially different with the urban residents in terms of regional culture and way of life as well as their demands on architecture, this difference is an embodiment of rural placeality. Through the in-depth research on Caiyuan village, we found that rural architecture is not just a single building; it is more of a whole system which has more specific and complex functional demands. In general, there are three kinds of functional demands: the production functional demands, life functional demands and ecological functional demands. In other words, rural architecture is a three-in-one architectural form with the combination of production, living and ecological function; these three aspects are interdependence, mutual influence, affecting the overall shape and use of rural architecture. Beautiful rural architecture is a combination of "3F-in-1" spatial and social demands of production, life and ecological function, which has a strong link among these three functions.

IOP Conf. Series: Earth and Environmental Science **63** (2017) 012036 doi:10.1088/1755-1315/63/1/012036

3. "3F-in-1" Model of Rural Architecture with Placeality

People gather in space and link with the place through communication. The interaction between human, human and space has gradually produced the local significance and symbol in a certain region. These constitute the text for others to interpret the placeality. A series of symbols which belong to the space reflect the placeality, and placeality communicate with people through the space, the environment and other symbols, so as to bring out continuous experience and all kinds of emotions, among which positive emotional experience can help to form individual recognition of the place [7]. The rural architecture is one of the most intuitive manifestations of placeality, human space to create the subjectivity spatial construction of human is an important way to construct placeality[8].

3.1 Production function of rural architecture

Rural economic activities are generally dominated by the primary industry, the rural life have inseparable relations with production, while the rural production often extends from the outdoors to the interior. In the rural construction, the production function is closely related with human's economic activities, in the traditional sense, the space function of the building should be in accordance with the production demand of the primary industry.

Compared with the urban residential buildings, where a variety of space configuration are mainly designed to meet the needs of life, the space configuration of rural residential building, are undoubtedly designed for realizing the production function. Specifically, it includes the production space system composed of people's production space, storage space for production tool and production-related economic crops and animals, and other functional elements. With the continuous development of the society and the adjustment and upgrading of the regional agricultural structure, the rural production mode has gradually changed, and the supporting facilities for production function in rural construction has improved so as to achieve human, animal, machine, storage separation. (Table 1)

Table 1. Spatial analysis of production function

Production Functional Space	Usage	Tradition method and problems	Reconstruction Strategy
Production space	Grain drying platforms	Artificial smoothened ground with poor drainage function.	Construct a flower bed and shading pergola
Storage space for means of production	Garage for car, motorcycle etc.	No fixed place	Combining with modern life demands of thief-proof and easy to management
	Equipment room (tractor, Harvesting equipment etc.)	Shabby timber set, not good for machine maintenance	Complete supporting facility
	warehouse	No fixed place, easy to be eroded by rats or buds	Complete supporting facility and create air network
	Places for fertilizers and pesticides	No fixed place, easy to take or use by mistake	Complete specific supporting facility
Space for livestock	Livestock House (cow, sheep, pig etc.)	Shabby timber set with poor smelly environment	Construct in leeward side, combine through corridor, complete ventilate and faeces collection and processing design

	Poultry House (chicken, duck, goose etc.)	Shabby timber set with smelly and noisy environment	Construct in leeward side, combine through corridor, complete ventilate and faeces collection and processing design
	House for Guarding Livestock (dog, cat)	No fixed feeding space, easy to cause pollution and hurt children from neighborhood	Build specific house for dogs or cats, get a fixed place for food feed and excretion

3.2 Life function of rural architecture
Living space of farmers and villagers is where the rural construction is directly related to human life. Compared with the living function of urban residential buildings, the rural buildings have more social interaction and ethnic consciousness because of the rural population composition, the nature of rural farming work, the rural customs and culture. In addition, the lifestyle of migrant workers, left-behind elderly women and children also lead to the diversity of rural living space in terms of the material living, entertainment and lifestyle, which should be fully taken into account in the reconstruction of rural building. (Table 2)

Table 2. Spatial analysis of life function

Life Functional Space	Usage	Tradition method and problems	Reconstruction Strategy
Resting Space	Bedroom	Functional diversification lead to poor indoor environment quality	Reasonably adjust the space to reduce the energy consumption of air conditioning
Communication Space	Function room like hall, greenhouse, room for table games	No fixed room, or use the hall for multi-function	Combining with modern lifestyle to separate courtyard with storehouse
Life Space	Front court, backyard	In disorder, use for both life and production, moody land and harden land coexist, with poor water discharge	Improve land and drainage design and afforest to optimize environment
	Kitchen	Various combustion mode coexist, cause cross contamination of food, has fire and health hazard	Combing traditional cooking bench with modern cooking methods
	Toilet	Inconvenience usage, without organized pollutant discharge, lead to breed of mosquitoes and serious domestic pollution	Design a modern bath system and improve the water supply and drainage system
	Drying space	Airing and drying clothes in the spot, has health hazard	Combining living and production space, divide a specific space for drying and airing

3.3 Ecological function of rural building
Ecological function is why the rural construction is closely related with the survival of the people. On one hand, the ecological function value reflects the harmonious symbiosis between the villagers and the regional climate, geographical conditions and natural environment. On the other hand, in the

process of local stratification, placeality is a sense of familiarity and belonging felt by the villagers themselves on the village. American geographer Seamon D proposed "place ballet", which is described as a combination of the "body ballet" and the "time-space routine" by him and refers to the behaviors people conducted rhythmically in accordance with the habit of normal conduct. "Space-time routine" refers to a group of habitual physical activities carried out by people over time (for example daily). Place ballet gives people access to the local characteristics from perspective of physical, experience and social construction. Specifically it reflects in the plane layout, envelope design and other specific aspects of rural construction. Meanwhile, the rural architecture can also reflect the social space of the rural villagers and the family relationship of the clan, such as spiritual needs and the needs of the blood lineage. The ecological feature of rural environment and social relations are important embodiments of rural placeality, and a projection in architectural space. (Table 3)

Table 3. Spatial analysis on the ecological function of village

Ecological Functional Space	Usage	Tradition method and problems	Reconstruction Strategy
Structure spatial ecologic function	building facades, roof	Constructed randomly without unified style and without characteristic	Make overall planning to reflect regional culture
	Doors and windows	Impede lighting and ventilating	Take sunshade and lighting into consideration, use screen window to stop buds from entering
	Water supply and drainage	Without systematic design, resulting in indoor and outdoor pollution	Take an overall consideration of water supply and drainage system to reduce pollution
	Electrical equipment (solar energy, black light bulb etc)	Not take into account	Combining with modern life demands, take the arrangement of solar energy equipment, black light bulb and air conditioning into consideration
Social Spatial ecologic function	Clan space	No fixed place, incense and photos are generally placed in the hall, influence the use of hall	Design a specific room with traditional clan layout to promote the construction of kinship in modern society
	Community communication space	No fixed place, resulting in disordered functional space	Combine with hall, greenhouse, and table game rooms to enhance rural community communication

3.4 The relationship between the value of the three function

The production, life and ecological function of village buildings should be organically integrated with one another, so as to create a local architecture system with good production function value, life function value, as well as the ecological function value. Based on this system, in the rural reconstruction process, the production, life and ecological functions should be taken as a system for comprehensive consideration. Especially in the process of urban and rural planning, it is necessary to meet the various needs of use and life, but also to protect the history of rural areas and the sustainable development of humanistic culture.

4. The "3F-in-1" Reconstruction Tactics for the Rural Reconstruction in Caiyuan Village

4.1 The retention of overall traditional spatial structure

IOP Conf. Series: Earth and Environmental Science **63** (2017) 012036 doi:10.1088/1755-1315/63/1/012036

Preserving and renovating the original spatial structure of villages(1) reserving and renovating the public spaces of villages, (2) transforming traditional buildings, (3) retaining main roads and natural spatial structures of villages, (4) preserving and repairing historical information elements such as old trees,(5) coordinating new buildings with traditional architectural style. The purpose of the above measures is to create a cultural diversity of rural space, so as to realize the sustainable inheritance of local culture. In addition, the existing places for recreational activities and production activities such as: square dance, drying crops, etc. should also be incorporated into the reconstruction of the local landscape, through the integration of existing public space, new functional areas can be built or expanded to meet the new functional demands.

4.2 Adaptation and function adjustment of rural buildings

The author carried out a door-to-door investigation and research on the housing quality, housing survey, household production and living demand in Caiyuan village in line with the "3F in 1" concept, divided the rural buildings into 7 types based on the specific construction status and demands(Figure1), and proposed the following strategies for local protection.

Type	Construction status	Thoughts for Reconstruction	Building plane	Building elevation
Common Gregarious type		Preserve both the production and living space in the courtyard, pay attention to the social and ecological relations among family members	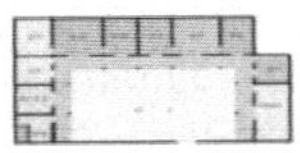	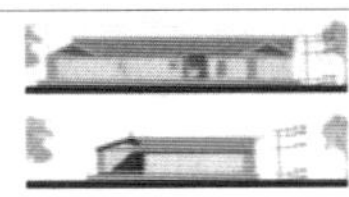
Crop-dominated type		Improve production and life space arrangement based on production function	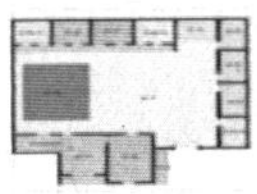	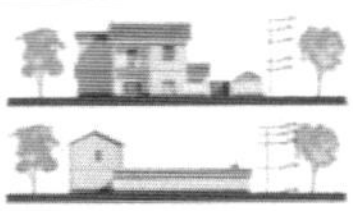
Aquaculture type		Maintain a balance between life and social ecology based on its production method	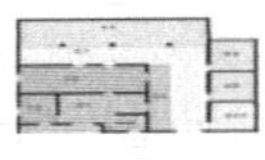	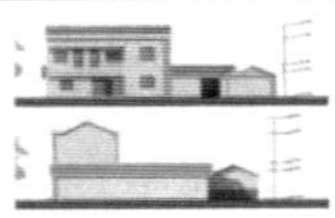
Rich and gregarious type		Take life and ecological function as the major design focus	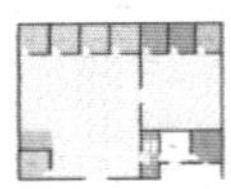	
Traditional spatial type		Enhance part of production and life function on the basis of traditional construction	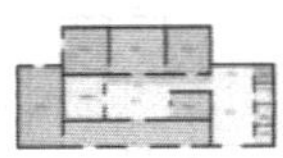	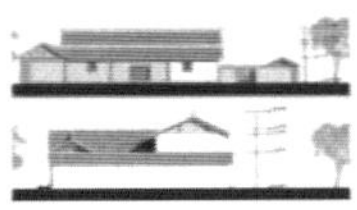
Improved small house type		Pay attention to the reconstruction of production and life space on the basis of traditional layout	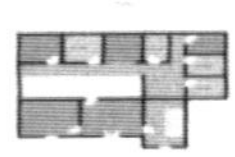	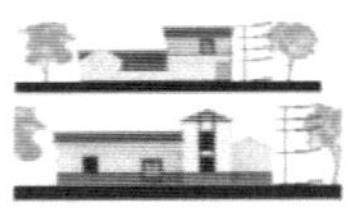
Non-agriculture type		Social ecological function is the major function of this type, some life function need to meet, no production space needed	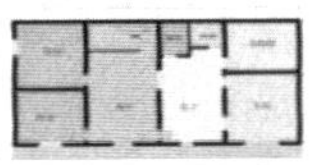	

Figure1. Adaptive functional adjustment of rural buildings(Source: Drew by the

4.3 Continuation of rural social cultural

The village is a living community composed of social networks such as blood relationship, kinship, clan system and geography. Although the standard of living and way of life are gradually urbanized in the present context, production and lifestyle of village still have some fundamental difference with that of the city. The transformation and reconstruction of Caiyuan village should be carried out around the three functional demands of local residents. The layout and decoration of family space should respect the rural life habits and the idea of ethnic group, local non-traditional rational surgery reconstruction method should be adopted for the conservation and extension of rural social ecological relations and placeality.

4.4 Focusing on local perception

Local perception is an important dimension of placeality, and it is an indispensable factor to realize the subjectivity of environment. "Nostalgia" involves space and place, and is closely related to human emotions [9]. Local perception involves the dimensions of body, experience and society. The familiarity of body with the place and the human nature embodied in it should be respected; both the old and new life experience should be preserved [10]. Jollification, united, clan, rich products, life style of rice fields are the common cognition of people in Caiyuan village. In the course of building a beautiful village, government policy and planners as external forces need to combine with the concept of "3F-in-1", attach importance to the local community's perception of rural memory and promote the local planning and reconstruction.

5. Conclusion

The reconstruction of beautiful countryside is not a "dressing" project, but a reconstruction of rural local meaning in the contest of globalization and modernization; it is the manifestation of production development, well-off life and good ecological performance. The overall and architectural transformation of Caiyuan village is based on local sustainable social construction; it will produce new local significance. In the context of urbanization development, in the beautiful rural reconstruction, we need to understand the rural life culture, respect the demand of agricultural production, and constantly explore and meet the demands of rural residents for living environment. This study attempts to carry out an in-depth and detailed social survey from the perspective of placeality, and analyzes the transformation of rural areas before and after the reconstruction.

In a certain sense, the reconstruction of beautiful rural village is to emphasize the existing rural production, life and ecology and meanwhile achieve the advancement of rural building with the times, in order to fundamentally build a rural living environment conducive to human life and production activities. In this context, the exploration of rural reconstruction has more practical significance.

6. Reference

[1] Qiu Tong, Xu Qiang Engineering Practice of Yingdong Ecological Village Construction In Chongming Island *J Green Building* 2012 **6**

[2] Xue Kaiqiang, Wen Jiangang, The Reconstruction Idea of Traditional Viliage in The Central Region of China *J Suzhou University of Science and Technology (Engineering)* 2016 **3**

[3] Zhang Tingwei, Changes of Urban Spatial Structure and Its Driving Mechanism in China in the 1990s *J Urban Planning* 2001 **25**

[4] Zhang Xinxin, Literature Review of Rural Living Environment Research *J Urban Geography* 2015 **14**

[5] Duan Yifu, My view on humanistic geography J translated by Zhicheng Zuo Yiou Proofread by Zhou Shangyi *Progress in Geography* 2006 **25** 2

[6] Pan Chaoyang, The geographical composition of the Placeality of Great Lakes region- the geographical interpretation from historical dimension *J Geographical Research Report (Taiwan)* 1996 **25**

[7] Tang Lin, Ling Hong, Chen Chen, Local, Landscape and Cultural Identity: An Analysis of Cultural Geography of Old City Reconstruction in China *J Journal of Hubei Correspondence University* 2014 **13**

[8]Seamon D. Body- subject, time- space routines, and place- ballets[M]/Buttimer A, Seamon D(eds.). The Human Experience of Space and Place. London: Croom Helm, 1980

[9] Tuan Y F. *Space & Place: The Perspective of Experience* （Minneapolis: University of Minnesota Press） 1977 P17

[10] Zhou Shangyi and Cheng Zhifen, Evaluation on the Spatial Morality and Local Morality of "Nostalgia" *J Human Geography* 2015 **6**

IOP Conf. Series: Earth and Environmental Science **63** (2017) 012037 doi:10.1088/1755-1315/63/1/012037

The Application of Downdraught Cooling in Vernacular Skywell Dwellings in China

H Xuan [*1] **and A M Lv**[1]

[1]Department of Architecture, Shanghai Jiao Tong University, Shanghai, China 200240

xuanh@sjtu.edu.cn

Abstract. Traditional skywell dwellings in the hot climate regions of China represent an important cultural heritage. Achieving indoor comfort meeting occupants' expectations, can contribute to the preservation of this unique traditional architecture. Improvement of ventilation and indoor temperatures through natural, sustainable and low impact solutions is an opportunity in achieving building thermal comfort in these traditional dwellings. The existence of skywells provides a good opportunity for the incorporation of downdraught cooling with minor interventions, and thus by avoiding extensive ductwork, saving energy and improving indoor temperatures, it can contribute to the preservation of traditional dwellings. Applicability of downdraught cooling, the history of traditional ventilation solutions, layout and space features of skywell dwelling are discussed and the way of incorporating downdraught cooling as an alternative to air-conditioning into these buildings is investigated.

1. Introduction

Today's worldwide conventional cooling systems are based on air-conditioning, which is recognized throughout the world as a major source of Building Energy Consumption. The extensive use of conventional air conditioning is a source of energy supply problems especially significant during summer peak loads in the USA, India and China and in other economies as well. Due to electricity disruption problems, furthermore high investment loads in expensive electricity back-up generating equipment, alternative refrigerating solutions as well as more radical alternatives to air conditioning are in the focus of research. Various systems, such as Evaporative coolers, techniques exploiting night sky radiation, for example roof ponds and Downdraught cooling (DC) showed encouraging results in hot climate conditions to become viable sustainable alternatives to conventional air conditioning. Downdraught cooling (DC) systems differ from conventional cooling systems in that the primary mechanism for distribution of cool air is either by buoyancy or wind assisted natural ventilation. In the past decade downdraught cooling has been proven as a feasible alternative and a number of pioneering buildings implementing this novel technique have emerged around the world such as the Torrent Research Centre in India, the Stanford Ecology Centre in the USA and the Habitat Research & Development Centre in Namibia to mention a few. [1] The focus of most of the research was to incorporate this technology in either new built buildings or to implement it during refurbishment, mostly in offices or other public building. However due to the fact that extensive ductworks characteristic for conventional air conditioning are extent, downdraught cooling can be a good alternative in vernacular architecture and in historic buildings, where the preservation of the original interior is important.

2. Historical background

Before modern mechanical means for obtaining thermal comfort appeared, people in the hot and dry or humid climate areas, such as the Mediterranean and the Middle East, were needed to develop ways to cool their houses using natural sources of energy. [2] Traditional ways use different techniques (ground cooling, wind towers, cooling based on evaporation of water, fountains and courtyard house) [3], [4]. The tradition of cooling in the areas mentioned, was sophisticated and highly developed as early as the 15th Century BC. A cardinal component of achieving thermal comfort in the past was ensuring adequate natural (passive) ventilation. For example, Middle Eastern and Mediterranean country houses are well-known for their cooling properties. Constant air movement is achieved through temperature difference between a small, well shaded and cool courtyard and a large and hot one. The enhancement of air movement through ventilation towers has its roots also in the past. In Iran and the countries of the Gulf, a specific type of ventilation tower was developed, which functioned as wind catcher and in conditions with little or no wind the shaft acted as an air stack after heated up by the sun. Natural ventilation was in many cases combined with the cooling effect of evaporation. One of the widest spread methods in traditional architecture was the use of fountains. They can be found in arab houses as well as in southern Europe. Other examples include the mashrabiya, a cantilevered space with an opening covered by lattice, where small water jars were placed [2] or the mazaria cooling jars used in rural areas of Upper Egypt [5]. As Hassan Fathy [2] emphasizes that traditional solutions in architecture should be researched and understood in more detail, and they should be adapted to be suitable for modern requirements.

China is a country with many ethnic culture, it has a long history and a vast territory. During the thousands of years of its history an extensive amount of different types of residential building have evolved. Over the recent decades, due to extensive urbanization and its promise to accomplish a better living environment, modern architecture became more widespread and is replacing traditional dwellings. Traditional vernacular dwellings are part of the cultural heritage and the ecological thinking and technical solutions used are a significant knowledge source for the future development of sustainable, climate and terrain adaptive new buildings.

The adaptability to modern life and comfort of a traditional vernacular dwelling must be improved. People due to the technological development become more and more accustomed to very stable and comfortable indoor conditions. Ventilation is an important measure of thermal comfort. However traditional ventilation techniques alone are mostly not able to achieve the expected level of comfort, as also shown in the study of Duan, Ford and Lau [6], thus improvements should also focus on enhancing indoor ventilation conditions, which is an area what needs to be researched in more detail. One typical traditional design element for ventilation and lighting purposes in China is the skywell, which seems suitable to an easy incorporation of downdraught cooling without major interventions, which is the focus of this study. By understanding these techniques and combining them with new sustainable solutions the results may satisfy occupants requirements, while preserving the cultural heritage and saving energy.

3. Downdraught cooling in China

Downdraught cooling is an energy and cost saving alternative to conventional air conditioning. It relies on the effect of gravity on a body of cold air to create a downdraught, and thus to move the air from the source of cooling within the building. [1] Downdraught Cooling types are differentiated according to the source of cool air, which can be as 'passive' or 'active'. When cooling of the air is achieved through the evaporation of water, it is labelled as Passive Downdraught Evaporative Cooling or 'PDEC'. In warm and humid conditions, Active Downdraught Cooling (ADC) can be the alternative to conventional air conditioning. Cooling occurs by using chilled water cooling coils or panels, meaning the strategy relies on mechanical cooling. In areas with mixed climate conditions during the summer months hybrid downdraught cooling can provide the solution. A hybrid downdraught cooling system (HDC) combines both 'passive' and 'active' downdraught cooling techniques. This cooling strategy uses evaporative cooling in hot and dry climate conditions but the chilled water cooling coils take over in warm and humid conditions. [1] The climate zone classification and climate applicability maps of China (Figure 1) provide a primary assessment about the opportunity in China of using downdraught cooling as an alternative or a complement to conventional air conditioning systems.

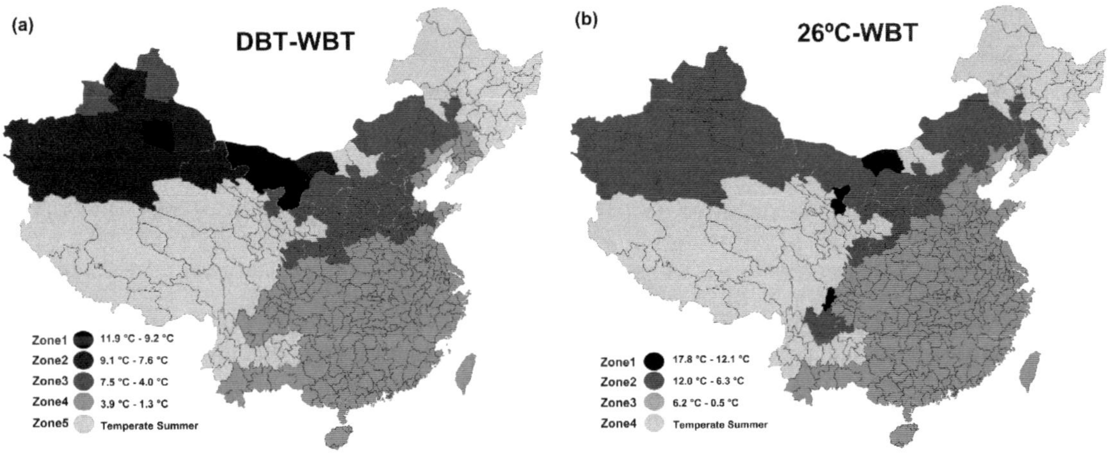

Figure 1. Maps of integrating cooling degree-hours. (a) Difference Outdoor Dry Bulb Temperature with Outdoor Wet Bulb Temperature integrates cooling degree-hours for June –August. and (b) Difference of Indoor Dry Bulb Temperature with Outdoor Wet Bulb Temperature integrating cooling degree-hours for June –August (Source: [7])

Based on the maps seven climate zone for downdraught cooling were defined with suggested cooling strategies. The three climate zones (dry, moderate humid and high humid) identified by the WBT depression were divided into two sub-zones with the difference of cooling degree hours (hot and warm).

In the hot climate zones PDEC, HDC and ADC integrated with other passive cooling strategies were proposed to address the cooling load. While in the warm climate zones, because the cooling load is not as high, the combination of downdraught cooling and natural ventilation was proposed. The only places where the use of downdraught cooling is not recommended are the temperate summer zones on the maps. Here, natural ventilation and high thermal mass can be relied on to meet the cooling load. The maps clearly show that downdraught cooling can be applied in a significant part of China and thus can represent a significant energy-saving potential. [7]

4. Vernacular architecture and ventilation strategies in China

China's different regions have different natural conditions and cultural background, which is reflected by diverse vernacular dwellings all over the country. In northwest regions of China, Gansu, Xinjiang and the Ningxia provinces, the traditional building design responds to the hot and dry climate. In Xinjiang province, the traditional houses have a strong response to climate and are using various techniques to encourage cooling. solar shading, thick adobe wall, small windows, courtyard for ventilation cooling and evaporative cooling. The cooler air from the courtyard circulates across the surrounding rooms, where the occupants appreciate both the lower temperature and the air movement and exhaust from the light well called 'A Yi Wang', which functioned as summer living and communication space, providing day lighting and ventilation. (Figure 2.) Buildings in the areas of Guanxi, Guizhou, and Yunnan with hot and humid climates, are typically open skeleton structures with ventilated, high-ceilinged rooms. Building materials are lightweight materials such as bamboo, fibres or leaves to ensure ventilation and make the best possible use of wind. Floors and shaded terraces are raised above the terrain to protect against flooding, moisture and small animals and to provide ventilation through the floor. Houses may also be placed on water to make maximum use of the cooling caused by evaporation.

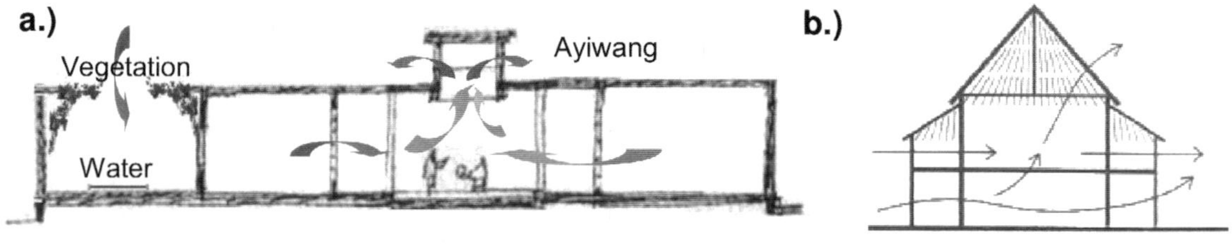

Figure 2. (a) Ventilation strategy of Ayiwang, (b) Ventilation strategy and of an open skeleton structure [8]

These two examples show the differences of adaptation to climate, but there are also some common similarities in building layouts deeply rooted in culture and design philosophy. Courtyard houses can be found all over the country and are a common feature of Chinese vernacular architecture. It has a 2000-year old history and the philosophy of this layout can be traced back to the Tao principles and the Ming t'ang, the calendar house [9]. Plans are usually symmetrical and rectangular. The core of the layout is the courtyard, with rooms around it on three or four sides, but climate conditions and differences the way of life in different regions have led to a large variety of plans and elevations. From the northeast to the southwest, the courtyard area gradually becomes smaller and are known as sky well. Sky well houses have relatively closed courtyard formed by closely spaced buildings two to three storeys high. The skywell has a rather small opening to the sky through the roof space. Types of skywell dwellings, varying in layout, can be seen in Figure 3. These enclosures help in temperature regulation and in assisting natural ventilation. Northern courtyards face the south direction and are more open, this allows the exposure of the building to the sun and keeps the cold northern winds out. Southern sky wells are smaller and their functions include the collection of rain water from the roof tops, limiting the amount of sunlight and also serve as vents for rising hot air, which draws cool air from the lower stories of the house and facilitates the exchange of cool air with the outside. The effect of skywells on ventilation can be enhanced further, or with some additional design elements they offer the possibility to introduce some type of downdraught cooling.

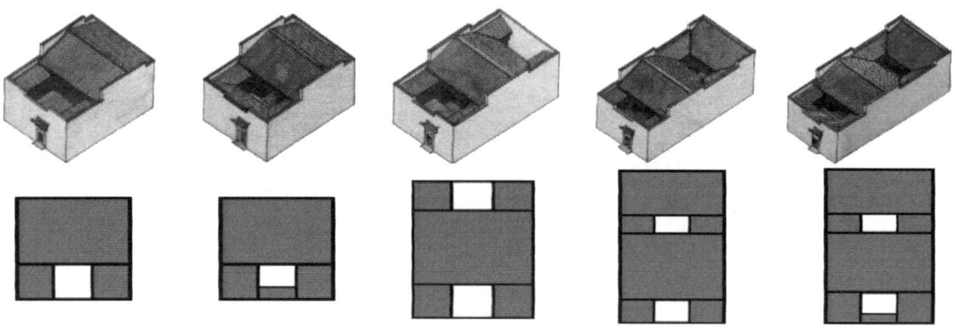

Figure 3. Skywell dwelling diagrams and plans (Source of illustration: [8])

5. Downdraught cooling integration

Studies of building typology and morphology undertaken to assess the applicability of downdraught cooling identified three levels of intervention required on existing buildings to be able to apply downdraught cooling, Minor intervention, Intermediate intervention and Major intervention [1]. The crucial factor in the feasibility of integrating downdraught cooling to existing buildings is the existence or the possibility to add a light well. (Figure 4)

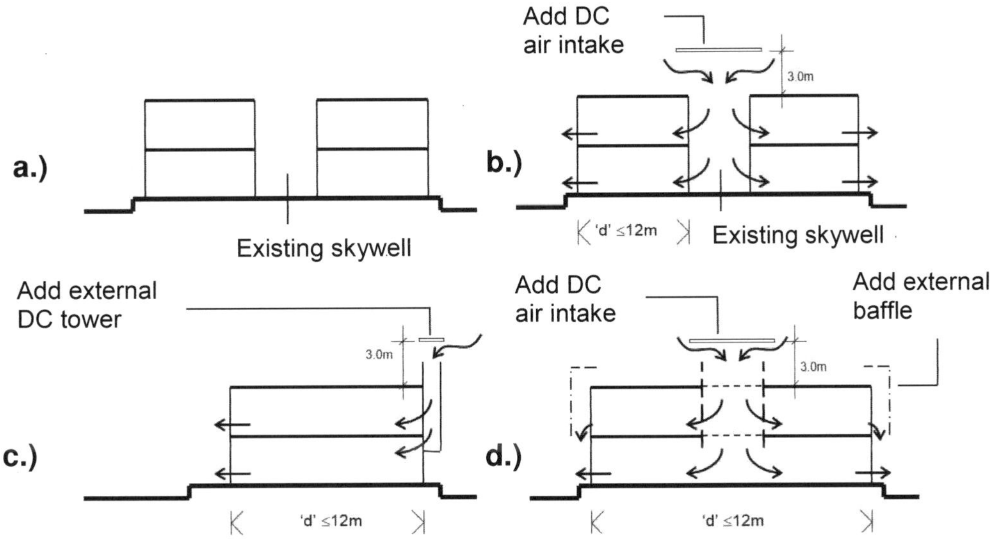

Figure 4. Downdraught cooling intervention options: (a) Existing building. (b) Minor intervention, (c) Intermediate intervention, (d) Major intervention (Source: [1])

According to the definition in the research Minor intervention will be applied where light wells or shafts could be utilized as DC towers, (Figure 4.b) which in this case is always present in the skywell houses. However existing skywells have certain requirements to fulfil in order to allow the installation of downdraught cooling with only minor intervention. Dimensions should not exceed 3-4 meters and the distance of the light well to an external wall should not exceed 12 meters. Rooms should be connected to the skywell through an opening (windows or doors) and opposite side of the skywell there should be openings for exhaust, for example manually operated windows. [1] Most of the skywell houses will fulfil these requirements, but individual assessment is necessary to establish accurate intervention strategies. Figure 5 shows the proposed interventions and airflow strategy for a typical skywell house.

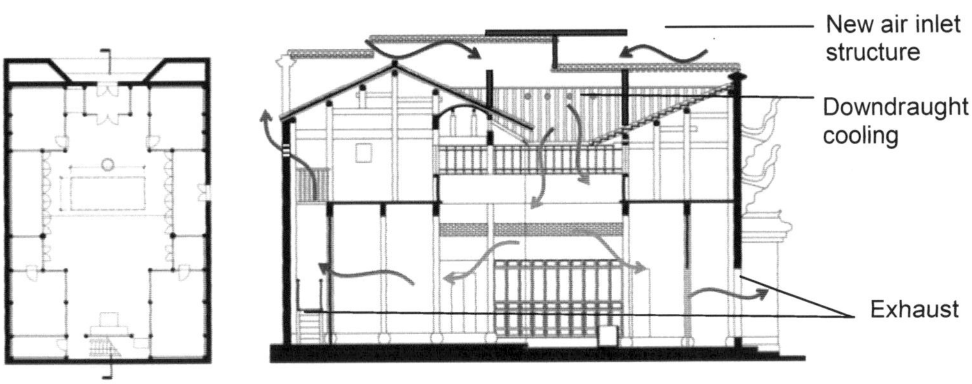

Figure 5. Downdraught cooling strategy in a vernacular skywell house

An air inlet structure can be designed using steel and glass to minimize visual impact and to clearly separate old building from modern extensions, following recent trends preferred in historical building conservation. The air inlet should be designed considering the shape and angle of the existing roof structures, and if possible using it to enhance driving air towards the cooling source. The design could easily incorporate features for collecting rainwater and thus not losing this function of the skywell. Some type of movable shading can also be added, to allow the control of solar heat gain. Front entrances, windows and staircases can be used as exhausts, but in very closed structures opening air vents on main walls should be considered. The source of cooling should be chosen according to climate conditions, based on the applicability maps and climate classification, but in the case of installing water nozzles, height of the skywell should be not less than 3 meters to allow evaporation.

6. Conclusion

Skywell dwellings represent an important heritage of Chinese architecture and thus should be preserved, however improving them is necessary to provide thermal comfort for the occupants. Downdraught cooling is a feasible alternative to conventional air conditioning in a significant area of China. General structure and plan features of skywell dwellings indicate that by minor intervention the application of Downdraught cooling is a feasible strategy to address the need for cooling and ventilation. Energy savings and conservation of original indoor design are among the benefits of such cooling strategies. However further research should be done, and the theory should be tested by empirical studies.

7. Acknowledgenemt

This study was carried out as part of a research funded by the National Natural Science Foundation of China (Grant No. 51308332).

8. References

[1] Ford B, Schiano-Phan R. and Francis E eds. 2010 *The architecture & engineering of downdraught cooling: a design sourcebook.* (UK: PHDC Press)
[2] Fathy H 1986 *Natural Energy and Vernacular Architecture* (University of Chicago Press)
[3] Bahadoori M N 1978 *Passive Cooling Systems in Iranian Architecture* (Scientific American) 238.
[4] Konya A 1980 *Design Primer for Hot Climates by Allan Konya* (Architectural Press, London)
[5] Cain A, Afshar F, Norton J and Daraie M R 1976 Traditional Cooling Systems in the Third World. *The Ecologist*, 6 (2), pp.60-65.
[6] Duan Z, Lau B and Ford B 2012 The Thermal Comfort of Vernacular Skywell Dwellings in South-eastern China, *PLEA2012 – 28th Conference*. Opportunities, Limits & Needs Towards an environmentally responsible architecture Lima, Peru 7-9 November, 2012.
[7] Xuan H and Ford B 2012 Climatic applicability of downdraught cooling in China, *Architecture Science Review,* UK, DOI:10.1080/00038628.2012.717687
[8] Wang Q J and Tan Y P 2005 *Pictorial illustration of Chinese Traditional Architecture*, (Waterpub press, China) (in Chinese).
[9] Bouillot J 2008 Climatic design of vernacular housing in different provinces of China, *Journal of Environmental Management* 87, p.287-299

IOP Conf. Series: Earth and Environmental Science **63** (2017) 012038 doi:10.1088/1755-1315/63/1/012038

Theoretical Study of Mercury Species Adsorption on CaO(100) Surface

Chong Zhang*, **Yu Liu, Li Zhao, Min-Shu Cui and Qiang Lu**

National Engineering Laboratory for Biomass Power Generation Equipment, North China Electric Power University, Beijing 102206, PR China

jl-zc@163.com

Abstract. As an important air pollutant, mercury produced by power generation can be captured by CaO incoal-firedfly ash. In order to understand the interaction mechanism of CaO with mercury in HCl atmosphere, the density functional theory method is utilizedto investigate the adsorption of mercury species on the fly ash components. The CaO(100) surface is built to study the stable adsorption configurations of Hg^0, HCl, HgCl and $HgCl_2$ on all possible adsorption sites. The results show that CaO(100) surface cannot absorb Hg^0 directly, but the intermediate HgCl can be adsorbed on the surface stably, and then continue to combine with a Cl radical to produce $HgCl_2$, which is easy to be removed.

1. Introduction

Mercury in coal-fired flue gas from power generation can cause global air pollution [1]. Previous studies showed that fly ash from the combustion of fossil fuel had a certain adsorption and oxidation ability to mercury [2-4]. Typically, fly ash is consisted of unburned carbon, CaO, MgO, Fe_2O_3, CuO, MnO_2, Al_2O_3 and SiO_2 and their metal salts [5]. The mechanism of heterogeneous interaction between fly ash and mercury is very complex. The difficulty of mercury removal in coal-fired flue gas is the elimination of elemental mercury which is insoluble in water and difficult to be captured by existing pollution control devices [6]. Previous studies have found that the unburned carbon in fly ash and activated inorganic components are the keys of adsorption and oxidation of Hg^0[7]. Some researchers regarded fly ash as a whole to study the mechanism of the oxidation of Hg^0 and found that the oxidation of Hg^0 by fly ash followed the Mars-Maessen mechanism, and the lattice oxygen in the fly ash played the role of the oxidant [8-10]. But up to now, there is no identical conclusion on the adsorption and oxidation mechanism of Hg^0 on the chemical composition of fly ash.

In order to investigate the detailed mechanism of Hg^0 adsorption on CaO(100) surface, first-principles calculations based on the density functional theory and the periodic slab models are used to study the adsorption of mercury species as well as the adsorption and dissociation of HCl.

2. Computational details

2.1. Model

The CaO(100) surface is selected as it is the most stable low-index facet which can well represent the property of CaO. A four-layer CaO(100) slab is cleaved from the perfect crystal with the topmost two layers fully relaxed and the bottom two layers fixed to model the CaO(100) surface. A 2×2 super cell of the unit cell is performed containing 8 Ca cations and 8 O anions for each surface layer. 12 Å vacuum layer is built above the surface to eliminate the interference of the imaging CaO(100) surface, as is shown in Figure 1.

2.2. Computational methods

All DFT calculations are carried out using DMol^3 package in Materials Studio software in this study. The generalized gradient approximation (GGA) with Perdew–Burke–Ernzerh of (PBE) functional is selected for calculating exchange and correlation functional. The molecular orbitals are expanded by double numerical basis set with polarization functions (DNP). All electrons are treated with DFT Semi-core Pseudopots (DSPP). 5.5 Å atomic radius cutoff is selected with a smearing value of 0.005 Ha to accelerate convergence. 3×3×1 Monkhorst–Pack k-point sampling is selected for Brillouin zone integration during geometric optimization calculation.

For the adsorption of mercury species on CaO(100) surface, the adsorption energy E_{ad} is defined as:

$$E_{ad} = E_{sys} - E_{ads} - E_{sur} \quad (1)$$

WhereE_{sys} is the system total electronic energy after adsorption; E_{ads} is the electronic energy of adsorbate; E_{sur} is the electronic energy of the surface.

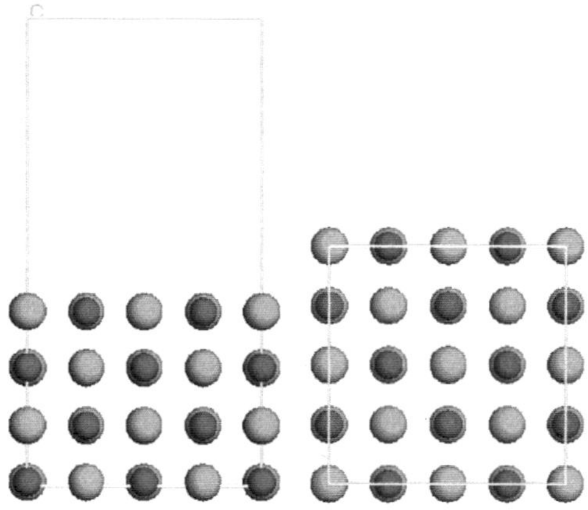

Figure 1. CaO(100) surface. Green and red balls represent Ca and O atoms, respectively.

3. Results and discussion

3.1. Hg^0 adsorption on CaO(100) surface

Adsorption of Hg^0 is the premise of its oxidation on CaO(100) surface. There are four possible adsorption sites for Hg^0 on the surface, Ca site (1A), O site (1B), bridge site (1C) and hollow site (1D), respectively. The optimized configurations ofHg^0 adsorption on each site are shown in Figure 2. The adsorption energies and geometric parameters are given in Table 1.The adsorption energy results show that the stability of Hg^0 adsorption is in the order of 1B < 1C < 1D < 1A. The adsorption energy of Hg^0 adsorption on O site is relatively lower, and Hg^0 is closer to the surface with a distance of 3.017Å, indicating that Hg^0 is more likely to be bound to the surface O atoms. However, the adsorption energy of each configuration is positive, that is, Hg^0adsorption on CaO(100) surface is physisorption, which shows that Hg^0 cannot be adsorbed directly by CaO(100) surface.

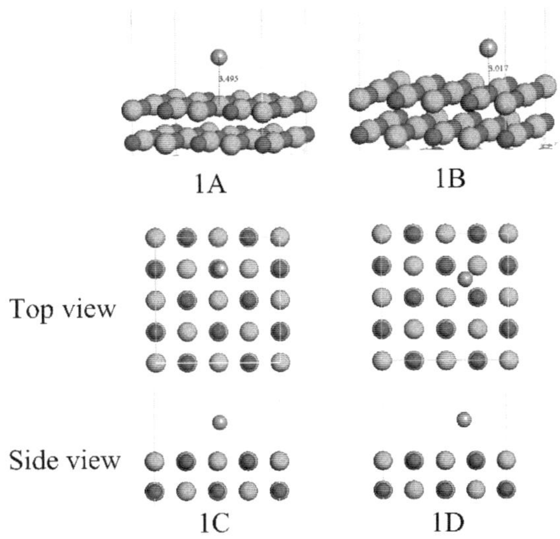

Figure 2. Optimized configurations ofHg^0 adsorption onCaO(100) surface.(The grey balls represent Hg atoms. The same below.)

Table 1. The adsorption energies and geometric parameters of Hg^0 adsorption on CaO(100) surface.

No.	Hg-O (Å)	Hg-Ca (Å)	E_{ads} (kJ/mol)
1A	-	3.495	23.192
1B	3.017	-	20.666
1C	3.045	3.794	20.776
1D	3.549	3.668	22.280

3.2. HCl adsorption on CaO(100) surface

HCl in coal-fired flu gas can take part in the Hg^0 catalytic oxidation reaction, while whether HCl can affect the Hg^0 oxidation on the fly ash composition is still not clear. The optimized configurations ofHCl adsorption on each site are shown in Figure 3. The adsorption energies and geometric parameters for HCl adsorption on the surface are given in Table 2.The adsorption energy results showthat the stability of HCl adsorption is in the order of 2B < 2C < 2A. The adsorption energy of 2A and 2C is positive, and the HClmolecule doesn't interact with the surface, indicating that these two configurations are physisorption only. The most stable structure for HCladsorption on the surface is 2B with an adsorption energy of -46.50 kJ/mol. In 2B, H atom of HCl interacts with surface O and the bond length of H-Cl elongates from 1.291 Å to 1.980Å. The length of new-formed H-O bond is 0.999Å, and the distance between Cl and Ca is 2.974 Å. After the dissociative adsorption of HCl, the Cl radical doesn't adsorb on the surface, instead, it releases to the gas phase, which can be concluded that Cl radical may combine with Hg in the gas phase to form HgCl, an intermediate of oxidation reaction.

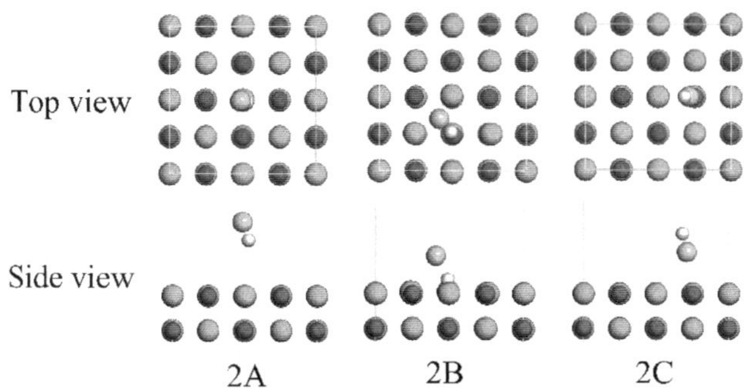

Figure 3. Optimized configurations of HCl adsorption onCaO(100) surface.(The white and grey-green balls represent H and Cl atoms respectively. The same below.)

Table 2. The adsorption energies and geometric parameters of HCl adsorption on CaO(100) surface.

No.	H-O (Å)	H-Ca (Å)	Cl-O (Å)	Cl-Ca (Å)	H-Cl (Å)	E_{ads} (kJ/mol)
2A	-	3.776	-	-	1.292	26.157
2B	0.999	2.609	-	2.974	1.980	-181.126
2C	-	4.452	2.868	-	1.306	12.807

3.3. HgCl adsorption on CaO(100) surface

HgCl is a possible intermediate of oxidation reaction of Hg^0. So, it is necessary to calculate the adsorption of HgCl on CaO(100) surface. The optimized configurations of HgCl adsorption on each site are shown in Figure 4. The adsorption energies and geometric parameters for HgCl adsorption on the surface are given in Table 3.The adsorption energy results show that the stability of HgCl adsorption is in the order of 3B < 3C < 3A. The adsorption energies of these three configurations are negative, indicating that HgClis chemically adsorbed on CaO(100) surface. The most stable structure for HgCl adsorption on the surface is 3B with an adsorption energy of -108.905 kJ/mol. In 3B, HgCl adsorbs on the surface horizontally, and Hg atom is close to the surface O atom, Cl atom is close to the surface Ca atom, Hg-Cl bond elongates from 2.495 Å to 2.948 Å, and the distance between Hg and O is 3.002 Å, the distance between Cl and Ca is 2.768 Å. The chemisorption of HgCl on CaO(100) surface provides the possibility for the combination of Cl and HgCl to form $HgCl_2$. However, in 3C, the Hg-Cl bond breaks, the distance of Hg and Cl elongates from 2.495 Å to 3.315 Å, Cl atom approaches to the surface, which indicates there is possibility for HgCl dissociation and Hg^0re-release to the gas phase.

IOP Conf. Series: Earth and Environmental Science **63** (2017) 012038 doi:10.1088/1755-1315/63/1/012038

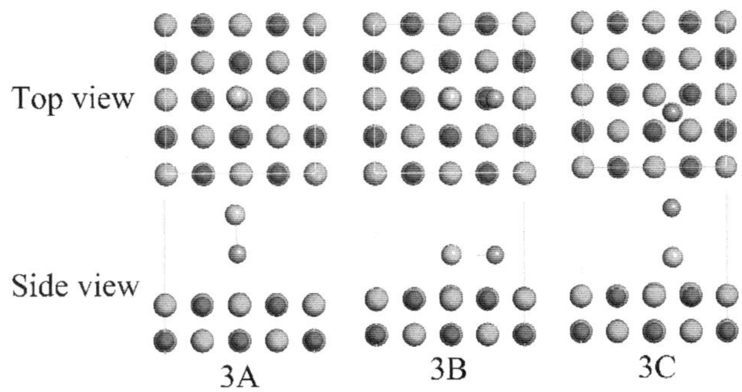

Figure 4. Optimized configurations of HgCl adsorption on CaO(100) surface.

Table 3. The adsorption energies and geometric parameters of HgCl adsorption on CaO(100) surface.

No.	Hg-O (Å)	Hg-Ca (Å)	Cl-O (Å)	Cl-Ca (Å)	Hg-Cl (Å)	E_{ads} (kJ/mol)
3A	-	3.316	-	-	2.621	-50.833
3B	3.002	-	3.92	2.768	2.948	-108.905
3C	-	-	3.219	2.950	3.315	-99.878

3.4. $HgCl_2$ adsorption on CaO(100) surface

The oxidation of Hg^0 is to transform Hg^0 to Hg^{2+} which is easy to remove. In the Hg^0 oxidation process with HCl, $HgCl_2$ is the final product of the reaction. The optimized configurations of $HgCl_2$ adsorption on each site are shown in Figure 5. The adsorption energies and geometric parameters for $HgCl_2$adsorption on CaO(100) surface are givenin Table 4.The adsorption energy results show that the stability of $HgCl_2$ adsorption is in the order of 4A < 4C < 4B. The adsorption energy of 4B is positive, showing it is physisorption. The $HgCl_2$ adsorption in 4C is very weak chemical adsorption with an adsorption energy of -5.729 kJ/mol. In 4A, the angle of Cl-Hg-Cl changes greatly from 180° to 147.476°, Hg atom interacts with the surface O atom forming a 2.288 Å Hg-O bond, while the Cl-Hg bonds have no significant elongation. The adsorption energy of 4A is -81.812kJ/mol, much greater than that of 4C, but it is still a weak chemical adsorption. These results suggest that $HgCl_2$ can desorb from CaO(100) surface and then be captured and removed by the downstream WFGD system. But at the same time, it is still possible for $HgCl_2$ to stay on the surface and be captured by the dust removal devices together with fly ash.

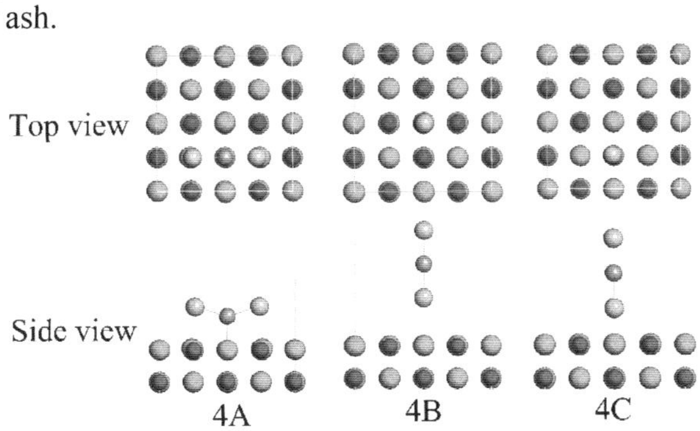

Figure 5. Optimized configurations of$HgCl_2$ adsorption onCaO(100) surface.

Table 4. The adsorption energies and geometric parameters of $HgCl_2$ adsorption on CaO(100) surface.

No.	Hg-O (Å)	Hg-Ca (Å)	Cl-O (Å)	Cl-Ca (Å)	Hg-Cl (Å)	Cl-Hg-Cl (°)	E_{ads} (kJ/mol)
4A	2.288	-	-	3.049	2.450	147.476	-81.812
4B	-	-	-	3.415	2.423	179.703	7.821
4C	-	-	2.559	-	2.432	179.903	-5.729

4. Conclusions

The detailed adsorption mechanism of Hg on CaO(100) surface is investigated by a systematic density functional theory study and the periodic slab model. Hg^0 is physically adsorbed on CaO(100) surface, while HCl is dissociatively absorbed, and HgCl and $HgCl_2$ is chemically adsorbed. After the dissociation adsorption of HCl, the Cl radical releases to the gas phase, which indicates that Cl radical may combine with Hg in the gas phase to form HgCl, and then HgCl may continue to combine with a Cl free radical to form $HgCl_2$. $HgCl_2$ can either desorb from CaO(100) surface or stay on the surface and can be removed by the downstream pollutant control devices under both conditions, hence to realize the removal of Hg^0.

5. Acknowledgments

The authors thank the Beijing Natural Science Foundation (3172030), Beijing Nova Program, and Fundamental Research Funds for the Central Universities (2016YQ05, 2016MS55) for financial support.

6. References

[1] Lu W F, Tian Y, Zhan J M and Xie M T2012 *Journal of Environment and Health***29** 761
[2] Chen L, Duan Y, Zhuo Y, Yang L, Zhang L, Yang X, Jiang Y and Xu X2007*Fuel***86** 603
[3] Duan Y F, Jiang Y M and Yang L G 2008 *Proceedings of the CSEE* **32**1
[4] Meng S,DuanY, Huang Z, Wang Y and Yang L2009*Proceedings of the CSEE***20**66
[5] Huang H and Luo J2010 *Proceedings of the CSEE***S1**70
[6] Ren J, Zhou J, Luo Z, Cen K 2002 *Acta Scientiae Circumstantiae***22** 289
[7] Yang J, Zhao Y and Zhang J 2014*Journal of Chinese Society of Power Engineering***05**337
[8] Jiang Y, Duan Y and Yang X 2007 *Journal of Southeast University***03** 436
[9] Meng S, Duan Y and Huang 2011 *Boiler Technology***05** 70
[10] Zhao Y C, Zhang J Y and Liu J 2009 *Chinese Science Bulletin***21**3395

Analysis of Influence Factors of $PM_{2.5}$ in Chengdu Based on VAR Model

Luo Mingzhi[1]

[1]School of Tourism,Sichuan University, No.24 South Section 1, Yihuan Road, Chengdu 610065, China

luomingzhi_123@163.com

Abstract. Air pollution and smog are the serious harms to public health and has attracted public attention. Based on the vector auto-regressive (VAR) model, we analysed the influence factors of $PM_{2.5}$ in Chengdu, investigated the effect of other kinds of air pollutants and meteorological factors onthe $PM_{2.5}$ by using the methods of generalized impulse response function, variance decomposition analysis, Granger causality test and therelated daily data from December 1, 2013 to November 14, 2016 in Chengdu city to the empirical study. The resultsshow that the influence factors of $PM_{2.5}$ were stable;the increase of nitrogen dioxide, ozone,precipitation and temperature difference led to the increase of $PM_{2.5}$ concentration while the increase ofthe wind speed, PM_{10}, sulphur dioxide and carbon monoxide resulted in the decrease of $PM_{2.5}$ concentration.Climate conditions,nitrogen dioxide and ozone are Granger causes for $PM_{2.5}$.It is suggestedthat the key for the control of $PM_{2.5}$ must be based on the cause and formation rules of $PM_{2.5}$.A further study on nitrogen dioxide and ozone may play an important role in finding out the real source and formation reasons of $PM_{2.5}$.

1. Introduction

With the rapid development of economy, air pollution problems are more and more prominent. Together with smog, atmospheric pollutant $PM_{2.5}$ also attracts public attention. $PM_{2.5}$ in the atmosphere is a kind of particulate that can be breathed into the lung. It is very harmful to human body, which can lead to diseases such as bronchitis, asthma, as well as an important factor that causes smog and dropped visibility.

Since $PM_{2.5}$ has become the focus of attention of the society, the domestic and foreign scholars made deep investigation and extensive researches on it.Chow J C et al. [1] and Houthuijs D et al.[2] did researches in $PM_{2.5}$ and PM_{10}, in which they found that $PM_{2.5}$ accounts for large proportion in PM_{10}.Vega E et al. [3] used the data of Mexico from year 2000 to 2002 years to analyse the time and spatial variation characteristics of $PM_{2.5}$ concentration, which pointed out that in industrial area and daytime, $PM_{2.5}$ is of higher concentrations.Tran H N Qand Molders N[4] did a research on the relationship between $PM_{2.5}$ and meteorological conditions with 10-years data of $PM_{2.5}$ in Alaska Fairbanks, and found out that different weather conditions have different effects on $PM_{2.5}$ concentration.HuangFangfanget al. [5]did research into the and the relationship between $PM_{2.5}$ andmeteorological factorsduring 2013-2014 in Beijing,which pointed out $PM_{2.5}$ concentration was significantly higher in winter and significantly higher in the night than day. Yang Leiet al.[6]analysedtheconcentration characterization of $PM_{2.5}$ and its impact factors in national road ofZhengzhou, and found out that traffic flow contributed most to the $PM_{2.5}$ mass concentration;relative humidity and windspeed contributed to the reduction of the $PM_{2.5}$ mass concentration.LiuJinpeiet al.[7]used VAR model to analyse the dynamic relationship between $PM_{2.5}$ and its influence factors in Xi'an city, which pointed out theincrease of carbon monoxide, sulphur dioxide, ozone and the weather led to the increase of $PM_{2.5}$ concentration.

By literature analysis, we can see the existing researches mainly analyse the harm, composition of $PM_{2.5}$ and its relationship with PM_{10}, time and spatial characteristics and its relevance between meteorological factors from the static perspective. But there are very few researches on the mutual effects on $PM_{2.5}$ and other atmospheric pollutants and meteorological factors. Therefore, based on the dynamic system constructed by $PM_{2.5}$, other air pollutants and meteorological factors, this paper studies the influence factors of $PM_{2.5}$ of Chengdu city with Vector Auto-regression model (VAR model). It focuses on the study of the influence of other air pollutants and meteorological factors on $PM_{2.5}$ in order to provide corresponding suggestions on comprehensive control of air pollution.

2. Empirical Methods

2.1. VAR model

VAR model is a system with some equations where all variables are treated as dependent variables,which is usually used for predicting interconnected time sequence and analysing the impact of stochastic disturbance on variation system. There is one equation for each variable as dependent variable. Each equation has values for all variables included as dependent variables, including the dependent variable itself. Since there is no explanatory variable included in the model simultaneously, the model is in a reduced form. So equations have the same form as long as they have the same variables on the right side of the equation.

VAR models are also used for the analysis of causality. In general, the empirical analysis of meteorological data, the relationship of cause and effect is very difficult to determine. If we consider two variables x and y and see that these variables show a high correlation then we can say that they have an apparent tendency to move together, but in the absence of more information, cannot add more significantly in the direction of causality. The correct way to interpret these tests is to consider as graphical analysis that is required to see if the performance of a variable precedes another.

2.2. Data

In order to comprehensively reflect the changes of $PM_{2.5}$, in our paper we have considered eight variables, which are as follows: PM_{10}; Sulphur dioxide (SO_2); Nitrogen dioxide (NO_2); Carbon monoxide (CO); Ozone (O_3); Temperature difference (T); Wind speed (FF) and Precipitation (RR). All data were collected on a daily basis from December 1, 2013 to November 14, 2016 in Chengdu, involving a total of 1080 observations. Through auto regression vectors model (VAR) we will try to discover the link between these variables to conclude which of the model equations is more important. Through various tests such as causality test of Granger we will try to determine which of the dependent variables is independent and which dependent, therefore to determine which of them is the cause or the consequence of other variables.

2.3. Stationarity Test

For the considered variables the stationarity testing was done with the Augmented Dickey - Fuller Test, the results are reported in table 1 and indicate that all the variables are stationary.

Table 1. Augmented Dickey - Fuller Test (ADF).

	Test for unit root in	t-Statistic	Prob.
$PM_{2.5}$	level	-10.1099	0.0000***
PM_{10}	level	-11.2068	0.0000***
SO_2	level	-7.1531	0.0000***
CO	level	-270.6010	0.0001***
NO_2	level	-11.5696	0.0000***
O_3	level	-4.7857	0.0001***
T	level	-17.7520	0.0000***
FF	level	-28.7673	0.0000***
RR	level	-29.1090	0.0000***

Notes: *, **, *** signifies the rejection of the null hypothesis (the presence of unit root orabsence of stationarity) at a significance level of 10%, 5%, 1%.

2.4. Lag Order Selection

The choice on the number of lags was done based on the evaluation criteria of informational content (Likelihood ratio test, Final prediction error, Akaike information criterion, Schwarz information criterion and Hannan-Quinn information criterion). The results are reported in table 2 and provide more support to lag 3 times for VAR model.

Table 2. The judgments of lag time for VAR model.

Lag	LogL	LR	FPE	AIC	SC	HQ
0	-29257.88	NA	4.18e+12	54.60240	54.64418	54.61822
1	-27047.04	4380.449	7.86e+10	50.62880	**51.04667***	50.78708
2	-26882.05	324.1150	6.72e+10	50.47212	51.26607	**50.77285***
3	-26776.52	205.5546	**6.42e+10***	**50.42634***	51.59638	50.86954
4	-26701.88	144.1272	6.50e+10	50.43821	51.98433	51.02386
5	-26650.76	97.85161	6.87e+10	50.49396	52.41615	51.22206
6	-26603.61	89.47360	7.32e+10	50.55710	52.85538	51.42765
7	-26519.84	157.5296	7.29e+10	50.55194	53.22630	51.56495
8	-26448.99	**132.0578***	7.43e+10	50.57087	53.62131	51.72633

Note: *indicates the lag order selected according to the standards.

2.5. VAR model construction

The VAR model can be written as:

$$BY_t = A_0 + \sum_{i=1}^{n} A_i Y_{t-i} + \varepsilon_t \quad (1)$$

Where Y_t is the $n*1$ vector $(Y_{1t}, Y_{2t}, \dots, Y_{nt})'$, $Y_t = \left(PM_{2.5t}, PM_{10t}, SO_{2t}, CO_t, NO_{2t,}O_{3t}, T_t, FF_t, RR_t\right)'$ ε_t is the $n*1$ vector of white-noise contemporaneously uncorrelated shocks $(\varepsilon_{1t}, \varepsilon_{2t}, \dots, \varepsilon_{nt})'$.

2.6. AR Root Test
We use AR characteristic polynomial root test to inspect the stability of VAR model system. The result shows that, all roots are less than 1, that is, locates within the unit circle, as shown in figure 1, which indicates that VAR model system is stable.

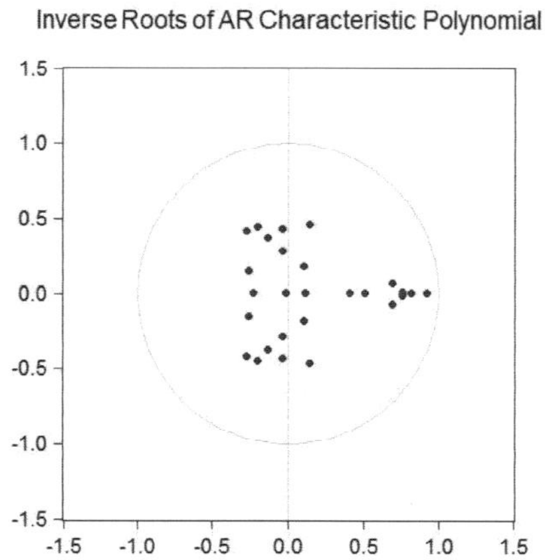

Figure 1. The verification of the VAR model stability.

3. Empirical Results

3.1. Impulse Response
Figure 2 shows the impulse responses produced by a one standard deviation shock to each of the series. The PM_{10}, SO_2 and CO shock cause $PM_{2.5}$ to decrease and continue to be negative through day ten. On the other hand, the NO_2, Precipitation and Temperature difference shock increase the $PM_{2.5}$ and continue to be positive through day ten. $PM_{2.5}$ initially increases after an O_3 shock, but the point estimates turn negative around day 4 and remain negative through day ten. Finally, the Wind speed shock causes $PM_{2.5}$ to decrease by day two and the point estimate continues to be negative only through day four.

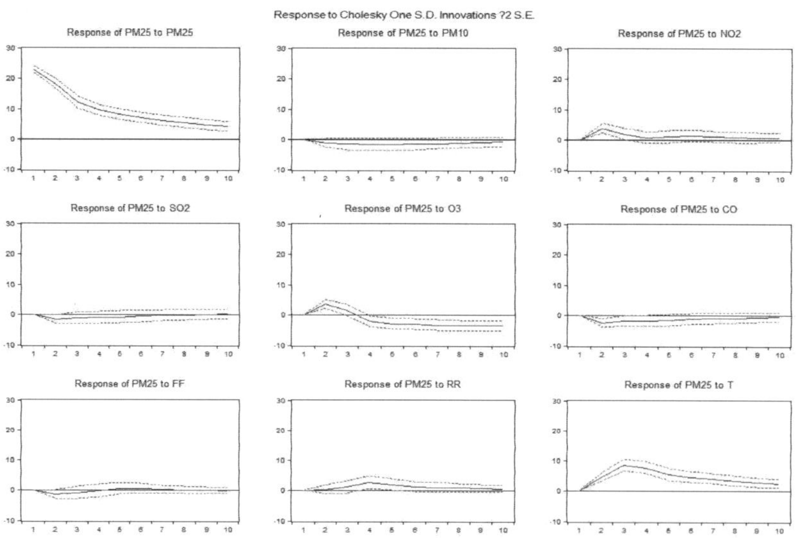

Figure 2. Impulse Response Functions.

3.2. Variance decomposition

According to table 3 at a time horizon of five days, the $PM_{2.5}$ variation is explained in proportion of 80.568% of personal innovations, 12.787% of shocks in the Temperature difference, 2.009% shock of O_3, 1.424% shock of NO_2, 1.166% shocks of CO. Also, the change in $PM_{2.5}$ is not significantly influenced by Wind speed. On a longer time horizon (ten days), the variation in $PM_{2.5}$ is explained in the proportion of 75.587% of personal innovations, 14.269% of shocks in the Temperature difference, 5.101% shock of O_3, 1.457% shock of NO_2, 1.166% shocks of CO.

Table 3. The variance decomposition of $PM_{2.5}$.

Period	S.E.	$PM_{2.5}$	PM_{10}	NO_2	SO_2	O_3	CO	FF	RR	T
1	22.926	100	0	0	0	0	0	0	0	0
2	30.252	93.255	0.152	1.608	0.3	1.394	0.739	0.179	0.008	2.365
3	33.927	86.926	0.319	1.617	0.365	1.297	0.883	0.199	0.119	8.275
4	36.361	82.631	0.499	1.458	0.402	1.507	1.021	0.177	0.627	11.678
5	37.937	80.568	0.641	1.424	0.421	2.009	1.116	0.179	0.856	12.787
6	39.09	79.133	0.752	1.46	0.417	2.532	1.152	0.196	0.9	13.458
7	39.998	77.923	0.827	1.471	0.405	3.165	1.167	0.193	0.933	13.915
8	40.697	76.981	0.871	1.465	0.392	3.829	1.175	0.186	0.957	14.145
9	41.247	76.224	0.893	1.46	0.382	4.482	1.174	0.181	0.964	14.24
10	41.695	75.587	0.905	1.457	0.375	5.101	1.166	0.178	0.962	14.269

3.3. Granger Causality Test

The Granger causality tests reported in table 4. Wind speed, Precipitation Granger causes $PM_{2.5}$ at the 10% level. Temperature difference Granger causes $PM_{2.5}$ at the 1% level. There is two-way granger causality relationship between O_3 and $PM_{2.5}$. And NO_2 Granger Causes $PM_{2.5}$.

Table 4. Granger Causality Test.

Null Hypothesis	Obs	F-Statistic	Prob.
PM_{10} does not Granger Cause $PM_{2.5}$	1077	0.401	0.752
$PM_{2.5}$ does not Granger Cause PM_{10}		1.663	0.173
O_3does not Granger Cause $PM_{2.5}$	1077	23.881	0.000^{***}
$PM_{2.5}$ does not Granger Cause O_3		7.207	0.000^{***}
NO_2 does not Granger Cause $PM_{2.5}$	1077	12.536	0.000^{***}
$PM_{2.5}$ does not Granger Cause NO_2		0.340	0.797
FF does not Granger Cause $PM_{2.5}$	1077	2.276	0.078^{*}
$PM_{2.5}$ does not Granger Cause FF		1.073	0.360
CO does not Granger Cause $PM_{2.5}$	1077	1.128	0.337
$PM_{2.5}$ does not Granger Cause CO		4.198	0.006^{***}
RR does not Granger Cause $PM_{2.5}$	1077	2.280	0.078^{*}
$PM_{2.5}$ does not Granger Cause RR		1.742	0.157
SO_2 does not Granger Cause $PM_{2.5}$	1077	0.056	0.983
$PM_{2.5}$ does not Granger Cause SO_2		10.772	0.000^{***}

T does not Granger Cause $PM_{2.5}$	1077	24.851	0.000***
$PM_{2.5}$ does not Granger Cause T		0.293	0.831

Notes: *, **, *** signifies the rejection of the null hypothesisat a significance level of 10%, 5%, 1%.

4. Discussions

Based on VAR model, this paper researched on the influencing factors of $PM_{2.5}$ in Chengdu through the impulse response function, variance decomposition and Granger causality test, conclusions and recommendations are as follows:

(1) From the perspective of impulse response function, PM_{10},CO and SO_2 have negative influence on $PM_{2.5}$, and O_3 has positive effects on $PM_{2.5}$ only in short-time, NO_2 has positive effects on $PM_{2.5}$ both in short-term and long-term perspective. Therefore, prevention and control of $PM_{2.5}$ should be taken together with that of NO_2. And in terms of climate conditions, only wind has a negative effect on $PM_{2.5}$ in the short term. Since Chengdu is located in the Sichuan basin and with less windy weather, it is difficult to control $PM_{2.5}$ through changing climate conditions.

(2) For the variance decomposition results, $PM_{2.5}$ variance contribution is the biggest. Therefore, the key for the control of $PM_{2.5}$ must be based on the cause and formation rules of $PM_{2.5}$. In addition, temperature difference, O_3 and NO_2 also have obvious influence on $PM_{2.5}$. On one hand, control of O_3 and NO_2 can decrease the harm of $PM_{2.5}$, on the other hand, temperature difference changes also can help to give more accurate warning for $PM_{2.5}$.

(3) Through the results of Granger causality test, climate conditions are Granger causes for $PM_{2.5}$. But in air pollutants, only NO_2 and O_3 are Granger causes for $PM_{2.5}$. Combined with the results of impulse response function and variance decomposition, NO_2 and O_3 have close relationship with $PM_{2.5}$. A further study on air pollutants as O_2 and O_3 may play an important role in finding out the real source and formation reasons of $PM_{2.5}$.

5. Acknowledgments

The study was supported by the Scientific Research Staring Foundation for the Young Teachers of Sichuan University (No. 2015SCU11031).

6. References

[1] Chow J C,Watson J G,Edgerton S A andVega E 2002 *Sci. Total Environ.***287** 177-201

[2] Houthuijs D, Breugelmans O, Hoek G, Vaskovi E, Mihalikova E, Pastuszka J S, Jirik V, Sachelarescu S, Lolova D, Meliefste K, Uzunova E, Marinescu C, Volf J, Leeuw F, Wiel H, Fletcher T, Lebret E and Brunekreef B 2001 *Atmos. Environment***35** 2757-71

[3] Vega E,Reyes E, Ruiz H,Garcia J,Sanchez G, Martinez G V and Gonzalez U 2004 *J. Air Waste Manage. Assoc.* **54** 786-98

[4] Tran H N Q andMolders N 2011 *Atmos. Res.***99**39-49

[5] Huang Fangfang, Li Xia, Wang Chao, Xu Qin, Wang Wei, LuoYanxia, Tao Lixin, Gao Qi, Guo Jin, Chen Sipeng, Cao Kai, Liu Long, Gao Ni, Liu Xiangtong, Yang Kun, Yan Aoshuang and GuoXiuhua 2015 *Plos One***10**1-17.

[6] Lei Yang, Wei Ran, Wang Shiyuan, Wang Yan, Du Liuliu and ShenHenggen 2015 *Nat.Envion.Pollut. Tech.***14**409-14

[7] Liu Jinpei, Wang Guanzhen, Chen Huayou, Xiao Peng and Du Bowen 2016 *J.Arid LandResour.Envion.***30** 78-84

IOP Conf. Series: Earth and Environmental Science **63** (2017) 012040 doi:10.1088/1755-1315/63/1/012040

The Distribution of Rock Salt and Gypsum in Sequence Stratigraphic Framework—A Case in Jialingjiang Formation in Chongqing

Caixa Gao[1, 2, a], Jun Yi[1, b], Junjiang Su[1,c], Yongqin Zhang[1,d], Bo Li[1,e]

[1]Chongqing Vocationan Institiute of Engineering, Chongqing, 402260, China;
[2]State Key Laboratory of Coal Resources and Safety Mining and Department of Resources and Earth Sciences, China University of Mining and Technology, Beijing, 100083, China

[a]gaocaixia_004@163.com, [b]yijun2366@yahoo.com.cn, [c]40107287@qq.com
[d]7541492@qq.com, [e]21289660 @qq.com

Abstract. The Jialingjiang Formation of the Lower Triassic in the eastern Sichuan Basin is subdivided into four members. The first and third members are featured with limestone, while the second and forth members which are clipped with gypsum and halite are characterized by dolomite and limy dolomite. At the same time, the second and forth members are the main geothermal reservoir in Chongqing area. Two sequence boundaries represented by regional unconformities and the lithology-lithofacies transition surface are recognized in the Jialingjiang formation based on outcrop and borehole data analysis. A total of two third-order sequences are subdivided in the Jialingjiang formation, and they are subcorrelated to the intervals of Member 1 to Member 2, Member 3 to Member 4, respectively. Each sequence is further subdivided into lacustrine transgresive system tract (TST) and highstand system tract (HST) according to variation in lithology and lithofacies. The sea level change is the main factor to control the sequence development. With the sea level change, the middle and late periods of HST in sequence are the favorable periods of the development of rock salt and gypsum.

1. Introduction

There was rich natural gas developed with Jialingjiang formation in the Lower Triassic in the east of Sichuan basin. At the same time, the Jialingjiang formation was rich of halite and gypsum and was important reservoir for geotherm in Chongqing in the east of Sichuan basin. Now, the predecessors have made a lot of achievements about sedimentary facies[1,7], gas accumulation mode[2], hydrocarbon sources[3], gypsum-salt rock[4], dolomite reservoir prediction[5], geochemistry characteristic[6], ichnofabric[7], lithofacies palegeography[8-10], sequence stratigraphy[9,10], and so on. On the base of detailed sedimentary facies and accurate stratigraphic classification, the sequence framework has been established including sequence and system boundaries based on outcrop and borehole data analysis for early Triassic in the east of Sichuan basin, so that discussing the distribution of halite and gypsum in the sequence framework.

2. Regional geological background

Sichuan basin is located in the east of Longmen mountains nappe zone, by the northern boundary of Chengba fault zone and Daba mountains obduction zone, the eastern boundary of Qiyaoshan fault and passive boundary fold belt in the south of Yangzi plate (Fig 1). The studied area was located in the east of Huayingshan fault. In the late Triassic, the study area had been developed into inland sea basin and the Jialingjiang formation featured by carbonatite had been sedimented. The Jialingjiang formation was subdivided into four members. The first and the third members were mainly filled with limestone, while the second and the forth members mainly filled with dolomite rock and limy dolomite and being the important reservoir for geotherm in Chongqing.

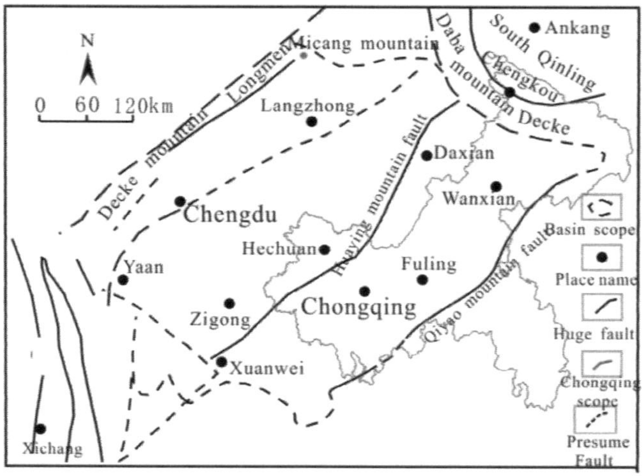

Figure 1. Sketch map showing structural outlines of Sichuan Basin

3. Sequence and system boundaries identification

Based on the Vail's theory from EXXEN Company, the distribution of sequence and system tract with Jialingjiang formation has been studied. The key sequence and system tract boundaries were as follows.

3.1. Sequence boundaries

A. Regional unconformity surface

At the end of the Feixianguan stage in early Triassic in Yangtze area, there was a massive regressive and regional exposed caused by corrosion function. It had been deposited until the stage of early Jialingjiang. The unconformity surface had been present in the study area, even larger area.

B. The top of Jialingjiang formation was the interface between Jialingjiang formation and Leikoupo formation.

3.2.The initial flooding surface and the maximum flooding surface

The Jialingjiang formation was continuous deposited. The sedimentary facies reversal surface was developed in the Jialingjiang formation, usually being the initial flooding surface and the maximum flooding surface.

4. Sequence division and lateral throughgoing

According to regional unconformities, sedimentary facies reversal surface, we have subdivided the Jialingjiang formation into 2 third-order sequences, and they are subcorrelated to the intervals of Member 1 to Member 2, Member 3 to Member 4. But, Jialingjiang formation had been developed differently in different areas influenced by later structures. Two comparison figures of the sedimentary facies and sequence stratigraphy from northeast to southwest were studied in order to understand the characteristics of the sequence stratigraphic framework (Fig 2).

4.1.Sequence I

Sequence I corresponds to the intervals of Member 1 to Member 2. The basal sequence boundary is the bottom of Jialingjiang formation and is a lithology reversal surface. This sequence includes transgressive and highstand system tract.

The transgressive system tract is developed basically in the whole region, and composed by limestone in open platform. The deposition range of TST is wide and deposition thickness is large. There are abundant fossils, such as mussel-shrimp, lamellibranch, proleg, cephalopoda and ammonite. On the whole, the sea water is continuous increasing and the rest of accommodation space is gradually increasing. This is the filling process of basal level rising and basin retrogradation.

The highstand system tract is also developed comparatively, and composed by dolomite rock, evaporite solution breccia, gypse in restricted platform and limestone in open platform. The halite and gypse are developed very well and thickly, such as the points at ZK5 and ZK12. In the middle and late of this system tract, the sea level descented and the accommodation space decrease progressively and slowly. When the rate of accommodation space decreasing was close to the rate of carbonate productivity and accumulation, the aggradation was dominant. The depositional sequences of shallowing-upward and shallow face were built, representing the dolomitization and gypsification of carbonate.

4.2. Sequence II

Sequence II corresponds to the intervals of Member 3 to Member 4. The basal sequence boundary is a lithology reversal surface between Member 2 and Member 3. This sequence includes transgressive and highstand system tract.

The transgressive system tract is developed basically in the whole region too, mainly composed of limestone in open platform and dolomite rock, argillaceous dolomite in restricted platform. The deposition range of TST is wide and deposition thickness is large. There are also abundant fossils, such as mussel-shrimp, lamellibranch, proleg, cephalopoda and ammonite, and so on. On the whole, the sea water is continuous increasing and the rest of accommodation space is gradually increasing. This is the filling process of basal level rising and basin retrogradation.

The highstand system tract was also developed comparatively, but the thickness is nonuniform. The lithologic association of this system tract is gypse, dolomite, evaporite solution breccias, argillaceous dolomite, and limy dolomite, cryptite in restricted platform and limestone in open platform. The halite and gypse are developed very well and thicker, such as the points at ZK5 and ZK12.In the middle and late of this system tract, the sea level descented and the accommodation space decreased progressively and slowly. When the rate of accommodation space decreasing was close to the rate of carbonate productivity and accumulation, the aggradation was dominant. The depositional sequences of shallowing-upward and shallow face were built, representing the dolomitization and gypsification of carbonate.

5. The distribution of halite and gypse in sequence framework

The distribution and system tract type of carbonate sequence are controlled by the change of relative sea level. The global eustasy and tectonic descent together control the change of relative sea level and accommodation space; further influence the type and distribution of system tract. Each sequence includes TST and HST in study area.

In the stage of TST sea level fast ascends, the open platform is mainly developed, followed by restricted platform. The sediment scope is wide and thicker in restricted platform. There are abundant fossil organism, such as mussel-shrimp, bivalve, gastropod, cephalopoda and ammonite. On the whole, the sea water is continuous increasing and the rest of accommodation space is gradually increasing. This is the filling process of basal level rising and basin retrogradation.

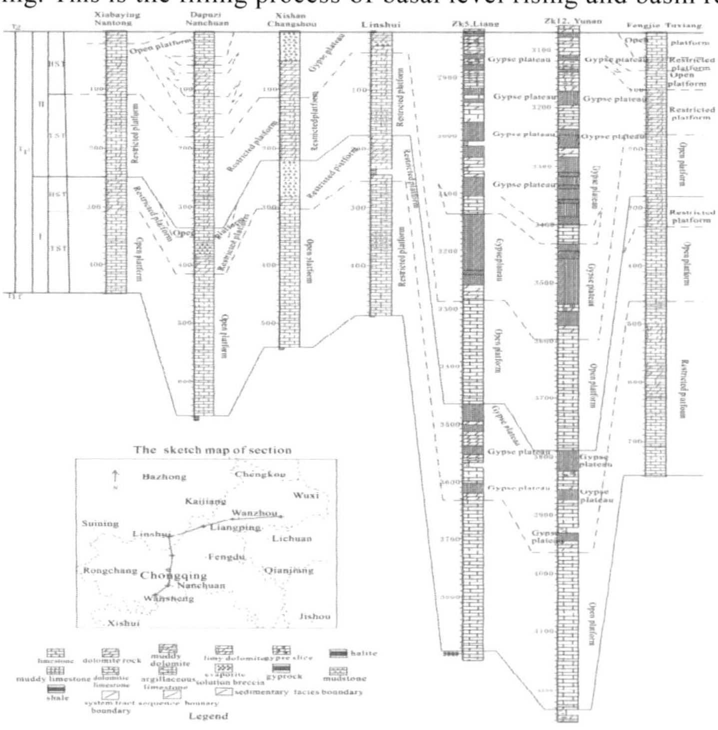

Figure 2. Columnar section showing sedimentary facies and sequence stratigraphy of the Early Triassic Jialingjiang Formation in Chongqing

In the stage of HST sea level fast descends, the restricted platform is mainly developed. Because the water in restricted platform is shallower in open platform and flat, the evaporation makes the water salty. As the relative sea level lightly fluctuates, large areas tableland will appear at the up and down of sea level, while tidal flat is developed well in restricted platform. The penecontemporaneous dolomitization will happen between carbonate mud and particle in the surface water which salinity is high and magnesium is rich. So, one of lithology characteristics is dolomite rock develops widely.

As the sea level descents, the accommodation space decreases progressively and slowly. When the rate of accommodation space decreasing is close to the rate of carbonate productivity and accumulation, the aggradation is dominant. The depositional sequences of shallowing-upward and shallow face are built, representing the dolomitization and gypsification of carbonate.

In a word, the sea level change is the most important influencing the sequence development and lithofacies paleogeography. As the sea level goes up and down, it is the favorable period for halite and gypse development in the middle and late of HST.

6. Summary

1) The regional unconformities and the lithology- lithofacies transition surface are recognized in the Jialingjiang Formation. A total of two third-order sequences are subdivided and they are subcorrelated to the intervals of Member 1 to Member 2, Member 3 to Member 4, respectively. Each sequence is further subdivided into lacustrine transgressive system tract (TST) and highstand system tract (HST) according to Variation in lithology and lithofacies.
2) The sea level change is the main factor to control the sequence development. With the sea level change, the middle and late periods of HST in sequence are the favorable periods of the development of rock salt and gypsum.

7. Acknowledgements

This study was supported by the National Natural Science Foundation of China (41030213), Land and Resources Survey, China (1212010633901), Chongqing Education Commission, Chongqing (2015ZX011) and Chongqing Vocational Institute of Engineering, Chongqing (KJB14).

8. References

[1]Zhong Yijiang, Chen Hongde, Zhu Lidong, etc, 2008, Sedimentary microfacies of the Lower Triassic Jialingjiang Formation in Shizhu, Chongqing, China [J]. Journal of Chengdu University of technology (Science & Technology Edition).35(6): 655~662

[2]Lu Zhengyuan, Luan Haibo, Lu Zonggang, etc, 2009, Study on the distal gas accumulation model of the Lower Triassic Jialingjiang Formation in the south of Sichuan Basin, China[J]. Journal of Chengdu University of technology (Science & Technology Edition). 36(6): 617~620

[3]Lu Zhengyuan, Zhao Luzi, etc, 2004, Study on hydrocarbon sources of Lower Triassic Jialingjiang Formation in south of Sichuan Basin, China[J]. Journal of Chengdu University of technology (Science & Technology Edition). 31(6): 720~722

[4]Zhangqi, Xuliang, Jin Xiaolin, etc, 2009, Sedimentary Environment Analysis on T_1j^4 Gypsum 2 salt Rock in Eastern Sichuan Basin[J]. Journal of Salt Lake research. 17(4): 1~5

[5]Ji Yuxin, Wei Xiucheng, Liu Chunyuan, etc, 2008, Dolomite Reservoir Prediction in the Jialingjiang No. 2 Member of the Tongnanba Area , Northeast of the Sichuan Basin , China[J]. Acta Geologica Sinica. 82(3): 353~359

[6]Li Yanjun, Li Qirong, Wang Yandong, etc, 2006, Geochemical research of hydrocarbon reservoiring in Lower Triassic Jialingjiang Formation in Luzhou palaeohigh[J]. Oil and Gas Geology. 27(3): 363~369

[7]Wangkun, 2010, Ichnofabrics and Their Sedimentary Significance from the Jialingjiang Formation in Longmendong Section, Mount Emei, Sichuan Province, China[D]. Henan Polytechnic University

[8]Tian Ruixiao, Wei Lingchen, 1985, Paleogeography and Facies of Jialingjiang period, Early Triassic in Sichuan Basin[J]. Experimental Petroleum Geology. 7(3): 218~226

[9]Lin Liangbiao, Chen Hongde, Zhu Lidong, etc, 2010, The sequence-based lithofacies-paleogeography of Jialingjiang Formation and Leikoupo Formation in eastern Sichuan Basin,China[J]. Journal of Chengdu University of Technology (Science and Technology Edition). 37(4): 446~451

[10]Hu Mingyi, Wei Guoqi, Li Sitian, etc, 2010,Characteristics of Sequence-based Lithofacies and Paleogeography and Reservoir Prediction of the Jialingjiang Formation in Sichuan basin[J]. Acta Sedimentologica Sinica. 28(6): 1145~1152

Structural Decomposition Analysis of China's Industrial Energy Consumption Based on Input-Output Analysis

X Y Huang[1], J Q Zhou[2, 1], Z Wang[1], L C Deng[3] and S Hong[1]

[1] School of Resource and Environmental Sciences, Wuhan University, Wuhan, PR China

[2] Zhejiang Environmental Monitoring engineering limited company, Hangzhou, PR China

[3] Center for Environmental Progress, Wuhan, PR China

Z Wang. E-mail: sinoo@whu.edu.cn

Abstract. China is now at a stage of accelerated industrialization and urbanization, with energy-intensive industries contributing a large proportion of economic growth. In this study, we examined industrial energy consumption by decomposition analysis to describe the driving factors of energy consumption in China. Based on input-output (I-O) tables from the World Input-Output Database (WIOD) website and China's energy use data from 1995 to 2011, we studied the sectorial changes of energy efficiency during the examined period. The results showed that all industries increased their energy efficiency. Energy consumption was decomposed into three factors by the logarithmic mean Divisia index (LMDI) method. The increase in production output was the leading factor that drives up China's energy consumption. World Trade Organization accession and financial crises had great impact on the energy consumption. Based on these results, a series of energy policy suggestions for decision-makers has been proposed.

1. Introduction

With increasing energy demands arising from China's industrialization, China has been the largest energy consumer in the world since 2010[1]. Coal is the major energy source in China, making up about 70% of China's total energy use [2-3]. As a consequence, China emits large volumes of green-house gases as well as dust, raising serious domestic and global concerns [4-5]. Industrial sectors are the biggest energy consumers, with over 70% of China's total energy consumption [6]. In addition, the compound growth rate of China's industrial added value increased dramatically from 1995 to 2011. Meanwhile, energy consumption increased greatly as well, with an average increase of 6.3% per year. Compared to major developed countries, China's energy utilization efficiency is still at a relatively low level.

Decomposition analysis is a useful tool for distinguishing the intrinsic driving forces of changes in energy efficiency. Index decomposition analysis (IDA) and structural decomposition analysis (SDA) are two common types of decomposition methods. IDA is a type of analytical framework using direct energy data. Numerous studies have established IDA frameworks to analyse the driving force for energy consumption and its environmental impact, including industry energy use [7], residential energy consumption [8-9], CO_2 emission [10-14], etc. SDA is based on input-output analysis, using both direct and indirect energy consumption data. SDA is also applied to analyse environmental impacts such as CO_2 emissions [15-16], material consumption [17], and footprint evolution [18]. Compared to IDA, SDA has its advantage to reveal indirect energy consumption concealed in the material flows among industrial sectors, therefore being widely and increasingly adopted for such studies. More details about the differences between IDA and SDA can be found in Ref. [19].

China is of particular interests to the international community due to its scale of economy, energy use and emissions. Many studies6, [20-21] have focused on China's energy intensity by tracing the data back to each sector and using national level statistics. In order to understand the rapid growth of China's energy consumption and energy productivity, a group of researchers [22-23] used production-theoretical decomposition analysis and IDA to investigate the driving force behind changing energy data. A study based on an energy input-output model and SDA analysis [24] found that the main challenge for China is that of the final demand structure. Previous SDA work used only gap years and failed to give continuous analysis on the driving factors of energy use. This paper proposes to use SDA and continuous data to reveal the driving factors of energy use change in China. The main method, including input-output analysis and the logarithmic mean Divisia index (LMDI) method, is described in Section 2, along with descriptions of the data sources and preliminary process. Section 3 describes the results. Section 4 contains a discussion of the implications of the results from Section 3.

2. Methods and Data

2.1. Methods

Input-output analysis was first introduced by Leontief [25]. It is widely used to study the interdependent relationships between sectors in an economic system, and to systematically analyse the internal complex transactions among various industries[26]. An economic-energy input-output (EEIO) table is a table in which the monetary energy flows in the original I-O table are replaced by physical energy flows. The formula can be expressed as follows:

$$c_{kj} = d_{kj} + \sum_{s=1}^{n} d_{ks} a_{sj} + \sum_{s=1}^{n} \sum_{l=1}^{n} d_{ks} a_{sl} a_{lj} + \cdots \tag{1}$$

$$d_{kj} = \frac{E_{kj}}{x_j} \tag{2}$$

Table 1 showed the implication of each variable. Equation (1) also can be expressed in a matrix form:

$$C = D(I - A)^{-1} = D \cdot L \tag{3}$$

where C and D represent the matrixes of c and d. $(I-A)^{-1}$ is known as the Leontief inverse matrix, and $(I-A)^{-1} = L$.

In this paper, the driving force of energy use in industry is decomposed into three elements by the LMDI method. The LMDI method is called the perfect decomposition technique that can process data equal to or smaller than zero and generate no residual [27]. The three decomposed elements are production effect, structure effect, and intensity effect. The total energy consumption is formulated as follows:

$$E_{i,t} = f(X_{i,t}, S_{i,t}, C_{i,t}) \tag{4}$$

By using LMDI methods, equation (4) can be derived into equation (5):

$$dE_{i,t} = \frac{\partial E_{i,t}}{\partial X_{i,t}} dX_{i,t} + \frac{\partial E_{i,t}}{\partial S_{i,t}} dS_{i,t} + \frac{\partial E_{i,t}}{\partial C_{i,t}} dC_{i,t} \tag{5}$$

which can be also expressed as follows:

$$\Delta E_t = E_t - E_o = \Delta E_x + \Delta E_s + \Delta E_c \tag{6}$$

Finally, the production effect, structure effect, and intensity effect of the change in energy use can be formulated as follows:

$$\Delta E_{i,x} = \sum_i \frac{E_{i,t} - E_{i,o}}{\ln E_{i,t} - \ln E_{i,o}} \ln(X_{i,t} / X_{i,o}) \tag{7}$$

$$\Delta E_{i,s} = \sum_i \frac{E_{i,t} - E_{i,o}}{\ln E_{i,t} - \ln E_{i,o}} \ln(S_{i,t} / S_{i,o}) \tag{8}$$

$$\Delta E_{i,c} = \sum_i \frac{E_{i,t} - E_{i,o}}{\ln E_{i,t} - \ln E_{i,o}} \ln(C_{i,t} / C_{i,o}) \tag{9}$$

Table 1. Implication of each variable in Section 2.1.

Variable	Meaning
c_{kj}	the complete inputs required from energy k for sector j to generate one unit value added
d_{kj}	the direct inputs required from energy k for sector j to generate one unit value added
a_{sj}	the direct inputs required from sector s for sector j to generate one unit value added
E_{kj}	the total inputs required from energy k for sector j
x_j	total amount of energy consumed in sector j
$E_{i,t}$	total amount of energy consumed by sector i in year t
$X_{i,t}$	total amount of economic output in sector i in year t
$S_{i,t}$	the proportion of total output for sector i in year t
$C_{i,t}$	the complete energy consumed coefficient of sector i in year t
ΔE_t	total effect from year 0 to year t
ΔE_x	production effect from year 0 to year t
ΔE_s	structure effect from year 0 to year t
ΔE_c	intensity effect from year 0 to year t

2.2. Data

All data for this analysis were obtained from open public databases. China's monetary I-O tables were obtained from the World Input-Output Database (WIOD), including 34 sectors. Industrial energy consumption data were obtained from China Energy Statistical Yearbooks, containing 36 sectors. We consolidated some of the industries into 21 sectors to coordinate the two different sectorial classifications and formed 17 EEIO tables (1995–2011).

3. Results

3.1. Energy efficiencies of different sectors

China's economy began its rapid growth in 1978, and its energy consumption also showed an analogous growth at the same time (Figure. 1). China's gross domestic product (GDP) has a significant positive linear relation with total energy consumption ($R2 = 0.946$, $P < 0.001$). However, there is no evidence that the inflection point of the environmental Kuznets curve (EKC) will appear in the near future [28].

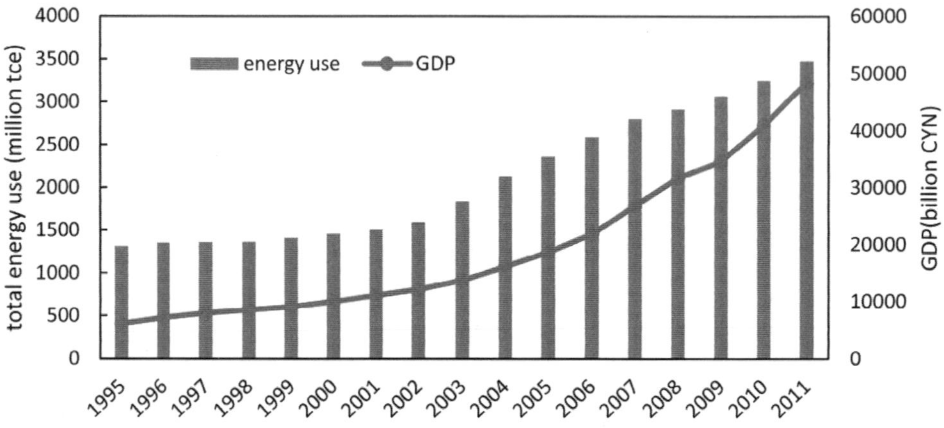

Figure 1. The changes in total energy use and GDP in China from 1995 to 2011.

Table 2 shows the statistics of all sectors. The value of DV (the difference in energy coefficient between 1995 and 2011) was positive in all sectors, indicating the total improvement of energy saved per unit output. The DV/AC values indicate that most industrial sectors had remarkable improvements in efficiency. Sector H20 had a relatively low DV/AC value, possibly because the sector is highly dependent on transportation. By contract, the improvement of energy efficiency with regard to vehicles and trains was not as significant as that of industrial sectors [29].

Table 2. The change in energy consumption coefficients of industrial sectors during the period from 1995–2011. AC: average energy coefficient; RAC: rank of the AC; DV: difference in energy coefficient between 1995 and 2011; DV/AC: ratio between DV and AC.

NO.	Industry	AC (tce/1000 USD)	RAC	DV (tce/1000 USD)	DV/AC
H1	Agriculture, Hunting, Forestry, and Fishing	0.432	21	0.632	1.46
H2	Mining and Quarrying	1.954	6	3.442	1.76
H3	Food, Beverages, and Tobacco	0.604	17	0.957	1.58
H4	Textiles and Textile Products	0.774	15	1.02	1.32
H5	Leather, Leather, and Footwear	0.54	18	0.72	1.33
H6	Wood and Products of Wood and Cork	0.697	16	0.907	1.30
H7	Pulp, Paper, Printing, and Publishing	1.003	11	1.33	1.33
H8	Coke, Refined Petroleum, and Nuclear Fuel	2.247	2	3.79	1.69
H9	Chemicals and Chemical Products	2.074	4	3.452	1.66
H10	Rubber and Plastics	1.096	9	1.498	1.37
H11	Other Non-Metallic Mineral	1.963	5	2.358	1.20

H12	Basic Metals and Fabricated Metal	2.351	1	3.343	1.42
H13	Machinery, n.e.c[a]	1.024	10	1.367	1.33
H14	Electrical and Optical Equipment	0.774	14	1.005	1.30
H15	Transport Equipment	0.898	13	1.148	1.28
H16	Manufacturing, n.e.c; Recycling	1.241	7	1.979	1.59
H17	Electricity, Gas, and Water Supply	2.176	3	3.292	1.51
H18	Construction	1.21	8	1.912	1.58
H19	Wholesale and Retail Trade, Hotels, and Restaurants	0.457	20	0.663	1.45
H20	Transport, Storage, and Post	0.938	12	0.678	0.72
H21	Others	0.533	19	0.813	1.53

[a] [not elsewhere classified]

3.2. Decomposition of China's total energy consumption

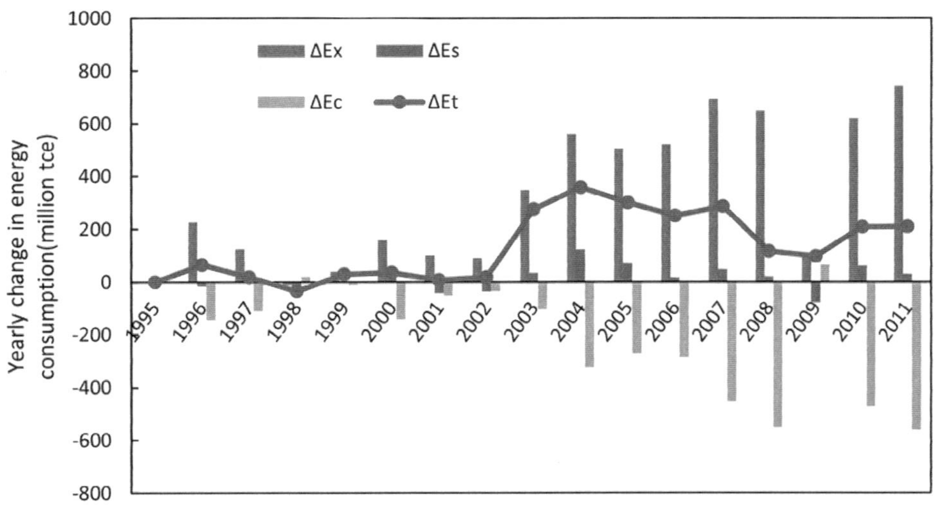

Figure 2. Decomposition of China's yearly energy consumption during the period from 1996 to 2011.

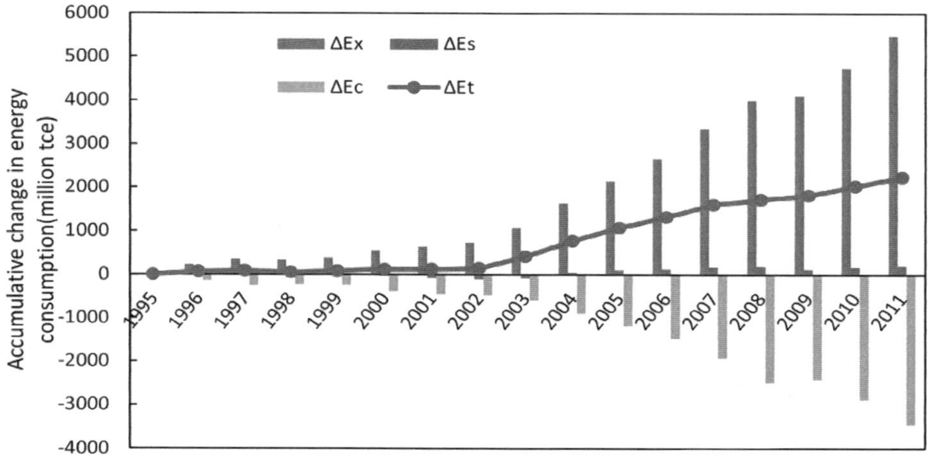

Figure 3. The accumulative effects during the period from 1996 to 2011 (Base year: 1995).

As to the production effect, its variation was rather small before 2002 as Figure 2 and Figure 3 depicted. During this period, the production effect contributed positively to the total energy consumption, except in 1998. Possibly, the outbreak of Asian financial crisis in 1998 caused decline in exports, which in turn influenced the industrial outputs. As the consequence, the production effect presented a negative contribution in 1998, equivalent to -6.32 million tce. During 2003-2008, the outputs of industrial sectors dramatically increased with an annual growth rate at 18.3% due to China's entry into the World Trade Organization (WTO), which created a large foreign demand for industrial products and manufacturing capacities from China. Consequently, the production effect also increased significantly as Figure 2 showed. The 2008 crisis caused the recession of outputs and the production effect (-8.27 million tce). Generally, the production effect was the largest effect and contributed positively to the energy consumption increase.

Structure effect was insignificant compared to the other two effects. As a result, energy consumption relatively concentrated to the energy intensive sectors, and the structure effect became unfavourably positive. This trend was also very clear in Figure 3. The accumulative structure effects kept larger than zero, which reflected China failed to transform its economy structure into an energy saving one as it promised.

Energy intensity effect contributed negatively to the total energy consumption in general during 1996-2011. Suppressed production demand and excessive energy supply might desert their renovation projects and investments for energy saving technologies.

3.3. Sectorial energy consumption decomposition

The contribution of each effect for industrial sectors' energy consumption was diverse. Table 3 showed the effects of contributions and their rankings at sectorial level. The structure effects were positive together with the production effect in these sectors, implying their share for economy output enlarged in the past 16 years and contributed the largest part for the undesirable structure change. H1 was a special case with negative accumulative total effect.

The accumulated production effect of each sector was positive and played a major role on the energy consumption growth. Besides, the rank of accumulated production effect presented a high positive unsupervised correlation with the rank of total energy consumption. That means high energy intensive sectors often had high production effect, implying the economy growth based on the expansion of large industrial energy consumers. Sectors with high accumulative production effect basically also had very large accumulative energy intensity effect in absolute term. Two reasons might explain this phenomenon. One is, the elimination of backward production capacity especially the energy intensive technologies indirectly promoted the energy efficiency of industrial sectors. The other is, the newly built production capacity driven by the expanding needs after 2002 often used the new technologies and devices with high energy efficiency, guaranteed by the policy makers through the strict examination and approval system for high energy sectors. However, the accumulative energy intensity effect of each and all sectors could not compensate the increment of the accumulative production effect. H1 had the closest two effects, with a gap about 11.10 million tce, or 8.4% of the accumulative production effect. Comparatively, H12 had the largest absolute gap about 510.29 million tce, and H20 had the largest relative gap about 61.2%. Thus, to achieve the favourable decoupling status between energy consumption and the economy, much still can be do to minimize the gaps for each industrial sector.

The accumulative structure effects for industrial sectors were much complicated. Half of sectors had positive effect. Unfortunately, these sectors were mostly energy intensive, such as H12, H17, H2, H8, and H9. These five sectors only consisted of 19.55% of the economy output, but contributed 206.6% of the accumulative structure effect. In other words, these undesirable expansions were biased towards energy intensive sectors, and lead to energy consumption increase. Other sectors had negative accumulative structure effect, mainly due to their shrinking share of total economy outputs. The structure change of H1 together with the energy efficiency improvement had cancelled out all the increment of the production effect. As a result, H1 became the only sector that had negative total effects.

Table 3. Accumulated effects of each sector (million tce; TE: total effect; PE: accumulative production effect; SE: accumulative structure effect; IE: cumulative energy intensity effect; R: the ranking of each effect)

NO.	TE	R	PE	R	SE	R	IE	R
H12	63515.12	1	131036.54	1	12485.80	1	-80007.22	1
H17	33508.35	2	51909.32	3	10990.06	2	-29391.02	4
H9	30595.97	3	77241.58	2	4839.97	5	-51485.59	2
H2	20681.59	4	47221.07	4	8420.60	3	-34960.08	3
H8	15870.03	5	34131.94	6	6479.90	4	-24741.81	6
H20	14786.71	6	33115.67	7	-5490.54	20	-12838.42	8
H21	10757.88	7	24364.60	8	647.33	8	-14254.04	7
H18	5674.39	8	9767.00	14	1514.56	6	-5607.17	14
H14	5030.46	9	7384.23	16	894.78	7	-3248.54	19
H13	4679.18	10	10804.84	12	-94.50	12	-6031.16	13
H11	4207.08	11	44032.03	5	-10867.87	21	-28957.08	5
H15	3538.05	12	6811.92	17	189.53	9	-3463.41	16
H10	2557.54	13	5864.05	18	65.50	10	-3372.01	17
H4	2446.51	14	12296.64	11	-1943.92	18	-7906.21	11
H19	2203.74	15	10448.08	13	-1071.15	15	-7173.19	12
H3	2040.23	16	12350.40	10	-1128.62	16	-9181.55	10
H7	1532.23	17	7557.28	15	-1206.80	17	-4818.25	15
H6	1098.50	18	2397.27	20	3.31	11	-1302.08	20
H16	714.70	19	4484.98	19	-440.70	14	-3329.59	18
H5	48.75	20	709.70	21	-176.71	13	-484.23	21
H1	-2078.54	21	13276.70	9	-3188.82	19	-12166.42	9

4. Discussion

There is still room to improve in catching up with developed countries. Basically, these sectors could reduce their energy use by innovating their production processes through cleaner production audits. In addition, updated energy saving technologies, such as waste heat conversion technology, heat storage technology, and high temperature air combustion (HTAC) technology, have great potential for application in these sectors and for reducing their energy use in terms of efficiency improvements [30-32]. Further policymaking and funding should encourage these energy intensive sectors to decouple their energy use and economy output.

The industry structure effect exhibited a weakening trend in comparison to the production effect and intensity effect. Two industries including non-metallic minerals, and transport, storage, and post contributed a large share of favourable structure effect. These two sectors reached –108.68 and –54.91 million tce of the cumulative structure effect, respectively. In contrast, China's economy is still highly reliant on high energy intensive sectors such as metal-processing, electric power, and mining, whose structure effects caused an unfavourable rise in the total energy consumption growth. Therefore, more attention should be focused on adjusting the industrial structure. Low energy intensive industries such as H21 and H19 should be carefully supported to further establish an energy saving economic structure.

Although there was no difference in decomposition results, regardless of the type of energy used, it is obvious that China's energy consumption has been dominated by coal, which has caused terrible carbon and dust emissions. However, China's international pledge to the United Nations Framework Convention on Climate Change (UNFCCC) is to achieve a peak of carbon dioxide emissions around 2030, making its best efforts to peak at an early date, to lower carbon dioxide emissions per unit of GDP by 60% to 65% of the 2005 level, and to increase the share of non-fossil fuels in primary energy consumption to around 20%. The general public also has a vision set out to improve the environment. There is no denying the fact that China urgently needs to optimize and upgrade its energy structure, and to use more clean energy, such as solar power, hydropower, and nuclear power [33].

5. Conclusion

China's energy utilization efficiency greatly improved over the study period. The chemical industry has shown the most marked change. The structural decomposition results showed that an increase of the production effect, as the dominant driving force, had been raising the consumption of energy significantly after 2002. The energy intensity effect also grew impressively in terms of absolute value, and cancelled out most of the production effect. The structure effect was the least dominant effect; however, it was revealed that China's industry structure adjustment had an unfavourable result. Sectorial decomposition indicated that the increment of production effect and the structural effect mainly came from the energy intensive sectors, although these sectors also contributed the largest share of decrement of energy intensity effect. Future policymaking should focus on the energy intensive sectors through technology and device renovation, and economic structure readjustment.

6. Acknowledgments

This work was supported by National Natural Science Foundation of China (Grant No. 41101567), and the Youth Chen-guang Project of Science and Technology of the City of Wuhan (Grant No. 2014070404010199).

7. References

[1] Zeng L, Yu Y and Li J 2014 China's Promoting Energy-Efficient Products for the Benefit of the People Program in 2012: Results and analysis of the consumer impact study Appl. Energy vol 133:22–32.

[2] Long R, Wang H and Chen H 2015 Regional differences and pattern classifications in the efficiency of coal consumption in China J. Clean. Prod. vol 112: 3684–91.

[3] Shealy M, Dorian J and P 2010 Growing Chinese coal use: Dramatic resource and environmental implications Energy Policy 38: 2116–22.

[4] Intergovernmental Panel on Climate Change (IPCC) 2016 IPCC Guidelines for National Greenhouse Gas Inventories Cambridge University Press Cambridge.
[5] Solomon S, Plattner G, Knutti R and Friedlingstein P 2009 Irreversible climatechange due to carbon dioxide emissions. Proceedings of the National Academy of Sciences vol 106(106): 1704-9.
[6] Liao H, Fan Y and Wei Y M 2007 What induced China's energy intensity to fluctuate: 1997-2006? Energy Policy vol 35: 4640–49.
[7] Zhang M and Song Y 2015 Exploring influence factors governing the changes in China's final energy consumption under a new framework. Nat. Hazards vol 78: 653–68.
[8] Nie H and Kemp R 2014 Index decomposition analysis of residential energy consumption in China: 2002–2010 Appl. Energy vol 121: 10–19.
[9] Xu X Y and Ang B W 2014 Analysing residential energy consumption using index decomposition analysis Appl. Energy vol 113: 342–51.
[10] Dong Y, Ishikawa M, Liu X and Wang C 2010 An analysis of the driving forces of CO2 emissions embodied in Japan–China trade Energy Policy vol 38: 6784–92.
[11] Ouyang X and Lin B 2015 An analysis of the driving forces of energy-related carbon dioxide emissions in China's industrial sector Renew. Sustain. Energy Rev vol 45: 838–49.
[12] Xu X Y and Ang B W 2013 Index decomposition analysis applied to CO2 emission studies Ecol. Econ vol 93: 313–29.
[13] Yao C, Feng K and Hubacek K 2015 Driving forces of CO2 emissions in the G20 countries: An index decomposition analysis from 1971 to 2010 Ecol. Inform vol 26: 93–100.
[14] Yuan R and Zhao T 2015 Changes in CO2 emissions from China's energy-intensive industries: A subsystem input-output decomposition analysis J. Clean. Prod. vol 117: 98–109.
[15] Akpan U S, Green O. a, Bhattacharyya S and Isihak S 2015 Effect of Technology Change on CO 2 Emissions in Japan's Industrial Sectors in the Period 1995–2005: An Input–Output Structural Decomposition Analysis Environ. Resour. Econ vol 61: 165–89.
[16] Wu C, Huang X, Yang H, Lu Q, Xu G, Li L and Li J 2015 Embodied carbon emissions of foreign trade under the global financial crisis: A case study of Jiangsu province, China. J. Renew. Sustain Energy vol 7: 043-110.
[17] Wenzlik M and Eisenmenger N and Schaffartzik A 2015 What Drives Austrian Raw Material Consumption?: A Structural Decomposition Analysis for the Years 1995 to 2007 J. Ind. Ecol. vol 19: 814–24.
[18] Yang Z, Liu H, Yang T and Xu X 2015 A path-based structural decomposition analysis of Beijing's water footprint evolution Environ. Earth Sci vol 74: 2729–42.
[19] Su B, Ang B W 2012 Structural decomposition analysis applied to energy and emissions: Some methodological developments Energy Econ vol 34: 177–88.
[20] Chai J, Guo J E, Wang S Y and Lai K K 2009 Why does energy intensity fluctuate in China? Energy Policy vol 37: 5717–31.
[21] Ma C and Stern D 2008 China's changing energy intensity trend: A decomposition analysis. Energy Econ vol 30: 1037–53.
[22] Du K and Lin B. Understanding the rapid growth of China ' s energy consumption : A comprehensive decomposition framework. Energy 2015; 90: 570–577.
[23] Du K, Huang L and Yang Z 2014 Understanding industrial energy productivity growth in China: a production-theoretical approach Energy Effic vol 8: 93–508.
[24] Xie S C 2014 The driving forces of China's energy use from 1992 to 2010: An empirical study of input-output and structural decomposition analysis Energy Policy vol 7: 401–15.
[25] Miller R E, Blair P D 2009 Input – Output Analysis Cambridge University Press Cambridge.
[26] Zhou J Q, Wang Z, Niu B B and Hong S 2015 Global Environmental Input-output Research Trends During 1900-2013:A Bibliometric Analysis Fresenius Environmental Bulletin vol 24(5B): 1996–2004.
[27] Ang B W 2004 Decomposition analysis for policymaking in energy: Which is the preferred method? Energy Policy vol 32: 1131–39.

IOP Conf. Series: Earth and Environmental Science **63** (2017) 012041 doi:10.1088/1755-1315/63/1/012041

[28] Yu Y N and Chen L L 2012 Research progress of the Environmental Kuznets Curve (EKC)in China Ecol. Environ. Sci vol 21: 2018–23.
[29] Song X, Hao Y and Zhu X 2015 Analysis of the Environmental Efficiency of the Chinese Transportation Sector Using an Undesirable Output Slacks-Based Measure Data Envelopment Analysis Model Sustainability vol 7: 9187–06.
[30] Price L, Levine M D, Zhou N, Fridley D, Aden N, Lu H, McNeil M, Zheng N, Qin Y and Yowargana P 2011 Assessment of China's energy-saving and emission-reduction accomplishments and opportunities during the 11th Five Year Plan Energy Policy vol 39: 2165–78.
[31] Zhou Q, Yabar H, Mizunoya T and Higano Y 2015 Exploring the potential of introducing technology innovation and regulations in the energy sector in China: a regional dynamic evaluation model J. Clean. Prod. vol 112: 1537–48.
[32] Schaffel-Mancini N, Mancini M and Szlek A, Weber R 2010 Novel conceptual design of a supercritical pulverized coal boiler utilizing high temperature air combustion (HTAC) technology Energy vol 35: 2752–60.
[33] Yong J Y, Klemeš J J, Varbanov P S and Huisingh D 2015 Cleaner energy for cleaner production: Modelling, simulation, optimisation and waste management J. Clean. Prod. vol 111: 1–16.

Review and Future Research Directions about Major Monitoring Method of Soil Erosion

Yue Li[1, 2], Xiaoyong Bai[*1, 2], Yichao Tian[1, 2] and Guangjie Luo[1, 2]

1. State Key Laboratory of Environmental Geochemistry, Institute of Geochemistry, Chinese Academy of Sciences, Guiyang 550002, China;
2. Puding Karst Ecosystem Observation and research station, Anshun, Guizhou, 561000, China

baixiaoyong@126.com

Abstract. Soil erosion is a highly serious ecological problem that occurs worldwide. Hence, scientific methods for accurate monitoring are needed to obtain soil erosion data. At present, numerous methods on soil erosion monitoring are being used internationally. In this paper, we present a systematic classification of these methods based on the date of establishment and type of approach. This classification comprises five categories: runoff plot method, erosion pin method, radionuclide tracer method, model estimation, and 3S technology combined method. The backgrounds of their establishment are briefly introduced, the history of their development is reviewed, and the conditions for their application are enumerated. Their respective advantages and disadvantages are compared and analysed, and future prospects regarding their development are discussed. We conclude that the methods of soil erosion monitoring in the past 100 years of their development constantly considered the needs of the time. According to the progress of soil erosion monitoring technology throughout its history, we predict that the future trend in this field would move toward the development of quantitative, precise, and composite methods. This report serves as a valuable reference for scientific and technological workers globally, especially those engaged in soil erosion research.

1. Introduction

Soil erosion monitoring is important in determining the soil erosion rate and upholding soil and water conservation [1]. Along with the gradual progress of soil erosion research, international studies have shown considerable attention on the proper choice of soil erosion monitoring method to be used in particular aspects of investigation [2] .To date, several works are focusing on soil erosion [3]. Yoo, Kyungsoo [4] studied soil formation and soil erosion by using the method of geochemical mass balance. Kairis, Orestis [5] emphasized the importance of using an efficient land management model to prevent soil erosion in the desert portions of rural areas in Crete. Research on soil erosion provides a scientific basis and theoretical guidance for soil erosion monitoring and aids in developing new technology and methods for this purpose [6].

Several soil monitoring methods have been established, each having its own advantages and disadvantages as well as specific conditions for use. Soil erosion monitoring has been performed for over 100 years, but a deep understanding of the involved theories, methods, and applications has been difficult to achieve. The history of the development of international mainstream technology and methods of soil erosion monitoring has been seldom discussed. Furthermore, a comprehensive comparison of these approaches, including the enumeration of their advantages and disadvantages, has been rarely reported. Current research statuses and future trends in the development of these methods have been highlighted infrequently. In this regard, this article examines the runoff plot method and erosion pin technique, two methods used in the quantitative observation of soil erosion. Furthermore, this report discusses and summarizes the development of soil erosion monitoring, including the gradual improvement of the radionuclide tracer method, model estimation, and "3S" technology. It also touches on the soil erosion monitoring techniques common in the international mainstream and reviews the development history, research status, main advantages and disadvantages, and future development trend and direction of each method. This paper aims to provide a valuable reference for research on soil erosion and its related surface processes. It also seeks to formulate plans and measures for preventing and controlling water and soil erosion.

Throughout its history, soil erosion monitoring has involved techniques that developed gradually from crude to precise and semi-quantitative to quantitative. These approaches also progressed from outdoor to indoor monitoring forecast and simulation and from small-scale single-slope to large-scale regional comprehensive monitoring. Methods for soil erosion monitoring can be classified based on the date of establishment and type of approach. This classification gives rise to five major categories. Other techniques of soil erosion monitoring are basically derived from these five types and are being continuously enhanced for global soil erosion monitoring (Figs. 1 and 2).

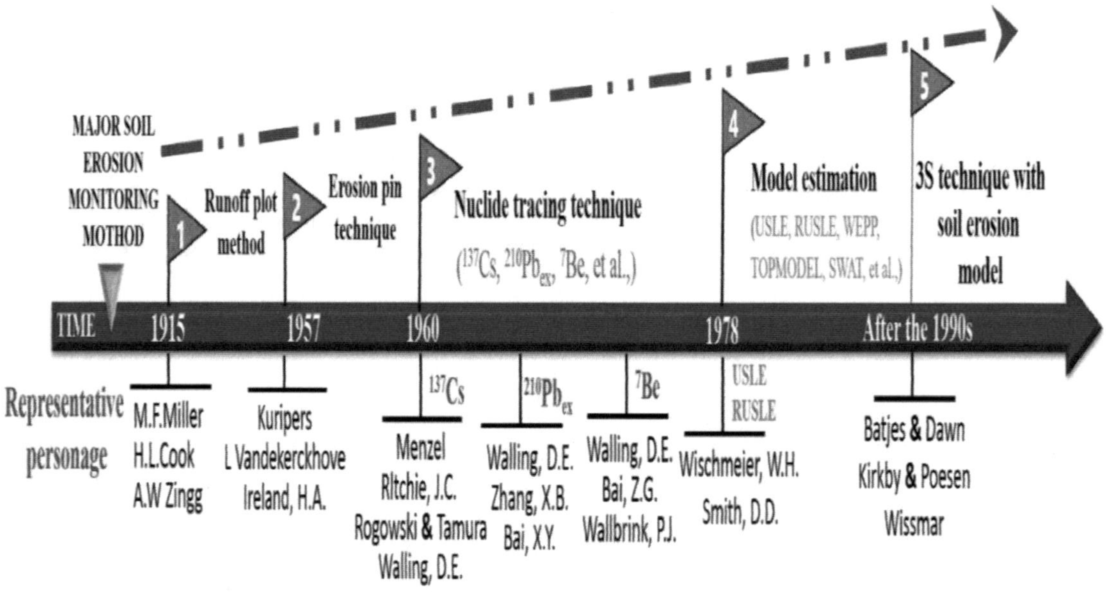

Figure1. The developing process chart of the method of soil erosion monitoring

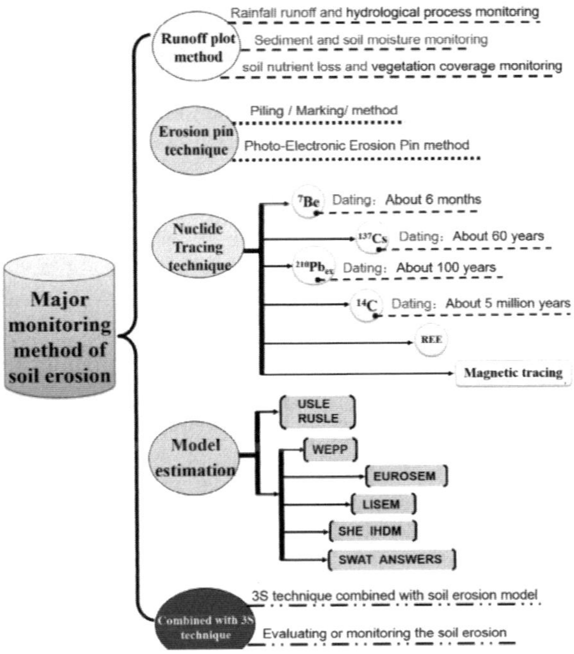

Figure2. The classification chart of the method of soil erosion monitoring

2. Runoff plot method

2.1. Historical review and current research

Soil erosion monitoring was first conducted using runoff plots [7]. The runoff plot method was first used in 1915 by the Missouri Agricultural Experiment Station of America. M. F. Miller and colleagues investigated the effect of crops and crop rotation on soil erosion and used the data to create a runoff plot [8]. The runoff plot method is used in experiments on soil and water conservation. Studies on soil and water conservation usually employ unique ground observation methods. H. L. Cook analysed numerous runoff plots and proposed three major factors that influence soil erosion [9]. This work launched the development of soil erosion forecasting technology and laid a good foundation for follow-up study. After nearly 20 years of exploration and research, the runoff plot method gained increasing attention and application. In 1940, A.W. Zingg simulated runoff plots and applied rain conditions in the field to study the relationships among slope length, slope and building terrace, and soil erosion rate and slope [10]. Subsequently, D. D. Smith added soil and water conservation measures as well as crop factors to simulation studies, results of which established the universal soil loss equation (USLE). This work also further promoted the progress of research on soil erosion and ground monitoring methods, the basic determination of soil erosion, and the monitoring of the rule of soil erosion.

The development of runoff plots in the mid-twentieth century greatly enhanced awareness and understanding of the loss of soil and water during soil erosion. In 1971 and 1978, American scholars Wischmeier and Smith (1965) studied 65000 rains, and observed and analysed data from 8250 and 2500 small watershed erosion areas for a year [11]. The data collected from these investigations were used to formulate the USLE. After the start of the 21st century, the runoff plot method continued to be used and promoted, as well as developed and challenged. For instance, Peng et al applied the Karst slope to six land and vegetation runoffs under the conditions of fixed, continuous field observation of through fall and groundwater level changes. Findings of the study revealed that the surface runoff coefficient is characterized by exponential function variation with changes in rainfall; the surface runoff coefficient is also easily produced with a growth type of frequent transformation [12]. Cao et al (2015) studied the runoff plot method and simulated rainfall in the Chinese southern forest of Pinus massoniana by using a runoff simulation after rainfall and soil erosion. The results of the study helped in building a model to predict soil erosion and explore the factors influencing this phenomenon [13].

Overall, the runoff plot method developed over the course of nearly 100 years and served as the means for determining the link between soil erosion and slope. Moreover, in-depth analysis of runoff plots provided a collection of data for understanding the occurrence and development of soil erosion.

2.2. Future research direction

The runoff plot method, a soil erosion monitoring approach that applies both traditional and classical means, is anticipated to follow three major trends for future development. First, the trend of model construction will be described by the location of the main existing situation and will be based on dynamic and timely observation. Second, the modernization and diversification of monitoring technology and equipment will gradually continue to be scientific and reasonable. Third, the accurate transition from rough to precise quantitative determination will be achieved, and the discipline will continue to expand. Research results and practical applications will be considered closely, demonstrated, and promoted.

3. Erosion pin technique

The erosion pin method is first proposed by Kuripers[14]. It is a simple and feasible approach for soil erosion monitoring through general survey. It is also a classic method for gully erosion monitoring. L Vandekerckhove investigated soil erosion in 46 banks of the Guadalentin and the Guadix basin in Spain by using the erosion pin method. The findings of the study revealed that the gully head retreat rate difference is mainly due to regional differences in rainfall and gully wall crack rifting activities [15]. Sun G.H et al. used the erosion pin technique to determine the eroded quantities from six gully erosions in Ledu County, Qinghai Province. According to the study, gully erosion is influenced by vegetation coverage at a greater degree than the slope and K [16]. In addition, the erosion pin method is simple, practical, economical, and widely used in the dynamic monitoring of river morphology. The earliest reference to the study on river morphology is Wolman's work on gully erosion in Ireland [17]. Herein, the study of bank collapse was introduced by measuring the length of the earth's surface regularly to reflect the depth of soil erosion. Since then, the erosion pin method has been playing a crucial role in erosion monitoring in riverbanks [18]. Along with the advances in monitoring technology, some scholars have successfully modified and improved the erosion pin method. In 1989, Lawler used photo-electronic erosion pin (PEEP) to monitor the intensity and frequency of soil erosion and accumulation in riverbanks [19]. This modified technique greatly reduced the repetitive use of the traditional erosion pin method in multiple field observations. As a result, the monitoring efficiency was improved and a more scientific and convenient method was formulated for the study of soil erosion.

3.1. Advantages and disadvantages

The erosion pin method entails short-term soil erosion observation. Hence, its advantages are evident, particularly, its suitability for field monitoring, low cost, the lack of need for many related facilities, high precision, and simplicity and ease of operation. However, this method also has some limitations. These limitations include the following: (1) low degree of automation, small range of observation, tedious work outside the industry, arduousness of long-term observations, and difficulty of locating the erosion pin in small ground heights; and the (2) need for close-contact measurement of the height of the pin exposed (buried) after erosion, high susceptibility to human interference in sensitive areas around the erosion pin, the limitations posed by environmental and human activities, and many other factors. The PEEP method also has several shortcomings, such as the loss of data when the detector becomes covered by snow or vegetation and during high-intensity turbulence.

3.2. Future research direction

The erosion pin (pile) method has become the bottleneck of research on soil erosion. It provides a convenient method for investigating bank collapse. Moreover, with related technological development and innovation, the PEEP method has greatly promoted the application and improvement of the erosion pin technique. Overall, the erosion pin method gradually enriches soil erosion research and combines with other technologies to complement and enhance investigations. Its trend of development involves its improvement from simple dynamic observation to explore erosion characteristics and the surface roughness caused by different areas of erosion and deposition. In this regard, the erosion pin method plays a significant role in the dynamic monitoring of the initial stage of gully development and the control of soil erosion.

4. Radionuclide tracer method

4.1. ^{137}Cs tracer method

^{137}Cs is an important radioactive isotope. Its soil distribution was first detected during the late 1950s [20]. This radioactive isotope has an extremely poor ability for self-removal from the soil and hence was exploited for soil erosion monitoring [21]. In the early 1960s, Menzel (1960) was the first to study the relationship between soil erosion and radionuclide deposition and migration [22]. Subsequently, Rogowski (1965) and Tamura (1970) first applied the 137 Cs method to study soil erosion by measuring runoff, soil erosion, and 137 Cs loss. Through this process, the exponential relationship between soil erosion and 137 Cs loss was elucidated [23]. Since then, the ^{137}Cs tracer method has been widely used in soil erosion monitoring and research. ^{137}Cs tracer technology rapidly developed and became the major means for monitoring soil erosion, determining soil erosion and sedimentation rates, quantitatively analysing soil net loss, and other applications in the field. Furthermore, numerous models of ^{137}Cs migration and soil erosion of soil profiles have been established. These models can be divided into two types, namely, empirical and theoretical.

4.1.1. Empirical model. Ritchie (1974) was the first to establish a quantitative relationship between the rate of ^{137}Cs loss and the amount of soil erosion [24]. Since then, several researchers have established a linear estimation model of the logarithmic form of the soil erosion rate and the ^{137}Cs loss rate [25], which is the basic form of ^{137}Cs models as follows:

$$Y = \alpha X^{\beta}, \qquad (1)$$

The formula includes the annual soil erosion amount (t/hm2·a), the percentage of soil ^{137}Cs loss, and the undetermined coefficient. Subsequently, Canadian researchers Elliot and Campbell (1984) used the model created by Ritchie to calculate the amount of soil loss from a particular farmland [26]. The generated formula is shown below:

$$S_t = (S_{t-1} + F_t - E_t \times CT)\, K, \qquad (2)$$

Where S_t is the ^{137}Cs area (Bq/m^2) at the end of the concentration of T, F_t is the t ^{137}Cs settlement in the T of the year (Bq/m^2), Et is the annual erosion rate (kg/m^2), CT is the plough layer soil ^{137}Cs concentration (Bq/kg), and K is the ^{137}Cs attenuation coefficient (0.977).

In 1989, Zhang X.B proposed the following simplified formula:

$$X = X_0(1 - H/H)^{N-1963}, \qquad (3)$$

Where X is the concentration ^{137}Cs in the area of study (Bq/m^2), X_0 is the background value of ^{137}Cs (Bq/m^2), H is the depth of the plough layer (CM), H is the thickness of the annual soil loss (CM), and N is the year of soil sampling. In 1997, Zhang X.B considered a past nuclear explosion distributing ^{137}Cs into the soil to show the effect of the enrichment and separation of grain erosion in a farmland and modified the formula accordingly [27].

In 1990, Elliott (1990) proposed the following estimation model based on the loss of ^{137}Cs in non-tillage soil [28]:

$$Y=\alpha\beta^{X}, \qquad (4)$$

Where Y is the loss of soil during erosion (kg/hm^2·a), X is the loss of ^{137}Cs from the soil, and α and β are undetermined coefficients.

4.1.2. Theoretical model. The theoretical model is mainly established through research on the model of profile distribution and mass balance. The earliest mass balance model was proposed by Kachanoski. Subsequently, many other researchers further explored the model and suggested different forms. Meanwhile, its basic form is shown as follows [29]:

$$S_t = (S_{t-1} + T_t - E_t)\,k\ (t = 1, 2, 3, \ldots, N), \quad (5)$$

Where S_t and S_{t-1} represent T and T−1 year's total ^{137}Cs amount in the soil profile (Bq/m^2), F_t is the total settlement of ^{137}Cs in year t (Bq/m^2), E_t is the soil erosion loss in year t (Bq/m^2), K is the radioactive decay constant (0.977) of ^{137}Cs, and N = M – 1954 (M = sampling year).

The establishment of the ^{137}Cs model laid a solid foundation for soil erosion monitoring through the qualitative or quantitative analysis of ^{137}Cs spatial distribution. For instance, Simpson (1976) investigated the deposition in the lower reaches of the Hudson River and found that ^{137}Cs levels vary with different positions and depths in the connecting estuary [30]. This work provided basic data to monitor soil erosion and migration. In Canberra, Australia, Wallbrink (1994) determined that ^{137}Cs content is greater than that in the slope toe [31]. C. Alewell et al. (2013) used the ^{137}Cs tracer method to evaluate and analyse soil erosion in mountain grassland [32]. Whereas H. D. Leckie et al. (2015) studied the wind erosion of Basin Mackenzie with the same technique [33].

Overall, the emergence and development of the ^{137}Cs tracer method greatly contributed to the establishment of the soil erosion estimation model and the development of new technology and methods for soil erosion monitoring. The ^{137}Cs tracer method has also become the most thorough and refined method for the quantitative study of soil erosion, monitoring of soil loss, and investigation of sediments since the 1970s.

4.2. Advantages and disadvantages

The ^{137}Cs tracing method can provide information on soil erosion and deposition that cannot be obtained through traditional means. This technique not only determines the erosion and deposition of the specific particle and the source of the erosion and sediment but also demonstrates the spatial distribution of soil erosion and movement as well as the formation age of different levels of soil. Moreover, the amount of soil erosion of about 40a can be estimated with low cost and short cycles. Evidently, it has become an indispensable method in soil erosion monitoring.

The limitations of the ^{137}Cs tracer method are as follows. (1) ^{137}Cs has a half-life of 30.12 years, which is suitable for the macro estimation of medium and long periods of erosion. (2) Only a small proportion of the ^{137}Cs content remains in seriously eroded areas, especially in the Loess Plateau, with a widely covering steep slope gully erosion zone. In these cases, the application of the ^{137}Cs tracer method has regional limits. (3) The earliest ^{137}Cs subsidence occurred in 1954. Hence, calculations on earlier times of deposition cannot be performed. (4) The spatial variability of ^{137}Cs deposition is relatively large. However, in this method, ^{137}Cs settlement is assumed to be fixed, which may account for some inaccuracies in ^{137}Cs tracer studies.

4.3. Future research direction

The ^{137}Cs tracer method has been used to study and monitor from the starting time of soil erosion. However, its many advantages and rapid development foresee great potential in four aspects. In particular, (1) the ^{137}Cs tracer method should be used to improve the soil erosion estimation model and its application to a large area. Furthermore, a database of the background values of ^{137}Cs in different regions must be established to obtain the spatial variations of soil erosion and provide a scientific basis for soil erosion monitoring and research. (2) The ^{137}Cs tracer method is also implicated in multidisciplinary aspects. For instance, the study of soil erosion can be extended to the investigation of its effect on the water environment in terms of resultant pollution. In this regard, the relationship between soil pollutants and specific water environmental conditions involved in sediment release can be determined. (3) The complete decay of ^{137}Cs from the nuclear fallout will soon be achieved; hence, the development of a new isotope to replace ^{137}Cs represents the current trend. The distribution of ^{210}Pb in soil profiles is similar to that of ^{137}Cs and is consistent with the rule of soil movement. Therefore, ^{210}Pb appears as the best alternative for ^{137}Cs in soil erosion monitoring applications.

5. $^{210}Pb_{ex}$ tracer method

5.1. Future research direction

The $^{210}Pb_{ex}$ tracing method is used in the earliest research on sedimentation rate. It is valuable in determining the basin erosion rate and lake sedimentation rate and their relationship with time [34]. To date, no model has yet been established for the $^{210}Pb_{ex}$ estimation of soil erosion. The ^{137}Cs model remains the most commonly used model for weight approximation and mass balance [35].

In recent years, some researchers have begun to explore the use of $^{210}Pb_{ex}$ to study soil erosion rate. Walling (1999) investigated the possibility of using $^{210}Pb_{ex}$ to trace soil erosion in the United Kingdom and proposed the quantitative model equation for erosion rate [36]. The potential of the $^{210}Pb_{ex}$ tracer method for estimating long-term soil erosion rates was also suggested. Subsequently, Zhang X.B (2003) conducted an in-depth study on the depth distribution of $^{210}Pb_{ex}$ in the soil profile of China and the United Kingdom. As a result, a stable-state model of agricultural land erosion rates was formulated, but the actual application in research and the reliability of the method are still to be verified at the time. Porto, P (2013) used both ^{137}Cs and $^{210}Pb_{ex}$ tracer methods to investigate and analyze the sediments in a small river basin in southern Italy [37]. X. Y. Bai (2013) combined the ^{137}Cs and $^{210}Pb_{ex}$ tracer techniques to study soil erosion in the Karst depression and explore the effects of land use changes on soil erosion in a small river basin in the Karst region [38].

5.2. Advantages and disadvantages

The $^{210}Pb_{ex}$ tracer method can distinguish the changes in atmospheric particles and human causes of trace elements, the reconstruction of pollution sources, and the history of river deposition and erosion in the past 100 years. However, its limitations include complex sample processing, high accuracy requirements, and difficulty in obtaining the flux of deposition for a particular year.

5.3. Future research direction

$^{210}Pb_{ex}$ technology can study 100 years of soil redistribution and hence provides a way to compensate for the defects of the soil redistribution rate in a short period. The key in the future development of this technique is to further improve the quantitative relationship between the amount of $^{210}Pb_{ex}$ loss and soil erosion. Furthermore, combining $^{210}Pb_{ex}$ technology with ^{137}Cs, ^{7}Be, and other isotopes can facilitate better understanding of soil erosion and help establish a soil erosion prediction model that can improve soil erosion monitoring.

6. ^{7}Be tracer method

6.1. Historical review and current research

^{7}Be is used as a tracer for soil erosion research and monitoring. However, its use is relatively recent and its application has not yet been deeply explored. Bai Z. G. (1997) led the study on the seasonal variation of ^{7}Be in the Karst area. In this work, the potential of using ^{7}Be to study soil erosion was discussed, and a quantitative model for estimating the soil erosion rate of the 7Be tracer was proposed [39]. Consequently, the study laid a solid foundation for the future applications of the ^{7}Be tracing method. Subsequently, Walling (1999) used ^{7}Be to study the seasonal erosion of agricultural land and the effect of ploughs on soil erosion by the ^{137}Cs method. On the basis of the characteristics of the ^{7}Be distribution profile in soil, the Walling model was proposed. Blake (2002) used the ^{7}Be tracer technique to study the erosion, slope erosion rate, distribution, and migration of fine sediments [40]. Zhang Q.W (2014) estimated the rill erosion amount and relative contribution rate via the ^{7}Be tracer method, thereby providing a novel method and means to understand further the mechanism of soil erosion and the development of soil erosion prediction models [41]. With the development of tracing technology, scholars have begun to combine ^{7}Be with other radionuclides to study and monitor soil erosion. Wallbrink (1996) used ^{137}Cs and ^{7}Be to investigate the relative contribution of different erosion types in soil erosion. Their work proved that ^{7}Be is an ideal means for tracing shallow sources of topsoil [42]. Burch (1998) showed the initial source of sedimentary soil from the soil profile based on the deposition of ^{137}Cs in soil and ^{7}Be activity, and then inferred the possibility of erosion [43] .G. Matisoff (2005) used the $^{210}Pb_{ex}$ and ^{7}Be tracer methods to develop and improve the measurement of suspended sediment age [44]. Therefore, the ^{7}Be tracer method has gradually become a novel tool for the study of soil erosion, with a very important role in soil erosion monitoring.

6.2. Advantage and disadvantage

The ^{7}Be tracer method has background values that are easy to measure and not subject to site constraints. Given the advantages of this convenient and simple method, the spatial distribution of soil erosion rate and erosion deposition can be well applied to the soil erosion rates in a short term or in specific erosion events (sub rainfall). The method can also be applied to evaluate soil erosion under a particular intensity of land use, thereby providing an important basis for the monitoring and control of soil erosion. However, the application of ^{7}Be tracing still has some problems. For example, the shallowness of ^{7}Be distribution complicates sampling. Moreover, a relatively perfect and simple ^{7}Be erosion tracer model has yet to be established, and research progress is relatively slow. The ^{7}Be tracer method is suitable for low and medium intensity soil erosion events, but its accuracy for high strength soil erosion events is poor.

6.3. Future research directions

The existing model was established without considering the spatial distribution characteristics of ^{7}Be. Therefore, the temporal distribution of precipitation, the temporal distribution of erosion, and the redistribution of rainfall runoff should be considered in the future to strengthen research on the distribution patterns of ^{7}Be in the soil profile and establish a quantitative model with a wide range of applications. ^{7}Be could reflect the effects of soil erosion factors. Therefore, the scope of applications of the ^{7}Be tracer method can be broadened to explore the comprehensive effect on specific small watersheds based on the hydrological and meteorological conditions of soil erosion. The operability of ^{7}Be tracing in soil erosion determines the necessity and importance of designing a novel experimental scheme to study the migration characteristics of ^{7}Be with rainfall runoff. The application of the ^{7}Be tracer in soil erosion research started relatively late; thus, the related quantitative models are few, and their precision needs to be tested. Therefore, research on the ^{7}Be tracer quantitative model should be strengthened. In addition, single radionuclide tracing cannot meet the needs of soil erosion research. Hence, the tracing of ^{7}Be and other radionuclide compounds has become an important direction of future research.

7. Magnetic tracer method

7.1. Historical review and current research

As early as 1986, magnetic tracer technology has been applied to soil particles to determine the surface coverage of soil erosion movement [45]. Caitheon (1993) and Ventura (2001) proposed the feasible use of magnetic tracers as a new method to understand soil erosion [46].

The application of the magnetic tracer technique in soil erosion research has two main aspects. One aspect is the use of environmental magnetism to trace sediment sources. Australian scholar Caitcheon (1993) proposed the use of environmental mineral magnetism to trace sediment sources. Within a certain range of geological and climate conditions, the magnetic minerals of river sediments represented by the magnetic parameter set are stable in space and time [47]. The relationship between magnetic parameters shows the characteristics of the collected magnetic minerals. Therefore, the tracers can be used to study soil erosion. Subsequently, Delong (1998) used changes in soil magnetic susceptibility to investigate the redistribution of long-term soil erosion [48]. Their results showed that the soil erosion and redistribution of the region near the sedimentary region can be estimated based on the variation in magnetic susceptibility distribution. Jordanova (2013) used the magnetic tracer method to study the formation of dark red soil and its mineralogical properties [49]. The other aspect is the use of sediment magnetism to indicate environmental changes in the basin. Dearing (1981) used the magnetic tracer method in Peris Lake because of the excessive grazing caused by soil erosion and explained the history of land use change in the river basin [50]. Meanwhile, Franciskovic-Bilinski (2014) used the magnetic tracer method to discuss the geochemical and mineralogical characteristics of sediments in Croatia and Slovenia [51].

The magnetic tracer method can reflect the history of land use pattern, vegetation succession, and soil erosion in a watershed. It can also identify the soil distribution and the erosion rate for a certain period. Therefore, this method can be used to provide a theoretical basis for soil erosion prediction and monitoring, and a history of the development of small watersheds.

7.2. Advantages and disadvantages

The magnetic tracer technique has obvious potential advantages. First, the measurement of the magnetic parameters is simple, rapid, and does not involve destruction. The use of the conventional magnetic method can meet the analysis requirements of a large number of samples. Second, the magnetic measurement instrument can be used to directly determine various magnetic parameters. It requires a simple operation and an easy-to-carry instrument, making measurements in the field and laboratory convenient. However, the magnetic properties and the depth of soil erosion or deposition are difficult to determine via the magnetic tracer method. Thus far, no reasonable quantitative model exists for the magnetic tracer method.

7.3. Future research direction

The magnetic tracer method has greatly progressed in the study of soil formation, the classification of soils, and the quantitative description of the evolution, occurrence, and development of erosion. With the continuous progress of science and technology, the study of soil erosion by soil magnetism will be a hot topic in soil science. Therefore, the development trend of the magnetic tracer method in soil erosion research is mainly reflected in the following aspects. (1) The magnetic contribution rate and the separation of soil erosion are quantified by exploring the species, grain shape, and characteristics of magnetic minerals in the soil to completely understand soil erosion on slope and establish a theoretical soil erosion prediction model. (2) The test results of the total magnetism can be incorporated in the empirical or theoretical model to separate magnetic signals from different sources and thus improve soil erosion prediction and monitoring accuracy. (3) A quantitative model of soil magnetic parameters and soil erosion rate is gradually established to improve the accuracy and precision of soil erosion prediction and provide reliable data for the effective prevention and control of soil erosion. (4) The magnetic tracer and other tracer techniques (e.g., ^{137}Cs, ^{7}Be, ^{210}Pb, and REE) were combined for compound tracing to explore the evolution of sediment sources, the different patterns and intensities of erosion, and the spatial differentiation of the basin.

Despite the sensitivity of magnetic minerals to environmental change, the use of magnetic tracers is a novel method to study the spatial distribution of soil erosion and the evolution of erosion patterns. The variable interpretation of magnetic parameters, the effects of soil processes and human activities on magnetic properties, and the total magnetic contribution rate require further research. Magnetic tracers in soil erosion research have broad prospects for improvement with the development of testing methods and the completion of basic theory.

8. Model estimation method

8.1. Historical review and current research

Soil erosion models are indispensable to quantitatively study soil erosion worldwide. The development of a soil erosion model can be divided into three stages: empirical statistical model, physical process model, and distributed model.

(1) Stage I: In 1877, German scientist Ewald Woolly began the construction of a soil erosion model [52]. By the 1960s, the USA USLE for the quantitative analysis of soil erosion was basically empirical. Empirical models of soil erosion are mainly based on the USLE. Wischmeier and Smith (1978) analysed the runoff and sediment data from all over the USA with the empirical USLE [53]. The RUSLE is a representative model. The soil empirical model remains widely used to date. For example, G. S. Pradeep (2015) used the AHP and RUSLE models to estimate the annual soil loss of ghats in southern India [54]. Shi Z.H (2004) combined RUSLE and GIS to evaluate the level of soil and water conservation planning in the Three Gorges area of China [55].

(2) Stage II: With the study of the mechanism of soil erosion, many scholars have found numerous limitations and shortcomings in the use of previous empirical models. During a 1985 workshop in Lafayette, Indiana, USA, a new generation of soil erosion prediction model that can replace the empirical model was created. The advent of a new generation of water erosion prediction model called the WEPP occurred in 1987. This model provides an up-to-date description of the physical process of water erosion parameters. The WEPP model can simulate soil erosion, non-regular steep slope, and soil, tillage, and management measures by calculating the temporal and spatial distribution of soil erosion and predicting the movement of sediment in the slope and basin. The WEPP model reflects the applicability and ductility of the temporal and spatial distribution of erosion and sediment; thus, numerous scholars still use this method. For example, R. E. Brazier (2000) used the WEPP model to assess the uncertainty of a soil erosion model based on physical processes in the UK and the USA [56]. B. Saghafian (2015) applied the WEPP model to determine the runoff and sediment sources in a forest watershed [57].

To describe hydrological processes, the distributed model has been recently developed into the SHE and IHDM models based on the traditional lumped conceptual model and the physical process model. The SWAT model was gradually constructed to simulate watershed management of soil erosion and the deposition effect during and after heavy rain with rainstorm catchment characteristics [58]. These models are widely applied in soil erosion research and monitoring. However, the parameters of the distributed model need to be evaluated. Therefore, numerous semi-distributed models such as TOPMODEL have been used [59].

(3) Stage III: In the 1990s, the soil erosion model was integrated with GIS to survey global and regional soil erosion, understand the ecological effects of soil erosion, and analyze soil erosion dynamics. Batjes and Dawn (2003) combined global earth science data and USLE with RUSLE to quantitatively assess soil erosion on a global scale [60]. Kirkby (1998) and Poesen (1996) used GIS technology for the dynamic monitoring of soil erosion in a small watershed [61]. Aiello (2015) combined GIS technology and RUSLE3D to evaluate the soil erosion in the southern Italy basin [62]. Zhu (2015) used the USLE model to evaluate soil erosion in the Danjiangkou reservoir area. Further improvement of GIS and its applications is necessary to facilitate the scientific, modern, and quantitative research and monitoring of soil erosion [63].

8.2. Advantages and disadvantages

The main advantages of the empirical model (and RUSLE USLE) are the following. (1) The formula is concise and the meaning of each factor is clear. (2) The calculation method of the factor has been basically mature and the parameters are easy to obtain for the continuous improvement and perfection of the model. (3) After several years of verification and testing, the accuracy of the model meets the needs of the application. The model is widely recognized and used in the calculation of soil erosion. However, the empirical model of this series also has some limitations. (1) The limited factors in the model cannot completely explain the complex and changeable phenomena of sediment yield and sediment flow in the river basin. (2) The model is highly regional and difficult to promote because it is based on observation data. (3) Simulation of soil erosion and sediment transport is difficult to perform.

The physical process model is based on the study of the processes and mechanisms of soil erosion. (1) To simulate soil erosion, sediment yield calculation can be realized. (2) The physical process model provides more profound scientific theory and higher adaptability in different regions than the empirical model. Although the physical model greatly compensates for the defects of the empirical model, this approach also has some shortcomings. (1) The physical mechanism of soil erosion is relatively complex and unclear. Some parameters in the physical process model are still dependent on the empirical model. (2) The large range of the study area is the major obstacle that hinders the use of the model because of the exacting demand of the model parameters. (3) The structure of the physical process model is complex and may change because the form has not been unified.

At present, GIS technology can be combined with the experience and physical models. The use of GIS spatial data management and analysis in the study of soil erosion prediction and evaluation has great advantages but still has some problems. For example, GIS will generate errors in the overlay and data operations when the nonlinear operation contains great errors. If no changes occur in the time and space of the response factor or if the acquisition of the factor itself has errors or uncertainty and cannot reflect the scale and space–time characteristics, then the combination with GIS technology will also produce large errors.

8.3. Future research directions

Future research should focus on theoretical analysis, particularly on the quantitative research and theoretical development of an erosion prediction model based on the erosion factor. The factor analysis model has been developed from a single-factor analysis to a comprehensive factor analysis, which is suitable for different regions. GIS is dynamic and capable of powerful spatial analysis; thus, the combination of GIS and soil erosion models has more advantages than the traditional soil erosion model. Present and future research should focus on GIS technology. Moreover, GIS is particularly useful for the description and quantitative distribution of parameters. GIS with spatial-scale conversion methods and models as technical support can maximize data, maps, and remote sensing data to quantify the sloping field system of erosion background conditions and the intensity of simulated and realistic soil erosion.

9. Conclusion

The application and research direction of soil erosion monitoring methods have achieved great progress. These monitoring methods are set up in different environments, terrains, and scales, but none of the methods have general applications. Each method has inherent advantages, disadvantages, and regional characteristics. Therefore, future research on soil erosion monitoring methods should include exploration of the processes, mechanisms, prediction models, and environmental effects soil erosion. Moreover, researchers should focus on the advantages of each method and gradually improve the current defects to develop these methods in the quantitative, accurate, crossover, and composite directions. Development of these methods is indispensable for extensive soil erosion research.

10. Acknowledgements

The authors gratefully thank for the financial support provided by the auspices of National Key Research Program of China (No. 2016YFC0502300, 2016YFC0502102, 2014BAB03B00), National Key Research and Development (No.2014BAB03B02), Agricultural Science and Technology Key Project of Guizhou Province of China (No. 2014-3039), Science and Technology Plan Projects of Guiyang Municipal Bureau of Science and Technology of China (No. 2012-205), Science and Technology Plan of Guizhou Province of China (No. 2012-6015)

11. References

[1] Launiainen S, Katul GG, Lauren A, Kolari P (2015) Coupling boreal forest CO2, H2O and energy flows by a vertically structured forest canopy-Soil model with separate bryophyte layer. ECOL. MODEL 312(24):385-405.

[2] Ozsoy, Gokhan, Aksoy, Ertugrul (2015) Estimation of soil erosion risk within an important agricultural sub-watershed in Bursa, Turkey, in relation to rapid urbanization. *ENVIRON.MONIT. ASSESS* 187(7), 4653.

[3] Garcia-Ruiz JM, Begueria S, Nadal-Romero E, Gonzalez-Hidalgo JC, Lana-Renault N, Sanjuan Y (2015) A meta-analysis of soil erosion rates across the world. *GEOMORPHOLOGY* 239, 160-173.

[4] Yoo K, Fisher B, Aufdenkampe A, Klaminder J (2015). The geochemical transformation of soils by agriculture and its dependence on soil erosion: An application of the geochemical mass balance approach. *SCI. TOTAL. ENVIRON* 521, 326-335.

[5] Kairis O, Karavitis C, Salvati L, Kounalaki A, Kosmas K (2015). Exploring the Impact of Overgrazing on Soil Erosion and Land Degradation in a Dry Mediterranean Agro-Forest Landscape (Crete, Greece). *ARID LAND RES. MANAG* 29(3), 360-374.

[6] Ludwig JA, Wilcox BP, Breshears DD, Tongway DJ, Imeson AC (2005). Vegetation Patches and runoff-erosion as interacting ecohydrological processes in semiarid landscapes. *ECOLOGY* 86(2), 288-297.

[7] Zhu TX (2013) Spatial variation and interaction of runoff generation and erosion within a Semi-arid, complex terrain catchment: a hierarchical approach. *J. SOIL. SEDIMENT* 13(10):1770-1783.

[8] Miller MF (1926). Waste through soil erosion. *Journal Am. Soc. Agron* 18, 153-160.

[9] Florens JP, Mouchart M, Rolin JM (2013) ASSESSING THE PERFORMANCE OF A SPATIALLY DISTRIBUTED SOIL EROSION AND SEDIMENT DELIVERY MODEL (WATEM/SEDEM) IN NORTHERN ETHIOPIA. *Land Degradation & Development*, 24(2):188–204.

[10] Zingg AW (1940) Degree and length of land slope as it affects soil loss in runoff. *Agricultural Engineering* 21, 59-64.

[11] Wischmeier WH, Smith DD (1965) Predicting rainfall-erosion losses from cropland east of the Rocky Mountains[R].*USDA Agricultural Handbook* No.282.

[12] Peng T, Wang SJ, Zhang XB (2008) RESULTS OF PRELIMINARY MONITORING OF SURFACE RUNOFF COEFFICIENTS FOR KARST SLOPES. EARTH AND ENVIRONMENT. Earth & Environment 36(2), 125-129.

[13] Cao LX, Liang Y, Wang Y, Lu HZ (2015) Runoff and soil loss from Pinus massoniana forest in Southern China after simulated rainfall. *Catena* 129, 1-8.

[14] Zhang JJ, Lu FC, Zhao JG (2013) Present Situation and Prospects of the Research in Monitoring Technology for Hillslope Soil Erosion. *Zhejiang Hydrotechnics* (6), 43-45.

[15] L Vandekerckhove, JP, D Oostwoud Wijdenes, G, Gyssels (2001) Short term bank gully retreat rates in Mediterranean environments. *Catena* 44, 133-161.

[16] Sun YH, Wang T, Zhao CC, et al (2009) Contributing ratio of impact factor on gully erosion in Loess hilly and gully area of Qinghai province. *Ecology and Environmental Sciences* 18(4), 1402-1406.

[17] Wolman MG (1959) Factors influencing erosion of a cohesive river bank. *American Journal Science* 257, 204-216.

[18] Stephenson R, Lambeck K (1985) Erosion-isostatic rebound models for uplift: an application to south-eastern Australia. *Geophysical Journal International* 82(1):31–55.

[19] Lawler DM (1989) Some New Development in Erosion Monitoring: The Potential of

Optoelectronic Techniques [R]. *School of Geography, University of Birmingham Working* Paper 47, 44.

[20] Du P, Walling DE (2011) Using ^{137}Cs measurements to investigate the influence of erosion and soil redistribution on soil properties. *APPL. RADIAT. ISOTOPES* 69(5), 717-726.

[21] Honda M, Matsuzaki H, Miyake Y Maejima Y, Yamagata T, Nagai H (2015) Depth profile and mobility of ^{129}I and ^{137}Cs in soil originating from the Fukushima Dai-ichi Nuclear Power Plant accident. *J. ENVIRON. RADIOACTIV* 146, 35-43.

[22] Haghiri F (1969) Effects of Soil and Crop Management Practices on the Removal of Strontium90 by Plant Uptake, Leaching, Runoff and Erosion1. *Agronomy Journal* 61(5):793-796.

[23] RItchie JC, McHenry JR (1990) Application of radioactive fallout ^{137}Cs for measuring soil erosion and sediment accumulation rates and Patterns. A Review. *J. Environ. Qual* 19, 215-233.

[24] Ritchie JC, Spraberry JA, McHenry JR (1974) Estimating soil erosion from the redistribution of ^{137}Cs. *Soil Sci. Soc. Am. Proc* 38(1), 137-139.

[25] Walling DE, He QP, Blake W (1999) Use of ^{7}Be and ^{137}Cs measurements to document short and medium term rates of water induced soil erosion agricultural land. *WATER RESOUR.RES.* 35(12), 3865-3874.

[26] Elliott GL, Campbell BL, Loughran RJ (1990) Correlation of erosion measurements and soil Caesium-137 content. *APPL. RADIAT. ISOTOPES* 41(8), 713-717.

[27] Feng MY, Wen AB (2002) Research Progresses on ^{137}Cs Measurement in Soil Erosion in China. *J .SOIL WATER CONSERV* 16(2), 61-64.

[28] Lowrance R, McIntyre S, Lance C (1988) Erosion and deposition in a friend forest system estimated using Caesium) 137 activity. *J Soil Water Conserve* 43(2), 195-199.

[29] Kachanoski RG, de Jong E (1984) Predicting the temporal relationship between soil Caesium-137 and erosion rate. *J. Environ. Qual* 13(2), 301-304.

[30] Simpson HJ, Olsen CR, Triver RM and Williams SC (1976) Man-made radionuclides and sedimentation in the Hudson river estuary. *Science* 194:1979-1982.

[31] Wallbrink PJ, Olley JM and Murry AS (1994) Measuring soil movement using ^{137}Cs implications of reference site variability. *ISAH* 1-9.

[32] Alewell C, Meusburger K, Juretzko G, Mabit L, Ketterer ME (2014) Suitability of 239+240PU and Cs-137 as tracers for soil erosion assessment in mountain grasslands. *CHEMOSPHERE* 103, 274-280.

[33] Leckie HD, Almond PC (2015) Evidence of prehistoric wind erosion of the Mackenzie Basin, South Island, New Zealand: an assessment based on 137Cs and Kawakawa-Oruanui tephra. *SOIL RESEARCH* 53(1), 56-66.

[34] Porto P, Walling DE, Callegari G (2013) Using ^{137}Cs and $^{210}Pbex$ measurements to investigate the sediment budget of a small forested catchment in southern Italy. *HYDROL. PROCESS* 27(6), 795-806.

[35] Dörr H (1995) Application of ^{210}Pb in soils. *J. PALEOLIMNOL* 13, 157-168.

[36] Walling DE, He Q (1999) Using of fallout lead-210 measurements to estimate soil erosion on cultivated land. *Soil Sci. Am. J* 63, 1404-1412.

[37] Porto P, Walling DE, Tamburino V et al (2003) Relating caesium-137 and soil loss from cultivated land. *Catena* (53), 303-326.

[38] Bai XY, Zhang XB, Long Y, Liu XM, Zhang SY (2013) Use of 137Cs and 210Pbex measurements on deposits in a karst depression to study the erosional response of a small karst catchment in Southwest China to land-use change. HYDROL PROCESS 27(6), 822 829.

[39] Bai ZG, Wan X (1997) GEOCHEMICAL SPECIATION OF ^{7}Be, ^{137}Cs, ^{226}Ra and ^{228}Ra IN SOILS OF THE KARST REGION, SOUTHWESTERN CHINA AND THEIR EROSION TRACE. *ACTA SCIENTIAE CIRCUMSTANTIAE* 17(4), 407-411.

[40] Blake WH, Walling DE, He Q (2002) Using cosmogenic beeryllium-7 as a tracer in sediment budget investigations. *Geografiska Annaler* 84(2), 89-102.

[41] Zhang QW, Dong YQ, Li F, Zhang AP, Lei TW (2014) Quantifying detachment rate of eroding rill or ephemeral gully for WEPP with flume experiments. *J. HYDROL* 519, 2012-2019.

[42] Wallbrink PJ, Murray AS (1996) Soil loss using the inventory ratio of excess Lead-20 to Cesium-137 *Soil Sci. Soc. Am. J* 60, 1201-1208.

[43] Burch GJ, Barling RD, Branes CJ et al (1998) Detection and prediction of sediment sources in catchments use Be-7 and Cs-137. *Hydrology and Water Resources Symposium* 44, 146-150.

[44] Matisoff G, Wilson CG, Whiting PJ. The ^{7}Be/^{210}Pbxs ratio as an indicator of suspended sediment age or fraction new sediment in suspension. *Earth Surface Processes and Landforms*, 2005, 30(9):1191–1201.

[45] Dearing JA, Morton RI, Price TW, et al (1986) Tracing movements of topsoil by magnetic measurements: two case studies. Phys. *Earth Planet. Inter.* 42, 93-104.

[46] Chaparro MAE, Suresh G, Ramasamy V, Sinito AM (2013) Magnetic studies and elemental analysis of river sediments: a case study from the Ponnaiyar River (Southeastern India). *ENVIRON. EARTH SCI.* 70(1), 201-213.

[47] Caitcheon GG (1993) Sediment source tracing using environment almagnetism: A new approach with examples from Australia. *HYDROL. PROCESS.* 7(4), 349-358.

[48] Jong ED, Nestor PA, Pennock DJ (1998) The use of magnetic susceptibility to measure long-term soil redistribution. *Catena*, 32(1):23-35.

[49] Jordanova N, Jordanova D, Liu QS, Hu PX, Petrov P, Petrovsky E (2013). Soil formation and mineralogy of a Rhodic Luvisol-insights from magnetic and geochemical studies. *GLOBAL PLANET. CHANGE.* 110, 397-413.

[50] Dearing JA, Elner JK, Happey-Wood CM (1981) Recent sediment in flux and erosional processes in a Welsh upland lake-catchment based on magnetic susceptibility measurements. *QUATERNARY RES.* 16, 356-372.

[51] Franciskovic-Bilinski S, Scholger R, Bilinski H, Tibljas D (2014) Magnetic, geochemical and mineralogical properties of sediments from karstic and flysch rivers of Croatia and Slovenia. *ENVIRON. EARTH. SCI.* 72(10), 3939-3953.

[52] Meyer LD (1984) Evaluation of the universal soil loss equation. *J. SOIL. WATER. CONSERV.* 39:99-104.

[53] Wischmeier WH, Smith DD (1978) Predicting rainfall erosion losses. *USDA Agricultural Handbook* 537.

[54] Pradeep GS, Krishnan MVN, Vijith H (2015) Identification of critical soil erosion prone areas and annual average soil loss in an upland agricultural watershed of Western Ghats, using analytical hierarchy process (AHP) and RUSLE techniques. *ARAB. J. GEOSCI.* 8(6), 3697-3711.

[55] Shi ZH, Cai CF, Ding SW, Wang TW, Chow TL (2004) Soil conservation planning at the small watershed level using RUSLE with GIS: a case study in the Three Gorge Area of China. *CATENA* 55(1), 33-48.

[55] Brazier RE, Beven KJ, Freer J, Rowan JS (2000) Equifinality and uncertainty in physically based soil erosion models: Application of the glue methodology to WEPP-the water erosion prediction project-for sites in the UK and USA. *EARTH SURF. PROC. LAND* 25(8), 825-845.

[57] Saghafian B, Meghdadi AR, Sima S (2015). Application of the WEPP model to determine sources of run-off and sediment in a forested watershed. *HYDROL.PROCESS* 29(4), 481-497.

[58] Kumar S, Mishra A, Raghuwanshi NS (2015) Identification of Critical Erosion Watersheds for Control Management in Data Scarce Condition Using the SWAT Model. *J. HYDROL. ENG.* [SI]: 20(6).

[59] StockermBD, Spahni R, Joos F (2014) DYPTOP: a cost-efficient TOPMODEL implementation to simulate sub-grid spatio-temporal dynamics of global wetlands and peatlands. *GEOSCI. MODEL. DEV* 7(6), 3089-3110.

[60] Dawen Y, Kanae S, Oki T (2003) Global potential soil erosion with reference to land use and climate change. *HYDROL PROCESS* 17, 2913-2928.

[61] King D, Fox DM, Daroussin J (1999) Upscaling a simple erosion model from small areas to a large region. *NUTR. CYCL. AGROECOSYS* 50, 143-149.

[62] Aiello A, Adamo M, Canora F (2015) Remote sensing and GIS to assess soil erosion with RUSLE3D and USPED at river basin scale in southern Italy. *CATENA* 131, 174-185.

[63] Zhu MY (2015) Soil erosion assessment using USLE in the GIS environment: a case study in the Danjiangkou Reservoir Region, China. *ENVIRON. EARTH. SCI* 73(12), 7899-7908.

A Review on New Technological Progress for Beneficiation of Refractory Phosphate Ore in China

Xu Dong , Siqing Liu, Yanqing Yao, Hailin Liu and Yi Pei

Kunming University of Science and Technology, Kunming City, Yunnan Province, China

451184428@qq.com

Abstract. The world is rich in phosphate resources, but the vast majority of the resources are of middle and low grade ores, and it is difficult to obtain high grade concentrate at a high recovery by using traditional separation methods. In this paper, the distribution and flotation processes of typical phosphate ores in China are introduced. Considering the current situation in phosphate ore beneficiation both home and abroad, the new trend of processing technology and reagents in China are summarized. Flotation is still regarded as the main processing method in phosphate ore beneficiation for a certain time in near future.

1. Introduction

According to the US Geological Survey (USGS) statistics, the world's phosphate reserves was 49.75 billion tons in 2015, and China has 13 billion tons. China's phosphate ore resources ranked the second in the world, but since 2006, many experts and scholars pointed out the current situation of phosphate ore resources, and issued an urgent appeal. In particular, the rational exploitation of China's phosphate ore resources and their sustainable utilization have been given a great deal of attention [1]. The next few years, China's major mining operation of the phosphate ore will focus on low-grade phosphate ore, which contains a large number of gangue minerals, such as silicon, magnesium and so on. Desiliconization of magnesium has been a problem of phosphate ore beneficiation. Mineral processors have been committed to the desilication and de-magnesium to improve the grade of the concentrate by applying new process and new reagents

2. Phosphate ore resources and beneficiation process

China's phosphate ore resources are mainly distributed in Yunnan, Guizhou, Sichuan, Hubei and Hunan provinces, reserves account for about 74% of the national in total[2]. Although China is rich in reserves of phosphate ore, the phosphate ores are mainly of low-grade and lean type, which are difficult to process, and the easy-to-process ores are in quite a small number, and majorities are of poor availability [3] and refractory. In the past, the man beneficiation processes of phosphate ores were heavy medium separation technology, roasting digestion process and chemical beneficiation which are high in costs and low in efficiency.

2.1. Flotation process

2.1.1. Direct flotation process. Fang Shixiang[4] and other inventors disclosed the method of a low-grade mixed refractory colloidal phosphate in a positive way. The monomer of mixed type phosphate ore is firstly dissociated. After adding positive sodium carbonate, sodium silicate and other direct flotation regulators and fatty acid anion collectors, and then positive roughing flotation, positive cleaning floatation and positive scavenging flotation are applied tailings of the roughing flotaion and tailings of the cleaning flotation were combined to the positive scavenging flotation. When the mixed collophane is in the calcium-silica type, the concentrates of the positive scavenging flotation then returned to the positive roughing flotation, which forms a closed circuit operation; when the mixed collophan is of silica-calcium type, the concentrates obtained by scavenging flotation can be combined with the concentrates of cleaning flotation directly to the next reverse flotation operation. This invention simplifies the flotation process, is of simple operation, and greatly improve the recovery rate and the yield of concentrate, which is a kind of flotation method that is easy to operate, efficient to remove impurities and suitable to promote for the low-grade refractory mixed collophane.

2.1.2. Reverse flotation process. Liu Yuntao[5] invented a coarse-grain reverse flotation method for phosphate ore, which is characterized in that the low grade phosphate ore is crushed and grinded, and flotation feed size is of -0.074mm less than 50%, and then put it into the flotation tank with the pulp slurry concentration of more than 50%, pulp temperature between 10℃ and 30℃, and then regulator--inorganic acid and mixed reverse flotation collector were added to, supplemented with water for reverse flotation operation, in which the tank product is low magnesium phosphate concentrate. The inventive method reduces the final concentrate magnesium content and improves the quality of the concentrate. The process has advantages of being simple in flow, good in separation performance and high in removal rate of magnesium.

Wang[6] and others improved the reverse flotation process of Yunnan Phosphating Group An increase of four sets of $120m^3$ flotation cell, which were arranged in parallel series for rougher flotation, making the roughing time increased from 14min to 17min.This makes the roughing process time much longer, reducing the pressure of the scavenging. Therefore, it stabilizes the system process, reducing the grade of P_2O_5 in the tailings and in the meanwhile, increasing the recovery.

2.1.3. Flocculation-flotation process. Wang Qiao[7] applied the flocculation to the sorting of phosphate ore. After series of experiments, the study shows that some flocculants involving Y3, F-08, N+09 and 850 can promote the flotation performance obviously. Of all the experimented flocculants, the flocculation F-08 has a great influence on the phosphate ore of Jinning, Yunnan Province. When the dosage of F-08 is 30g/t, the concentrate grade is 24.54% and the recovery is 85.30%, with beneficiation efficiency of 19.55%. Grade of the product increased by 4.17% than the grade of raw ore, and increased by 0.55% than that without adding flocculation. The recovery rate is 4.5% higher and the beneficiation efficiency is 3.95% higher than that without adding flocculation. The experiments results prove that this method is efficient in recovering the phosphate ore.

2.1.4. Direct-reverse flotation process. Fu Kewen[8] and others found that there are many problems using direct flotation to sort Dayukou open-pit mining of phosphate ore, as the concentrate foam is high in viscosity, the bubble is not easy to remove, and the concentrate is difficult to settle, the mass fraction of MgO in the concentrate is too high. Therefore, on the basis of laboratory tests, a new SP reagent was selected as an inhibitor of phosphate minerals, and a new technology of direct-reverse flotation was used to de-magnesium for the and rough concentrate. The results show that when the P_2O_5 of the raw ore is 17.90% and the MgO is 4.28%, concentrate of the P_2O_5 and MgO can be obtained at 31.62% and 0.74%, respectively, with the recovery of 81.35% MgO.

2.1.5. Reverse-direct flotation process. Huang Qi-mao[9] and others optimized the reagent system of the flotation according to ore properties of a low-grade silica-calcium phosphate ore in Hubei better selection indicators obtained by using reverse-direct flotation process are as follows: When the raw ore is of 17.09% P_2O_5 and the mass fraction of MgO is 5.29%, the obtained phosphate ore concentrate is 29.03% P_2O_5, with the recovery of 78.22%, and the mass fraction of MgO is 0.71%. This study provides a reference for the development and utilization of a low-grade silica-calcium collophanite.

2.1.6. Double-reverse flotation process. Li Songqing[10] and others studied the effect of the collector of BK425 for the double-reverse flotation, on a high-magnesium high-alumina phosphate ore in Guizhou. After two-staged grinding, the raw ore achieved the grinding fineness of -0.074 mm 75%. And phosphate concentrate assaying 33.97% P_2O_5, 0.73% MgO, with the recovery of P 73.71% was obtained by using closed-circuit process of the double-reverse flotation, and the sum of the sesquioxide ($Al_2O_3 + Fe_2O_3$) was 2.23%.

A medium-low grade collophane in Yunnan is a mixed type of phosphate ore, Wang Canxia[11] and others separated the phosphate ore minerals from the gangue minerals effectively by double-reverse flotation, which is based on the ore properties. When the grinding fineness is -0.074mm 88.36%, phosphate concentrate assaying 28.66% P_2O_5, 0.73% MgO, and 19.56% SiO_2, with the recovery of P 79.65% can be obtained.

2.2. Other methods

Ma Xiaoqing[12] invented a method that using medium mechanical processing to de-magnesium from the collophane ore. They added regulators and collectors of phosphate mineral flotation to the slurry of the broken silica-calcium phosphate ore then diluted with water, to conduct the flotation of phosphate mineral in the direct flotatio system. After de-reagents and de-sliming, the concentrate enters the agitating tank to conduct the de-silication in the reverse flotation system to reversely float the silicious gangue minerals. The slurry of the rough concentrate of floatation after concentration enters the mechanical scrubbing system by adding a number of media to carry out series of scrubbing and desliming. The concentrate after filtration, and drying then becomes qualified phosphate concentrate for sale. This method proved to be more economical and practical by adopting. The method of mechanical de-magnesium which eliminates the need for grinding and reagents, simplifies the flotation process, reduces the cost and improves efficiency, and the obtained concentrate is low in MgO content and high in grade.

Li Ji[13] and others developed a method for the removal of silicic acid salts in low-grade phosphate ore. After magnetic separation, phosphate ore tailings were grinded to fineness of 90% -0.074mm and then some adjusters, collectors and silicic acid metal salt inhibitors were added into the pulp, by adopting the flotation flowsheet of one-stage roughing and two-stage cleaning, to achieve high-grade phosphate ore concentrate products. The invention realizes the efficient separation of telling the silicate metal salts apart from the associated low-grade phosphate ore, and the obtained phosphate ore ore containing over 34% P_2O_5, with the recovery more than 75%, less than 1.0% MgO, less than 2% $Al_2O_3 + Fe_2O_3$ content. The produced high-quality phosphate concentrate can be used as raw material for production of phosphate fertilizer, which proves that this method is a feasible mineral processing technology for utilization of a low-grade phosphate ore.

3. New reagents for the beneficiation of phosphate ores

The traditional flotation processes can be divided into 5 categories, they are: direct flotation, reverse flotation, direct-reverse flotation, reverse-direct flotation and double-reverse flotation. Generally, in the direct flotation, sodium carbonate is used as a regulator, sodium silicate as a inhibitor; in the reverse flotation, sulfuric acid as a regulator, phosphoric acid as a inhibitor[14]. Separation of magnesium and silicon in phosphate ore has long been a difficult problem, and the traditional flotation method is not ideal enough to separate them.

3.1. Collectors

Zhang Ying[15], developed a new type of de-magnesium collector of FC-2B that can be applied to flotation of phosphate ore through series of separation tests on phosphate ore and dolomite, flotation test of the actual mineral results show that, FC-2B collector can effectively separate phosphate ore and dolomite. The raw ore assays 19.16% P_2O_5, 4.67% MgO, and the obtained concentrate assays 28.54% P_2O_5, with the recovery of 80.40%; 4.67% MgO, with the exclusion rate of 90.52%. And through infrared spectrum test, it is proved that the collector has a strong chemical adsorption on the surface of dolomite, and it keeps a better selectivity to dolomite.

Li Chengxiu[16] developed a new type of collector EM-LP-01 for direct flotation of collophane, with oleic acid, linoleic acid and linolenic acid-based (ratio of 3: 2: 1) mixed acid saponification and the SDS and OP-10 were mixed in 1: 1 to make the synergist. Then the mixed oil fatty acid soap and the combined synergist were mixed in 100: 20. This collector was applied to a silica-calcium collophane ore from Guizhou Wengfu Plant by adopting direct-reverse flotation process, under the condition of normal temperature After desilicication direct flotation of "one-stage roughing one-stage scavenging one-stage cleaning" and de-magnesium reverse flotation of "one-stage roughing one-stage scavenging", concentrate assaying 30.86% P_2O_5 with the recovery of 89.57%, 0.77% MgO can be obtained. The new agent has a strong ability to selectively capture, and its solubility and dispersion are relatively good under the condition of room temperature.

3.2. Regulators

3.2.1 Modifying reagent. Ma Xianlin[17] invented a new regulator, using the mother liquor of production of sodium fluoride as pH adjustment reagent. The effects of mother liquor with different compositions on the residual MgO content in phosphate concentrate was studied. And the effects of H_2SO_4 and H2SiF6 in the mother liquor on the de-magnesium of phosphate ore were preliminary analysed. At the same time, the influence of the residual MgO content of phosphate concentrate that using the different proportions of mother liquor of villiaumite instead of sulphuric acid was studied. The results show that the fluoride mother liquor certainly has de-magnesium effect, and it can replace 50%~75% the consumption of sulphuric acid.

3.2.2 Depressant. Jiang Zhensheng etc. [18] invented a carbonate inhibitor. The inhibitor is an organic high molecular compound that contains a large amount of polar groups, such as carboxyl group, hydroxyl group, methoxy group, carbony l group and amino group. The method has the advantages of simple process and no pollution. This inhibitor can effectively reduce the content of MgO in phosphate concentrate and COD in tailings. It also can reduce the dosage of regulator Na_2CO_3 in the process of direct phosphate flotation, which has good economic and environmental benefits.

3.2.3 Other regulators. Li Yaoji etc. [19] invented a method that using phosphoric acid residue as flotation regulator of phosphate ore. The residue of industry produced phosphoric acid is used as the raw material, through the pump the residue can be sent to the flotation regulator tank directly. After the flotation, phosphate flotation index can reach the requirements of dihydrate production of phosphoric acid. The invention has the advantages of simple process and applicability, and realizes the phosphate chemical mineralization coupling with phosphorite ore dressing, which has great significance in terms of resources and environment protection.

Zhao Jing, Chen Hongjian[20] found thatin the mineral processing of phosphate ore (especially calcium or silica-calcium collophane, adding a small amount of TZ can significantly reduce the

IOP Conf. Series: Earth and Environmental Science **63** (2017) 012043 doi:10.1088/1755-1315/63/1/012043

viscosity of the flotation foam, improve the liquidity of the foam and significantly shorten the flotation time. Hence, the grade of phosphate concentrate can be significantly improved and it has a far-reaching impact on industrial production.

4. Conclusion

The phosphate ore reserves in China rank the second in the world, but the high-grade phosphate ore is deficient, which cannot meet the need of agriculture and other industries. The development and utilization of low-grade phosphate ores will last a long time for mineral processors to develop the new process with low cost and high efficiency. At present, the comprehensive utilization of phosphate ore resources in China has made certainly achievements by mineral processors. Innovation on technological process for the refractory ore utilization should be conducted to produce the high grade phosphate concentrate of higher added value, and flotation will still be the leading technology for a long time. At the same time, developing more efficient flotation reagents and applying accurate and reasonable process to match with the ore properties become more and more important. Moreover, for the low grade and refractory phosphate ore beneficiation, more efforts should be paid on research of low cost and efficient beneficiation process.

5. References

[1] Xiucheng X 2014 *China's phosphate ore resources can be mined* (Mining and Metallurgical Engineering)

[2] Yongfu Y 2015 *Phosphate beneficiation progress and existing problems* (Mining and Metallurgical Engineering)

[3] Xiaohong S 2015 *China Phosphate Resource Demand Forecast* (Acta Geoica Sinica)

[4] Shixiang F 2014 *In the low-grade mixed refractory phosphate ore is floating branch flotation method* (Chemical Minerals & Processing)

[5] Yuntao L 2015 *A Method for Desilication of Coarse Grain Reverse Flotation for Phosphate Ore* (Mining and Metallurgical Engineering)

[6] Chong W 2013 *Single reverse flotation process of phosphate ore phosphate optimization* (Phosphate Fertilizer and Nitrogen Fertilizer)

[7] Qiao W 2015 *Positive flotation test of Jinning Phosphate Ore in Yunnan Province* (IM & P Chemical Mineral Processing)

[8] kewen B 2013 *Dayukou phosphate ore phosphate positive and negative flotation of new technology research and industrial application* (Chemical Minerals & Processing)

[9] Qimao H 2010 *Study on the Flotation Process of a Calcium Phosphate Ore in Hubei Province* (Chemical Minerals & Processing)

[10] Songqing L 2016 *Guizhou a high-magnesium high-alumina phosphate ore double anti- flotation drop miscellaneous test study* (Mining and Metallurgy)

[11] Canxia W 2015 *Yunnan in a low-grade phosphate ore double reverse flotation test* (Chemical Minerals and Processing)

[12] Xiaoqing M 2015 *A mechanical treatment of mechanical processing of magnesium phosphate dephosphorizing ore beneficiation* (Phosphate Fertilizer and Nitrogen Fertilizer)

[13] Ji L 2013 *A low-grade phosphate ore metal silicate method of prolapse* Mining and Metallurgical Engineering)

[14] Fengting Z 2016 *Flotation process of different types of phosphate ore in Dianchi area* (Phosphate and Compound Fertilizer)

[15] Ying Z 2014 *Experimental Study on New Type Magnesia Collector FC-2B Floatation Phosphate Ore in Haikou, Yunnan Province* (Chemical Minerals and Processing)

[16] Xucheng L 2014 *Difficult selection of phosphate ore flotation new pharmaceutical agent test and mechanism of action* (Phosphate Fertilizer and Nitrogen Fertilizer)

[17] Xianlin M 2015 *Application of sodium fluorosilicate mother liquor in phosphate ore flotation* (IM & P Chemical Minerals Processing)

[18] Zhensheng J 2013 *A carbonate inhibitor and its preparation method and application* (IM & P Chemical Minerals Processing)

IOP Conf. Series: Earth and Environmental Science **63** (2017) 012043 doi:10.1088/1755-1315/63/1/012043

[19] Yaoji L 2014 *A Method for Phosphate Flotation Regulator by Phosphoric Acid Residue* (Mining and Metallurgical Engineering)

[20] Jing Z and Hongjian C 2011 *Effect of a new regulator on beneficiation of a siliceous colloidal phosphate ore* (IM & P Chemical Minerals Processing)

Analysis on the Climate Change Characteristics of Dianchi Lake Basin under the Background of Global Warming

Zhenyu Yu[1,2,a)], Yi Luo[1,2,3,b)] , Kun Yang[1*,2,3,c)] and Qiongfei Deng[1,2,d)]

[1]College of Information, Yunnan Normal University, Kunming 650500, China
[2]The Engineering Research Center of GIS Technology in Western China, Ministry of Education, Kunming 650500, China
[3]College of Tourism and Geography, Yunnan Normal University, Kunming 650500, China

yangkunynnu@163.com

Abstract. Based on the data published by the State Statistical Bureau and the weather station data, the annual mean temperature, wind speed, humidity, light duration and precipitation of Dianchi Lake in 1990 ~ 2014 were analysed. Combined with the population The results show that the climatic changes in Dianchi Lake basin are related to the climatic change in the past 25 years, and the correlation between these factors and the main climatic factors are analysed by linear regression, Mann-Kendall test, cumulative anomaly, R/S and Morlet wavelet analysis. Population, housing construction area growth and other aspects of the correlation trends and changes in the process, revealing the population expansion and housing construction area growth on the climate of the main factors of the cycle tendency of significant impact.

1. Introduction

Global warming is a scientific issue which has always been the focus of attention of scholars at home and abroad. Climate change is the combined result of the internal processes of nature and the continual changes in human activities to atmospheric constituents and land use. Global climate change will bring disasters such as glacier melting, permafrost melting, coral reef dying, sea level rise, ecosystem change, aggravation of droughts and floods for mankind and ecosystem[1] . According to the IPCC, the global average surface temperature has risen by 0.5 °C over the past century [2]. World Meteorological Organization in 2007 announced that from 1998 to 2007 is the warmest decade recorded since then have been announced in 2009, 2011, 2012, respectively, the corresponding range of the warmest year, the global warming trend will continue [3]. "Second National Assessment Report on climate change" that China's climate warming rate may be 0.5~0.8℃/100a, higher than the global warming rate[4]. Plateau climate has obvious characteristics, for global warming under the study has a certain representation. Dianchi Lake is located in the Yunnan-Guizhou Plateau, north-south distribution, is the sixth largest freshwater lake, the lake about 1886 meters above sea level, an area of about 330 square kilometers, the average depth of about 5 meters, are semi-closed lakes, only the southwest of Haikou As the outlet, known as the plateau pearl, in the subtropical plateau southwest monsoon climate zone, climate change is mainly affected by the southwest monsoon and tropical continental air mass alternating control [5]. At the same time, Dianchi Lake is the main body of water for irrigating, transferring livestock and accepting in Kunming. It is also the basic condition of urban development, and plays an important role in climate regulation in Dianchi Lake Basin [5].

In recent years, the Dianchi Lake ecological environment has been seriously damaged, the climate of the basin had a certain impact, the issue of attention by the relevant departments. Therefore, it is necessary to study the climate change in the Dianchi Lake basin, which can provide data support for the relevant departments to protect and manage the environment and prevent the disaster caused by climate change. Wen H et al [6] studied the aquatic biota structure and water pollution factors and their spatial distribution characteristics in the Rivers of the Rivers by means of factor analysis and other methods based on the data of 29 aquatic organisms and water pollution in Dianchi Lake. Liang Z Y et al [7] based on the 1998-2010 Dianchi Lake monitoring stations, water quality indicators of the time series to STL and other methods of water quality trends in time and space characteristics of the identification and identification. Li Z Y et al [8] used the principle of conservation of mass to construct the simulation system of the pollution control effect of the old Yunliang River and calculated the pollution load into the Dian River. Chen C et al [9] used day-to-day sounding data from 1958 to 2013 in Wuhan, and used climatological tendency and wavelet analysis to reveal the relationship between periodic and aperiodic mean wind speed Sexual changes. Qi D L [10] used the basic meteorological data from 14 meteorological stations from 1964 to 2013 in the Sanjiangyuan area of Qinghai Province, and analyzed the temporal and spatial variation and trend of evaporation in the Sanjiangyuan area by using linear propensity estimation and Mann-Kendall method. Zhou D et al [11] used the linear trend, Morlet wavelet analysis, Mann-Kendall and other methods to analyze the influence of the temperature, precipitation and sunshine duration on 468 meteorological stations on the continent from 1961 to 2010, variation characteristics of each climate element. Ding W R [12] studied the effects of temperature, precipitation, precipitation and regional climate change in the Erhai Lake area from 1951 to 2014 on the basis of linear trend estimation, Mann-Kendall trend test, Morlet wavelet analysis and R/S analysis. Based on the previous research [13-16], and with reference to the above research team research methods and ideas, this paper studies and analyzes the climatic factors (temperature, wind speed, precipitation, duration of light) in the Dianchi Lake basin under the background of global warming on a long time scale , Humidity and so on), taking into account the human factors such as the total population at the end of Kunming and the housing construction area, to explore the causes of climate change and provide data support for further decision-making of related departments.

2. Data Selection and Methods

2.1. Data Selection

From 1990 to 2014, the population of Kunming and the floor space of the building were derived from the total population at the end of the corresponding year in Yunnan Statistical Yearbook, the area of housing construction and the floor space, average temperature, sunshine duration and precipitation from the China Statistical Yearbook, the average annual wind speed, relative humidity data from China Meteorological Data Network and the World Weather Web site published data (https://www.wunderground.com).

2.2. Method Introduction

The main research methods include linear regression [17-19], significance test [19], correlation coefficient [20], Mann-Kendall test [21], cumulative anomaly [22], Pettitt mutation test [23], R/S analysis [24] and Morlet wavelet analysis [4, 7].

2.2.1. Correlation analysis. The four factors, such as total population, light duration, annual minimum temperature and construction area, did not pass the normal distribution test. In order to facilitate the comparative analysis, this paper used Spearman correlation coefficient to describe the correlation between the factors and the linear relationship. The Spearman correlation coefficient, also known as the Spearman rank correlation coefficient, is used to estimate the correlation between two variables, x and y, where the correlation between variables can be described using monotonic functions. If the two variables do not have the same two elements, then when one variable can be expressed as a good monotonic function of another variable, that is, the two variables have the same trend of change, r between two variables can reach 1 or -1. The formula is:

$$r = \frac{\sum_{i=1}^{N}(x_i-\bar{x})(y_i-\bar{y})}{\sqrt{\sum_{i=1}^{N}(x_i-\bar{x})^2\sum_{i=1}^{N}(y_i-\bar{y})^2}} \quad (1)$$

Where r is the Spearman correlation coefficient, x_i and y_i represent the ith sample value of the x variable and the y variable, respectively, and N is the number of sample values. Spearman correlation coefficient$(-1 \leq r \leq 1)$ is generally divided into:

- $|r| < 0.4$ for low linear correlation.
- $0.4 \leq |r| < 0.7$ for significant correlation.
- $0.7 \leq |r| < 1$ for highly linear correlation.

2.2.2. Cyclical analysis Wavelet analysis is a powerful statistical tool, the earliest used in signal processing and analysis. In geosciences, various meteorological factors, hydrological processes and material exchange between ecosystems and the atmosphere can be seen as periodic signals with time, so the wavelet analysis method is also applicable to the field of geology, And the complex time structure of geo-science process.

A continuous wavelet transform of a discrete time series $x_n(n=1,\ldots,N)$ having an equal time step δt may be defined as the wavelet functionψ_0 scale and the convolution of x_n at the conversion:

$$W_n^X(s) = \sqrt{\frac{\delta t}{s}}\sum_{n=0}^{N-1} x_n \cdot \varphi^*\left[\frac{(n'-n)\delta t}{s}\right] \quad (2)$$

Where * is the conjugate complex number, N is the total number of data in the time series, and *(δt/s)1/2* is a factor used to normalize the wavelet function, so that the wavelet function has unit energy at each wavelet scale *s* .

Morlet wavelet not only has no orthogonality, but also by the Gaussian adjustment of the index complex wavelet.

$$\varphi_0(t) = \pi^{-1/4}e^{i\omega_0 t}e^{-t^2/2} \quad (3)$$

Where t is the time andω_0 is the dimensionless frequency. Whenω_0=6, the wavelet scale s is almost equal to the Fourier period (λ=1.03s), so the scale term and the periodic term can be substituted for each other. It can be seen that Morlet wavelet has a good balance between time and frequency localization.

$|W_n^X(s)|^2$ is defined as the wavelet power spectrum, which expresses the magnitude of the time series fluctuation in a given wavelet scale and time domain. Since the Morlet mother wavelet is a complex wavelet, $W_n^X(s)$ is also a complex number and its complex part can be interpreted as local phase. The wavelet power spectrum is averaged over a period to obtain the wavelet full spectrum:

$$W^2(s) = \frac{1}{N}\sum_{n=0}^{N-1}|W_n(s)|^2 \quad (4)$$

Wavelet full spectrum can show the unbiased and uniform estimation of real power spectrum of time series. Since the full spectrum of the wavelet can show the background spectrum measurement, the peak of the local wavelet can be verified, so the periodic fluctuation characteristics and intensity of the time series can be clearly discerned by the wavelet full spectrum.

3. Interannual variability of meteorological factors and human factors

In the climate of global warming, the temperature of Kunming in the past 25 years is on the rise, the trend of temperature is 0.391℃/10a, the global temperature tendency is 0.12℃/10a (1951 ~ 2012) [4]. Which is higher than the global trend of temperature, with an average value of 15.9℃. As shown in Figure 1(a), the linear correlation coefficient is 0.28, and the trend of α=0.01(P=0.006) is significant.

The annual temperature changes have a certain fluctuations; the range did not exceed 2℃. Using the 3-year sliding average method, it can be seen that the fluctuation range of temperature has a tendency to increase. The annual maximum temperature has passed the significant test of α=0.01(P=0.3099), the upward trend is significant and the annual minimum temperature does not pass α=0.1 significance test, the decline trend is not significant.

As shown in Figure 1(b), the UF values are all greater than 0, indicating an upward trend in temperature and a significant confidence level fluctuation at the level of α=0.05 after 2003, indicating a significant growth trend. UF and UB have only one intersection, and the cumulative anomaly, Pettitt method of seeking the same mutation time, all in 1997, the cumulative anomaly shown in Figure 1(c).

The Hurst exponent (H=0.73, H>0.5) of the annual mean temperature indicates that the memory of the time series is higher than that of the random walk, that is, persistent and the temperature will continue to rise in the future time series. In the Morlet wavelet analysis, there are two types of periodic changes in the temperature evolution process: 0 ~ 9a, 10 ~ 17a. There are cold-warm-cold oscillations on the scale of 10 ~ 17a, and the periodic variation is very stable throughout the analysis period, and it is global. The shocks appearing on the scale of 0 ~ 9a are unstable, as shown in Figure 1(d). 10 ~ 17a is weak in energy, but the period distribution is obvious, almost throughout the study time domain (1990 ~ 2014), and the period is local and obvious, as shown in Figure 1(e). There are three obvious peaks in the temperature evolution time series, corresponding to 5a, 14a and 22a in turn, and the maximum peak value is 14a, which shows that the periodic oscillation of this time scale is the strongest and is the first major cycle of annual mean temperature change. 5a and 22a As the second and the third main cycle, as shown in Figure 1(f), 1997 is the second main cycle and the first major cycle of the junction point, that is, the mutation, the three cycles of the fluctuations in the control of the entire time domain Variation characteristics of annual mean temperature.

The Hurst index (H_{max}=0.56, H_{min}=0.66) of extreme temperature indicates that the extreme temperature time sequence has a certain random walkability, that is, the persistence is weak, and the significance test shows that the temperature extremum, and the trend is difficult to judge.

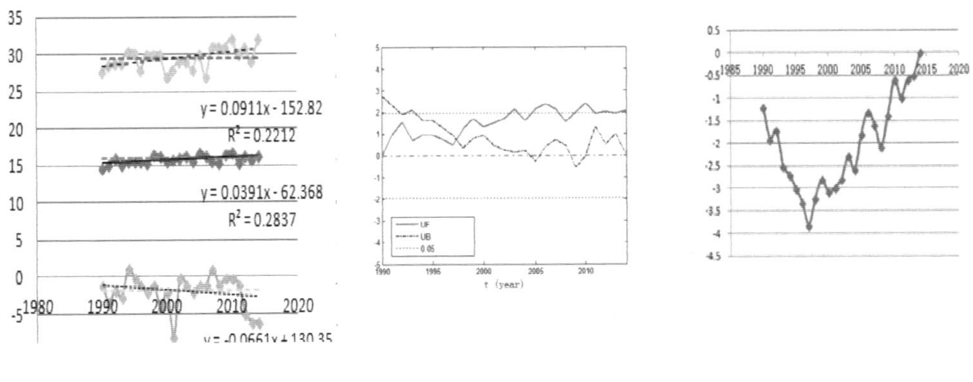

(a). Factor linear regression graph

(b). M-K test chart

(c). Cumulative anomaly

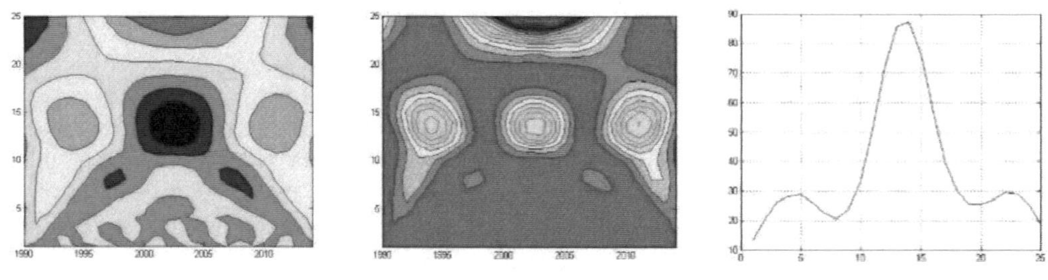

(d). Real-time time-frequency distribution

(e). Square of wavelet coefficients

(f). Wavelet variance map

Figure 1. Temperature analysis

Several other factors were analyzed in the same way, as seen in the Conclusions and Discussion section.

4. Multi-factor Composite Analysis

In multivariate linear regression analysis, multiple variables are contrasted and corresponding relationships are found. Usually, the method is time-consuming and difficult to compare with each variable, making the analysis more difficult. To this end, the study uses the scatter matrix to compare the relationship between the various variables, and use Spearman correlation coefficient that the correlation between the various factors. The scatter plot matrix and the Spearman correlation coefficient between the two factors are shown in Figure 2. The diagonal is the factor variable name, the lower left triangle is the scatter plot of the corresponding factor, and the upper right triangle is the correlation coefficient of the factor. The larger the absolute value of the correlation coefficient is, the larger the space occupied by the graph is, the stronger the covariant trend is between the variables, and the negative correlation coefficient indicates the negative correlation between them.

Among them, 7 groups of variables (air temperature-humidity, wind speed-population, construction area-population, construction completion area-population, construction area-wind speed, building completion area-wind speed, construction area-building completion area) passed the significance test of α=0.001, indicating that the linear correlation of the variables is strong, and the sample r value is greater than 0.7. 9 groups of variables (temperature-precipitation, humidity-precipitation, duration of light-precipitation, duration of light-population, duration of light-wind speed, duration of light-construction area, duration of light-construction area, precipitation-construction area, precipitation-building The completion area) did not pass the significance test, indicating that the linear correlation is small, the significance of linear regression test shown in Table 1. With the population growth, the construction area and completion of the corresponding increase in area to meet the living, office and other needs; the contrary, the construction area of the growth, the construction area of the building area, but also for the population growth provides the necessary conditions.

IOP Conf. Series: Earth and Environmental Science **63** (2017) 012044 doi:10.1088/1755-1315/63/1/012044

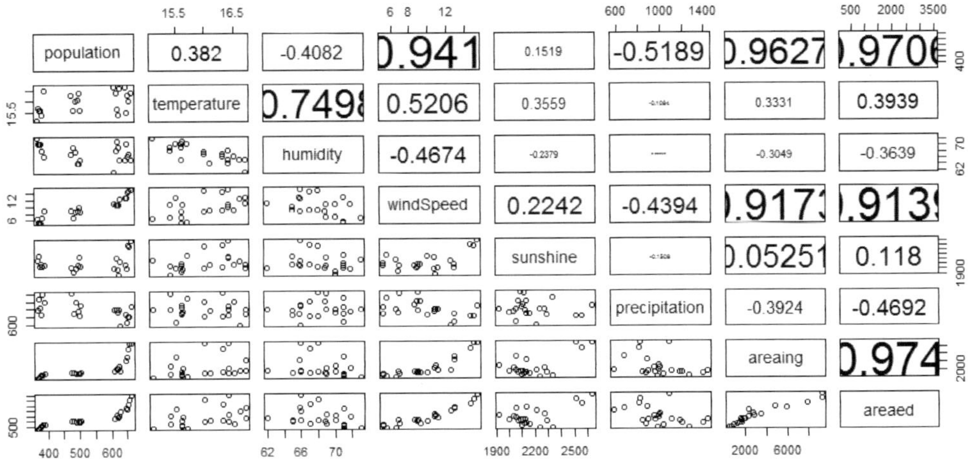

Figure 2. A matrix of scatter plots

Table 1. Significance test

Factor	Significance test	Spearman correlation coefficient	Correlation test	Mutation time
Temperature	0.0060	0.4856	0.0139	1997
population	8.80E-16	1	3.20E-07	2002
humidity	0.0017	-0.5968	0.0016	1994
Wind speed	1.90E-10	0.8921	2.12E-09	2007
Sunshine duration	0.0211	0.3177	0.1219	2013
precipitation	0.0825	-0.3615	0.0765	2006
Building construction area	2.25E-08	0.9738	2.68E-16	2010
Completed area of housing construction	3.22E-13	0.9792	<2.2e-16	2009

5. Conclusion and Discussion

Based on the above experimental analysis, we can see that there are some correlations among the eight factors, such as temperature, humidity, light duration, precipitation, wind speed, population, housing construction area and building completion area, as shown in Table 2.

Table 2. Factor correlation

	Temperature	**Humidity**	**Sunshine duration**	**Precipitation**	**Wind speed**	**Population**	**Building construction area**	**Completed area of housing construction**
Time scale	10~17a	10~17a	10~17a	10~17a	11~25 a	10~17a	4~22a	4~22a
Variety	Cold-warm-cold	Wet-dry-wet	Long-short -long	More-less-more	Small-large-small	Fast-slow-fast	More-less-more	More-less-more
The first main period	14a	14a	13a	13a	19a	15a	18a	18a
Future trends	rise	fall	rise	fall	rise	rise	rise	rise
Mutation Year	1997	1994	2013	2006	2007	2002	2010	2009

The change rate of humidity is 1.27%/10a after the year of mutation (1994), and it is about 8%/10a before 1994, about 6 times of the mutation. The temperature increase rate after 1998 is 0.134℃/10a, and the growth trend before 1997 is 1.1℃/10a, which is much higher than the warming rate of global warming, about 8 times of that before the abrupt change. The change trend of the two trends is obviously slowed down after the mutation year, and the change of the humidity is more than that of the air temperature. It can be concluded that the change of humidity affects the temperature change to a certain extent. Precipitation has a trend of increasing 31.32mm/10a before the year of 2006 and declining trend since 2006 (52.03mm/10a), while the population has a tendency of increasing. The rate of change produced 19.51 million people/10a changes, indicating that the population growth rate has been greatly improved, it is concluded that the decrease in humidity affect the temperature rise at the same time, the sharp increase in the population in the context of precipitation was significantly reduced trend, Followed by a significant increase in wind speed and duration of light before the abrupt year.

There were five periods of temperature, humidity, light duration, precipitation, and population and so on, and the first principal period of climatic factor was 13-14 years, and the first principal period of population factor was 15 years. From 2000 to 2002, the period from 2003 to 2004 was characterized by low temperature, high humidity, and long duration of light and high precipitation. The period from 2003 to 2004 was characterized by high temperature, low humidity, short illumination time and little rainfall. Low temperature, humidity, light for a long time, precipitation and more periods. Precipitation increased, and humidity increased, the first major cycle of precipitation than the reduction of the humidity of 1a, indicating that precipitation cycle ahead of the humidity, the wind continued to increase on the basis of the precipitation cycle will be further reduced, resulting in increased rainfall period, With the rise in temperature, which increased the drought, making the climate worse.

Population, housing construction and completion of the area to a large extent by the more serious human intervention, such as the promulgation and implementation of relevant policies and regulations, so discuss the relevance of which there are some limitations, but with the population growth, To meet the needs of people living space, the city will, housing construction area also increased, resulting in dramatic increase in impermeable surface, and have continued a clear upward trend, which the population in 2002 mutation in 2004 produced a sharp rise. Precipitation in 2006 produced a mutation, since then 8 years were less than 25 years of average precipitation; the mutation of wind speed is in 2007, after the growth rate was significantly higher than the previous growth rate; temperature and humidity since 2004 Fluctuation range is obviously increased. Housing construction and completion of the area were in 2010 and 2009, respectively, mutation in 2010 and 2011, a dramatic rise in light duration from 2011, the growth trend in 2013, a mutation. In summary, population and housing construction area of the sharp rise in the climatic factors of the cycle and tendencies have had a significant impact.

6. References

[1] GE Q S, ZHENG J Y, HAO Z X 2015 *J. Journal of Geographical Sciences*. Progress and Prospects of Integrated Study on Climate Change in Asia over the Past 2000. **70** 355-363

[2] WANG S W, YE J L 1995 *J. Chinese Journal of Atmospheric Sciences.* Analysis of Global Warming over the Last 100 Years. **19** 545-553

[3] United Nations News 2007 *J. United Nations News*. WMO: past decade is the warmest decade ever

[4] DING Y H, WANG H J 2015 *Conf. Special Report of the Annual Meeting of the Chinese Meteorological Society* (China: Tianjin)

[5] YAO P, ZHAO F Z, XIA H Y 2009 *J. Chinese Journal of Environmental Science*. Climatic Characteristics of Dianchi Lake Watershed. **28** 8-10

[6] WEN H, CAI J L, et al 2011 *J. Chinese Journal of Environmental Science*. **31** 69-80

[7] LIANG Z Y, LIU Y, SHENG H, et al 2014 *J. Chinese Journal of Environmental Science*. **34** 754-62

[8] LI Z Y, XIE P C, DU P F 2016 *J. Chinese Journal of Environmental Science*. A Simulation Study on Pollution Control of Old Transport Grain River in Kunming City. **10** 1-16

[9] CHEN C, MENG D 2015 *J. Resources and Environment in the Yangtze*. **S1** 30-37

[10] QI D L, LI X D, XIAO H B, et al 2015 *J. Resources and Environment in the Yangtze*. Analysis of the Variation of Evapotranspiration and Its Influencing Factors in Sanjiangyuan Area in Recent 50 Years. **24** 1613-20

[11] ZHOU D, ZHANG B, LI X Y, et al 2014 *J. Resources and Environment in the Yangtze*. **23** 549-58

[12] DING W R 2016 *J. Resources and Environment in the Yangtze*. Characteristics of climate change in the Erhai Lake region. **25** 599-605

[13] LUO Y, SHI Y B, QU L L, et al 2014 *J. Chinese Journal of Scientific Instrument.* Study on High Altitude Temperature Detection Method Based on Harmonic Analysis and Alternating Current. **35** 721-9

[14] LUO Y, YANG K, SHI Y, et al 2014 *J. Sensors & Actuators A Physical*. Research of radiosonde humidity sensor with temperature compensation function and experimental verification. **218** 49-59

[15] XU Y, YANG K, LUO Y, et al 2015 *J.* Cyanobacteria bloom monitoring and early warning system based on GIS and WSNs - a case study in Dianchi Lake. **2015** 1-8

[16] CHEN J, YANG K, ZHU Y, et al 2015 *J.* Analysis of the relationship between land surface temperature and land cover types — A case study of Dianchi Basin. **2015** 1-6

[17] HELBOK R, KO S, SCHMIDT J M, et al 2011 *J. Stroke; a journal of cerebral circulation*. **42** 1534-39

[18] YU W K, YAO X R, LIU X F, et al 2015 *J. Chinese Physics B*. **24** 340-5

[19] PERKINS S E, ALEXANDER L V 2013 *J. Journal of Climate*. **26** 4500-17

[20] LIU J T, FANG S W 2015 *J. Chinese Journal of Environmental Science*. **12** 3707-13

[21] WANG L, LIU D D, LI T Y, et al 2015 *J. Hydrology*. **4** 85-90

[22] GAO G, HUANG C Y 2001 *J. Advances in Atmospheric Sciences*. **18** 718-32

[23] CHEN L X, ZHU C W, WANG W, ZHANG P Q 1998 *J. Advances in Atmospheric Sciences*. **18** 623-38

[24] BING L F, SHAO Q Q, LIU J Y 2012 *J. Journal of Geographical Sciences*. **2**

IOP Conf. Series: Earth and Environmental Science **63** (2017) 012045 doi:10.1088/1755-1315/63/1/012045

Reflections on Beijing Rural Energy Policies and New Rural Construction ——Using Digital Information Means

WANG ZHI

Institute of Agricultural Scientech Information, Beijing Agriculture and Forestry Academy of Science, Beijing 100097.

angelwang1009@163.com

Abstract. In recent years, the Beijing construction in accordance with the philosophy of environmental protection, solid depth to carry out a series of policies and measures, such as rural energy structure optimization and adjustment; rural residential seismic energy-saving; village green landscaping. If displaying and recording movement states and regulars of the beautiful countryside construction through this information process, It will greatly strengthen regulation of the relationship between new rural people, to realize system optimization, make urban and rural become conducive to human survival and sustainable development of space.

1. Introduction

Digital City is the use of spatial information to build virtual platforms, including urban natural resources, social resources, infrastructure, cultural, economic and other information about the city, in order to obtain and load up the digital form, so as to provide a wide range of aspects of government and society service[1-3]. Through advanced means of information, support Beijing new rural planning, construction, operation, management and emergency, can effectively improve the government management and service levels, improve the new rural management efficiency[4], conserve resources, promote the development of the beautiful countryside.

2. Beautiful countryside construction is an important basis for the construction of ecological civilization and beauty of China

Chinese capital of Beijing, an area of 16,400 square kilometres in rural areas accounted for 1.5 million square kilometres, if rural development is not good, ecological Beijing will be no barrier, there will be no space environment, and sustainable development will be on the lack of foundation [5].

3. Major policies and measures to promote the construction of beautiful countryside

We recognize the importance and urgency of the beautiful countryside construction, but also, perceive that the complexity and difficulty to build a beautiful countryside. In recent years, Beijing launched a series of livelihood projects, including farmhouse seismic energy-saving projects, risk village and insurance households from mountain relocation project, Clean Air Initiative in rural areas reducing coal, Agricultural non-point source pollution control projects, Rural Ecological Environment Construction Engineering, and so on, rural environment continues to improve, rural infrastructure and public service facilities gradually improved, the rural energy structure was further optimized, farmer housing conditions and income levels are rising[5,6]. New Rural Construction of Beijing has been built a beautiful countryside and laid a good foundation. At the same time we should also see that although the new rural construction of Beijing has made remarkable achievements, yet, with the urgent needs of the majority of farmers, and the requirements of building world-class city, there is still a big gap, but also need to be improved in the following areas..

3.1. Rural energy structure adjustment and optimization

In recent years, the Beijing government has been to actively promote the use of new energy sources, new technologies in rural areas, and started to implement the " Reducing Coal and Replaced Coal, Clean air" Initiative. Preliminary accounting, in 2013 heating season, the Beijing rural areas completed 2954 tons of sulphur dioxide emissions, nitrogen oxides 3512 tons emissions, and 7397 tons of carbon dioxide emissions, accounting for 22% of the city's rural areas, 4.3 million tons of total coal-fired. but there are still nearly 80% of farmers are still using low-grade coal [5], as well as a large number of public facilities in rural areas, enterprises and institutions were not clean energy transformation. Therefore, we should continue to promote the rural areas "Reducing Coal and Replaced Coal, Clean air" Initiative. Through the joint action of government, market and society, at a rate of about 1 million tons per year, with a further three years to complete rural clean alternative, while by replacing heating coal to electricity, sending natural gas to home, as well as using liquefied petroleum gas, appliances solar, biomass, air energy, geothermal energy utilization, etc., to realize gasification of rural areas for cooking, heating and clean [7, 8].

Through the survey of rural areas in Beijing found that the residents of the current application of the main heating technologies are electric heating, heat pump heating, gas heating, clean coal heating and solar heating in five ways[5]. According to the survey data, combined with the application of the Beijing area and product technology maturity, the existing mainstream rural areas in Beijing can be used for rural use of clean energy technologies in Table 1.

Table 1. Commonly used in rural areas heating technology

Serial number	Heating mode				
	category	Heat source (priority)	Auxiliary heat source (alternate)	Terminal	
1	Heat pump	Low temperature air source heat pump	——	Floor heating	Low temperature hot water terminal
2		Ground source heat pump	——	Floor heating	
3	Solar composite heating category	Solar hot water heating	Clean coal boilers	Floor heating	
4		Solar hot water heating	Gas boiler	Floor heating	
5		Solar hot water heating	Low temperature air source heat pump	Floor heating	
6		Solar hot water heating	Ground source heat pump	Floor heating	
7		Solar hot water heating	Electric heating	Floor heating	

<table>
<tr><td>8</td><td>Gas category</td><td>Gas boiler</td><td>——</td><td>Radiator / floor heating</td><td rowspan="4">High temperature / low temperature hot water terminal</td></tr>
<tr><td>9</td><td>Coal-fired class</td><td>Clean coal heating stove</td><td>——</td><td>Radiator / floor heating</td></tr>
<tr><td>10</td><td rowspan="4">Electric heating</td><td>Electric hot water boiler</td><td>——</td><td>Radiator / floor heating</td></tr>
<tr><td>11</td><td>Heat storage type electric boiler</td><td>——</td><td>Radiator / floor heating</td></tr>
<tr><td>12</td><td>Carbon crystal plate class</td><td>——</td><td>——</td><td rowspan="2">——</td></tr>
<tr><td>13</td><td>Energy storage electric heating</td><td>——</td><td>——</td></tr>
</table>

3.2. Rural Residential Seismic energy-saving projects

Beijing government started farmhouse seismic energy-saving projects since 2007, it has supported 380,000 rural residents in the implementation of energy-saving residential earthquake. Through the implementation of seismic energy-saving projects, farmers housing winter indoor temperature has increased by 4 to 6 degrees, saving more than 30% of coal-fired[9,10]. In summer, obviously feeling the cool interior, indoor temperature is relatively low outside temperatures 5 to 6 degrees [11]. According to statistics, only this thermal insulation has an annual saving of coal-fired nearly 55 million tons, a significant reduction in carbon dioxide emissions, effectively improve air quality. According to surveys, there are still 530,000 rural residents in Beijing have seismic energy-saving requirements. Therefore, to further increase the Farmhouse seismic energy-saving efforts, using of 5 years, at a rate of about 100,000 a year, it will complete the city's Farmhouse seismic energy-saving task.

3.3. Sewage treatment works in rural areas

In recent years, new rural construction in Beijing, more than 500 villages have been built sewage treatment facilities in 1010, but due to complex process, facilities scattered layout, unfocused, factory network is not complete, the high cost of operation and maintenance, management operation is not in place and other reasons, rural sewage treatment facilities, 30% did not run or intermittent operation. In Beijing rural area, annual output of sewage is more than 100 million tons, but, treatment rate is only about 50%. Large amounts of sewage is discharged directly [12]. With the rapid development of urbanization and rural economic and social, supply and demand contradictions in rural city sewage treatment facilities have become increasingly prominent. Therefore, the reality requires us to rural sewage treatment as an important part of the construction of the beautiful countryside. For facilities that have been built to run on to assess the situation, according to local conditions to transform and strengthen the operation and maintenance, and ensure that the newly built sewage treatment facilities operating normally, the basic objectives of rural sewage discharge standards may successfully achieve in 2020.

4. Beijing 's rural energy demand forecast

4.1. Analysis of energy - using factors

4.1.1. Farmers own needs. With the development of rural economy, farmers' concept of energy use is gradually changing. The higher income level of farmers, the more emphasis on the convenience and comfort with energy, and then health, the last is the economy. Therefore, in cooking, the farmers are more willing to use the cleanliness of liquefied petroleum gas or biogas, rather than coal. In the heating, according to affordability, farmers will focus on electricity, solar and other clean energy heating.

4.1.2. The Impact of Urbanization Policy With the acceleration of urbanization in the rural areas and the implementation of ecological resettlement, the rural energy structure has also been greatly affected.

(1) Heating will be converted into centralized heating. The Beijing municipal government proposed that by 2020 the city will make about two-thirds of the farmers into urban residents, the living conditions of farmers will be dispersed from the living, into a centralized way to live, therefore, the corresponding heating will focus on the form heating mode change, which will help save energy consumption of heating [13-15].

(2) Farmers' choice of energy will become more commoditized and cleaner. Due to the concentration of living forms and the difficulty in obtaining biomass resources such as straw, farmers will tend to favour commercialized energy sources for energy selection [13-15]. With the popularization and cleaning of public facilities such as heat and gas, the proportion of the use of energy will be a substantial increase.

(3) Housing construction will become more energy-saving. With the process of urbanization, the new farmers will focus on energy-saving housing and comfort [16], thus directly reduce the heating load and reduce the use of heating energy.

(4) Energy use will become more low-carbon and comfortable [17]· Urbanization directly promote the increase of farmers' income, and with the construction of new houses, solar energy and geothermal energy and other low-carbon new energy and renewable energy in the proportion of domestic energy consumption will gradually increase.

4.2. Potential evaluation of coal adjustment in rural energy structure

In the heating, through the existing energy-saving transformation of rural households, the annual heating coal can be saved about 62.76 million tons of standard coal (in accordance with the energy-saving transformation to save 40% of the heating coal to calculate). On this basis, by 2020, 952,000 farmers will live in the cities and towns of residence, so that the original decentralized heating will transform into a central heating. This way can reduce heating coal-burning 94,100 tons of standard coal per year.

In addition, for the remaining decentralized heating farmers, the Government can actively promote solar heating, electric heating and other new green energy-saving technologies, which can reduce the heating coal-fired 25.10 million tons of standard coal.

From the adjustment potential, based on full use of existing resources and actively promote green energy-saving technology, respect for heating and cooking, the total coal pressure reduction will be more than 1.1451 million tons of standard coal per year.

5. Some thoughts on rural energy policy and new rural construction in Beijing

In the capital, to promote the digital city construction process, the author has deeply thinking for rural energy policies and beautiful countryside construction as follows:

First, the establishment of Beijing rural information service platform and the beautiful countryside of information resource library. It will cover many aspects of rural infrastructure, rural economy, industry characteristics, population health, culture, education, grass-roots organizations and other resources for effective integration of information and data, and establish a cover Beijing, connecting Beijing, Tianjin city, county, township and village "digital village" information network platform.

The second is the establishment of "Beijing Digital Village" platform; its function should have a strong practical and advanced. It can form in a variety of vivid visual image, data, icons and text, etc., it can provide information on all aspects of rural work for the government at all levels in Beijing, so as to provide a reliable basis for the scientific and democratic decision-making of new rural construction. Beautiful countryside construction in every measure can be demonstrated and information management, the establishment of relevant information platform. Moreover, with the continuous improvement platform, which will provide reliable first-hand information for all agricultural sectors and investment units, it lay a solid foundation for the economic development of Beijing Rural Development and Investment, the same time, the Present Situation can be shown by the platform the beautiful countryside as a window, it can be to promote the rural areas this emerging e-commerce business model, so as to promote the exchange of agricultural products and accelerate rural economic development has laid a good foundation.

The third, pay attention to the establishment of several information platform, focusing on people's livelihood and ecological protection. The main measures "optimal adjustment of rural energy structure, agricultural nonpoint source pollution control, rural residential seismic energy-saving projects, sewage treatment works in rural areas" and other beautiful countryside construction are to be monitored and managed through the information platform to maximize for the construction of ecological civilization provide effective protection, as well as the national new rural construction to enhance digital play a leading and exemplary role..

6. Acknowledgement

This research was financially supported by the Beijing Municipal Rural Work Committee Technology Services Special Fund.

7. References

[1] Meng xiang-lin.Strategic studying for Tangshan City's economic integeration in JingJinJi region based on "Double Parents+Double Sons" theory[J]. CITY, 2011（4）.
[2] HE Ling, WANG Jun,DONG qian. Agricultural development model of regional economic integration of Beijing City, Tianjin City and Hebei Province[J]. JIANGSU AGRICULTURAL SCIENCES, 2011, 39(2).
[3] Digital City Network. http://www.4d3s.com/
[4] http://baike.baidu.com/view/2282762.htm.
[5] WANG Zhi. Thoughts on Beijing's Long-Term Rural Infrastructure Management and Perspective of the Government to Effectively Perform Their Duties. Mechanical Engineering and Control Systems. 2016,PP:490-493
[6] CHEN Zhang, YAN Ping, SHAO Wei. Beijing, Tianjin Economic Contacts[M].DATA, 2011, (3).
[7] Yang X, Jiang Y, Yang M, et al. Energy and environment in Chinese rural housing: Currentstatus and future perspective. Front. Energy Power Eng. China 2010, 4(1): 35-46.
[8] Ho-Lin H, Murray M, Cohen T, Colijn C, et al. Effects of smoking and solid-fuel use on COPD, lung cancer,and tuberculosis in China: a time-based, multiple risk factor, modeling study. Lancet, 2008, 372:1473-1483.
[9] Al-Homoud M S. Performance characteristics and practical applications of common building thermal insulation materials. Building and Environment, 2005, 40: 353–366.
[10] Omakli K, Yuksel B.Optimum insulation thickness ofexternal walls for energy saving. Applied Thermal Engineering,2003,23:473-479.
[11] Erlandsson M. Energy and environmental consequences of an additional wall insulation of a dwelling. Building andEnvironment,1997,32(2):129-136.
[12] WANG Zhi, ZHOU Lian-di, LIU Shi-rong, et al. The Effect on Abating Pollution from Farmland Runoff by Using Biogas Manures[J]. JOURNAL OF ANHUI AGRICULTURAL SCIENCES, 2009, 37(10).
[13] Ahmad.A, Saini JS, Varma HK. Effect of geometrical and thermophysical characteristics of bed materials on the enhancement of thermal performance of packed bed solar air heaters. Energy Conversion and Management, 1995, 36(12): 1185-1195.

[14] Hegazy AA. Performance of flat plate solar air heaters with optimum channel geometry for constant/variable flow operation. Energy Conversion and Management,2000, 41(4):401-417.
[15] Varol Y, Oztop HF. A comparative numerical study on natural convection in inclined wavy and flat-plate solar collectors. Building and Environment, 2008, 43(9): 1535-1544.
[16] Varol Y, Oztop HF. Buoyancy induced heat transfer and fluid flow inside a tilted wavy solar collector. Building and Environment, 2007, 42(5): 2062-2071.
[17] Wang Gehua, Chen Yanbini, Shi Deming. The effect assessment of integrated rural energy development and the application of Grey System Method. Transactions of the CSAE.1992(9):60-66.

IOP Conf. Series: Earth and Environmental Science **63** (2017) 012046 doi:10.1088/1755-1315/63/1/012046

Application of Entropy Method in River Health Evaluation Based on Aquatic Ecological Function Regionalization

Yan-ting SHI [1], Jie LIU [1], Peng WANG [1,2], Xu-nuo ZhANG [1], Jun-qiang WANG [1] and Liang GUO [1,2,*]

[1] School of Municipal and Environmental Engineering, Harbin Institute of Technology, Harbin 150090, China;
[2] State key Laboratory of Urban Water Resource and Environment, Harbin Institute of Technology, Harbin 150090, China;
[*] Corresponding author at: School of Municipal and Environmental Engineering, Harbin Institute of Technology, Harbin 150090, China; State key Laboratory of Urban Water Resource and Environment, Harbin Institute of Technology, Harbin 150090, China.

E-mail address:guoliang0617@hit.edu.cn

Abstract. With the implementation of water environment management in key basins in China, the monitoring and evaluation system of basins are in urgent need of innovation and upgrading. In view of the heavy workload of existing evaluation methods and the cumbersome calculation of multi-factor weighting method, the idea of using entroy method to assess river health based on aquatic ecological function regionalization was put forward. According to the monitoring data of songhua river in the year of 2011-2015, the entropy weight method was used to calculate the weight of 9 evaluation factors of 29 monitoring sections, and the river health assessment was carried out. In the study area, the river health status of the biodiversity conservation function area (4.111 point) was good, the water conservation function area (3.371 point) , the habitat maintenance functional area (3.262 point) , the agricultural production maintenance functional area (3.695 point) and the urban supporting functional area (3.399 point) was light pollution.

1. Introduction

Since 2008, the Ministry of Environmental Protection in the water pollution control and control of major science and technology projects carried out the research of aquatic ecological function regionalization, and in the "eleventh five-year" and the "twelfth five-year" topics, the special research construct the new ideas of aquatic ecological function regionalization system[1-2]. Through the management of different regionalization units using "zoning, grading, classification, staging" management mode, to ensure the ecological health of the whole basin and the normal function of aquatic ecology[1,3-4]. The establishment of river health rating system based on aquatic ecological function regionalization is helpful to the accurate management of river management. In the river health assessment, the establishment of index system is one of the most important part[5-6]. The evaluation of index weights is reasonable or not will directly influence the evaluation results. In the numerous method of index weighting, due to the actual objectivity of the monitoring data, it is considered that the objective weighting method——entropy method——is more suitable for the river health evaluation study [7-9].

The concept of entropy originated from thermodynamics, it was more early proposed by the German physicist Boltgman and Clausius in 1864 to describe the physical state of the system. Later, American mathematician, cybernetics and information theory founder Shannon put forward a broader information entropy based on it, as a measure of uncertainty[10].Li Peiyue[11] etc. applied the entropy method to groundwater quality evaluation, and the method is simple, reasonable, reliable and intuitionistic. It can be used for groundwater quality evaluation and worthy of popularization. Guo Xiuying[12] gave an objective entropy method to determine the weight of each interval number index, which is the objective weighting of the index in the multiple interval decision making with unknown weight information. Zou Zhihong[13] etc. applied the fuzzy evaluation factor entropy weight method to the Three Gorges reservoir water quality evaluation, that the evaluation results are reasonable, the calculation process is simple. This paper attempts to apply the entropy method to the river health assessment based on aquatic ecological function regionalization. The Songhua River Basin is taken as an example to evaluate the river health.

2. The basic theoretical process of determining the weight of Index by entropy method

Entropy is a measure of the degree of system disorder in information theory, it also can measure the amount of the effective data provided information [14]. Therefore, entropy can be used to determine the weight. In river health evaluation, through the calculation of "entropy" to determine the weight, is based on the difference between the various monitoring indicators to determine the weight of each indicator.

Specific calculation steps are as follows:

There are m evaluation indicators, n evaluation objects, the formation of the original data matrix is $X=(x_{ij})_{m\times n}$. For a certain index i, the greater the difference of the index value, the greater the role of the indicator in the comprehensive evaluation. If the index values of all indicators are equal, then the indicator in the comprehensive evaluation of almost no effect[15].

Step 1: For calculating entropy weight when m water samples (i=1, 2,..., m) are taken to evaluate the water quality and each sample is analyzed for "n" quality parameters (j=1, 2,...,n) , according to observed data, eigenvalue matrix X can be constructed.

In order to eliminate the influence caused by the difference of different quantity grades and different units, data pretreatment must be put into force[11].The matrix is normalized as follows:

$$R=(r_{ij})_{m\times n} \quad (1)$$

Where r is the standard value of the j-th evaluation object on the i-th evaluation index, $r_{ij}\in[0,1]$.

The standardization process is as follows:

For the efficiency type, the construction function of normalization is:

$$r_{ij}=\frac{x_{ij}-\min_j\{x_{ij}\}}{\max_j\{x_{ij}\}-\min_j\{x_{ij}\}} \quad (2)$$

And for the cost type, the construction function of normalization is:

$$r_{ij}=\frac{\max_j\{x_{ij}\}-x_{ij}}{\max_j\{x_{ij}\}-\min_j\{x_{ij}\}} \quad (3)$$

Step 2: Define the entropy

In the evaluation index of m indicators, n evaluation objects, the entropy of the i-th index is defined as

$$H_i=-k\sum_{j=1}^{n}f_{ij}\ln f_{ij}, i=1,2,\ldots,m \quad (4)$$

Where $fij = r_{ij} \backslash \sum_{j=1}^{n} r_{ij}, k = 1/\ln n$, when $f_{ij} = 0$, make $f_{ij} \ln f_{ij} = 0$.

Step 3: Define the entropy weight

$$w_i = \frac{1 - H_i}{m - \sum_{i=1}^{m} H_i} \quad (5)$$

Where $0 \le w_i \le 1$, $\sum_{i=1}^{m} w_i = 1$.

3. A case study of river health evaluation based on entropy method

3.1. Compute objects and their data sources

In this paper, the monitoring sections of the Songhua River Basin, which belong to the five aquatic ecological function regionalization, are selected to study the river health comprehensive evaluation and the entropy method for the five years monitoring data of the Songhua River Basin. The water quality monitoring data used were from environmental monitoring bulletin.

3.2. The screening of calculated indicator factor

In the original monitoring data, there are 32 monitoring indicators, a primary screening are operated on them. Remove the auxiliary indicators and indicators of high undetection rate in monitor, including: water temperature, flow, pH, sulfate, volatile phenol, copper etc.

3.3. Evaluation factors and evaluation

After screening, 9 indexes of dissolved oxygen, permanganate index, five day BOD, ammonia nitrogen, chemical oxygen demand, total nitrogen, total phosphorus, fluoride and fecal coliform were selected as evaluation factor. Namely, $U = \{DO, COD_{Mn}, BOD_5, NH_3\text{-}N, COD, TN, TP, F^-, \text{m}\}$. According to the national surface water environmental quality standard (GB3838-2002), the water quality level is set to 5, so the evaluation set is $V = \{\text{I}, \text{II}, \text{III}, \text{IV}, \text{V}\}$.

3.4. Raw data processing

For the individual data that the original data are not monitored, the data are supplemented by the cubic polynomial interpolation method, taking into account the whole data. For the data below the detection limit in the monitoring, use the detection limit value of the monitoring indicators to assignment.

3.5. Weight calculation of partition index based on entropy method

According to the relationship between the monitoring points and the five levels of aquatic ecological function regionalization, the monitoring points were classified into five levels of aquatic ecological function regionalization, the results shown in Figure 1. In this paper, the name of the section is replaced by number.

The entropy weight formula is used to normalize the data to obtain entropy and weight. Take the monitoring data in August 2012 as an example. For the agricultural production maintenance functional area, the data are normalized to obtain the matrix as shown in Table 1.

The index entropy and the weight of the function partition are obtained as shown in Table 2.

Figure 1. corresponding monitoring points of aquatic ecological function regionalization

Table 1. Standardization of data

Monitoring section	DO	COD_{Mn}	BOD_5	NH_3-N	COD	TN	TP	F^-	Fecal coliform
D1	0.000	0.861	1.000	0.560	1.000	0.796	0.813	0.630	0.922
D2	1.000	0.466	0.707	0.691	0.591	0.757	0.761	0.752	0.000
D3	0.387	0.284	0.523	0.871	0.240	1.000	0.926	0.817	1.000
D4	0.610	0.000	0.000	0.000	0.000	0.000	0.523	0.000	0.395
D5	0.518	0.257	0.640	0.748	0.909	0.514	0.074	0.569	0.977
D6	0.518	0.257	0.627	0.789	0.909	0.107	0.074	0.589	0.967
D7	0.860	0.375	0.747	0.597	0.877	0.803	0.875	1.000	0.993
D8	0.845	0.672	1.000	0.661	0.877	0.843	1.000	0.728	0.993
D9	0.415	1.000	0.333	1.000	0.130	0.253	0.415	0.467	0.984
D10	0.549	0.358	0.653	0.773	0.455	0.026	0.000	0.386	0.892
D11	0.671	0.324	0.813	0.696	0.195	0.408	0.188	0.549	0.917

Table 2. Index entropy and weight value

Monitoring indicators	DO	COD_{Mn}	BOD_5	NH_3-N	COD	TN	TP	F^-	fecal coliform

Entroy	0.942	0.909	0.944	0.955	0.895	0.871	0.867	0.946	0.951
Entroy weight	0.081	0.126	0.078	0.063	0.146	0.179	0.184	0.075	0.068

All data from 2011 to 2015 are calculated in accordance with the above method of index weight, the average value of the index weight obtained as the recommended weight value for each monitoring indicator. From the obtained weight we can see: the total phosphorus, total nitrogen two indicators of water nutritional status of the index weight value is larger, should be concerned. This is also related to the monitoring time is summer of August.

3.6. Weight calculation of monitoring section based on entropy method

The weight value of the monitoring section of each aquatic ecological function area is calculated in the same way as the index weight calculation method, and the weight of the monitoring section of five water ecological function zones is calculated. The weight of monitoring indicators and the monitoring section are sort out in Table 3:

Table 3. Recommended weight value of monitoring indicators and monitoring section

Functional area	Monitoring indicators	Recommended weight	Monitoring section	Recommended weight
The biodiversity conservation function area	DO	0.153	A1	0.204
	COD_{Mn}	0.107	A2	0.098
	BOD_5	0.110	A3	0.172
	NH_3-N	0.099	A4	0.527
	COD	0.116		
	TN	0.120		
	TP	0.110		
	F^-	0.086		
	fecal coliform	0.100		
The water conservation function area	DO	0.104	B1	0.195
	COD_{Mn}	0.124	B2	0.037
	BOD_5	0.107	B3	0.565
	NH_3-N	0.099	B4	0.204
	COD	0.129		
	TN	0.141		
	TP	0.102		
	F^-	0.092		
	fecal coliform	0.103		
The habitat maintenance functional area	DO	0.119	C1	0.017
	COD_{Mn}	0.109	C2	0.025
	BOD_5	0.102	C3	0.003
	NH^3-N	0.100	C4	0.895
	COD	0.111	C5	0.033
	TN	0.105	C6	0.026
	TP	0.102		
	F^-	0.131		
	fecal coliform	0.121		
The agricultural production maintenance functional area	DO	0.134	D1	0.028
	COD_{Mn}	0.171	D2	0.039
	BOD_5	0.113	D3	0.029
	NH_3-N	0.082	D4	0.567
	COD	0.109	D5	0.048
	TN	0.093	D6	0.059
	TP	0.102	D7	0.030
	F^-	0.113	D8	0.045
	fecal coliform	0.083	D9	0.060

			D10	0.056
			D11	0.040
The urban supporting functional area	DO	0.105	E1	0.206
	COD_{Mn}	0.091	E2	0.366
	BOD_5	0.135	E3	0.176
	NH_3-N	0.107	E4	0.251
	COD	0.106		
	TN	0.146		
	TP	0.107		
	F^-	0.101		
	fecal coliform	0.102		

3.7. Comprehensive river health assessment

According to the standard limit value of the surface water environmental quality standard[16] (GB3838-2002) to score the evaluation index, the scoring standard was :{ class Ⅰ, class Ⅱ, class Ⅲ, class Ⅳ, class Ⅴ} corresponds to{5, 4, 3, 2, 1}. According to the results of the above calculation, the monitoring data of Heilongjiang Province in January-May 2016 were comprehensively evaluated. The sections used in the evaluation in Heilongjiang Province and evaluation results are shown in Table 4:

Table 4. The health assessment results of the Songhua River Basin

The biodiversity conservation function area		The water conservation function area		The habitat maintenance functional area		The agricultural production maintenance functional area		The urban supporting functional area	
Section	Score	Section	Score	Section	Score	Section	Score	Section	Score
A1	4.169	B3	3.371	C5	3.150	D3	3.373	E1	3.788
A2	4.003			C6	3.346	D7	3.786	E2	3.355
A3	4.174					D8	3.786	E3	3.171
						D9	4.234	E4	3.358
						D10	3.492		
						D11	3.242		
Score	4.111	Score	3.371	Score	3.262	Score	3.695	Score	3.399

The grading standards of water eco-environmental quality is shown as follows:{5, 4, 3, 2, 1} corresponds to {excellent, good, slight pollution, moderate pollution, heavy pollution}.The river health status of the biodiversity conservation function area was good, the water conservation function area, the habitat maintenance functional area, the agricultural production maintenance functional area and the urban supporting functional area was light pollution. This is consistent with the water quality status of the Songhua River basin reported in the monthly water quality report. Moreover, due to the assignment of the partition weights, the indicators with higher pollution in the monitoring have received more attention.

4. Conclusion

(1) The result of weighting by the entropy method shows that the weights of the total nitrogen, dissolved oxygen and chemical oxygen demand in the five aquatic ecological function regionalization are relatively large, indicating that the information given by the monitoring data is large, also from the side indicates a serious degree of pollution. Through the comprehensive evaluation results can be seen, the overall river health status of the Songhua River is slight pollution. In terms of their ecological functions, the functional areas with high water quality requirements for maintaining biodiversity are in good health, while those for urban support functional areas are slightly polluted. This also fit in the basic requirements and laws of aquatic ecological function regionalization.

(2) Compared with the current "Surface Water Environmental Quality Assessment Method", river health assessment based on aquatic ecological function regionalization is more targeted to the river

evaluation than that of the entire river basin or the whole tributary. The evaluation combined with water demand of regionalization ecological functions providing a scientific basis for reasonable pollution control strategy.

(3) Compared with other methods, the entropy method based on data can objectively weigh the situation of monitoring data, and eliminate the influence of human subjective factors on the weight. And the entropy method can be calculated by MATLAB is very simple to achieve, simplifying the calculation steps to reduce the workload of the evaluation.

(4) The results show that entropy method is a more effective method, and it is reasonable in river health comprehensive evaluation.

5. Acknowledgement

This research was supported by Natural Science Foundation of China (71471050).

6. References

[1] Meng Wei, Zhang Yuan, Zhang Nan, etc. Study on aquatic ecological function regionalization and water quality target management in the river basin[J]. Acta Scientiae Circucnstantiae, 2011, 31 (7) : 1345-1351.

[2] Huang Yi, Cai Jialiang, Zheng Weishuang, etc. Advances in the research on aquatic ecological function regionalization and partitioning methods[J]. Chinese Journal of Ecology, 2009, 28 (3) : 542-548.

[3] Meng Wei. Technology and demonstration of total amount control of water pollutants in river basins[M]. Beijing: China Environmental Science Press, 2008..

[4] Lei Kun, Meng Wei, Qiao Fei, etc. Study and application of the technology on waterquality target management for control unit[J]. Chinese Journal of Engineering Science, 2013, 15 (3) : 62-69.

[5] Wang Tao, Wei Yani, Qian Hui. Influence of different weights on fuzzy comprehensive evaluation method for water quality[J]. South-to-North Water Transfers and Water Science & Technology, 2010, 8 (2) : 87-90.

[6] Wang Shenglong, Zou Zhihong. Application of information entroy-based grey clustering method on water quality assessment[J]. Mathematics in Practice and Theory, 2012, 42 (21) : 83-89.

[7] Wang Tiefeng, Pan Xiaohui. An application on entropy weight of fuzzy mathematics in water quality assessment[J]. Shanxi Architecture, 2010, 36 (14) : 359-360.

[8] Vahab Amiri, Mohsen Rezaei, Nasim Sohrabi. Groundwater quality assessment using entropy weighted water quality index (EWQI) in Lenjanat, Iran[J]. Environ Earth Sci (2014) 72: 3479-3490.

[9] Wu Jianhua, Li Peiyue, Qian Hui, etc. On the sensitivity of entropy weight to sample statistics in assessing water quality: statistical analysis based on large stochastic samples[J]. Environ Earth Sci (2015) 74: 2185-2195.

[10] Chen Qiyun, Qiu Wanhua, Relation entropy and transferable think of aggregation on group decision making[J]. Systems Science and Engineering, 2002, 11 (1) : 11-18.

[11] Li Peiyue, QianHua, Wu Jianhua. Application of set pair analysis method based on entropy weight in groundwater quality assessment- a case study in dongsheng city, Northwest China[J]. E-Journal of Chemistry, 2011, 8(2), 851-858.

[12] Guo Xiuying. Improvement of entropy method for determining weights of interval[J]. Statistics and Decision Making, 2012 (17) : 32-34.

[13] Zou Zhihong, Sun Jingnan, Ren Guangping. Study and application on the entropy method for determination of weight of evaluating indicators in furySynthetic evaluation for water quality assessment[J]. Acta Scientiae Circucnstantiae, 2005, 25 (4) : 552-556.

[14] Meng Qingsheng. Information theory[M]. Xi'an：Xi'an Jiaotong University Press, 1989: 19-36

[15] Qiu Wanhua. Management decision and applied entropy[M]. Beijing: Machinery Industry Press, 2002: 193-196
[16] State Environmental Protection Administration, State Administration of Quality Supervision, Inspection and Quarantine. GB3838-2002Surface water environmental quality standard[S]. Beijing: China Environmental Science Press, 2002.

IOP Conf. Series: Earth and Environmental Science **63** (2017) 012047 doi:10.1088/1755-1315/63/1/012047

Reflective Insulation for Energy Conservation in South East Asia

Khar San Teh[1,3], David W Yarbrough[2,4], Chin Haw Lim[1,5] and Elias Salleh[1,6]

[1]Solar Energy Research Institute (SERI), Universiti Kebangsaan Malaysia (UKM), Bangi, Malaysia
[2]R & D Services, Inc., Cookeville, TN, USA
[3]Corresponding author, Master Degree Candidate at SERI, UKM
[4]V. President of R&D Services, Inc.
[5]Senior Research Fellow at SERI, UKM
[6]Principal Research Fellow at SERI, UKM until November 2016; he has since moved to International Islamic University Malaysia as Professor of Architecture from December 2016.

ks.teh@sanmiguelwoven.com

Abstract. Thermal resistances have been measured for attic spaces insulated with reflective insulations. Three test units located in Malaysia were instrumented to provide heat flux and temperatures for the calculation of time-average RSI-values (RSI is representing R-value in SI units). The RSI for attics with enclosed reflective air spaces were in the range 2-3 $m^2{\cdot}K/W$ while the uninsulated attics averaged about 0.4 $m^2{\cdot}K/W$. The RSI-values determined in this project were for heat-flow down, the predominant heat-flow direction for attic spaces in Equatorial regions. The observed thermal resistances due to the installation of the reflective insulation results in an 80-90% annual decrease in the heat transfer across the ceiling. This reduces utility usage for air conditioned units and improved comfort for occupants. The research demonstrates the use of transient data for the determination of thermal insulation performance and usefulness of enclosed reflective air spaces for thermal resistance.

1. Introduction

Residential and commercial buildings in the countries located in South East Asia are excellent candidates for the use of reflective insulation in below-roof applications. The building energy consumption for air-conditioning systems in these countries is cooling-load dominated. This energy is costly and contributes to air pollution when fossil fuels are used for electricity generation. Reflective insulation makes use of low-emittance surfaces to reduce heat flow across open air spaces. When the air spaces are enclosed to prevent flow of air through the space, thermal resistances, R-value Systeme International or known as RSI-values ($m^2{\cdot}K/W$), can be determined either by measurement or calculation.

A reflective insulation material has one or more surfaces with infrared emittances less than 0.1 resulting from the use of a metallic foil or film usually aluminum. The low-emittance surface is normally bonded to a substrate to facilitate handling and installation. These insulations materials usually have RSI- values from 0 to 0.25 $m^2{\cdot}K/W$ based on the thickness and type of substrate material. The primary RSI-value attributed to a reflective insulation system, RIS, which is the reflective insulation material plus the adjacent enclosed reflective air space is provided by the air space. Heat transfer across an air space occurs because of thermal radiation, conduction, and convection. In cooling dominated regions, the heat flow direction is mostly downward across regions between the roof and the ceiling. This means that free convection is absent or minimal in many roof assembly applications. If the effective emittance of the air space is in the range 0.03 to 0.05, then the radiative component of heat transfer is small. The effective emittance, E, for parallel plane surfaces depends on the emittances, ε_1 and ε_2, of the surfaces emitting and receiving radiant energy is given by equation (1) [1]. Heat transfer by radiation across an air-space is directly proportional to E.

$$E = 1/(\frac{1}{\varepsilon_1} + \frac{1}{\epsilon_2} - 1) \quad \text{where } 0 < E < 1 \tag{1}$$

With heat flow by convection and radiation largely absent, conduction remains as the major mechanism for heat flow across an air space. The thermal conductivity of air at 23°C is about 0.026 W/m·K [2]. This means that if convection and radiation are absent, then a 25-mm air space will provide RSI of 0.96 m^2·K/W. For air spaces with temperature differences less than 10°C, convection is predicted to be absent for air spaces less than 60 mm for heat flow down, 20 mm for horizontal heat flow, and 13 mm for heat flow up [3]. RSI of 0.96 for a 25-mm is an upper limit for air-based insulations are known as rock wool, fiberglass, expanded polystyrene, and reflective insulation.

2. Reflective insulations

Laboratory measurements of the thermal resistance for a RIS are made using a hot-box facility for testing building assemblies [4,5]. A hot-box test can provide RSI-values and U-values for a building element such as a RIS under steady-state conditions with specified boundary temperatures. Hot-box data are widely used for product labeling or comparisons. Additional product characterization can be obtained from field data that involve actual weather conditions. RISs have RSI-values that depend on heat-flow direction because the air in an enclosed space can exhibit convection when the heat flow direction is not downward. Figure 1 contains calculated RSI-values obtained using ISO 6946 for a single air space with an average air-space temperature of 25°C, ΔT of 6°C and an effective emittance of 0.03 [6]. Results for up, horizontal, and down show the variation of RSI with heat-flow down are much greater than the other heat flow directions when the air spaces are greater than about 30 mm. Enclosed air spaces can be partitioned with a reflective insulation to provide two or more enclosed reflective air spaces in series that provide substantially more thermal resistance than the same space with one low-emittance surface facing a single air space.

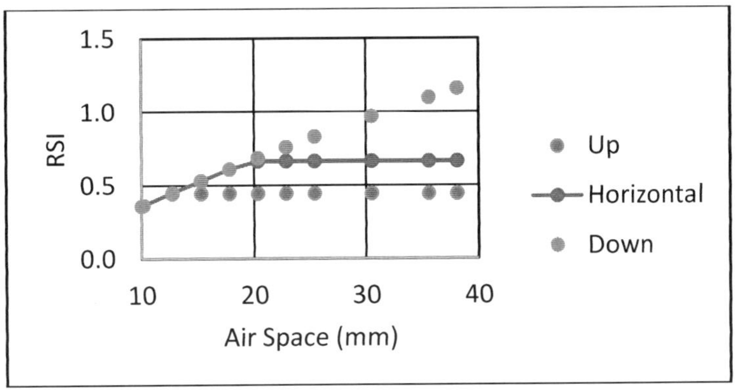

Figure 1. Calculated RSI-values for a single air space.

Thermal data obtained from reflective assemblies installed in structures (buildings) exposed to the environment provide transient performance data that can be more realistic than steady-state laboratory data. As a result, a study of field performance was undertaken.

3. Roof constructions

Concrete roof tiles are commonly found in South East Asia residential buildings and are well-suited for the use of RSIs. A layer of reflective insulation installed below battens creates an enclosed reflective air space below the roof tiles. This reflective air space provides thermal resistance and decreases the heat flow into the attic region. Metal roofing is also used in both residential and commercial applications. Reflective insulation installed below metal panels with a low-emittance surface facing downward provides thermal resistance and a reduction of heat flow into the structure.

4. The hut project

Figure 2. Picture of the three test huts.

The hut project pictured in Figure 2 was carried out to study the performance of reflective insulation under actual building and environment conditions. Three identical huts were built in an open area in Melaka, Malaysia and subjected to natural weather conditions. The dimensions of the huts are 2.2 m (width) x 2.5 m (length) x 3.3 m (height). The distance between the huts is 1.9 m. The huts face West to avoid shading.

The walls, including the attic area, are constructed using hollow metal frames and covered on the outside with 3-4 mm cement board and on the interior with 12 mm gypsum board. The floor cavity is faced with 12-mm plywood on the top and bottom. The wall and floor cavities contain 100 mm thick mineral wool with density of 80 kg/m^3 to reduce heat gain or heat loss through the walls and floor. As a result, temperature changes in the attic and interior of the huts are mainly affected by the roof insulation. The attic assembly for the tile roof huts uses a gable-roof design. Reflective insulation with low-emittance surfaces on both sides was laid on the rafters. Battens were used to create an enclosed reflective air space between the reflective insulation and the roof tiles. The attic enclosure was completed by fascia boards installed around the eaves. Figure 3 shows the attic region from the inside with the ceiling removed. A single pitch-roof design is used for the metal-roof structures. Reflective insulation and/or rock wool insulation is placed or draped above the battens with roofing panels installed above the insulation. The bottom side of the reflective insulation material creates an enclosed reflective air space.

Figure 3. Wall Insulation Material

5. Instrumentation

The huts were instrumented with a pyranometer, thermocouples, heat flux transducers and data loggers. The pyranometer was placed outdoor on top of the middle hut to record the irradiance during the day.

Type-K thermocouples with precision ±1°C were positioned to monitor roof, ceiling, and reflective insulation temperatures. The surfaces being monitored represented the bounding surfaces for the enclosed reflective air spaces. Each hut has six thermocouples underneath the roof covering, three thermocouples attached below the insulation material, and three thermocouples on the top of the ceiling. Each hut contained a heat-flux transducer adhered to the ceiling to determine the heat flux across the ceiling. Temperature and heat flux data were recorded in order to calculate RSI for the attic assemble.

6. Calculate RSI from transient data

Temperatures, heat flux and irradiance were recorded at two-minute intervals for 10 days. Figure 4 shows an example of the average temperature for roof tiles (red line), woven foil (blue line) and ceiling (grey line) in the 10-day period. Figure 5 compares temperatures for the test huts for a 24-hour period. Test Hut 2 in this case did not have insulation in the attic space. Test huts 1 and 3 were insulated. As expected, the ceiling temperatures for Test Huts 1 and 3 are significantly lower than the temperatures in Test Hut 2 during the day.

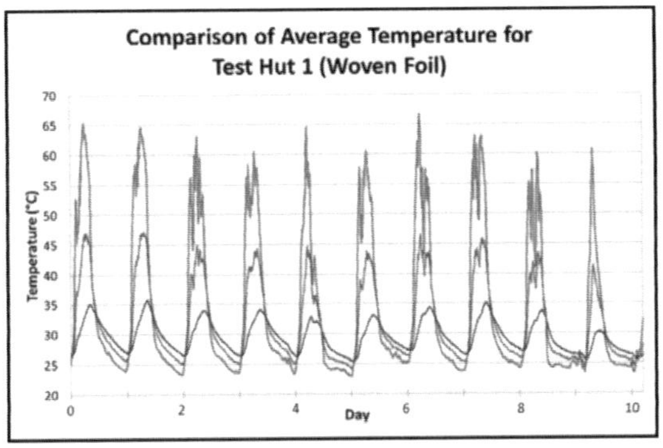

Figure 4. Average temperatures for roof tiles, woven foil and ceiling.

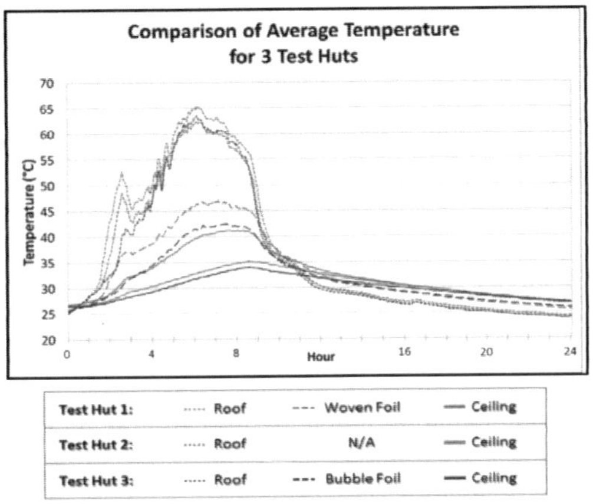

Figure 5. Comparison of average temperature for three test huts.

RSI_i in this study is the total thermal resistance of two enclosed attic spaces as indicated by equation (2). The heat fluxes for the upper attic region were obtained from the transducer mounted on the ceiling using equation (3). RSI_{block} is the average of ten consecutive two-minute readings with RSI_i calculated using equation (4). Equation (5) is for a running average of RSI-values.

$$RSI_i = RSI_A + RSI_B \tag{2}$$

$$Q_A = Q_B \cdot \left(\frac{Ceiling\ area}{Roof\ area}\right) \tag{3}$$

$$RSI_{block} = \textstyle\sum_i \left\{\frac{(T_1 - T_2)}{Q_A} + \frac{(T_2 - T_3)}{Q_B}\right\}_i \cdot \Delta t_i / \sum_i \Delta t_i \tag{4}$$

$$RSI_{running-average} = \frac{\sum_n \{RSI_{block}\}_n}{n} \tag{5}$$

Where,

RSI_A is for the air space between the roof and the insulation material
RSI_B is for air space between the insulation material and the ceiling
T_1 is the average temperature of the roof
T_2 is the average temperature of the insulation material
T_3 is the average temperature of the top of the ceiling
Q_A is the heat flux across the roof

Q_B is the heat flux across the ceiling
Δt is the time interval
RSI_{block} is the average of ten RSI_i values

This reduction in heat flow translates to a reduction in electrical use when a conditioned space is maintained. Using the average CDD_{25} of 1311 from the data taken from www.degreedays.net and an air-conditioning coefficient of performance of 3, the estimated annual savings for a change in attic RSI from 0.4 to 2.4 is 109 kWh/y·m^2. The area in the result is for the ceiling.

7. Attic air temperature

Figure 6 shows attic air temperature results for Phase I. The results show that the attic with insulation is has attic air temperatures for a ten data period that are consistently lower than those of the uninsulated attic. The blue curve in the figure represents the uninsulated hut while the red and green curves are for huts containing reflective insulation that form enclosed reflective air spaces. Maximum attic temperatures during this ten-day period were reduced by about 10°C as a result of the reflective insulation.

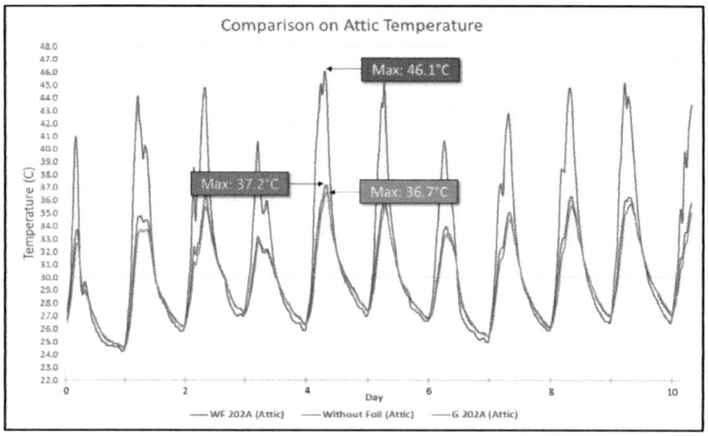

Figure 6. Results for attic air temperature.

8. Results for RSI

RSI values for three sets of tests are provided in table 1. Test hut 2 for all three tests showed RSI of about 0.40 m^2·K/W. The RSI for this configuration given in Table K2 in AS/NZS 4859 for heat flow down across an un-ventilated attic with high-emittance surfaces is 0.28 m^2·K/W.[7] The RSI for the insulated attic spaces are in the range 2 to 3 m^2·K/W. The percent reduction in ceiling heat flux (heat added to the occupied region of the hut) was found to be 80 to 90% using equation (6).

$$HFR\ (\%) = \left(1 - \frac{0.4}{RSI}\right) \cdot 100 \tag{6}$$

This reduction in heat flow translates to a reduction in electrical use when a conditioned space is maintained. Using the average CDD_{25} for S.E. Asia (data take from www.degreedays.net) and an air-conditioning coefficient of performance of 3, the estimated annual savings for a change in attic RSI from 0.4 to 2.4 is 109 kWh/y·m^2. The area in the result is for the ceiling.

Table 1. RSI for the hut project.

Phase	Type of roof	Air space	RSI (m^2·K/W)		
			Test Hut 1	Test Hut 2	Test Hut 3
I	Concrete tile	25mm	2.37 [a]	0.37 [b]	2.93 [c]
II	Concrete tile	50mm	2.16 [a]	0.40 [b]	2.69 [c]
III	Clay/Concrete tile [d]	25mm	2.26 [a]	0.40 [b]	2.40 [a]

[a] Woven foil
[b] Without foil
[c] Bubble foil
[d] Concrete tile for Hut 1, clay tile for Huts 2 & 3

Table 2. Maximum attic air temperature.

Phase	Type of roof	Air space	Temperature (°C)		
			Test Hut 1	Test Hut 2	Test Hut 3
I	Concrete tile	25mm	37.2 [a]	46.1 [b]	36.7 [c]
II	Concrete tile	50mm	38.3 [a]	45.7 [b]	36.6 [c]
III	Clay/Concrete tile [d]	25mm	36.7 [a]	43.8 [b]	36.0 [a]

[a] Woven foil
[b] Without foil
[c] Bubble foil
[d] Concrete tile for Hut 1, clay tile for Huts 2 & 3

Table 2 shows the maximum attic air temperature for all the phases. It shows that there is about 9°C difference for roof with and without insulation. The data in table 2 indicate that attic air temperature for clay tiles (without insulation) is cooler than concrete tiles.

9. Conclusions

RIS are effective as attic insulation in S.E. Asia where the predominant heat-flow direction is downward. Measured RSI for typical attic applications of reflective products showed results in the range 2 to 3 $m^2 \cdot K/W$. Measured thermal resistances for attics with reflective insulation are comparable to RSI for mass insulations in the same structure. Ceiling heat-flux reduction for attics with reflective insulation was determined to reduce ceiling heat flux by greater than 80% relative to an uninsulated attic. Small-scale field measurements using custom-built test huts can provide thermal resistance values from transient heat flux and temperature data.

10. Acknowledgements

The authors acknowledge and appreciate the contributions of Dr. Patrick Loi Suok Tee and Mr. Yam Kah Wei to the preparation of this paper and the research project. Their skill and experience has enhanced the research. The authors are also grateful to Solar Energy Research Institute (SERI), Universiti Kebangsaan Malaysia and San Miguel Yamamura Sdn Bhd for the data obtained from the joint research collaboration, without which this research would not have been possible.

11. References

[1] Siegel, Robert and Howell John R 1972 *Thermal Radiation Heat Transfer* (Kuala Lumpur: McGraw Hill Book Company) pp 240-1
[2] Stephan K and Laesecke A 1985 The thermal conductivity of fluid air *J. Phys. and Chem. Ref. Data* 14 (1) 227-234
[3] Yarbrough D W 2010 *Materials for energy efficiency and thermal comfort in buildings* Matthew R Hall (New York: CRC Press) pp 311-3
[4] ISO 8990 *Thermal Insulation Determination-steady-state Thermal Transmission Properties-calibrated and Guarded Hot Box* (Geneva: International Organization for Standardization)
[5] ASTM C1363 2014 *Standard Test Method for Thermal Performance of Building Materials and Envelope Assemblies by Means of a Hot Box Apparatus* vol 04.06 (Annual Box of ASTM Standards) pp 798-849
[6] ISO 6946 2005 *Building Components and Building Elements - Thermal Resistance and Thermal Transmittance - Calculation Method* (Geneva: International Organization for Standardization) Annex A and Annex B

IOP Conf. Series: Earth and Environmental Science **63** (2017) 012048 doi:10.1088/1755-1315/63/1/012048

Reinvestigation of the Dynamic Tidal Power Dams and their Influences on Hydrodynamic Environment

Dong Shao[1,*], Weibing Feng[1], Xi Feng[1] and Yinfeng Xu[2]

[1] College of Harbor Coastal and Offshore Engineering, Hohai University, Nanjing, 210098, China

[2] Nanjing Hydraulic Research Institute, Nanjing 210029, China

shao20130905@163.com

Abstract. Dynamic tidal power (DTP) system is known as an efficient method to exploit the tidal power. Large storage of tidal power in the Yellow Sea and the Bohai Sea along the Chinese coastline is revealed by previous studies. In the consideration of the local environment, combination of smaller DTP dams located at three attractive positions in this area is investigated. Their efficiency is compared with that of larger DTP dams working singly. The influences of the triple smaller dams are also discussed together with those of the single larger dams placed at the same locations.

1. Introduction

As a new form of exploitation of the tidal power, the dynamic tidal power system (DTP) becomes popular and interests many researchers of this domain. Different from other methods, for example, constructing a barrage across an estuary or the mouth of a bay to form a basin or installing submerged small devices (fixed or floating) in open water to extract kinetic power from the currents, the DTP may have less important ecological and environment impact and a good efficiency. The concept of dynamic tidal power exploitation was firstly introduced by Hulsbergen et al.[1], which consists of creating a long dam-like structure perpendicular to the coastline. The DTP dam would interfere with local tidal dynamics and lead to water level differences on both sides of the dam providing considerable water head over the dam. It is known from the early work of the researchers like Hulsbergen et al.[1] and Adema and Hartsuiker [2] that the DTP dams may be located where the form of the coastline is prominent in order to obtain a good efficiency. For the sites of the DTP dams along Chinese coastline, the ideal ones include the those in the Yellow sea, the Taiwan Strait and Qiongzhou Strait. In the area of the Yellow sea, the capes of Liaoning Peninsula and Shandong Peninsula together with the prominent part of northern Jiangsu coast present the locations which store relatively large DTP potential.

However, although the DTP dams are not barrages which form a basin with the coastline, they are nevertheless very huge structures and furthermore, in the pursuit of larger water head, they may extend very far into the open sea. From the DTP investigations of the Yellow sea, it can be seen that sometimes a large DTP dam at the cape of Liaoning Peninsula or Shandong Peninsula may occupy one third or even half the opening of the Bohai Bay, which may cause extreme changes of the local hydrodynamic conditions within the Bohai Bay and furthermore the ecological and environmental system there. As an alternative solution, smaller structures may be placed at the locations with large dynamic tidal potential instead of very large ones. The stem of the T-form DTP dams may be reduced to make it occupy less space and the bar of the T-form dams may be removed since its construction can become very difficult and costly in relatively deep water (as the position of the bar is far from the coast). Although this solution with smaller dams may render each dam a lower efficiency (a smaller difference of water level on both sides of the dam comparing with a larger T-form dam), this disadvantage may be compensated by a combination of several small dams with less impact on the local environment system than a huge dam.

In this study, combinations of small DTP dams located at three attractive positions (triple small DTP dams) in the Yellow Sea and the Bohai Sea (A: the prominent part of northern Jiangsu coast, B: the cape of Shandong Peninsula and C: the cape of Liaoning Peninsula) are investigated. Their efficiency is compared with single larger T-form dams placed at the same position. As for sure that smaller dams may have less impact on the area close to them than larger ones, in this study, more attention will be paid to the influences on the hydrodynamic environment at a certain distance from the dams within the area of Yellow Sea and Bohai Sea.

2. Numerical model

As the water depth of the Yellow sea and Bohai sea is small comparing with the horizontal scale in this study, a two dimensional tidal model ADCIRC[3][4] is applied. The governing equations include the primitive continuity equation

$$\frac{\partial \zeta}{\partial t}+\frac{\partial UH}{\partial x}+\frac{\partial VH}{\partial y}=0 \tag{1}$$

the primitive momentum equation

$$\frac{\partial U}{\partial t}+U\frac{\partial U}{\partial x}+V\frac{\partial U}{\partial y}-fV=-g\frac{\partial}{\partial x}\left[\frac{P_s}{g\rho_0}+\zeta-\alpha\eta\right]+\frac{\tau_{sx,winds}+\tau_{sx,waves}-\tau_{bx}}{\rho_0 H}+\frac{M_x-D_x}{H} \tag{2}$$

$$\frac{\partial V}{\partial t}+U\frac{\partial V}{\partial x}+V\frac{\partial V}{\partial y}+fU=-g\frac{\partial}{\partial y}\left[\frac{P_s}{g\rho_0}+\zeta-\alpha\eta\right]+\frac{\tau_{sy,winds}+\tau_{sy,waves}-\tau_{by}}{\rho_0 H}+\frac{M_y-D_y}{H} \tag{3}$$

with ζ the free surface departure from the geoid, H , the total water depth, $H=\zeta+h$ and h the average sea level from the bottom. U,V indicate the average velocities in the horizontal plan. $f=2\Omega\sin\phi$ is the Coriolis parameter with $\Omega=7.29212\times10^{-5} rad/s$. P_s represents the atmospheric pressure at the sea surface. η is the Newtonian tidal potential. τ_{bx},τ_{by} are the bottom stress components and $\tau_{s,winds},\tau_{s,waves}$ denote the surface stress components imposed by

IOP Conf. Series: Earth and Environmental Science **63** (2017) 012048 doi:10.1088/1755-1315/63/1/012048

the winds and the waves respectively. M_x, M_y are lateral stress gradients and D_x, D_y represent the baroclinic pressure gradients in the x and y direction.

The investigation area is a little larger than that of Hulsbergen et al.[1] for the Yellow sea. The land boundary is controlled by the wetting and drying parameters while the ocean boundary is defined by the tidal components. Thirteen short-period constituents were applied to define the open boundary including the principal lunar, principal solar, lunar solar, elliptical solar, elliptical lunar constituents and they were derived from the satellite altimeter data (Matsumoto et al.[5]) and the application of the components to the numerical simulation is feasible. The numerical model established was validated with the tidal data records from three stations on the Chinese coastline of the Yellow Sea. The simulation period was from the 25th Oct. 2015 to the 30th Oct. 2015. Following the results of the research of Hulsbergen et al.[1], for the single large DTP dams located at the three positions indicated below, the length of the stem of the T-form DTP dams is fixed at 50km and the length of the bar is 0.75 times that of the stem. As the efficiency may decrease much more than linearly when the dams become shorter, for the triple small DTP dams, the length of their stem is chosen to be 25km, half that of the large dams, with a bar always 0.75 times that of the stem, so that the total efficiency of the triple small DTP dams may reach that of a single large dam. The positions of the DTP dams and the water level recording locations on both sides of the dams are shown in figure 1.

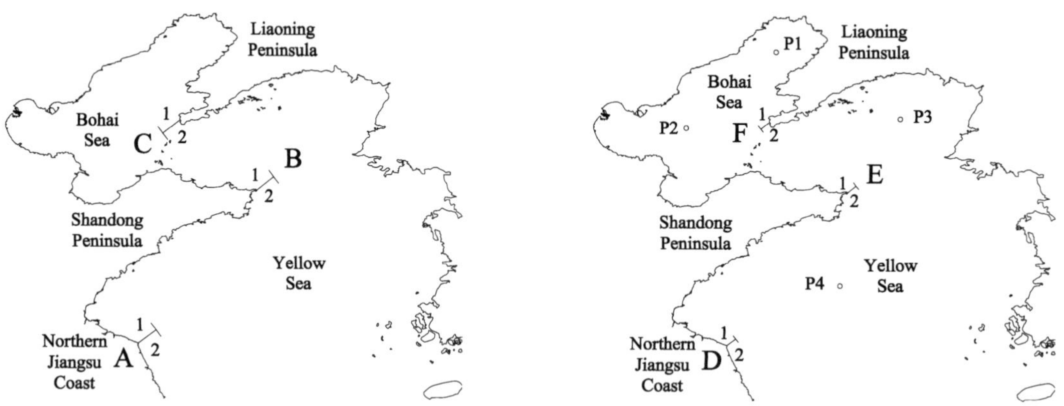

(a) The three single large DTP dams (b) The triple small DTP dams

Figure 1. Investigation area and positions of the three single large DTP dams A, B and C and the triple small dams D, E and F with the water level recording locations (1 and 2) on both sides of the dams (The observation points of the hydrodynamic environment are marked by small circle P1, P2, P3 and P4 in (b))

3. Results and discussion

Similar to the research of Hulsbergen et al.[1], the working efficiency of the DTP dams (predominantly the water level difference across the dams) is examined with the numerical model described in the previous section, including those of the three single large dams and that of the three small dams working in triple. The water levels at the observation locations on the two sides of the DTP dams are shown in figure 2 and figure 3 for each dam which give the water level differences at every moment. It can be observed that the water level differences at DTP dam A, B and C may reach 2.02m, 1.69m and 1.43m respectively. The water level differences for the three small dams working in triple are obviously smaller than those of the larger ones as their sizes are reduced. The efficiencies of the small dams at the prominent part of northern Jiangsu coast (D) and the cape of Shandong Peninsula (E) are slightly better than half of their larger counterparts and that of the small dams F at the cape of Liaoning Peninsula decreased more grievously (only one third of its counterpart). The triple small dams may have a total efficiency equivalent to that of a single large one and even slightly better. In fact, a good efficiency may be achieved conveniently by triple or even double small dams. The water level differences are listed in table 1.

Table 1. Water level differences of the large DTP dams and the triple small DTP dams

		Δh_{max} (m) for each dam	Total Δh_{max} (m)
Single DTP dams	large DTP A	2.02	2.02
	large DTP B	1.72	1.72
	large DTP C	1.5	1.5
Triple DTP dams	Small DTP D	1.17	2.75
	Small DTP E	0.92	
	Small DTP F	0.65	

As it is obvious that smaller DTP dams make less interference on the hydrodynamic environment within the area very close to them than larger ones, attention may be paid to the area relatively farther form the positions of the DTP dams to compare the impacts of single larger dam and triple small dams. Four points P1, P2, P3 and P4 are chosen for observation in both the Yellow Sea and the Bohai Sea as shown in figure 1. The water levels simulated at these points in different cases (with large dam A, B or C or with triple small dams D, E and F) are compared with those simulated without any DTP dams presented in the investigation area in figure 4.

It is revealed that the single large dams A and B may interfere very little with the area at a certain distance from them (the area relatively far from them) as they situate in relatively open water, especially the single large dam A at the prominent part of northern Jiangsu coast. However, it can be seen that the impact of the single large DTP dam C at the cape of Liaoning Peninsula is very important for the observation points within the Bohai Sea since it occupies a certain space right at the opening of the Bohai Sea (nearly one third of the width of the opening) and disturbs the movements of the inward and outward currents. For the small dams, as they work in triple, it is reasonable that they have slightly stronger influence on the farther hydrodynamic environment than a single large dam in relatively open water like the DTP dam A and B, but their impact is to some extent limited and much less important than that of DTP dam C.

IOP Conf. Series: Earth and Environmental Science **63** (2017) 012048 doi:10.1088/1755-1315/63/1/012048

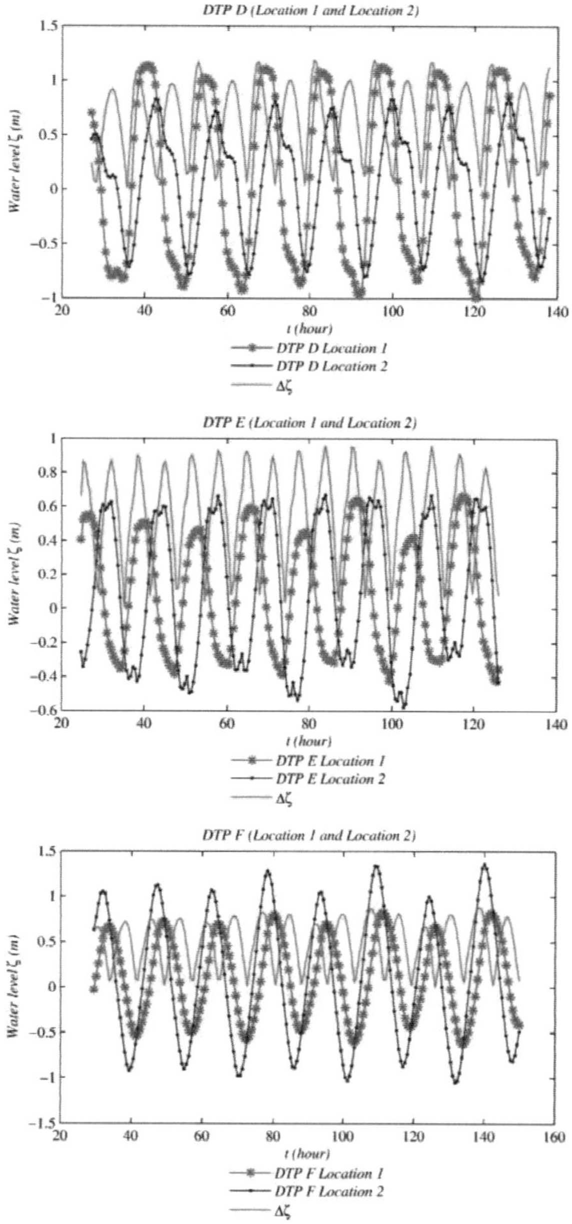

Figure 2. Water levels and their differences for each small DTP dams D, E and F working in triple

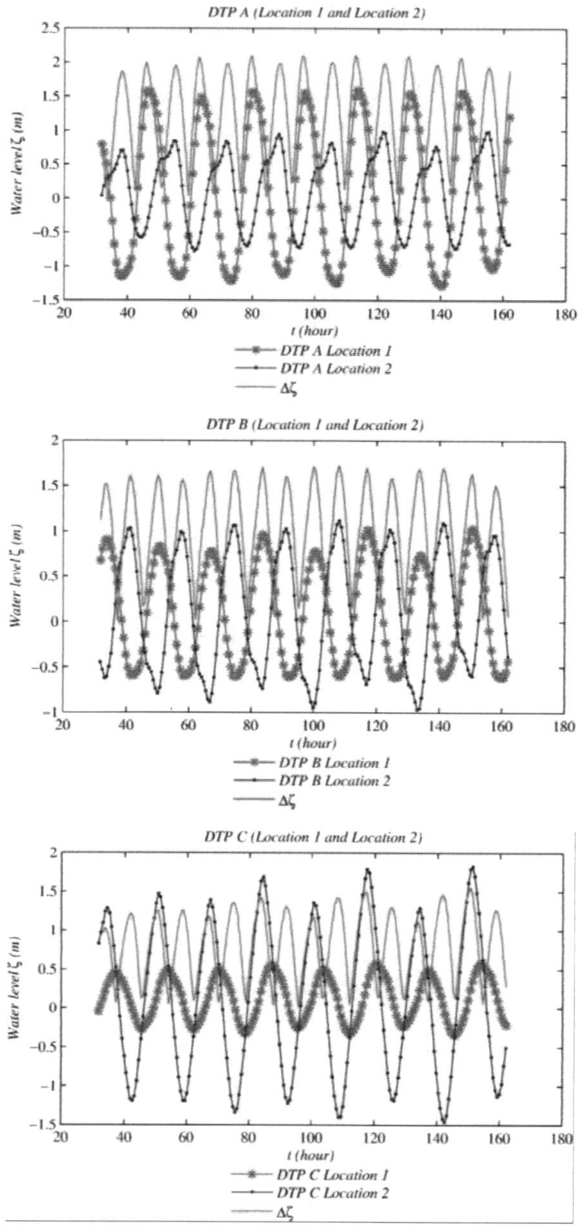

Figure 3. Water levels and their differences for each single large DTP dams A, B and C

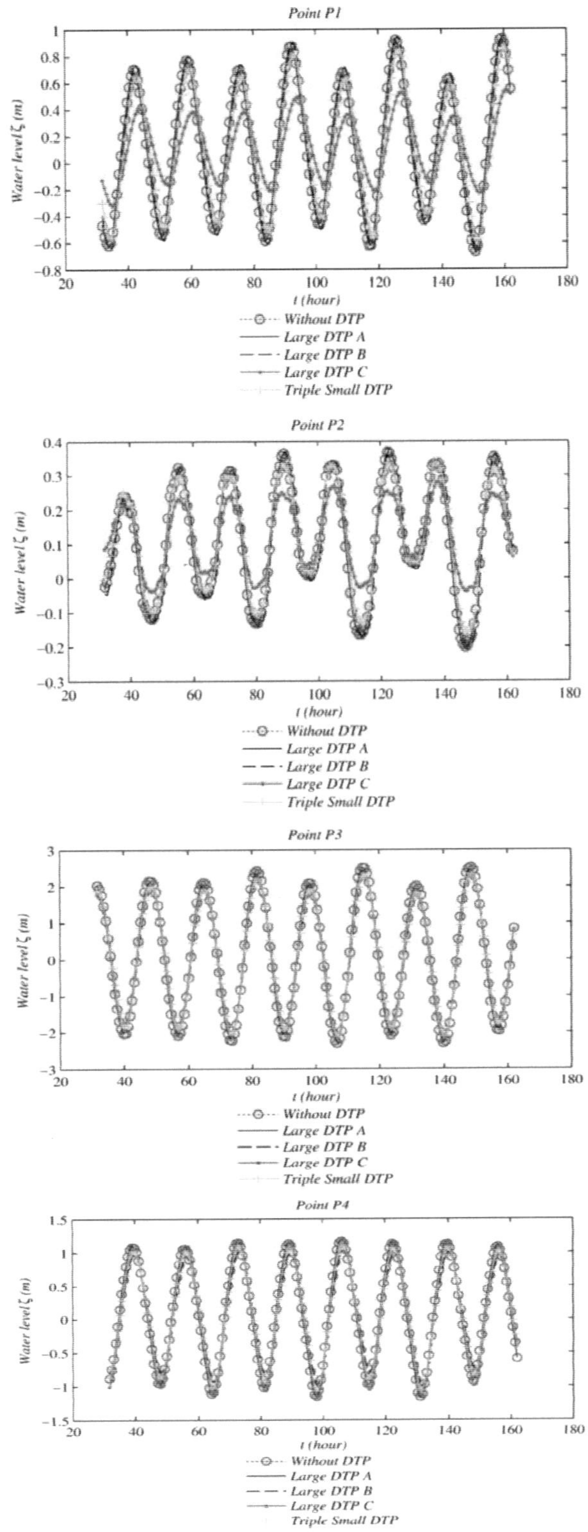

Figure 4. Comparison of the impacts of different DTP dams (with large dam A, B or C, with triple small dams D, E and F and without DTP dam)

4. Conclusion

The numerical investigation of the dynamic tidal power systems in the Yellow Sea and the Bohai Sea along the Chinese coastline for single larger DTP dams and triple smaller DTP dams reveal that combination of two or three small dams (half the size of a regular one) may reach or even exceed conveniently the efficiency of a single larger one working alone. While the small dams may not interfere very strongly with the area close to them, triple small dams have not much more impact on the hydrodynamic environment than one larger dam neither. For the larger DTP dams in the investigation area of this study, the dam at the cape of Liaoning Peninsula influences strongly the hydrodynamic environment within the Bohai Sea whereas the two others have very little impact.

5. Acknowledgement

This research was financially supported by the National Science Foundation of China (51279055).

6. References

[1] Hulsbergen K, Steijn R, Hassan R, Klopman G, Hurdle D. 2005. Dynamic tidal power (DTP). In 6th European Wave and Tidal Energy Conference, Glasgow, UK.

[2] Adema J, Hartsuiker G. 2010. Potential location for dynamic tidal power in China. Vol. 2571. Report No. C04021.

[3] Luettich Jr R A, Westerink J J, Scheffner N W. 1992. ADCIRC: An Advanced Three-Dimensional Circulation Model for Shelves, Coasts, and Estuaries. Report 1. Theory and Methodology of ADCIRC-2DDI and ADCIRC-3DL. COASTAL ENGINEERING RESEARCH CENTER VICKSBURG MS.

[4] Westerink J J, Luettich Jr R A, Blain C A, Scheffner, N W. 1994. ADCIRC: an advanced three-dimensional circulation model for shelves, coasts, and estuaries. report 2. user's manual for ADCIRC-2DDI (No. WES/TR/DRP-92-6-2). ARMY ENGINEER WATERWAYS EXPERIMENT STATION VICKSBURG MS.

[5] Matsumoto K, Ooe M, Sato T, Segawa J. 1995. Ocean tide model obtained from TOPEX/POSEIDON altimetry data. Journal of Geophysical Research: Oceans, 100(C12), 25319-25330.

Scale Effect Analysis of Urban Compactness Measurement Index Based On Grid

Lei Cao

School of Urban and Environment, Huaiyin Normal University, Huai'an, Jiangsu223300, China

lc@hytc.edu.cn

Abstract. Scale effect is one of the basic aspects of urban compactness. This paper takes the use of urban construction land as the breakthrough point, continuity degree and Gini coefficient as the indexes to analyse the scale effect of grid from the calculation through the different scales grid analysis of urban construction land extracted by ERDAS and ARCGIS respectively. The result showed that the selected indicators quantitively indicate the features of research area land, different scale effects of two indexes, big difference in continuity degree influenced by scale, and slight impact on Gini coefficient influenced by scale.

1. Introduction

As one of the key issues in the research of sustainable development, the sustainable development of cities has been paid much attention by governments and academia. Compact city is the concept based on urban sustainable development which aims at solving the problem of unordered city expansion. The concept of compact city first appeared in Holland and later was written in the urban development policy in England and America, and eventually become a widely accepted urban development strategy [1]. However, city is not more compact, the merrier. The form of compact city is under the great influence of population and building density. There will be problems in the aspects of urban communications, infrastructure and environmental capacity, if it is too much compact. And urban ecosystem and the quality of life of residents will be adversely affected [2]. The rationality of city compactness directly affects the efficiency of relevant strategies, therefore, the research of urban compactness has become a hotspot issue in the fields of geography and planning.

In this paper, the continuity degree and Gini coefficients are selected as the representative indicators of urban compactness taking Nanjing city of Jiangsu Province as an example. First, it extracts the building lot map of Nanjing from remote sensing images by using remote sensing software, then sets a certain scale gradient difference grid and calculates the index of urban compactness under different scale by using the method of grid analysis. The index of recognition of urban compactness is under the influence of grid scale through the comparison of two indexes analysis under different grid scale, which provides useful reference and powerful technical support for the research of urban compactness and city development plan.

2. Analysis method

2.1. Selection of structure index

There are different levels and aspects research about urban compactness. The measurement of urban compactness can be divided into five types, shape, scale, density, function and structure. And the index of structure is measured by the spatial distance and the internal relation of urban elements. There are two research areas. Firstly, the equilibrium degree of spatial distribution of urban elements, Gini coefficient can be selected. Secondly, the fragmentation degree of different components in city. It is different from shape, which emphasizes the overall situation of internal composition of research objects. There are some frequently-used indexes, splitting degree, effective grid size and urban layout continuity [3].

Therefore, this paper selects two compact indexes based on the land use structure, which are continuity degree and Gini coefficient to consider the degree of fragmentation and equilibrium degree of urban land respectively.

2.2. Calculation method of measure index

*2.2.1. Gini Coefficient*The Gini coefficient is derived from the measurement of the difference of income distribution in economics, and has been applied to measure the degree of urban development focusing on a small number of plots. Considering the global distribution of the whole building land, the Gini coefficient can effectively measure the equilibrium degree of urban construction land. The formula is as follows:

$$Gini = 0.5\sum_{i=1}^{N}\left|X_i - Y_i\right| \tag{1}$$

In the formula: Xi is the plot unit, i is construction land for the proportion of total construction land, Yi is the plot unit, i is area accounted for the proportion of the overall plot, N is the standard small plot of total number of i. The index is used to measure the density distribution of urban land, the greater the value, the greater the degree of imbalance.

*2.2.2. Continuity Degree*In order to measure the city development of continuous high density, some scholars choose residence or employment density as the threshold; set a proportion as the standard. If the functional land reaches the standard, it will be the continuous use, if it doesn't any continuous use. Marjo (2006) selects 80% as the boundary to distinguish between continuous and discontinuous city land, when he does the analysis of development mode of land use in 15 European cities. And this threshold conforms to our country city land use situation of heterogeneity after verification, which plays an important role in instructing the definition of city development mode [3]. Based on this research, the level of contiguous land unit construction can be calculated by the proportion of city construction land.

$$Continuity = \frac{\sum_{i=1}^{N}(if = D_i > 80\%, M = 1; otherwise \quad M = 0)}{N} \tag{2}$$

In the formula: Di is the proportion of construction land in the plot unit i. M=1, when Di>80%, in other cases, M=0. N is the total number of standard small plots of i. The value Continuity is between 0 and 1. The higher the level of city construction contiguous land is, if the value tends to 1, the small the relative degree of fragmentation.

2.3. Scale Effect of Grid

Spatial scale is to measure the space to carry out research and the scale effect is defined as changing of unit size, shape and direction, when spatial data polymerizes in geography [4]. This paper takes the continuity and Gini coefficient as the indexes by the calculation of grid analysis. In the process of grid analysis, it inevitably involves the grid size. Therefore, in the grid analysis, the scale effect on the implementation of what size of grid cell is able to express the specific geographical phenomenon. In the study of compact degree of city land use by the method of the grid analysis, not only can help select the appropriate grid size, but also can provide an important reference for the different indicators to screen effectiveness and adaptability [5].

This paper adopts the square grid size to partition the entire study area taking each unit block as a grid to calculate the proportion of construction land and compactness index. If the calculated results show that the size of grid computing and grid results are closely related, it can explain the measurement index with spatial scale effect. In addition, the characteristic difference between each index can be observed at different scales with selection of grid units of different sizes for the construction of compact measure, which, meanwhile, has specific impacts on each index of the grid scale. Therefore, the use of simple grid partition can effectively identify the effects of spatial scale and its influence.

3. Scale Effect Analysis of Compactness Measurement Index: a Case Study of Nanjing

3.1. Study area and data

Nanjing belongs to the "Yangtze River Delta" economic developed areas with superior natural conditions and dense population, which is in a rapid urban growth. Over the past 30 years, the scale of urban construction land has expanded more than 10 times in an average annual growth rate of 10%. The structure of urban land use in Nanjing is characterized by the relative concentration [6]. Meanwhile, Jiangning district, Pukou district and Liuhe district around the main city is in a rapid development recently, which has become a new hot spot for the development of Nanjing. This paper chooses Nanjing city as the study area, hoping that through the study of this typical area, on the one hand can reflect the practical significance of research on urban compactness index, on the other hand can be used to verify the grid scale effect index.

The remote sensing data of Nanjing uses seven bands and panchromatic band TM remote sensing images of Nanjing city in 2012 (from the GLCF website: http://glcf.umd.edu/). The map of administrative division of Nanjing city is from the Bureau of Surveying and Mapping Geographic Information in Jiangsu Province (http://www.jschj.gov.cn/). Remote sensing image processing software ERDAS 8.4 is used to complete the band synthesis, geometric correction and data tailoring of TM image data. Geographic information system software ARCGIS 9.3 is used to do the grid analysis and mapping by extracting the data of the distribution of city construction.

3.2. Grid analysis

For further research, this paper chooses 6 different grid scales according to the actual situation, which are 1000m, 2000m, 3000m, 4000m, 5000m, 6000m (Figure 1).

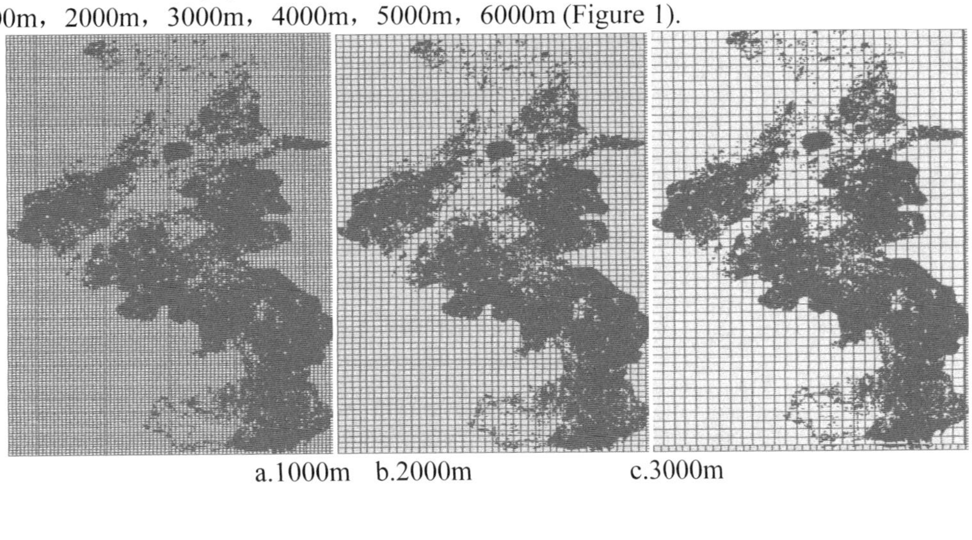

a.1000m b.2000m c.3000m

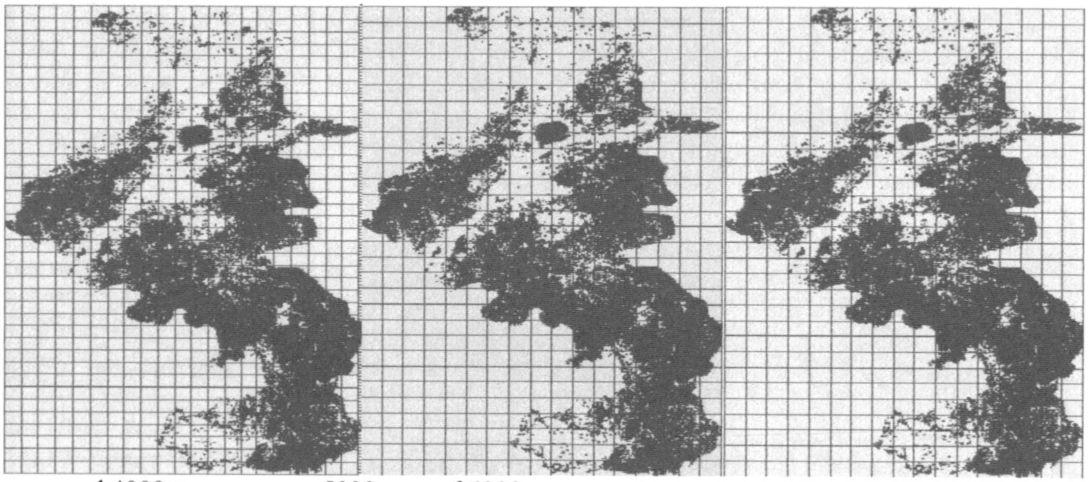

d.4000m e.5000m f.6000m

Figure 1. Grid overlay results atdifferent scalesFootnotes.

4. Result and Discussion

4.1. Result of Urban Compactness Measurement in Nanjing

The degree of continuity of construction land in Nanjing is still relatively high. However, the distribution of construction land is not balanced, and the construction land is still distributed in the central city. Nanjing has a long history of development. In early times, the relative developed city zone is in the intersection or collection of resources around the main city. Then a variety of economic information radiation has been generated through the traffic corridor, which brings a development of both sides of construction land along it. The lateral connection of the land on both sides of the line is strengthened, when the development of construction land is up to a certain extent, and he broken space is gradually filled [7]. Therefore, the Nanjing construction land around the city is in gathered distribution, which, however, is restrained by the hilly terrain and administrative boundary so that the city is in a north-south narrow form development [8].

Grid analysis at different scale can be done to the construction land of Nanjing, which has different properties. And the indicators will be changed by the change of scale, which directly confirms that there is obvious spatial effect in indexes of continuity and Gini coefficient(Table 1).

Table 1. The calculation results of two indexes at different grid scale.

	1000m	2000m	3000m	4000m	5000m	6000m
Continuity	0.256	0.134	0.095	0.074	0.032	0.013
Ginicoefficient	0.613	0.598	0.586	0.572	0.561	0.554

The degree of continuity of urban construction land in Nanjing is between 0.013 and 0.256 under different grid scales. The value is close to 0, the degree of continuity of urban construction land in Nanjing is relatively small and the degree of fragmentation is high. Although there is slight deviation compared with the visual perception of the remote sensing image of the construction land in Nanjing, it just shows it has scientific nature and accuracy. As for the case of value of degree of continuity of urban construction land in Nanjing approaches to 0, it might has something to do with the grid scale which is in spatial scale of Nanjing or not. And it needs further research about if its grid scale is relatively large. At the same time, continuity in different grid scale varies greatly. And the maximum value reached 18.7 times the minimum value, which means effect of spatial scale exists. The Gini coefficient of urban construction land in Nanjing is from 0.554 to 0.613, which shows that the spatial distribution of construction land in Nanjing is not balanced. It is consistent with the actual situation of

IOP Conf. Series: Earth and Environmental Science **63** (2017) 012049 doi:10.1088/1755-1315/63/1/012049

Nanjing urban land. The slightly variation in the Gini coefficient shows that there is no big influence on the grid scale calculation results for Gini coefficient, which means the Gini coefficient is not sensitive to the change of research scale. Based on the above two comparisons, it is found that the dependence of different indexes on the scale is different, the degree of continuity is affected by the scale, and the Gini coefficient is under slightly influence of the scale.

4.2. Scale Effect Analysis of Compactness Measurement

According to the analysis of different scale grid (Figure 1), it can be done from two different angles, which makes the result more meaningful.

*4.2.1. Selection of the grid scale of same size, analysis of two different index corresponding to the results*Under the grid scale of 6000m, the result of continuity is 0.013, reaching 0. It is presumed that is caused by the choice of small scale. The continuity is significantly affected by grid scale from the calculation results. The greater the grid is, the more map spot in unit grid of land is. And The amount of continuous construction in grid sharply reduces, which results a proportion decrease of continuous construction land. Therefore, the situation of value of continuity reaching 0 happened under 6000m grid system.

The value of Gini coefficient is 0.554 under 6000m grid scale. The larger the value is, the unbalanced the distribution of construction land in Nanjing is. And it is more reasonable. The value of Gini coefficient drops when the scale expands. The increase of grid size results a sharp reduction in the number of land units. The land use type is more diverse, which performance is the decrease of the value of Gini coefficient. The bigger the analysis scale is, the more balanced the distribution of land use is. Meanwhile, the Gini coefficient is slightly affected by grid scale. Its value only has become less than 10% from top to bottom.

In the process of measuring the urban compactness, scale effect of different indexes is different, or the sensitivity of scale effect of different indexes is different through the comparative analysis at the same scale.

*4.2.2. Research on structure under different grid scale respectively corresponding to two different indices.*With the increase of the grid scale, there is a downward trend of the size of continuity. But the magnitude of decline is relatively large. The value of the Gini coefficient goes down as the scale goes large. But the magnitude of decline is relatively gently (Table 2).

Table 2. The degree of index reduction at different grid scale.

Magnitude of Decline	2000m	3000m	4000m	5000m	6000m
Continuity	47.65%	29.01%	22.10%	56.70%	59.37%
Ginicoefficient	2.45%	2.00%	2.39%	1.92%	1.24%

There is a same trend between these two indexes, which is there is a downward trend, when the value of these two indexes go down as the scale goes large. There is a relationship between the size of the Gini coefficient and continuity and the scale of grid, and the closeness of the relationship between them is different. The continuity is close to grid scale more than the Gini coefficient. There is the same trend in change of index scale, but the stability of scale which shows the difference of land use is slightly different. And its scale effect is also different. The grid scale effect of continuity is much more obvious than of the Gini coefficient.

There is a particularly decrease in the continuity when in grid scale of 1000m and 2000m from Figure 1. With the increase of grid, the decline of continuity shows the trend of slowing down, because it is affected by the continuous threshold of land use. The change of scale leads to proportion decrease of construction land in grid and the total amount of continuous land sharply decrease. However, there is a difference when the scale up to 4000m. Therefore, it is helpful for the analysis of continuity index by choosing the right grid scale, so as to indentify the degree of fragmentation in city construction land accurately. However, there is steady decline with the change of scale in the Gini coefficient. And its scale of suitability can be selected according to the actual operating conditions. And it is easy for the use conditions. In the study of urban compactness, the effect of index under the influence of grid scale is different. They all have unique scale effect. Therefore, it is necessary to select the appropriate index, considering its corresponding scale effect, which makes results more reliable in actual process.

5. Conclusion

Based on the analysis of the grid, the Gini coefficient and the continuity reflect the characteristics of the spatial structure of the city, which respectively show the degree of continuity and equilibrium of city land use so as to facilitate the study of urban compactness.

The characteristics of the urban compactness space represented by the index have certain scale dependence, and the continuity and the Gini coefficient will decrease with the increase of the scale, showing a downward trend.

The scale effects of continuity and Gini coefficient are different. With the change of scale, the degree of difference is relatively large, so it is more important to choose the appropriate scale. The Gini coefficient is smaller than the scale, so it has better applicability.

In the actual research, it should pay attention to the scale effect of different index and select the appropriate grid scale to do the research to ensure the certain accuracy of work in order to study the urban compactness through grid analysis of the continuity and Gini coefficient.

This paper has some shortcomings, such as, there are only 6 different grid scales to do the calculation, which makes research limited. There is only one city as the sample, which lacks of comparison with other cities. Hope there will be further research in future.

6. Acknowledgments

This work was supported byThe Project of Philosophy and Social Science Research in Colleges and Universities of Jiangsu Province (2014SJB661), The Science and Technology Plan Projects of Huaian (HAS2014022-3) and National Natural Science Foundation of China (Grant No. 41271135).

7. References

[1] Lin L 2008 The compact strategy of the EU countries: examples from the UK and Netherlands *Urban Planning International***23** 106-116

[2] Shimou Y, Cheng W and Luocheng Z 2008 The influencing factors of resources and environments in the process of urbanization of China *Progress In Geography***27** 94-100

[3] Binyao L 1998 Urban space form of measurement method and its evaluation *Urban Planning Forum***3** 42-45

[4] Lin L and Shen Y 2005 Fundamental problem on spatial scale *Geomantic and Information Science of Wuhan University***30** 200-202

[5] Jing Y, Xiaohuan Y and Dong J 2010 The grid scale effect analysis on town leveled population statistical data spatialization*Journal of Geo-information Science***12** 40-47

[6] Ronghua M, Chaolin G, Yingxiang P, Xiaodong M and Chuangeng Z 2007 Urban spatial sprawl pattern and metrics in south of Jiangsu province along the Yangtze Rive*ActaGeographica Sinica***62** 1011-1022

[7] Suliu C 2012 The research of bidirectional organizational development character and design of urban morphology *Journal of TongjiUniversity(Social Science Section)* **23** 33-39

[8] Chuncai W 2007 The study on the interaction mechanism between urban transportation and urban spatial evolution *Urban Problems***6** 15-19

IOP Conf. Series: Earth and Environmental Science 63 (2017) 012050 doi:10.1088/1755-1315/63/1/012050

Green Development Status in Zhejiang Province and the City of Ningbo, China: Examination of Policies, Strategies and Incentives at Multiple Levels

Ali Cheshmehzangi, Wu Deng, Yun Zhang and Linjun Xie

Department of Architecture of Built Environment, The University of Nottingham Ningbo China, 199 Taikang East Road, Ningbo, 315100, China

Ali.Cheshmehzangi@nottingham.edu.cn

Abstract. China is undergoing the largest scale of urbanization in history and at an unprecedented pace. The construction and operation of buildings have inevitably brought severe pressures on resource conservation and environmental protection. China has initiated policies, strategies and financial incentive schemes at national level to address these issues. It is also seen that there is a growing interest in recent years at local government level in promoting green buildings. This paper will examine the current national policies, targets and standards and then discuss how these national initiatives are reflected at provincial and city level by taking Zhejiang Province and Ningbo City as case studies. A comparison between different levels of initiatives is conducted by reviewing incentive mechanisms, technological development and compliance requirements. It is concluded that the national initiatives may be not effective without local enhancement.

1. Introduction

China's urbanization is progressing rapidly at has gone through a phase of unprecedented pace in 1980s and 1990s. Currently, China's urbanisation is steady but still progressing at a large scale. Between 1991 and 2012, China's urban population has increased from 26.4% to 52.6% of the whole population, which signifies a major shift, doubling the number of urban residents in only two decades. The urban built areas are expanded from 12,856 km^2 to 45,566 km^2 over the same period, an increase of 3.5 times greater in about two decades National Bureau of Statistics [1]. The number of new buildings - and often large scale structures - in cities is increased significantly in these decades in order to accommodate the increased urban population. In return, the rural population has been perceptible in this process of transition, from a rural country towards an urbanising nation. For instance, as Li and Shui [2] argues, in recent years China has been adding about 1.7 billion square meters of new floor space on an annual basis. As estimated by the 'Building Energy Conservation Center (BECC)' of Tsinghua University, the annual rate of new construction in China equals the total amount of new building in all developed countries BECC [3]. In 2005 Shanghai alone added more space in the form of residential and commercial towers than exists in all of New York City [4]. Taking the annual addition to the residential building stock, the number has soared from 0.4 billion m2 to approximately 0.7 billion m^2, a growth rate of 69% in the last one decade [2].Such significant increase certainly puts major pressures on construction industry. The rapid pace of development often results in poor quality development that may not necessarily be sustainable. Now that China has gone through this rapid phase, a larger emphasis is expected to be given on quality of development to address issues of sustainability and green development. Cities in general and the built environment in particular, during their life-cycle, have a significant environmental impact both at local and global levels. The suggestions given on rapid expansion of urbanisation level has inevitably brought severe pressures on resource conservation and environmental protection in Chinese cities. This factor is highlighted significantly in recent research in the field. As a result, we now experience major urban pressures (and some are global) of increasing pollution, traffic and energy consumption in the urban areas. Major criticism is given on the pace and mode of development in most cases, which has resulted in mass development of low-quality or unsustainable development projects across major cities in China. These mainly include cities of tier 1, tier 2 and tier 3 that have mostly experienced rapid pace of development or are still facing that phase of growth and expansion. For instance, only eight of 74 major Chinese cities, have satisfied the national air quality standards in 2014 [5], which indicates a major nationwide challenge. Moreover, given the long lifetime of buildings, choices made today on the construction of buildings will have long-term effects influencing the overall environmental performance of cities for decades to come [6].

In this process of rapid development and urbanisation progress, buildings have become a major energy consumer in China. The most significant impacts are on building operational consumption that varies between 16% and 30% of the total national consumption based on various data (e.g. IEA [7], BEEC [3], Yang and Kohler [8], He et al [9]). If embodied energy used for manufacturing building materials is taken into account, the total building energy consumption may be around 40% of the national total energy consumption [10], cited in [8], which is relatively higher than just consideration of the operational energy consumption. However, this is not yet the energy consumption cap as further growth and urbanisation is under process. For instance, BEEC has estimated that the 2030 national total building energy consumption will be based on three variables that are: a) floor area per capital b) life-style changes; and c) technology evolution. The variables may not be so significant individually, but their combined effects would determine an estimated increase of minimum 14% (the best scenario) consumption increased from the building operation in 2030. The worst case scenario is the increase of 270% that is estimated based on the worst case scenario of all three variables combined together. This also implies there is huge potential to reduce energy use in building sector in China BECC [3]. It is therefore foreseen to comprehend more emphasis on initiatives, incentives and policies – and if not merely technologies – on this almost ever-growing sector of building construction.

1.1. Green Building Development Policies at the National Level.

At the national level, there is a list of programmes that are discussed in here.

Green Building Action programme – Announced in January 2013 by the state council, the GBAP has the clear objective of completing 10 billion square meters area of new green building construction and 20% of the new construction in the city meets the green building standard requirement till the end of 2015. Also since 2014, all of the construction invested by the government, affordable housing and large public building should implement the green building standard.

Notice on Implementation of Green Building Action in Protected Housing – promulgated in December 2013 by the Ministry of Housing and Urban-Rural Development (MOHURD), this specific programme requires a full understanding of the importance of green housing action in affordable housing. In this respect, since 2014 the municipal area of affordable housing, should implement green building action, at least meet one-star green building standards. During conducting annual plan of affordable housing projects, green building action requirements should be made clear. This is also required to strengthen publicity of affordable housing.

Planning for New Urbanization in China (or commonly as New-Type Urbanisation Plan) – Developed over the course of several years and announced in March 2014, the CPC Central committee and The State Council put together a list of key requirements on: 1) implementation of green building action plan; 2) improvement of the green building standards and certification system; 3) expansion of the scope of enforcement; 4) speeding up energy-saving transformation of the existing building; 5) development of green building materials; and 6) promotion of the construction industry as a whole.

Notice on Printing and Distributing the Action Plan for Energy Conservation, Emission Reduction and Carbon Reduction from 2014 to 2015 – Also announced by The State Council and in May 2014, the main requirement of the programme is to deeply develop the green building action plan. The public welfare buildings invested by government, large public buildings and affordable housing in cities should therefore fully implement the green building standard. Till 2015, Implementation rate of new urban construction which meets the green building standard should reach to 20%, and the area of new green building up to 3 billion square meters.

Notice on Promoting the Green Building Action in the Public Welfare Buildings Invested by Governmental and Large Public Buildings – Also developed by MoHURD and in October 2014, the requirements of the programme are to fully address the importance of government investment in public buildings and large public buildings to promote green building action. This is also required to encourage the use of industrialization construction model and the use of green building materials. As a result, the programme aims to strengthen management of the whole process of construction.

Opinions of the CPC Central Committee and the State Council on Accelerating the Construction – Developed jointly by the CPC Central Committee and The State Council and announced in April 2015, the programme requires vigorous promotion of green urbanization. During the process, this aims to strengthen the concept of energy conservation, development of low carbon and convenient transportation systems, and promote the construction of green ecological city.

1.2. Green Building Technical Standards and Strategies at the National Level.

In recent years, China has also developed several technical standards at the National Level.

Green Construction Evaluation Standard for Building Engineering (GB/T 50640-2010) – Lunched in October 2010, this standard includes eight categories of: 1) Construction preparation; 2) Construction site; 3) Foundation and foundation works; 4) Main structural engineering; 5) Construction and decoration works; 6) Insulation and waterproofing works; 7) Mechanical and electrical installation works; and 8) Demolition works.

Code for design of civil green building (JGJ/T 229-2010) – Developed in Oct 2011, the standard is applicable to design of the new built, alteration and extension of green building. It also includes general rules, terms, basic regulations, design plan, site and outdoor environment, water supply and sewerage, architectural design and indoor environment, HVAC and building electrical systems.

Evaluation Standard for Green Industrial Building (GB/T 50878-2013) – Lunched in March 2014, it includes an evaluation index of: 1) 1.Land saving and sustainable development; 2) Energy saving and energy utilization; 3) Water saving and water resources utilization; 4) Materials saving and materials resources utilization; 5) Outdoor environment and contaminant control; 6) Indoor environment and occupational health; 7) Operational management; and 8) Technology progress and innovation.

Green Building Inspection Technology Standard (CSUS/GBC 05-2014) – announced in July 2014, this standard includes comprehensive list of: 1) Outdoor environmental testing; 2) Indoor environmental testing; 3) Thermal performance test of enclosure structure; 4) HVAC system testing; 5) Water supply and drainage system testing; 6) Lighting and power supply and distribution system testing; 7) Renewable energy system testing; 8) Monitoring and control system verification; and 9) Building energy consumption and total energy consumption of air conditioning in heating.

Green Building Evaluation Standard (GB/T 50378-2014) – Developed in 2014 (second edition), this standard provides scope of application for various types of civil construction and public building. It also provides scope for Green Building Evaluation Index System, which highlights the following seven aspects of: 1.Land saving and outdoor environment; 2.Energy saving and energy utilization; 3.Water Saving and Water Resources Utilization; 4.Materials saving and Materials Resources Utilization; 5.Indoor environment quality; 6.Operation management; 7. Construction management. The evaluation indexes are then graded and the green building level is determined by the total score.

Green Evaluation Criteria for Existing Buildings (GB/T 51141-2015) – Developed recently in August 2016, the standard is provided to mainly supplement and enhance the green building evaluation standard system in China, The evaluation index is considered for seven categories of: 1) planning and architecture; 2) structure and materials; 3) HVAC systems; 4) building electrical systems; 5) water supply and sewage systems; 6) construction management; and 7) operational management.

Evaluation Criteria for Green Eco-City – currently at soliciting opinion stage, this standard comprehensively include: 1) Land use; 2) Ecosystem; 3) Green building; 4) Resources and carbon emissions; 5) Green transportation; 6) Information management; 7) Industry and economy; 8) Humanities; and 9) Technological innovation.

1.3. Green Building Incentive Schemes at the National Level.

At the national level, there is a list of incentive schemes at the National Level.

Green Building Action Program - Giving the finance incentives to buildings that reach the national green building evaluation standard: 45 RMB per square meter for two-star green building, 80 RMB per square meter for three-star green building.

Tax Policy - National ministries and commissions: the Ministry of Finance should consider and formulate preferential policies on taxes, encourage real estate developers to build green buildings, and guide consumers to buy green houses. For instance, Xiamen province aims to give back 20% deed tax to owners who brought the 2-star green house and 40% to those who brought the 3-star green houses.

Financial Policy – National ministries and commissions: Improve and optimize the financial services for green building, financial Institutions could give appropriate concessions on house loan interest rate to consumers who buy the green house. For instance, Jiangsu province gives loan amount can go up 20% for who buy two-star or above green housing using housing provident fund loans.

Land Policy - National ministries and commissions: Ministry of Land and Resources should consider and formulate policies which can promote green building development in land transfer. Ministry of Housing and Urban-Rural Development should consider and formulate policies in volume rate incentives. For instance, Fujian province proposes for green building district developed by real estate developer can awarded 1%, 2% and 3% of volume rate respectively according to one-star, two-star and three-star green housing.

2. Reflection of National Policies at Provincial and City Level

In China, the concept of green building was defined in the first edition of GBES in 2006, which elaborates that a green building should be 'energy-saving', 'water-saving', 'material-saving', 'land-saving' and 'environment-benign', summarised as "four savings and one benign". Generally the Chinese legal and policy systems for developing green buildings comprise four levels, including; 1) The Energy Conservation Law; 2) Renewable Energy Law; 3) Construction Law; and 4) Environmental Protection Law constitute the legal basis and give the general principles for developing green buildings. The State Council has issued regulations to enforce the laws and some provincial and local governments also have developed their own regulations for improving building environmental performance. Building design standards for various climatic zones have been issued by the Ministry of Housing and Urban-Rural Development (MoHURD). GBES is the first comprehensive system for whole building assessment and has been a national standard since its development as the assessment standard in China. Before GBES, there were fragmented assessments on various aspects of building environmental performance such as energy, material and water. As a result, policy and financial incentive schemes have been proposed at both national and local levels.

At the local level, some local governments have developed their own green building evaluation systems, which often have more rigid requirements on building environmental performance than national standards. A remarkable example is in Beijing, where they execute the new goal of 75% building energy efficiency since 2011 [11][12], compared to the national requirement of building energy saving of 50%. Similarly, two cities of Shanghai and Tianjin also require a minimum of 65%

energy saving for new buildings. Therefore, we can see a transition from national level to local level policies and initiatives.

2.1. Case Study Analysis: The Cases of Zhejiang Province and City of Ningbo
In this study, we also demonstrate the policies, technical standards, and incentive schemes at the Provincial level of Zhejiang Province first and the at the City level of Ningbo. Sub-sections 2.1.1 to 2.1.3 cover the provincial level, while 2.1.4 to 2.1.6 cover the city level.

2.1.1. Green Building Development Policies at the Provincial Level. At the Zhejiang provincial level, there is a list of programmes that are discussed in here:

Opinions of Zhejiang People 's Government on Actively Promoting the Development of Green Buildings – Developed in August 2011, the requirements of the programme are to mainly to enhance social awareness of energy conservation and form a green building development system in order to achieve the leaping development of transferring the energy-saving construction to green construction.

Notice of the General Office of the People's Government of Zhejiang Province on Deepening Printing and Distributing the Implementation of Promoting New-type Industrialization of Buildings and Promoting the Development of Green Buildings in Zhejiang Province – Developed in December 2014, the main requirement is to construction invested by the government, indemnificatory housing in Hangzhou and Ningbo cities, and large public buildings with area larger than 20,000 m^2 are required to meet green building standard, as well as implementing new building industrialization.

Three-year action plan of Green Building Development in Zhejiang Province – Developed in September 2015, this programme proposes to build new green construction. Green Design Standard for Civil Buildings in Zhejiang Province has been implemented on new civil construction within the province; government investment public welfare housing and large public buildings which reaches to 20,000 m^2 of area and are encouraged to meet the two-star or three-star green building standard. The programme also includes requirements to build new prefabricated building. The areas of industrialisation projects both completed and under construction, reach 1 million m^2 in 2015, 3 million m^2 in 2016 and 5 m^2 meter in 2017. It also includes energy saving renovation for existing building.

2.1.2. Green Building Technical Standards and Strategies at the Provincial Level. In recent years, Zhejiang Province has also developed two technical standards (Table 1).

Table 1. Two Technical Standards developed in Zhejiang Province

Green Building Evaluation Standard (DB33/T 1039-2007)	Started in 2008.8.1	Basically continuation the framework system of GB/T 50378-2006.
Code for design of civil green building (DB33/T 1092-2013)	Started in 2014.1.1	Basically continuation the framework system of JGJ/T 229-2010 .

2.1.3. Green Building Incentive Schemes at the Provincial Level. Also Zhejiang Province has proposed for one incentive scheme at the Provincial Level:

Financial subsidy - The only incentive scheme is the financial subsidy which proposes requirements according to the three-year action plan of Green Building Development in Zhejiang Province, including giving the financial incentive to those new construction industrialization projects, which have awarded national green construction two-star or three-star level. These are based on the Opinions on Accelerating the Development of Green Building in China enacted by Ministry of Finance and Ministry of Housing and Urban-Rural Development.

2.1.4. Green Building Development Policies at the City Level. At the City level, there is a only one major programmes that is discussed in here:

Ningbo Green Building Action Plan – Developed by Office of Ningbo Government, the policy action plan is for new built construction projects. From 2014, construction invested by government, indemnificatory housing in Ningbo, and large public buildings which area larger than 20,000 m^2 are required to meet green building standard. By the end of 2015, the area of new built green construction reaches 10 million m^2, accounting for 20% of the whole new construction area. Also the other requirement is the energy saving renovation for existing buildings. By the end of 2015, the area of completed energy saving renovation for existing building reaches to 1 million m^2, the renovation rate of high energy consume building up to 30%.

2.1.5. Green Building (Local) Technical Standards and Strategies at the City Level. In recent years, Ningbo city has also developed two local technical standards (table 2).

Table 2. Two Local Technical Standards developed in Ningbo City

Document name	Promulgation time	Content
Implementation Rules of Ningbo Green Building Evaluation (SS-01)	2014.7.1	Included in: 1. Land saving and outdoor environment 2. Energy saving and energy utilization 3. Water Saving and Water Resources Utilization 4.Materials saving and Materials Resources Utilization 5.Indoor environment quality 6. Operation management 7. Technological innovation for both residential building and public construction
Implementation Rules for Civil Green Building Design in Ningbo City	2016.10.1	Combination of Green Building Evaluation Standard (GB/T 50378-2006) and Green Building Evaluation Standard (DB33/T 1039-2007), summarize of Ningbo city green building practice experience and research results.

2.1.6. Green Building Incentive Schemes at the City Level. Also Ningbo City has proposed for one incentive scheme at the City Level (table 3).

Table 3. One Incentive scheme at the city level, Ningbo City

Notice on the adjustment of green building pre-sale conditions of commercial housing enacted by Ningbo City Housing and Urban Construction Committee	Through the establishment of special funds to support green building, both energy-saving and building renovation work. Encourage all localities to carry out green building volume rate of incentives and supporting policies.

3. Further Discussions: Green Development and A Cross-level Transitional Approach

China has initiated major policies, strategies and pilot projects at both national and local levels to address the key issues of resource and environmental challenges. These are heavily reflecting the construction sector and operation of buildings. Thus, China has developed a broad concept of development a "resource-conserving and environmental friendly society" in order to achieve a more harmonious development pattern. This seems to become a necessity as the estimated figures (some shown above) indicate major consumption challenges across the country. In this respect, China's focus is at both national and local levels in order to address these issues. The focus is therefore on three key aspects of: 1) low-carbon/green buildings; 2) new urban areas; and 3) retrofits for the existing contexts. Accordingly, there are key policy mechanisms implemented as incentives, which promote approaches to green building design. In recent years, we can witness a growing interest in green buildings and development in Chinese cities. According to Green Building Map [13], the following progress has occurred in less than one decade:

- 2098 projects have been certified by the Chinese Green Building Evaluation System (GBES) since it was launched in 2006. This includes 398 three-star certified buildings;
- 446 projects have been certified LEED certification;
- four projects are currently under BREEAM evaluation;
- There are also around 280 Chinese cities that have declared an ambition to develop "eco-city" or "low carbon city" China Society for Urban Studies [13][14].

More importantly, such large scale to implementation has deeply shapes the local construction industry, from design, material manufacturing, and construction to operational management. Therefore, the green development sector has become a valuable contributor to the local economy.

4. Conclusions

What is important to note and has been briefly demonstrated in this research paper is the policy transitions from national initiatives to the local scale. The examples of provincial level and city level policies, technical standards/strategies and incentive schemes demonstrate the importance of local enhancement for implementation. Also other city-level cases across the country show the importance of local enhancement in development of policies and strategies, if not financial incentive schemes.

Across various cases, it is visible to identity many Chinese cities that have allocated quantitative targets for energy saving and carbon emissions at the city or regional level. Green buildings are seen as a key area to achieve these targets and have become highlight of the green development plans in recent years. At the local level, there is still a major scope for policy improvement, such as incentive mechanism and coordination of various stages of green building development. These are yet to be addressed in the process of local enhancement based on national initiatives.

5. References

[1] National Bureau of Statistics, *China Statistical Yearbook 2013*, China Statistics Press, 2014.
[2] Li, J. & Shui B., A comprehensive analysis of building energy efficiency policies in China: status quo and development perspective. *Journal of Cleaner Production*, Vol. 90, pp. 326-344, 2015.
[3] Building Energy Conservation Center (BEEC), *China Building Energy Conservation Annual Report*, Tsinghua University, China Building Industry Press, 2009.
[4] Fernandez, J., Resource consumption of new urban construction in China. *Journal of Industrial Ecology*, Vol. 11, Issue 2, pp. 99-115, 2008.
[5] Ministry of Environmental Protection (MEP), official website of the Ministry of Environmental Protection, document on air quality of cities, available at : http://www.mep.gov.cn/gkml/hbb/qt/201502/t20150202_295333.htm, 2015. (Accessed 12 September 2016)
[6] Ye, L., Cheng, Z., Wang, Q., Lin, W., Ren, F., Overview on green building label in China. *Renewable Energy*, 53, pp. 220-229, 2013.
[7] IEA, *World Energy Outlook 2007*, International Energy Agency, 2007.
[8] Yang, W. & Kohler, N., Simulation of the evolution of the Chinese building and infrastructure stock. *Building Research & Information*, Vol. 36(1), pp. 1–19, 2008.
[9] He, B.J., Yang, M., Ye, Mou B., Zhou, Y., Overview of rural building energy efficiency in China. *Energy Policy*, Vol. 69, pp. 385-396, 2014.
[10] Li, Z. & Jiang, Y. (2006), Pondering over the situation of domestic generalized building energy consumption. *Architectural Journal*, Vol.7, pp. 30-33, in Chinese.
[11] Li, Y., Yang, L., He, B., Zhao, D., Green building in China: needs great promotion. *Sustainable Cities and Society*, Vol.11, pp.1-6, 2014.
[12] The State Council 2011, http://www.gov.cn/gongbao/content/2011.
[13] Green Building Map, http://www.gbmap.org:86/
[14] China Society for Urban Studies, *China low carbon and ecological cities annual report 2012*, China Building Industry Press, 2013.

Modeling of Energy Management Systems for Commercial Parks with Thermodynamic Equipment

Shi Shanshan[1, a], **Zhang Yu**[1, b], **Fan Liuyang**[2, c] **and Luo Tian**[2,d]

[1] State Grid Shanghai Electric Power Research Institute, China
[2] Shanghai Jiaotong University, China

[a]shishsh@sh.sgcc.com.cn, [b]zhangyu@sh.sgcc.com.cn, [c]404295262@qq.com, [d]pianist1993@qq.com.cn

Abstract. Commercial park energy management systems (CPEMS) can reasonably plan appliances' schedule of commercial tenants (CT) and lower their electricity purchasing cost. However, in the existing models, thermodynamic equipment like air conditioners and water heaters are not precise enough, failing to reflect the actual operating characteristics of the equipment. This paper presents an energy management system model including thermodynamic equipment. By coordinating the electricity consumption schedule of multiple CTs, CPEMS can reduce CTs' electricity purchase costs. In the demonstration example, electricity purchase costs of CTs are reduced and operators of CPEMS gain profit, proving the feasibility of the model.

1. Introduction

A commercial park (CP) has a function combining eating, accommodation, transportation, shopping and entertainment. With the development of China's economy, CPs' electricity consumption grow rapidly. On the one hand, various smart appliances usually controlled by smart devices like smart phones are being used gradually. On the other hand, with the development of photovoltaic and other distributed power technology together with cost reduction, more and more distributed photovoltaic power used by CTs (like restaurants, hotels, etc.) apply for access to the grid[1]. The importance of intelligent CP power consumption and demand side response technology are arousing scholars' attention.

In traditional power systems, user side is usually deemed as a simple power consumer. However, after the energy crisis in the 1970s, Europe and the United States launched a variety of electricity price systems, and began a research about demand side management (DSM)[2-5]. As the power demand grows, many scholars start to study energy management system (EMS)[6-10]. Through monitoring and collecting users' power consuming information, EMS can analyse users' consumption habits and add overall management and optimization to their power consumption. A user's load optimization model based on consumption efficiency was proposed in [6], which analysed the influence of different consumption habits in the model. Under the background of day-ahead and real-time power markets, a kind of mixed integer programming algorithm was proposed in [7], which focused on charging modes of electromobile with energy storing devices. The general model of EMS with changing electricity prices is discussed in [8], which compares the effects of solving high-dimensional optimization problems using heuristic algorithm and Q-learning algorithm respectively. The power's continuity and discreteness, transferability and intransferability, interruptibility and non-interruptibility of appliances are analysed in [10], which studied the modeling and optimization of EMS including appliances such as PVs, energy storing devices and electromobiles.

Papers above concluded that through optimizing users' power consumption, the peak-valley difference of power systems can be reduced effectively, and so does users' power costs. The existing researches achieved the modeling of CT appliances, but it's focused on electrical appliances (electromobiles, storage batteries, PV batteries, etc.). Actually, the behaviour of the thermodynamic equipment such as air conditioners and water heaters has played an important role in the whole CP's electricity consumption characteristics. More detailed model research should bc carried out.

This paper presents an energy management model with multiple CTs in a CP. Models of the transferable appliances; energy storage devices and thermodynamic equipment in CTs are built. The model is applied to a four-CT demonstration example, which shows that the model can achieve an optimal management of CT load, and the validity and feasibility of the model are verified.

2. Energy Management Model for Commercial Parks

2.1. Classifications and Characteristics of Appliances in Commercial Parks

According to the operating characteristics of CTs' appliances and control methods, CT appliances can be divided into four categories.

- **Non-transferable Appliances:** When the user gives a work order, the uncontrollable appliances must respond immediately without delay. Such appliances' start/stop and operating power are not controlled by the CPEMS. Such appliances are mostly lighting and entertainment appliances, whose start/stop has a strong randomness, varying with user's habits. Typical representatives are incandescent lamps, televisions and so on.
- **Transferable Electrical Appliances:** Users can set the working time of such electrical appliances in advance. In the time range set by the user, the CPEMS can be free to arrange the working time of electrical appliances. Further, such appliances can be divided into two categories as interruptible and uninterruptible. The difference between these two categories is whether the appliance can be interrupted when it's working. For example, the dishwasher can be set to work in a period of time between 19:00 and 22:00, but once the dishwasher starts to work, it cannot be interrupted until the dishwasher completes the job. On the contrary, take electromobiles for example, the user can set to finish the charging between 20:00 and 7:00, and the charging process can be interrupted for many times.
- **Energy Storage Devices:** Because the peak-valley hours and peak-valley generation amount of distributed power is different to that of CPs' load, clients can buy energy storage devices which will store or release electric power to take full advantage of distributed power. For CTs, considering that the typical peak value of CTs' load is of kW level, usually lithium batteries and lead batteries can be used as energy storage devices
- **Thermodynamic Equipment:** It can be seemed as a special kind of transferable appliances. Because of thermodynamic equipment like refrigerator, air conditioner and water heater has a temperature property, the change of internal and external temperature should be taken into consideration while modeling, and heat exchange process should be modeled. To ensure users' comfort, thermodynamic equipment is often set to work in a temperature range, which is usually set by users and varied with different living habits of users.

2.2. Modeling of CPEMS

In CPEMS, appliances' operation constraint conditions should be concluded according to different types of appliances' operation characteristics. The object is to minimize the cost of electricity purchase, and the optimal operation schedule of CTs' appliances should be made. In this paper, the time interval of CTs' appliances' working schedule is 15 minutes.

IOP Conf. Series: Earth and Environmental Science **63** (2017) 012051 doi:10.1088/1755-1315/63/1/012051

2.2.1. Distributed generator. For CTs, renewable energy sources like small size wind driven generators, PVs and biomass power generation can be used. Actually, PVs with solar panels are the most popular used clean energy. So in this paper, PV is modelled as an example of distributed generation.

Constraint condition of PV is:

$$0 \le p_{i,t}^{pv} \le p_{i,t}^{pv,\max} \tag{1}$$

Among the equation above, $p_{i,t}^{pv}$ is the ith solar panel's power of generation in the ith time interval. $p_{i,t}^{pv,\max}$ is the ith solar panel's maximum power of generation in the ith time interval, which can be obtained by making the solar panel work in Maximum Power Point Tracking (MPPT) mode. When CTs' smart grid and power grid operate together, SEMS can let PVs work in MPPT mode, selling extra electricity generated by PVs to power grid to gain more profit.

2.2.2. Non-transferable appliance. Non-transferable appliances are of strong randomness, greatly influenced by different users' living habits. They're the direct result of users' "electricity consumption modes". Based on users' history electricity consumption information, typical load curves of non-transferable appliances can be obtained through clustering algorithm in [11]. Let the ith CT's typical power of non-transferable appliances in the ith time interval be $EL_{i,t}^{nc}$.

2.2.3. Transferable appliance. Let the ith transferable appliance's working condition in i^{th} time interval be $u_{i,t}^{c}$. $u_{i,t}^{c}$ is a binary variable. A value of 0 stands for shutdown status, and value of 1 stands for operating status. Constraint condition of Transferable appliance is:

$$u_{i,t}^{c} = 0, \forall t < t_i^s \ or \ t > t_i^e \tag{2}$$

$$\sum_{t=t_0}^{t_{end}} u_{i,t}^{c} = D_i, \forall i \tag{3}$$

$$u_{i,t}^{c} - u_{i,t-1}^{c} - u_{i,\mathrm{k}}^{c} \le 0, \forall t \le k \le T_i^{on} + t - 1 \tag{4}$$

Among the equation above, t_i^s is the earliest start time of the ith transferable appliance, and t_i^e is the latest shutdown time of it. D_i is the time that the ith transferable appliance needs to finish working. t_0 and t_{end} respectively stands for SEMS's start time and end time for planning. T_i^{on} is minimum working time after transferable appliance booting.

Equation (2) indicates that transferable appliance can only complete its work within the time period specified by the user. t_i^s and t_i^e is specified by the user according to his or her own habits, and the user can change them at any time according to the need. For example, the user can take the washing machine as a transferable appliance which is opened to the operators of CPEMS, and specify the time that the washing machine can work between 18:00 to 22:00 every day. This schedule is informed to the CPEMS operator at the planning stage before day. One day the user needs to complete the laundry work between 10:00 to 12:00 for some reason. Then he needs to notify the CPEMS operators at real-time adjustment stage. The CPEMS operators will adjust the work plan to meet the user's demand for electricity. Equation (3) indicates that the start-up time of the transferable appliance in one day is equal to the time required to complete the work, thus ensuring the completion of the task. The value of D_i is determined by the appliance itself. Equation (4) is constrained by the minimum operating time of the transferred electrical appliance. As for uninterruptible loads, $T_i^{on} = D_i$,ensuring that electrical appliances cannot be stopped until it is finished after booting. As for interruptible loads, $T_i^{on} < D_i$, which means that the appliance can be started and stopped several times. T_i^{on} is decided by the electrical characteristics of the appliance itself.

2.2.4. Energy Storage Device. Energy storage device is not a general sense of appliances, and can be seen as the special equipment which can minimize the cost of power purchasing. Therefore, start and shutdown of energy storage devices with their power should be controlled by the CPEMS operators. Constraint condition of Energy storage device is:

$$soc_{i,t} = soc_{i-1,t} + \frac{p_{i,t}^{ch} \cdot \Delta t \cdot \eta_{ch} - p_{i,t}^{dis} \cdot \Delta t}{C_i} \tag{5}$$

$$soc_i^{\min} \le soc_{i,t} \le soc_i^{\max}, \forall i,t \tag{6}$$

$$soc_{i,end} \ge \varepsilon_i, \forall i \tag{7}$$

$$\sum_i p_{i,t}^{dis} \cdot \sum_i p_{i,t}^{ch} = 0, \forall t \tag{8}$$

Among the equation above, $soc_{i,t}$ is the state of charge (SOC) of the ith energy storage device in the ith time interval. $p_{i,t}^{ch}$ and $p_{i,t}^{dis}$ is charging power and discharge power of the ith energy storage device in the ith time interval respectively. η_{ch} is energy conversion efficiency of energy storage device. C_i is the volume of the ith energy storage device. ε_i is the ith energy storage device's minimum SOC at the end of the planning period. $soc_i^{\min}$ and $soc_i^{\max}$ stands for minimum and maximum charge state respectively of ith energy storage device.

Equation (6) is a constraint on the charge state of the energy storage system. For chemical battery type energy storage systems, the larger the depth of discharge will generally shorten the battery life, which is particularly evident for lithium-ion batteries. Therefore, the general battery type energy storage system usually set the maximum and minimum state of charge to protect the battery. $soc_i^{\min}$ / $soc_i^{\max}$ will be provided by the energy storage system manufacturer. Equation (7) ensures that at the end of the optimization period, the energy storage device maintains a certain amount of energy to ensure the ability to adjust for the next optimization period. Equation (8) constrains that a system with multiple energy storage systems cannot be in a state of charge and discharge at the same time. If part of the energy storage device is in the charging state, and the other part of the energy storage device is also in the discharge equipment, this is also a behavior of energy wasting. However, the formula (8) is essentially a quadratic constraint with multiple cross terms. This type of constraint will destroy the convexity of the problem and cause the problem to be difficult to solve. To solve this problem, flag variables f_t^{ch} is set. f_t^{ch} indicates whether the energy storage device can be recharged in the ith time period. A value of 0 means that it can be recharged , while value of 1 means not. Equation (8) can be re-expressed as:

$$\begin{aligned} &0 \le p_{i,t}^{ch} \le f_t^{ch} \cdot p_i^{ch,\max}, \forall i \\ &0 \le p_{i,t}^{ch} \le \left(1 - f_t^{ch}\right) \cdot p_i^{dis,\max}, \forall i \end{aligned} \tag{9}$$

2.2.5. Thermodynamic Equipment. The key to the modelling of thermodynamic equipment is how to describe the heat exchange process. Take air conditioning as an example, assume that the internal temperature of the room is T^r (ignoring the room temperature difference), outdoor temperature is T^{out}. The heat change in the room is caused by the heat exchange between the air conditioner and the room as well as the heat exchange between the room and the outside. The whole process can be described as:

$$\frac{dT^r(t)}{dt}=\frac{1}{M_{air}\cdot c}\left[\left(\frac{dQ(t)}{dt}\right)_{HVAC}-\left(\frac{dQ(t)}{dt}\right)_{loss}\right] \tag{10}$$

$$\left(\frac{dQ(t)}{dt}\right)_{HVAC}=\sum_i \eta_i^{co}\cdot p_i^{co}-\eta_i^{he}\cdot p_i^{he} \tag{11}$$

$$\left(\frac{dQ(t)}{dt}\right)_{loss}=\frac{T^r(t)-T^{out}(t)}{R_{eq}} \tag{12}$$

Among the equation above, M_{air} is the quality of the air in the room. c is heat capacity per unit mass of air at standard pressure. $p_i^{he}\,/\,p_i^{co}$ is the heating/cooling power of the ith thermodynamic equipment. η_i^{he}/η_i^{co} is the heating/cooling efficiency of the ith thermodynamic equipment. R_{eq} is the equivalent thermal resistance of the room at that time. $\left(\frac{dQ(t)}{dt}\right)_{HVAC}$ is the heat exchange between the air conditioner and the indoor unit per unit time. $\left(\frac{dQ(t)}{dt}\right)_{loss}$ is the heat exchange between the outdoor unit and indoor unit per unit time. It should be noted that c is a thermodynamic constant. M_{air} is related to the size of the room. For a CT, M_{air} can be regarded as constant. R_{eq} is related to the current room type and ventilation state. Assume that $p_{i,t}^{h}$ and T^{out} is a constant over a period of time. Solving the equations (10)-(12) and discretizing the results, a formula can be obtained:

$$T_{i,t+1}^{r}=R_{eq}\cdot(1-K_h)\left(\eta_i^{he}p_{i,t}^{co}-\eta_i^{he}p_{i,t}^{he}\right)+T_{i,t}^{r}\cdot K_h+(1-K_h)\cdot T_{i,t}^{out} \tag{13}$$

Among the equation above,

$$K_h=\exp\left(\frac{-t}{M_{air}\cdot c\cdot R_{eq}}\right) \tag{14}$$

Similar results have been reported in [9] [10], but they do not distinguish the heating and cooling power of air-conditioning equipment, which limits the application of the model. Generally, in order to ensure the comfort of the user, the user can set the upper and lower limits of the indoor temperature, the CPEMS operator will keep the indoor temperature within the range:

$$T_i^{r,\min}\le T_{i,t}^{r}\le T_i^{r,\max},\forall t \tag{15}$$

2.3. Operational Strategy of CPEMS

The user's distributed power generation system will meet the user's own needs of electricity preferentially, and the extra electricity can be selling to the power grid to gain more profit. If the PVs equipment and energy storage devices are not sufficient to meet the CT electricity demand, users need to purchase electricity from the grid. CPEMS operators need to ensure the power balance of CTs' smart grid by selling and buying electricity from the grid. Constraint condition of it can be described as:

$$p_t^{buy}-p_t^{sell}=\sum_i EL_i^{nc}+\sum_i u_{i,t}^{c}\cdot p_{i,t}^{c}+\sum_i\left(p_{i,t}^{co}+p_{i,t}^{he}\right)+\sum_i\left(p_{i,t}^{ch}-p_{i,t}^{dis}\right)-\sum_i p_{i,t}^{pv},\forall t \tag{16}$$

Among the equation above, $p_t^{buy}\,/\,p_t^{sell}$ is the electric power that CPEMS buys or sells to the grid in the tth time interval. This electric power is the sum of the power purchased and sold by the CTs and power grid who is responsible for the CPEMS operator. As a result, the CPEMS operator cannot simultaneously purchase and sell electricity. So Let f_t^g be a variable to show whether electricity can be purchased or not in t^{th} time interval.

Constraint condition of f_t^g is:

$$0 \le p_t^{buy} \le f_t^g \cdot p^{buy,\max}, \forall t \qquad 0 \le p_t^{sell} \le \left(1 - f_t^g\right) \cdot p^{sell,\max}, \forall t \tag{17}$$

Among the equation above, $p^{buy,\max}$ is the CPEMS operator's maximum purchase power, determined by the distribution network power supply capacity. $p^{sell,\max}$ is the CPEMS operator's maximum sold power, determined by the capacity of the CTs' smart grid's outlet inverter.

CPEMS operator's operating goal is to achieve the lowest cost of electricity purchase, under the premise that meeting the needs of multiple CTs' electricity demand. It can be described as:

$$\min \sum_t \left(p_t^{buy} \cdot c_t^{buy} - p_t^{sell} \cdot c_t^{sell} \right) \Delta t \tag{18}$$

Among the equation above, c_t^{buy} / c_t^{sell} is the real-time price at which the CPEMS operator buys / sells electricity to the grid in the tth time interval. Equations (1)-(17) are the constraints of the optimization problem. As a whole, the problem can be expressed as a mixed integer linear programming problem.

3. Demonstration Examples

3.1. Parameter Settings

The data presented in this paper is from a commercial park of a smart grid demonstration project. Four CTs in the commercial park are selected as model CTs with solar panels and energy storage devices installed.

According to the proposed CPEMS model, historical data of power consumption should be collected first during pre-research phases. After then, obtain the typical electricity consumption pattern of uncontrollable appliances by calculation. Such curves of the four CTs are shown in figure 1. The transferable load parameters, energy storage devices parameters and thermodynamic equipment parameters of the four CTs are shown in Table 1-3 of appendix.

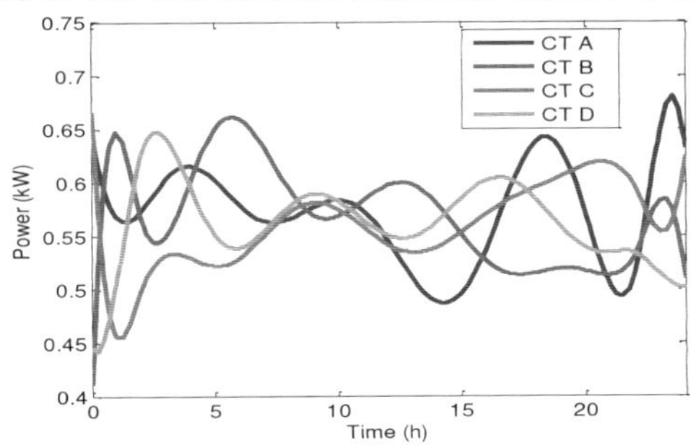

Figure1. The typical electricity consumption pattern of uncontrollable appliances

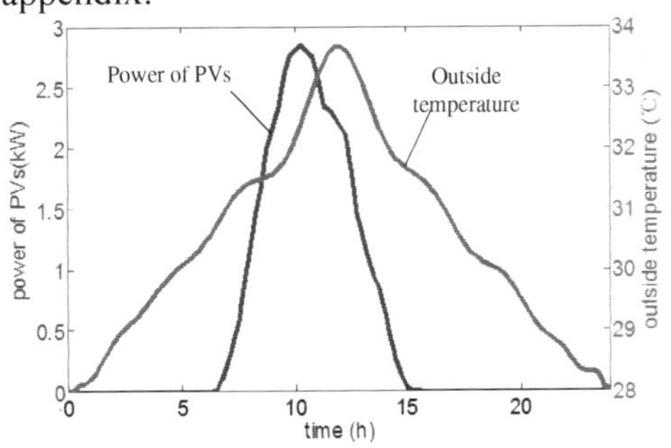

Figure 2. The outside temperature and solar power output

Exposed to the sunlight, solar panel's generation data and outdoor temperature conditions is shown in Figure 2. The outdoor temperature can be obtained from the local meteorological bureau, and the PV power can be calculated from the daylighting condition data. The electricity purchasing price is different in peak and valley hours. The time between 6:00 and 22:00 is peak hours, and the price is 0.56 Yuan/kWh. The time between 22:00 to 6:00 of the next day is valley period, and the price is 0.28 Yuan/kWh.

The demonstration examples use MATLAB 2014a and YALMIP [12] to complete the calculations, where the solution for mixed integer programming is done using IBM ILOG CPLEX 12.6. All calculations are done on a personal laptop with a 2.6G clock speed and 8G of memory with a processor model i5-4219M.

4. Result Analysis

In order to verify the effectiveness of the proposed energy management model, the following two situations are demonstrated and compared.

- **Situation 1:** No CPEMS participates. All the transferable load can begin to work at allowed working time. When the PV power is supplied to the user, the excess electricity is sold to the power grid. The electricity price is 0.3 Yuan/kWh for peak hours and 0.15 Yuan/kWh for valley period.
- **Situation 2:** The users make an agreement with the CPEMS operators and the operators will arrange the transferable load's and energy storage devices' operation. On the basis of purchasing power and selling electricity, the operators can reduce 7% of the electricity purchasing cost and give a subsidy of 7% of the electricity sales to the users, and subsidize the users' energy storage capacity according to 0.02 Yuan/(kWh·day).

In the two situations, the CTs' electricity costs are shown in Table 1. The operation of CPEMS in situation 2 is shown in Table 2.

Table1. Electricity costs in different situation (in Yuan)

Clients	Situation 1	Situation 2
CT A	2.46	1.14
CT C	3.50	2.31
CT C	5.61	3.03
CT D	4.23	2.02
Total	15.80	8.50

Table 2. The profit table of CPEMS operators (in Yuan)

Electric price without subsidy	Subsidy			Electric price with subsidy	Electricity Purchasing price from grid	Profit
	Electricity purchasing subsidy	Electricity selling subsidy	Energy storing subsidy			
15.35	4.094	1.973	0.780	8.50	7.92	0.58

It can be seen that in Situation 2 where the CPEMS participates, most of the users' electricity purchasing costs are significantly lower than that in Situation 1 due to the subsidy given by the CPEMS operator. In this case, CTs are willing to sign agreement with the CPEMS operator to obtain the subsidy. Also, in this situation, the CPEMS can reduce the total cost of purchasing power by optimizing the electricity plan of the four CTs, in spite that subsidizing the user increases the operating costs. Generally speaking, CPEMS is profitable. This shows that the operating mode in this paper can help users and CPEMS achieve a win-win situation.

5. Conclusions

This paper presents a multi-CT energy management model with thermodynamic equipment. The demonstration examples' result shows that under CPEMS mechanism, more PV power will be consumed locally and users' costs are reduced. At the same time, the CPEMS operator gains profit, achieving a win-win situation.

In this model, the effect of the uncertainty of the renewable energy is not taken into account. Using the mathematic tools such as robust optimization and stochastic programming to quantitatively consider the influence of PV forecast error on the specified electricity plan, and improving the robustness of the electricity plan is the direction of next research.

6. Acknowledgments

We thank the support by State Grid Technology Program (52094016000F).

7. Appendix

Appendix Table 1. Parameters of Transferable Appliances

Number	CT	Appliance name	Working time duration (min)	Earliest working hour	Latest working hour	Minimum working time duration(min)	Power (kW)
1	A	dehumidifier	40	09:00	19:00	35	1.5
2	A	disinfection cabinet	30	18:00	20:00	20	1.3
4	B	electric bicycle	45	1:00	8:00	20	0.8
5	B	dryer	70	1:00	7:00	50	0.9
6	B	dish-washing machine	45	18:00	22:00	30	0.5
7	B	sweeping robot	50	9:00	17:00	25	0.7
8	C	washing machine	60	11:00	23:00	50	0.8
9	C	electromobile	200	1:00	8:00	10	1.6
10	C	dish-washing machine	45	8:00	18:00	20	0.6
11	C	dehumidifier	85	09:00	19:00	10	0.4
12	D	washing machine	35	16:00	23:00	30	1.3
13	D	electromobile	250	1:00	9:00	25	1.4

Appendix Table 2. Parameters of Energy Storage Devices

Number	CT	Maximum charging power (kW)	Maximum discharging power (kW)	Charging efficiency	Capacity (kWh)	Minimum state of charge	Maximum state of charge	Initial state of charge
1	A	3	2.1	0.89	6	0.12	0.85	0.50
2	B	2.2	1.8	0.85	7	0.12	0.92	0.50
3	C	5	3.3	0.88	15	0.16	0.92	0.50
4	D	4.2	2.8	0.89	11	0.16	0.90	0.50

Appendix Table 3. Parameters of Thermodynamic Equipment

Number	CT	Equivalent thermal resistance	Initial temperature	Minimum setting temperature	Maximum setting temperature	Maximum cooling power (kW)	Maximum heating power (kW)
1	A	35	26	17	27	2.9	2.4
2	B	20	26	19	27	1.9	1.7
3	C	14	26	19	27	1.5	0.7
4	D	25	26	23	28	3.2	5.2

8. References

[1] TONG Huiming, DENG Lan, PAN Shuangshuang. Introduction of Zhejiang homes distributed photovoltaic power generation projects, network management and technical innovation [A]. China Electricity Council of Science and Technology Development Service Center.

[2] YAAGOUBI N, MOUFTAH H T. User-aware game theoretic approach for demand management[J]. Smart Grid, IEEE Transactions on, 2015, 6(2): 716-725.

[3] STEPHENS E R, SMITH D B, MAHANTI A. Game theoretic model predictive control for distributed energy demand-side management[J]. Smart Grid, IEEE Transactions on, 2015, 6(3): 1394-1402.

[4] ESTHER B P, KUMAR K S. A survey on residential Demand Side Management architecture, approaches, optimization models and methods[J]. Renewable & Sustainable Energy Reviews, 2016, 59:342-51.

[5] GHASEMI A, SHAYEGHI H, MORADZADEH M, et al. A novel hybrid algorithm for electricity price and load forecasting in smart grids with demand-side management[J]. IEEE Transactions on Power Systems, 2016.

[6] JINGHUAN M, CHEN H H, LINGYANG S, et al. Residential load scheduling in smart grid: A cost efficiency perspective [J]. IEEE Transactions on Smart Grid, 2016, 7(2): 771-84.

[7] JIN C, TANG J, GHOSH P. Optimizing electric vehicle charging with energy storage in the electricity market[J]. Smart Grid, IEEE Transactions on, 2013, 4(1): 311-320.

[8] LIANG Y, HE L, CAO X, et al. Stochastic control for smart grid users with flexible demand [J]. Smart Grid, IEEE Transactions on, 2013, 4(4): 2296-308.

[9] HUBERT T, GRIJALVA S. Modeling for residential electricity optimization in dynamic pricing environments [J]. Smart Grid, IEEE Transactions on, 2012, 3(4): 2224-31.

[10] ZHOU L, Modeling and optimal dispatch for residential load based on home energy management system under time-of-use pricing[J], Power System Technology, 2015, 02(39): 367-74.

[11] HINO H, SHEN H, MURATA N, et al. A versatile clustering method for electricity consumption pattern analysis in homes [J]. Smart Grid, IEEE Transactions on, 2013, 4(2): 1048-57.

[12] LÖFBERG J. YALMIP: A toolbox for modeling and opti-mization in MATLAB[C]//Computer Aided Control Sys-tems Design, 2004 IEEE International Symposium on. IEEE, 2004: 284-289.

IOP Conf. Series: Earth and Environmental Science **63** (2017) 012052 doi:10.1088/1755-1315/63/1/012052

Cooling Load Estimation in the Building Based On Heat Sources

Chairani[1,2], S Sulistyo[1] and Widyawan[1]

[1]Dept. of Electrical Engineering and Information Technology, Faculty of Engineering, Universitas Gadjah Mada, Grafika Street No. 2 Kampus UGM, Yogyakarta, 55281, (0274) 552305, Indonesia
[2]Informatics and Business Institute Darmajaya, Faculty of Computer Science, Z.A. Pagar Alam Street No. 93, Labuhan Ratu, Bandar Lampung, Lampung, 35141, Indonesia

chairani@mail.ugm.ac.id, selo@ugm.ac.id, widyawan@ugm.ac.id

Abstract. Heating, ventilation and air conditioning (HVAC) is the largest source of energy consumption. In this research, we discuss cooling load in the room by considering the different heat source and the number of occupancy. Energy cooling load is affected by external and internal heat sources. External cooling load in this discussion include convection outdoor/exterior using the DOE-2 algorithm, calculation of heat using Thermal Analysis Research Program (TARP), and Conduction Transfer Function (CTF). The internal cooling load is calculated based on the activity of the occupants in the office, a number of occupants, heat gain from lighting, and heat gain from electrics equipment. Weather data used is Surakarta weather and design day used is Jakarta design day. We use the ASHRAE standard for building materials and the metabolic of occupants while on the activity. The results show that the number of occupancies have an influence of cooling load. A large number of occupancy will cause the cooling load is great as well.

1. Introduction

The issue of global warming has become a particular concern in the world of research. To reduce the effects of global warming it have been developing of green building with the aim of energy saving [1]. For example, in the use of excessive cooling or heating of the building in addition to causing energy waste also causes inconvenience occupants. For saving energy, some researchers have also conducted research for occupancy detection [2][3], and natural lighting effects are also influenced by occupancy [4]. Setting the temperature in the room is affected by the cooling load. Cooling load is an expenditure of heat energy in the room to be cooled. Heat sources energy are affected by external and internal heat loads. External heat load is heat caused due to conduction, radiation and convection. Internal heat load is the heat caused by the occupants and electrical devices in the room [5]. External heat load include solar radiation transmitted through the glass, the solar radiation on the walls and roof, conduction in the room with irregularities through the walls, heat conduction and convection through the door and window glass due to the temperature difference, the heat caused by air infiltration due to the opening of the door and through gaps of the window. Internal heat load include heat from occupants, heat from lighting and electrical equipment, heat generated by other equipment [5]. Considering the importance temperature settings in smart building to support the objectives of green building, it is necessary doing research to determine the condition of heat in buildings for air conditioning. Heating load in the building is not only influenced by the number of occupancies, but other factors such as the condition of external and internal environment of the building. Furthermore, we also studied the effect of a number of occupancy for the cooling load in the room.

The purpose of this research was to determine the required cooling load of the building and to maintain an optimal state of the building considering with the number of occupancy effect. The reminder of the paper is organized as follows. In Section II, we provide the basic theory of the external cooling loads. While the internal cooling loads are presented in Section III. Before we conclude this research in Section V, we present and discuss our result in Section IV. Finally, in Section VI, we discuss the future work.

2. External Cooling Load

The external cooling load is affected by the heat transfer from/to the environment through the building envelope. It is also influenced by sensible heat load, latent heat load, and the total heat load. The sensible heat load is the heat of the temperature difference between the outside of the building to inside the room. Latent heat load generated by the phase change of water vapor contained in the air in the room. Total heat load is the total accumulation of sensible heat load and latent heat load. Heat balance method is the calculation of sensible cooling load method which apply the law of conservation of energy and principles of matrix algebra [5].

2.1. Outdoor/Exterior Convection

2.1.1. Outside Surface Convection Algorithm : DOE-2. The DOE-2 convection model is a combination of the MoWiTT and BLAST detailed convection models [6]. Total convection coefficient is:

$$h_{c,glass} = \sqrt{h_n^2 + \left[a\, V_z^b\right]^2} \tag{1}$$

Where $h_{c,glass}$ is convective heat transfer coefficient for very smooth surfaces (glass), h_n is calculated using Equation or Equation. Constants a and b are given in Table 1.

Table 1. A MoWiTT Coefficient (Yazdanian and Klems 1994) [6]

Wind Direction *(Unit)*	C_t $W/(m^2K^{4/3})$	a $W/(m^2K(m/s)^b)$	b -
Winward	1.143	0.286	0.286
Leeward	1.143	0.067	0.067

For less smooth surfaces, the convection coefficient is modified according to the equation:

$$h_c = h_n + R_f\left(h_{c,glass} - h_n\right) \tag{2}$$

Where R_f is the surface roughness multiplier (Table 2), h_c is surface exterior convective heat transfer coefficient, h_n is natural convective heat transfere coefficient.

2.2.2. Inside Surface Convection Algorithm : Thermal Analysis Research Program (TARP). Thermal Analysis Research Program (TARP) is a model developed by Walton. TARP is a comprehensive model for convection exterior by combining experimental flat plate (Sparrow et al) were correlated with ASHRAE [7]. The total convection coefficient is the sum of these components:

$$h_c = h_f + h_n \tag{3}$$

The forced convection component [8] is used:

$$h_f = 2.537\, W_f R_f \left(\frac{PV_z}{A}\right)^{1/2} \tag{4}$$

W_f is 1.0 for window ward surfaces or 0.5 for leeward surfaces, V_Z is local wind speed calculated at the height above ground of the surface centorid (m/s).

Table 2. Surface Roughness Multipliers [7]

Roughness Index	Rf	Example Material
3 (Medium Rough)	1.52	Concrete
4 (Medium Smooth)	1.13	Clear Pine
5 (Smooth)	1.11	Smooth Plaster
6 (Very Smooth)	1.00	Glass

If the temperature is no difference or a vertical surface, then used:

$$h = 1.31|\Delta T|^{\frac{1}{3}} \tag{5}$$

where $\Delta T = Air\ Temperature - Surface\ Temperature$, for ($\Delta T < 0.0$ and upward facing surface), OR ($\Delta T > 0.0$ and downward facing suerface) an enhanced convection correlation is used :

$$h = \frac{9.482|\Delta T|^{\frac{1}{3}}}{7.283 - |cos \sum|} \tag{6}$$

Where $\sum$ is the surface tilt angle. For ($\Delta T > 0.0$ and upward facing surface), OR ($\Delta T < 0.0$ and downward facing surface) an reduce convection correlation is used :

$$h = \frac{1.810|\Delta T|^{\frac{1}{3}}}{1.382 + |cos \sum|} \tag{7}$$

This research applies simulation on the energy plus software, using weather data Surakarta region. Latitude is -0.15 degrees, longitude is -78.48 degrees, the time zone is -5 hours, and elevation is 2812 meters. We use weather data in Surakarta region as the area most closely coordinates with the Jogjakarta region. The design day of Surakarta region is empty. We are using the design day of Jakarta region to completely of weather data Surakarta region. For summer design day, maximum dry blub temperature is 33.2^{0}c, the daily dry-bulb temperature range is 7.5 deltac, humidity condition type is a wetbulb with the utmost wetbulb or dewpoint at maximum with the maximum drybulb is 25.9^{0}c, barometric pressure is 101228.9 Pa, and sky clearness is 0.98. Design holiday using holidays in Indonesia region for one year. Surface inside and outside face temperature in Room21 at KPFT Building such as window1 can be seen in Figure 1.

2.2. Conduction Through The Walls

The process of conduction through the wall or the building envelope connecting the outside and the inside of the time heat balance. Conduction Transfer Function (CTF) is a function of time series that connecting the heat transfer conduction flux at a coating with a surface temperature and heat flux previous [9]. CTF for inside heat flux is:

$$q''_{ki}(t) = -Z_o T_{i,t} - \sum_{j=1}^{nz} Z_j T_{i,t-j\delta} + Y_o T_{o,t} + \sum_{j=1}^{nz} Y_j T_{o,t-j\delta} + \sum_{j=1}^{nq} \Phi_j q''_{ki,t-j\delta} \tag{8}$$

CTF for outside heat flux is :

$$q_{ko}^{"}(t) = -Y_o T_{i,t} - \sum_{j=1}^{nz} Y_j T_{i,t-j\delta} + X_o T_{o,t} + \sum_{j=1}^{nz} X_j T_{o,t-j\delta} + \sum_{j=1}^{nq} \Phi_j q_{ko,t-j\delta}^{"} \quad (9)$$

where X_j is outside CTF coefficient, j=0,1,...,nz, Y_j is cross CTF coefficient, j=0,1,...,nz, Z_j is inside CTF coefficient, j=0,1,...,nz, $\Phi_j$is flux CTF coefficient, j=1,2,...,nq, T_i is inside face temperature, T_0 is outside face temperature, $q_{ko}^{"}$ is conduction heat flux on outside face, $q^{"}$ is conduction heat flux on inside face.

3. Internal cooling load

The internal cooling load is affected by the heat sources in the building. The cooling load consists of three types that are internal heat gain from occupancy, heat gain from lighting, and heat gain from electric equipment.

3.1. Heat gain from occupancy

Heat gain occupancy consisting of latent and sensible heat gain. Latent heat gain and an estimate sensible heat gain emitted from the body according to age and activity (equation (10))[6].

$$S = 6.461927 + 0.946892\,M + 0.000255737\,M^2 + 7.139322T - 0.06279909TM + 0.0000589172\,T\,M^2 - 0.198550T^2 + 0.000940018T^2M - 0.00000149532\,T^2M^2 \quad (10)$$

LatenGain = MetabolicGain – SensibleGain, where T is air temperature (celsius), M is the metabolic rate (Watt), and S is a sensible gain (Watt) [6]. Simulations performed by the numbers of occupancy are 0, 10, 15, 25, 35 and 50. A period of 1 year in working time from 07:00 am to 16:00 pm. Criteria schedule is activity in the office. Holiday time for one year by the Indonesian state included in the design setting holiday. People laten gain rate in September and sensible heat gain rate in August can be seen in Figure 2 and Figure 3. People total heating energy in January can be seen in Figure 4.

3.2. Heat gain from electrical equipment

Electric equipment in the room21 at building KPFT UGM include LCD projector Sonny and speaker audio controls. LCD projector sonny amounted to two pieces. Each is having power 405 watts and the duration of the operation for 8 hours/day. Speaker audio controls amounted to two pieces. Each is having a power of 40 watts and duration of operation during 8 hours/day. The total heating rate of electric equipment and lighting can be seen in Figure 5. Energy used by electrical appliance turned into heat energy and worked energy. Energy is turned into heat energy. Heat energy is affected by latent heat, radiation, and convection. $f_{convected} = 1{,}0 - (f_{latent} + f_{radiant} + f_{lost})$, with $f_{convected}$ is a fraction of the heat is distributed by convection, f_{latent} fraction of the latent heat, $f_{radiant}$ fraction of the radiation heat, dan f_{lost} the fraction of energy is turned into work [10].

3.3. Heat gain from lighting

The lighting system in a building of which generate heat is the part that generates light, which is light. The energy consumed by the lighting system is transformed into heat energy in three section, namely the short-wave radiation, long-wave radiation, and convection. The three factions there that are distributed into the room (space fraction) and some are distributed to the ceiling so that it interacts with the air through the air system (plenum/return water fraction) [10]. $f_{convected} = 1{,}0 - (f_{ret} + f_{lw} + f_{sw})$ where$f_{convected}$ is a fraction of distributed heat by convection, f_{ret} is a fraction of the heat absorbed by the air through, f_{lw} is a fraction of long-wave radiation, and f_{sw} is a fraction of short-wave radiation [10].

4. Result

Total internal heating energy (joule) is count from zone lights total heating energy, zone electric equipment total heating energy, and zone people total heating energy. People Air Temperature is the zone temperature of people by Fanger Model Thermal Comfort. Fanger Model was developed using thermal sensation votes and Prediction Mean Vote (PMV) thermal sensation scale based on how energy lose (L) deviates from metabolic rate (M) in the following from $PMV = (0.303e^{-0.036M} + 0.028H - L$ [6]. Where H is internal heat production rate of an occupant per unit area (H=M-W). W is the rate of heat loss due to the performance of work (in the Fortran or Energy Plus variable name is WorkEff) [11]. Metabolic rates for office activities such as reading (seated), writing, typing, filing (seated), filing (standing), walking, and lifting (packing) can be seen in [10]. People air temperature in January can be seen. Of internal and external factors that have been discussed above, the cooling load estimates obtained in 1st August as follows.

5. Conclusion

From the above results, it can be concluded that in addition to the cooling load is influenced by external factors and internal factors, also affected by a number of occupancies. Figure 8 shows that the greater number of occupancy required cooling load will also be significant.

6. Future Work

Our next research shall be developing a knowledge base on the rule for conservation energy in the smart building.

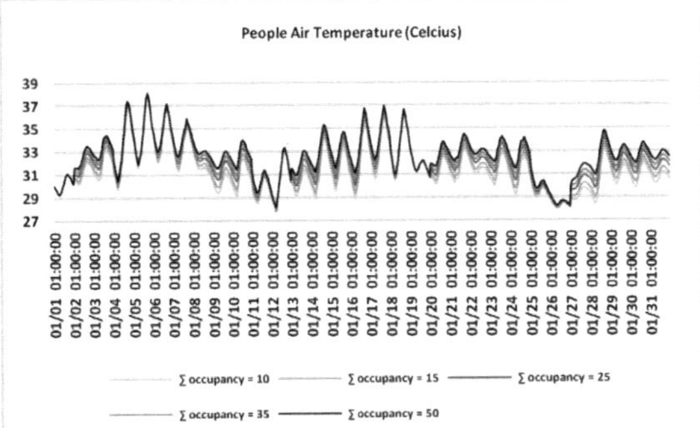

Figure 1. People air temperature in January (summer)

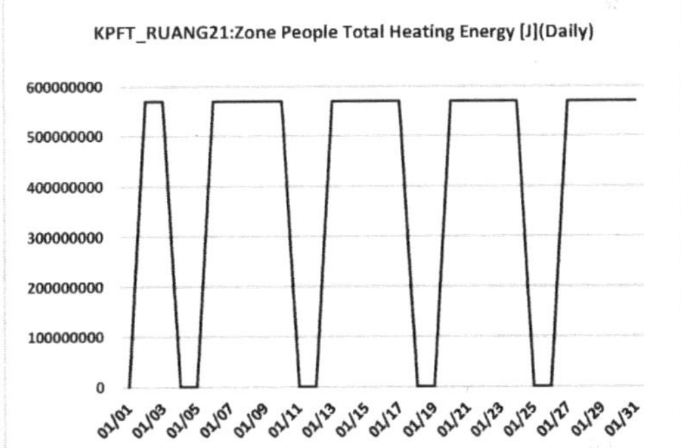

Figure 2. People total heating energy in January (summer)

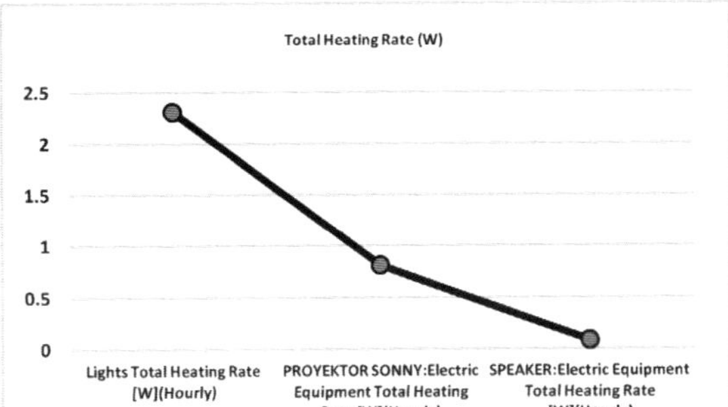

Figure 3. Total heating rate from electric equipment and lighting.

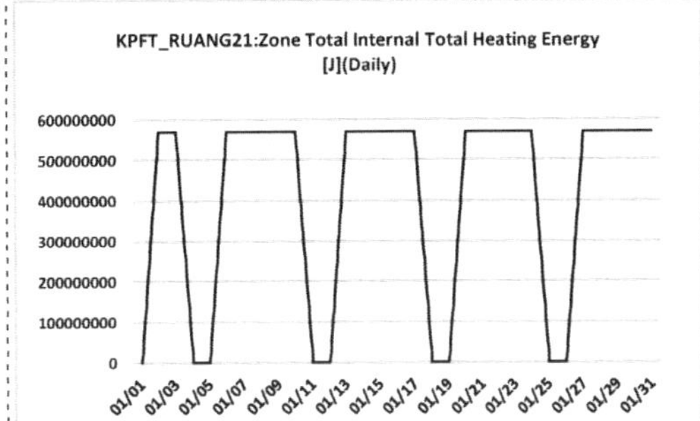

Figure 4. Total internal total heating energy in January (summer)

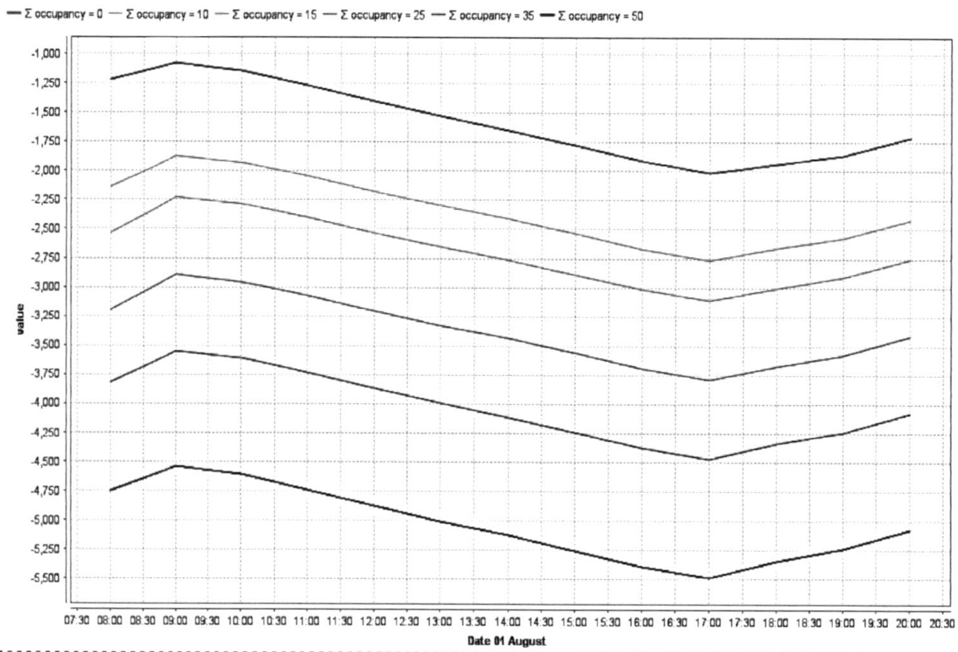

Figure 5. Predicted cooling load [W] – 1st August

7. Acknowledgments

Our thanks to Acoustics Laboratory at the Department of Physics, Universitas Gadjah Mada, for allowing us to information about weather data in Surakarta and Jakarta region.

8. References

[1] EIA 2003 *Washington US. Commercial buildings energy consumption survey (cbecs).* Technical report, US Department of Energy

[2] Pratama A R and Putra G D 2014 *An Infrastructure-less Occupant Context-Recognition in Energy Efficient Building. in IEEE, International Conference on Information Technology and Electrical Engineering (ICITEE)*

[3] Corna A, Fontana L, Nacci A and Sciuto D 2015 *Occupancy Detection via iBeacon on Android Devices for Smart Building Management. In Automation & Test in Europe Conference & Exhibition* 629–632

[4] Edwards L and Torcellini P 2002 *A Literature Review of the Effects of Natural Light on Building Occupants a Literature Review of the Effects of Natural Light on Building Occupants.*

[5] Hittle D 1981 *Calculating Building Heating and Cooling Loads Using the Frequency Response of Multilayered Slabs.*

[6] EnergyPlus Documentation 2016 *Engineering Reference* Version 8.6.

[7] Walton G N 1983 *Thermal Analysis Research Program Reference Manual* NBSSIR 83-2655. Thesis. National Bureau of Standards.

[8] Sparrow E M, Ramsey J W and Mass E A 1979 *Effect of Finite Width on Heat Transfer and Fluid Flow about an Inclined Rectangular Plate. Journal of Heat Transfer.* 101. p. 204.

[9] ASHRAE 2009 *ASHRAE Handbook Fundamentals (SI).* American Society of Heating, Refrigeration, and Air Conditioning, Engineers Inc. Atlanta. Georgia. America Serikat.

[10] EnergyPlus Documentation 2016 *Input Output Reference* Version 8.6

[11] Yazdanian M and Klems J H 1994 *Measurement of the Exterior Convective Film Coefficient for Windows in Low-Rise Buildings* ASHRAE Transactions. 100. Part 1. p.1087.

IOP Conf. Series: Earth and Environmental Science **63** (2017) 012053 doi:10.1088/1755-1315/63/1/012053

The Relationship Between Walkability and Environment Characteristics in Cold Region Cities: Case Study in Harbin

Yang Ye*, Teng Fei and Hongyuan Mei

School of Architecture, Harbin Institute of Technology. No 73 Huanghe Road, Architecture Design and Research Institute of HIT, Harbin, China
This work was supported by Major Project of Technology Research and Development Program of Hei Longjiang (GA15A501)

191775660@qq.com

Abstract. This study attempts to comprehensively and objectively understand whether the physical characteristic of urban space affect the walkability of Harbin city center. Besides, due to Harbin is located in the cold region, the temperature change a lot between winter and summer, this study also tried to find out whether the physical environment characteristics effect on walkability is different in winter and summer. Spatial feature and traffic management have been thought as the main determinate of walkability of urban space, however physical features and urban design details have been rarely mentioned. Yet, does physical quality deterioration of space decrease the walkability of urban center, does specific physical feature influence walkability differently in different season? To answer these question, users' perception toward the physical features of mix-used streets, have been examined in this study. 14 physical characteristic problems have been identified in the studied area based on the understanding of pervious researches. Through observations and questionnaire surveys, the physical characteristics of each case study were evaluated and the physical problems were discovered. Additionally, users' perception on the identified problems and their effects on walkability of the studied areas were found and defined, in both winter and summer.

1. Introduction

After new urbanism, TOD, and compact city theories were introduced to China, the domestic scholars start to focus their study on walking as a traffic mode and the quality of pedestrian space. At the same time, the high speed of cities sprawl and urban regeneration also make public space and activities in public space gain more attention than ever. More pedestrian friendly facilities were built in many Chinese cities, such as pedestrian bridges, pedestrian street and the pedestrian plaza, this reflects the governor's willing to improve walkability in urban center, however, the actual effect is far less than expected, unfriendly walking experience are still plagued pedestrians, walkability of urban space was still need to be promoted.

In previous studies, it was found that not only the public space, but also the social activity and social behaviour in public space can also affect the vitality of the public space. Moreover, the study indicates that the social problems in most public spaces are the source of the behaviour patterns of people in the form of public space. With the pass of time, the quality of public space will impact the social activities more and more. Improving the public space environment may not be able to solve the social problems, but at least could create a better social environment, and to prevent some social problems in the future.

There are such problems in the urban center of Harbin, not only in the traditional urban center, but also in new developed urban area: the walkability decreased, people choose walk and public transit as travel mode less and less. The main purpose of this research is to study the physical environment characteristics of Harbin, and tries to describe the impact of these characteristics on the walkability of urban center space. In order to answer these questions, the main research contents are as following:

- What kind of physical characteristics of urban space could affect walkability?
- What are the main characteristics of Harbin's urban center? What problems exist in physical environment in urban center space?
- What physical problems deteriorate walkability in Harbin urban center, does these physical problems affect on walkability differently in winter and summer?
- What kind of optimization strategy can effectively solve the physical problems, and enhance the walkability in urban center of Harbin?

'Walkability' is referred to and generally defined as "...the extent to which walking is readily available as a safe, connected, accessible and pleasant mode of transport". Livi and Clifton [1] (2004) also mention other friendliness terms including; functional, safety, aesthetic and destination as well as security, comfort and convenience, continuity, system coherence, and attractiveness. It is from the last decade of the 20-century when different evaluations took place based on walkable street agenda. These studies showed that street's walkability were influenced by spatial and traffic characteristics. In addition to spatial and traffic features, some studies examined the effects of other physical characteristics on walkability of urban center. For instance, Eyler [2] (2003) exam the landscape details such as shelter on transit station and lighting, which had provided safety and climate protection for pedestrian. Furthermore, Sallis [3] (1998) illustrated the visual damage and negative influence of commercial signage in traditional urban area. King [4] (2000) argued about the influences of seats, sculpture and planting on the walkability of an area and indicated that physical actives increase with increasing of streetscape elements. Also, Handy and Clifton [5] (2001) showed how friendly environmental design could enhance the walkability in six neighborhoods in Austin, TX. Overall, having sustainable ecology or a good quality environment is considered a major factor of walkability in cities.

Reviewed studies are based on valuations of urban space walkability from different perspectives. Yet, it is important to note that each research has only examined some of the affective factors of streets' walkability. This study attempts to identify the physical characteristics that affect the walkability of urban space, and try to examine how these physical characteristics influence walkability in new developed and traditional urban area. Accordingly, the scope of this research is limited to physical attributes, while the concept of walkability encompasses different social and functional aspects of urban space.

2. Research Method

2.1. The Studied Streets

This study selected two commercial areas located in the center of the city, both of the streets were high-accessibility. Choosing these two areas is based on the following reasons:

- Their physical characteristic could represent Harbin's urban center space
- This is a huge people flow, the number of visitor, no matter local people or tourists.
- They all have mixture functions of commercial office and residential

Shangzhi Street chosen in the study is from Youyi Road to Tiandi Road, with a total length of about 1.3 km, is representative of the traditional commercial area in Harbin. Shangzhi Street area near the Central Avenue – a famous commercial pedestrian street and the historic district in Harbin, this area is not only the office and residential for local people, but also popular in visitors. The buildings and space of this area have been regenerated several times, but there are still some problems. The area of Zhongxing Road is from Youjia Stree to Harbin Avenue, the total length of about 1.4 km. This area is the center of the New District of western Harbin, represent the new developed area emerging commercial and residential. The locations of two areas are showed in Fig1.

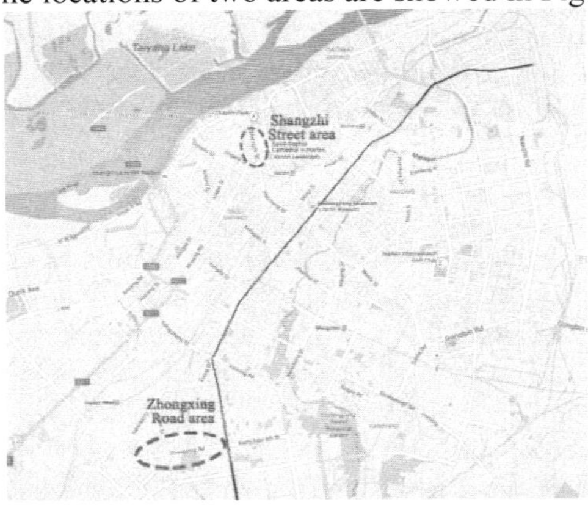

Figure. 1. Location of case studies in Harbin urban center

2.2. Research Process

This study combined a variety of research methods. In order to answer the main questions, both qualitative and quantitative approach were taken. The physical characteristics of the study areas were considered as variables. In this study, the first step was to identify physical problems of the study areas through observation. In order to describe the physical characteristics of the study areas accurately, the study areas have been observed in different times, including the daytime, evening, weekend, and special day of the year, including the hottest and coldest time in Harbin.

As the ultimate physical characteristics assessment was from the user's point of view, therefore, the study also used questionnaire, the questionnaire was designed for both cases. The purpose of the questionnaire was to explain user's viewpoint about existing physical problems of the urban center space, and how these problems affect walkability. The basic statistical information of the questionnaire survey included four categories: gender, age, race, and the purpose to the study of areas. The target respondents of the questionnaire are the people who know the study areas very well. In this study, the sample size is 150[6]. Through data analysis of the questionnaires, this study is focus on whether respondents agree with the identified problems in urban space in the study areas. Therefore, the respondents' opinions were set as three categories: "agree, neither agree or disagree, disagree". The identification of the physical characteristics of urban space impacts on the walkability on the urban space were analysed by applying descriptive statistics and Spearman correlation coefficient among the research variables - urban space walkability and liability as dependent variables and identified physical problems as the independent variables-urban space exists as independent variable – to get frequency distribution, percentages mean score of walkability. These mentioned quantitative data analysis were applied for the interpretation of total result, besides illustrating the identified physical problems' effects on the deterioration of the streets' walkability.

3. Results

After two phase of examining, the findings are presented as follows:

3.1. Visual Examination

The results show that most of the physical environment characteristics were not of satisfied condition. Although there were some of features, such as building density, mixed degree of functions, straight-line distance to the bus station and the accessibility of public transport, and streetlights status within the two regions showed the better situation. The following problems were found in both areas and presented as the result of the first phase of data collection: inadequate parking space, narrow sidewalk, inadequate of bike lanes, inadequate facilities for disabled people, inadequate public services and maintenance, inadequate seats and street furniture, inadequate shelter on transit station, inadequate shelter on sidewalk, inadequate lighting, irregular signs, inadequate planting and inadequate street decoration. In addition to the more common problems, some other problems, such as severe traffic congestion in Shangzhi street area were observed; low road network density and long block, space ratio (width to height) was too big were observed in Zhongxing Road area.

3.2. Questionnaire Survey

3.2.1. Basic Statistical Survey. Table 1 summarized people's attitude of the studied areas walkability. Table 2 shows users' agreement with the existence of these problems in both studied areas.

In fact, the results from the user's perspective reflect the significance of these problems. The results reveal the main difference in walkability of the two studied areas and of winter and summer. In the results of winter survey, there was less than half of the respondents believed that the area is walkable and livable, about 1/3 of the people did not agree with this viewpoint. Different from Shangzhi street, the study result of Zhongxing Road showed more than half of the users think that Zhongxing road does not have the required walkability, only 26% of users agreed this area is walkable and livable. In the results of summer survey, 50.7% of the respondents in Shangzhi street agreed that the area meet the requirement of walkability about 1/4 of the respondents disagree. In contrast, in Zhongxing road, only 32% of users believed that the area is a good place with for walkability, 42% of users think the walkability do not meet the requirement.

3.2.2.Correlation Analysis. Table 3 shows the findings of Spearman correlation analysis for both studied area in winter and summer. The results indicate that there is a negative correlation between walkability and of the studied areas and their physical problems, These results are shown in Table 3 which implies that the walkability of the studied areas are deteriorated by the increase of these physical problems.

Table 1. Users' attitude of the walkability of the studied areas

	Agree		Neither		disagree		Mean	Std. Deviation
	Frequency	Percent (%)	Frequency	Percent (%)	Frequency	Percent (%)		
Shangzhi street	62	41.3	45	30.0	43	28.7	1.80	0.830
Zhongxin road	39	26.0	31	20.7	80	53.3	2.27	0.851

Table 2. Users' attitude of the identified physical problems in the studied areas

The identified physical problems	Shangzhi street				Zhognxing road			
	Winter		Summer		Winter		Summer	
	Mean	Std. deviation	Mean	Std. deviation	Mean	Std. deviation	Mean	Std. deviation
Inadequate parking space	2.12	0.825	2.32	0.814	2.11	0.844	2.31	0.811
Heavy traffic congestion	2.25	0.817	2.28	0.836	——	——	——	——
Narrow sidewalk	2.02	0.805	2.31	0.811	2.14	0.784	2.34	0.784
Inadequate of bike lanes	1.82	0.844	2.11	0.856	2.37	0.857	2.10	0.883
Inappropriate space	——	——	——	——	1.82	0.712	1.94	0.821

Table 3. Correlation coefficient analysis: physical problems and walkability of the studied areas

The identified physical problems	Walkability			
	Shangzhi street		Zhognxing road	
	Winter	Summer	Winter	Summer
Inadequate parking space	-0.688**, P(0.000)	-0.612**, P(0.000)	-0.629**, P(0.000)	-0.465**, P(0.000)
Heavy traffic congestion	-0.583**, P(0.000)	-0.642**, P(0.000)	−	−
Narrow sidewalk	-0.498**, P(0.000)	-0.660**, P(0.000)	-0.529*, P(0.031)	-0.476**, P(0.000)
Inadequate of bike lanes	-0.212, P(0.128)	-0.304*, P(0.021)	-0.189, P(0.276)	-0.377*, P(0.017)
Inappropriate space proportion	−	−	-0.642*, P(0.014)	-0.521**, P(0.000)
Long block	−	−	-0.435*, P(0.021)	-0.072, P(0.380)
Inadequate facilities for disabled people	-0.217, P(0.088)	-0.139, P(0.091)	-0.163, P(0.105)	-0.081, P(0.323)
Inadequate public services and maintenance	-0.326*, P(0.031)	-0.397*, P(0.023)	-0.210*, P(0.041)	-0.169*, P(0.039)
Inadequate street furniture	-0.304, P(0.074)	-0.625**, P(0.008)	-0.246, P(0.083)	-0.467**, P(0.000)
Inadequate shelter on transit station	-0.598**, P(0.003)	-0.272, P(0.087)	-0.604**, P(0.002)	-0.281, P(0.081)
Inadequate shelter on sidewalk	-0.361*, P(0.033)	-0.193, P(0.058)	-0.474*, P(0.044)	-0.339**, P(0.000)
Inadequate lighting	-0.527**, P(0.000)	-0.358, P(0.060)	-0.623**, P(0.000)	-0.389*, P(0.033)
Irregular signs	-0.215, P(0.164)	-0.307, P(0.053)	-0.089, P(0.372)	-0.382*, P(0.026)
Inadequate planting	-0.521*, P(0.011)	-0.573*, P(0.018)	-0.472*, P(0.024)	-0.380**, P(0.000)
Inadequate streetscape elements	-0.588**, P(0.009)	-0.435**, P(0.091)	-0.509*, P(0.038)	-0.377**, P(0.000)

*P<0.05, **P<0.01

4. Discussion

In order to further understand the impact of the physical problems of Harbin on the walkability of the urban center space, the researchers make a comparison of the results of two case studies.

In the summer survey results, among all the identified problems, 12 of them had negative correlation. In the summer survey results, there are 11 identified problems show negative correlation in the two studied areas in total. Compared to the results of the summer survey, 9 problems show the same result, however, 3 problems do not show negative correlation.

When the researchers classified all the identified problems into different categories, they also find difference on impact of walkability of each category. Table 3 shows that people perceive all the physical problems about traffic conditions and the proportion of space could strongly affect the walkability in both studied areas in both winter and summer. Among summer survey results, the correlation coefficient of narrow sidewalks, heavy traffic congestion and inadequate parking space are ranked in the 1st, 2nd, and 4th in Shangzhi Street. In addition, inadequate of bike lanes also shows significant result. Meanwhile, in Zhongxing road survey, narrow sidewalks and inadequate parking space also show strongly correlation. The effect of inappropriate space proportion and long block on walkability could be attribute to user's visual and psychological feelings, lacking visual focus and losing interest and willing to walk when stay in such area. People perceive inadequate bike lanes in summer indicate user's willing to travel by bike, so cycling might be a supplement to current travel mode in summer.

In the facilities and streetscape problems categories, among summer survey results, people perceive that the inadequate public services and maintenance, inadequate streetscape elements, inadequate street furniture and inadequate planting are important factors which affect the deterioration of walkability in both studied area. In Shangzhi street, except inadequate street furniture, all the other problems are not as significant as traffic condition and space proportion problems, in Zhongxing road all these problems have similar correlation significant, what is more, irregular signs shows significant here, which may indicate this problem should gain more attention here than in Shangzhi street where the result of is different. In winter survey results, respondents thought that the walkability of the studied areas are deteriorated by inadequate public services and maintenance, inadequate planting and inadequate streetscape elements, and the impacts are similar to the results in summer survey. Inadequate facilities for disabled people show the same result as summer survey; however, inadequate street furniture and irregular signs show different results. This may indicated in winter time, people's willing to stay on street is not as strong as in summer, so the street furniture, such as seats, are not so important; but user's need of visual abundant are as same as summer, so they need more artificial streetscape elements, that could explain why irregular signs is accepted by users in winter and inadequate planting and inadequate is not as influential as summer.

Among all the identified problems, problems related to security and climate protection- inadequate shelter on transit station, inadequate shelter on sidewalk and inadequate lighting -show the strongest different results in summer and winter survey, and old & new developed areas. This may be because in summer the new developed area can not provide enough tree shade as the trees are all new and young ones, so people feel inadequate shelter when they walk on sidewalks, and as the space proportion is more empty, so the sightline density is lower on street, even the luminance on sidewalks is the same as Shangzhi street, people still more lighting at night to meet the need of safety.

In general, in the results of the comparative study of the winter and summer of two areas, the researchers find three major and common problems of the studied areas – inadequate parking space, narrow sidewalks, inadequate public services and maintenance, inadequate planting and Inadequate streetscape elements- which reveals poor urban environment characteristics are the main reason of the deterioration of the walkability in Harbin urban center area. The findings of this research reinforce the results of former studies, especially the research about walkability city [7], [8] (Appleyard, 1981; Appleyard &Lintell, 1972), which revealed traffic effects on the deterioration of streets' walkability. In addition, these results also confirmed the conclusions of other early studies such as Sauter and Huettenmoser 2008[9], on the Swedish streets, as well as the results of the study of Canadian contemporary citiesand research in California. Besides the impact of traffic condition, there are other research results shows the similar impact of physical characteristics on urban space's walkability. Although most of the results of this study confirm the results of previous studies, they do not corroborate all of them. For example, result of 2007 Portella's study [10] on urban visual pollution, irregular signs in traditional urban area shows significant, but in the studies in Shangzhi street, also traditional urban area, no significant correlation is discovered between streets' walkability and irregular signs in both winter and summer. The occurred contrast might be in relation to with people's unawareness of the issue's negative impacts. Despite the above-mentioned pervious researches have different approaches, but the comparison to them and discussion still revealed common results that verify the findings of this study.

5. Conclusions

The main purpose of this research was to examine whether the physical problems could influence walkability of urban center area, with a special focus on new and old urban center, in winter and summer. Since the physical environment of the two studied area could represent the physical characteristics of most part of Harbin, the recommended strategies could be applied to enhance the walkability of all Harbin urban center area.

5.1. Improve traffic condition and space proportion.

First of all, clear all the unnecessary facilities and parking vehicles on sidewalk, set street vendors at place where the sidewalks are wider than needed, give the sidewalks back to pedestrian. Secondly, provide more parking spaces, set the road parking, underground parking spaces to solve the car-parking problem. Thirdly, control of traffic congestion in traditional urban center such as only allows public transportation but not private cars. Furthermore, establish bicycle system, including bike lanes, bike parking place and maintenance facilities, reinforce the connective of bike travel and transit travel, especially in summer. Last but not least, redesign or regenerate urban space with space proportion problems, such as use plant or other streetscape elements to change space proportion, and add thoroughly pedestrian road in long block if possible.

5.2. Improve street facilities and streetscape elements

Street facilities and streetscape elements problems show little difference in the two studied areas in winter and summer, therefore, the improvement should consider priority to improving the common problems, and then adjust according to the differences. Firstly the governor should strengthen the maintenance and cleaning of the public facilities. Secondly, pay attention to the construction maintenance of facilities for the disabled, in Harbin's example, not only pay attention to set blind tracks, but also crosswalk ramps, building entrance ramp and other facilities for wheels. Thirdly, increase street furniture such as seats, from not only function but also visual abundant, and improve the aesthetic aspects of the urban space, leading to more activities on streets, especially in winter.

5.3. Improve climate protection and safety elements

The results of the research reveal climate protection and safety elements' affect on walkability is very different in winter and summer, and also different in new developed and traditional urban area. So the improvement scheme should be determined by different space types in different seasons. In traditional urban area, set temporary facilities might be a better way to optimize walkability. Meanwhile in the new developed area, the problem of climate protection and safety could deteriorate walkability in both winter and summer, therefore more permanent facilities should be a more reasonable way to optimize space's walkability.

6. References

[1] Clifton, K. J., & Livi, A. D. (2005). Gender Differences in Walking Behavior, Attitudes About Walking, and Perceptions of the Environment in Three Maryland Communities.

[2] Eyler, A. A., Matson-Koffman, D., Young, D. R., Wilcox, S., Wilbur, J. E., & Thompson, J. L., et al. (2003). Quantitative study of correlates of physical activity in women from diverse racial/ethnic groups: the women's cardiovascular health network project summary and conclusions. American Journal of Preventive Medicine, 25(3 Suppl 1), 93-103.

[3] Sallis, D., Nichols, J., F., Sarkin, & J., et al. (1998). Physical activity patterns of college men and women assessed by the tritrac r3d accelerometer over seven days. Medicine & Science in Sports & Exercise. [4]

[4] King, A. C., & Malone, B. R., et al. (2000). Promoting physical activity in rural communities: walking trail access and effects. American Journal of Preventive Medicine, 18(3), 235-241.

[5] Handy, S. L., & Clifton, K. J. (2001). Local shopping as a strategy for reducing automobile travel. Transportation, 28(4), 317-346.

[6] Dooley, K. (2001). Social research methods. In 4 th ed. Upper Saddle River, NJ.

[7] Appleyard,D.,&Lintell,M.(1972). The environmental quality of city streets: the residents' viewpoint. Journal of the American Institute of Planners, 38(2), 84-101.

[8] Appleyard, D., Gerson, M. S., & Lintell, M. (1981). Livable streets, protected neighborhoods. University of California Press.1.

[9] Sauter, D., & Huettenmoser, M. (2008). Liveable streets and social inclusion. URBAN DESIGN International, 13(2), 67-79.

[10] Portella, A. A. (2007). Evaluating commercial signs in historic streetscapes: the effects of the control of advertising and signage. Oxford Brookes University(2007).

Institute of Physics
Dirac House, Temple Back
Bristol BS1 6BE UK

ISSN: 1755-1307
ISBN 978-1-5108-4038-6

CURRAN ASSOCIATES INC.
proceedings.com